TROPICAL ANIMAL HEALTH

by

HORST S.H. SEIFERT
Professor of Animal Health in the Tropics at
the Georg-August University in Göttingen, Germany

Kluwer Academic Publishers
Dordrecht / Boston / London

Library of Congress Cataloging-in-Publication Data

Seifert, Horst S. H.
    [Tropentierhygiene. English]
    Tropical animal health / Horst S.H. Seifert. -- 2nd ed.
      p.   cm.
    Includes index.
    ISBN 0-7923-3821-9 (hardcover : alk. paper)
    1. Domestic animals--Diseases--Tropics. 2. Veterinary tropical
  medicine. 3. Animal health--Tropics.   I. Title.
  SF724.S4513 1996
  636.089'6'00913--dc20
                                                              95-40167

ISBN 0-7923-3821-9

This edition is published by arrangement with
Gustav Fischer Verlag, Jena. Original copyright
© Gustav Fischer Verlag Jena, 1992.
Title of the German original: Horst S. H. Seifert, Tropentierhygiene.

Published by Kluwer Academic Publishers
P.O. Box 17, 3300 AA Dordrecht, The Netherlands.

Kluwer Academic Publishers incorporates
the publishing programmes of
D. Reidel, Martinus Nijhoff, Dr W. Junk and MTP Press.

Sold and distributed in the U.S.A. and Canada
by Kluwer Academic Publishers,
101 Philip Drive, Norwell, MA 02061, U.S.A.

In all other countries, sold and distributed
by Kluwer Academic Publishers Group,
P.O. Box 322, 3300 AH Dordrecht, The Netherlands.

Printed on acid-free paper

Technical Centre for Agricultural and Rural Cooperation ACP-EU*

The ACP-EU Technical Centre for Agricultural and Rural Cooperation (CTA) operates under the Lomé Convention between Member States of the European Union and the African, Caribbean and Pacific States.

CTA collects, disseminates and facilitates the exchange of information on research, training and innovations in the spheres of agricultural and rural development and extension for the benefit of the ACP States.

To achieve this, CTA commissions and publishes studies; organizes and supports conferences, workshops and seminars; publishes and copublishes a wide range of books, proceedings, bibliographies and directories; strengthens documentation services in ACP countries; and offers an extensive information service.

Headquarters: Agro Business Park 2, 6708 PW Wageningen, The Netherlands

Postal address:
Postbus 380, 6700 AJ Wageningen, The Netherlands
Telephone : +31 317 – 467100
Fax   : +31 317 – 460067
Telex   : +44 30169 CTA NL

---

* Distribution of this English edition was made possible through the kind cooperation of the Technical Centre for Agriculture and Rural Cooperation ACP-EU.

Contents

Part III – Animal Health Management Adapted to the Tropical Animal Production System

Preface to the German Edition

This book, *Tropical Animal Health*, describes the problems of animal diseases in the tropics, in the tropical environment, and in relation to particular production systems. In Part I, those basic scientific facts of the special host defence mechanism and of the host-pathogen relationship in the tropics, which hardly play any part in animal husbandry in temperate climates, are explained. Of special importance are the resistance mechanisms of autochthonous breeds and in contrast to them, the high susceptibility of exotic breeds in the tropics. It is explained how immuno- and chemoprophylaxis can be used as instruments for animal health measures if they are adapted to the socio-economic and ecological conditions of both the tropics and developing countries. Scientific details of immunology are presented as far as they are necessary to understand the epizootiology of tropical diseases and diagnostic techniques for recognizing tropical diseases as well as the execution of prophylactic measures.

Vector-borne diseases are the disease complexes most difficult to control since they are bound to the tropical environment, thanks to the biology of their vectors. Therefore, a special chapter has been dedicated to the description of biology and eradication of vectors of vector-borne diseases. The extent of the description varies according to the importance of the specific vector. The acaricides, insecticides and alternative methods used to control vectors are discussed in detail. The author has tried to present a world-wide picture, but it is not possible to cover every aspect completely.

Part I, in principle, is a new edition of the book *Basics of Tropical Animal Health* which was published in 1983 by Göttinger Beiträge zur Land- und Forstwirtschaft in den Tropen und Subtropen. When no sources are cited, the information has been taken from this publication.

In Part II, aetiology, occurrence, pathogenesis, clinical symptoms and pathological lesions as well as the diagnosis, treatment and control of animal diseases which have an important bearing on economics in the tropics are described. In contrast to the usual text books, diseases are divided into the epizootiological complexes of

- vector-borne diseases,
- soil-borne diseases,
- contact diseases and
- plant poisoning.

Part II is supposed to carry on from the textbook by Mitscherlich and Wagener (1970) *Tropische Tierseuchen und ihre Bekämpfung*. The two disease complexes, soil-borne diseases and plant poisoning, have been added. The detailed information presented by Mitscherlich and Wagener has been used, especially in the description of clinical symptoms and pathological lesions.

Since aetiology, epizootiology and prophylaxis of endoparasitoses in the tropics are similar to identical invasions which occur in temperate climates, they are not included in Part II but rather taken into account within the framework of disease management in Part III.

All diseases of domestic animals are principally discussed. Since typical tropical diseases mostly occur in extensive ruminant production systems in the tropics, diseases of this complex play a most important part. Infectious diseases of pigs and poultry under intensive management conditions are identical with those in temperate climatic regions; factorial diseases are practically nonexistent in production systems of the tropics.

Part III describes the animal production systems in the tropics with their particular ecological, technical and socio-economic factors which might either favour or hinder disease control. On the basis of these preconditions, concepts of animal disease prevention and strategies of disease control within the different production systems are presented.

It is the aim of this integrated presentation of problems of animal health in the tropics and measures of prevention to be a guide to students of veterinary medicine, human medicine, and agriculture who are interested in working in the tropics. Last but not least, the book is supposed to help those many foreign students studying in Europe, and especially all experts who are engaged in the practical work of animal production in the tropics.

The results of much research done by my institute, especially from PhD theses, have been incorporated in this book and are documented in the respective chapters. I am grateful to Prof. Dr. Dr. h. c. E. Mitscherlich for allowing the use of material from his book. I would like to thank my collaborator Prof. Dr. H. Böhnel for reviewing the manuscript and his many suggestions along with Mrs. U. Sukop for her help in the preparation of the illustrations.

Göttingen 1992   H. S. H. SEIFERT

Preface to the English Edition

Thanks to the collaboration between the publisher of the first German edition, G. Fischer, Jena, Germany, Kluwer Academic Publishers, Dordrecht, The Netherlands, and the Technical Centre for Agriculture and Rural Co-operation in Wageningen, The Netherlands, it has been possible to publish this book in English, thus making it available to a much broader circle of readers. Basically, it is a translation of the first German edition, but has been partly brought up to date to incorporate where necessary, recent scientific information. This also applies to the data about pesticides which are used for vector control, but especially it applies to Part II where new findings about aetiology and disease prevention have been incorporated. In particular, the chapter on soil-borne diseases has been improved incorporating information from Seifert (1995) and Böhnel (1995). Furthermore, the taxonomy of viral diseases has been brought up to date.

Many suggestions made by reviewers of the German edition have been followed up. I am thankful to all of my critics for this constructive help. I decided not to act upon the suggestion to include a chapter on "epidemiology" because I believe that its theoretic scientific background and its methods are too general and do not present any specific techniques applicable to the tropical environment. Nor do I believe that reproductive diseases belong in this book, as has been suggested.

The references cited at the end of a sentence within a paragraph refer to the information immediately preceding. When author(s) have been cited at the end of a paragraph, it refers to the entire paragraph. If no reference is made, the source is Seifert (1992).

The English version was written by myself – and typed by my wife – and has been corrected by Mr. Ian Macleish B.A. He was a great help and I thank him very much indeed. I apologize for "germanisms" which still remain and would also like to mention that the writing of the medical terms, though I tried to follow one principle, has become a mixture of those gathered from English, American, Australian and South African scientific literature. Throughout the book, the terms epizootiology/epizootiological have been used instead of epidemiology/epidemiological because following older German tradition, I believe that these are the correct terms when dealing with animal diseases. "*Epo*" in Greek means "over" and "*Demos*" are the people, thus epidemiology hardly seems to be the appropriate definition in connection with animals.

Finally, I would like to mention that this book is last but not least the result of having worked for 40 years as a microbiologist exclusively in tropical countries with problems of animal diseases, and at the same time being responsible for organizing and carrying out schemes for disease control as well as of vaccine production.

Göttingen, 1995   H. S. H. SEIFERT

Part I – Scientific Background to the Problems of Animal Health in the Tropics

1. Introduction

Breeding techniques for increasing animal production in the tropics, and particularly the import of high-performance breeds, make it necessary to improve rearing, feeding and methods of animal health, otherwise the results obtained from these measures do not justify financial investment and can even end in failure, causing economic damage to the livestock farmer and to the economy of the country concerned. It is essential, therefore, to work out animal production projects in collaboration with all the various disciplines and specialized fields involved. These related fields must also be represented in discussions with partners in the developing countries, since a high-performance animal from an industrial country is frequently regarded as the only form of assistance for the expected increase in production in these countries.

If breeders on the one hand are to be required to combine improved breeding techniques with improved husbandry, feeding and hygiene, the veterinarian and animal health specialists on the other must also be asked to tie their measures in with the progress made in other specialized fields of animal production and not just deal with a health complex or a single major disease in isolation, however important it might be. Just as there are examples of failures in animal production in the tropics because of the introduction of one-sided breeding measures, such failures have also occurred as a result of one-sided animal health programmes. Animal health schemes adapted to the different production systems of animal husbandry in the tropics must take into account the peculiarities of

- the structure of the respective production system;
- the different breeds and their adaptability to the respective environment;
- the ecology of the region;
- the socio-economic situation of the owner;
- the infrastructure of the region;
- the economic and political situation of the country;
- the economic importance of animal production.

It must be the goal of an adapted animal health scheme to develop a strategy which, by taking into account the above-mentioned premises, will be able to produce an adequate animal health situation under justifiable economic conditions. The balance between infectious agent and host, or between agent, vector and host should only be changed if the development of the production system is able to provide the necessary technical prerequisites.

Basic prerequisite knowledge for such regionally and structurally adapted planning of an animal health scheme has to include information about

- the aetiology of diseases, especially regarding local variants and strains of the infectious agent;
- the epizootiology of diseases;
- the potential vector and pattern of transmission of the pathogen;
- the reservoirs of disease;
- the diagnostic facilities;
- the effective means of the prophylaxis of local diseases;
- the particular susceptibility or resistance of local and exotic breeds to local diseases.

The concept of animal health should provide systematic technical know-how which is coordinated with local climatic conditions and the production cycle of the particular production system. Planning of animal health schemes is basically the coordination of measures for management, rearing, breeding and feeding as well as immunoprophylaxis and chemoprophylaxis, taking into account the above-mentioned preconditions. It must be an integral part of the overall management of the production system. Animal health used as a collateral or emergency measure will never obtain optimal results and does not fulfil the special requirements of the structure of the different production systems in the tropics and the disease situation which is generally extreme in these regions.

The collaboration with the local veterinary administration presents a specific problem within bi- or multilateral economic and technical cooperation as well as in projects carried out by private enterprises or individual donors. The central veterinary service very often still runs along the lines of the system of the former colonial administration. Because of insufficient technical, financial and intellectual resources, they are not able to carry out their task in the field. Because of their conception, these institutions are also not in a position to fulfil the requirements of the animal holder, neither within the structure of the respective animal production system, nor within the development of animal production. Vaccination campaigns which are centrally planned often fail to understand the specific requirements of the production system and the specific local ecological factors, and thus are unable to lead to efficient disease prevention. Though better conditions exist in South America than for example in Africa because of the dominance of private initiatives – with the possible exception of Chile – even there the collaboration between government services and animal holders in the diagnosis and prophylaxis of infectious diseases does not function satisfactorily. This probably is one of the reasons for the failure to control FMD.

Figure 1 tries to present all the factors which act in combination with measures for disease prevention and which have to be coordinated accordingly. It is the aim of this presentation to point out that centrally planned measures

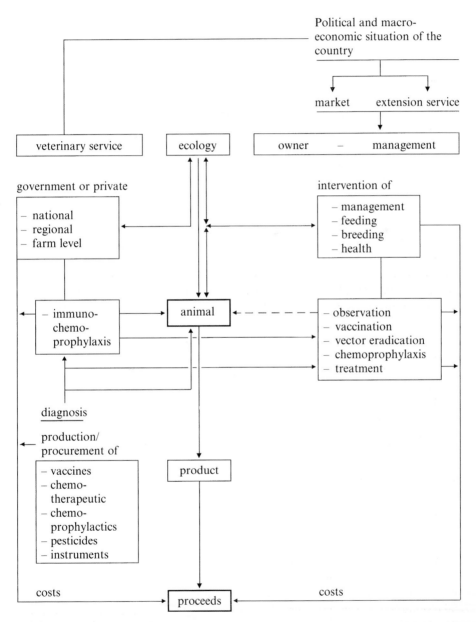

Figure 1. The interaction of factors relevant to animal health in the tropics.

which do not take into account the specific requirements of a region or of a specific production system cannot be successful. At the same time it ought to be pointed out that in a tropical production system, it is the owner or the keeper of the animal who is not only responsible for supervising his animal, but also generally responsible for carrying out therapeutic and prophylactic measures.

2. Host Defence Mechanisms against Infection

It is well known that autochthonous breeds of cattle in comparison with exotic high production breeds have a much higher level of resistance not only to infectious diseases but also, for instance, to plant poisoning in the tropical environment. For a better understanding of the concepts presented in Part III, genetically determined resistance mechanisms, i.e. those typical for a specific breed, as well as acquired host defence mechanisms must be discussed in detail.

So far, empirical criteria have been used for the selection of breeding partners or as selection criteria in a tropical production system. The criteria rely on the subjective experience of the animal holder who wants to obtain as much production as possible under the hygienic conditions of the environment, and also within given economic restrictions. In such cases, productivity gain is usually overestimated in relation to the actual economic result.

Efficient vaccines can be produced against most contact and soil-borne diseases. The immune response produced by these vaccines on the other hand depends on the condition, productivity and production system within which the animal is kept, and on its genetic disposition. As far as vector-borne diseases are concerned, genotypically fixed traits of resistance are mostly responsible for the survival of an individual animal or breed within the tropical environment, if the management of the farm is not able to stop the threat of infection through appropriate measures (vector-control).

Both humans and animals live, after having been born at least, in a microbial environment and have to combat potential pathogens (viruses, bacteria, fungi, protozoa, metazoic parasites).

The formation of the environment in the broadest sense (biosphere, production requirements) may increase or decrease the pressure of these factors on the organism; if the production requirements exceed the capacity of production of the organism, the health of the organism may be affected as a consequence.

Beside the pathogenic (contagiosity, pathogenicity, virulence) and environmental factors, the capacity of the host defence is also responsible for whether the signs of a disease appear or not. For its defence against infection, the organism possesses closely interrelated defence mechanisms which are varied in form according to how highly the organism is developed. This is referred to as a tiered system of host defence (Rolle and Mayr 1993).

The difference between humoral and cellular systems of host defence is characteristic for more highly organized animals; fish are the lowest form capable of producing specific antibodies.

In medical literature in general, and medical and veterinary literature written in English dealing with host defence mechanisms, a differentiation is made between acquired and inherited immunity. In my opinion, it is not clear enough when using the definition "inherited immunity" that this host defence mechanism is unspecific and, as will be described in detail later, is breed-specific amongst animals. This is especially important for animal production in the tropics. In order to present a clear differentiation of these definitions they are divided into

–   **resistance**, which is a congenital and hereditary trait of the organism used for defending itself against infectious, toxic, allergic and neoplastic antigens, and is composed of many clearly definable mechanisms. This, on the one hand, makes up the organism's constitution in the broadest sense to be able to form passive resistance. On the other hand, there are humoral and cellular factors which can first block an invading antigen, break it down and eliminate it, then set specific immunity mechanisms in motion and finally assume important catalytic functions in the formation of immune defence. During phylo- and ontogenesis they appear much earlier than specific mechanisms;
–   **immunity** is a defence mechanism acquired either actively or passively which can combat only that particular antigen against which it was formed. Immunity is specific against different pathogens, as well as their antigens and toxins. Its systems are not continuously active and are only active if unspecific mechanisms are unable to control the invading pathogen. Under those conditions, the organism has to fight the specific invading pathogen although for the most part, antibodies only act in conjunction with humoral and cellular resistance factors. Resistance factors as constitutional characteristics are of primary importance for the capacity of an individual to combat infectious agents (Fig. 2).

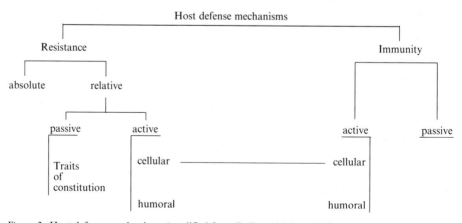

Figure 2. Host defence mechanisms (modified from Rolle and Mayr 1993).

   Resistance and predisposition have a close relationship. Predisposition in general means susceptibility of the organism; it is contrary to resistance and has no function as long as the trigger mechanism for disease can be kept away from the organism or as long as the resistance system is kept intact when a challenge is present. Reduced resistance enhances predisposition. Ecological influences and stress through production may lower the level of resistance and thus enhance predisposition.

   General, constitutional, and unspecific mechanisms first become active when there is an interaction between a host organism and pathogenic living factors. As they are constantly active, they do not challenge the organism (skin, mucosa, ciliated epithelium). A large part of the process of infection is determined this way. If on the other hand an infection appears because of inflammation, the

RESISTANCE

Absolute ———————————————————— Relative

Passive (constitution) —————————————— Active

cellular ——————— humoral

Skin:
Properties, arrangement and forms of
sweat glands, body surface, colour
and pigmentation

Phagocytosis: Lysozyme
Macrophages   Lysine
Microphages   Complement
                    Opsonin
                    Properdin

Adaptive physiological mechanisms:
Heat tolerance (limits of range of comfort)   Inflammation and   Unspecific
basal metabolism, fever, changes in pH,   products of   anti-infectious
enzymatic processes, electrostatic charges,   inflammation   and antitoxic
capacity for crude fibre metabolism,   serum properties
resistance to thirst, fat reserves, level of
production

Exocrinum:
Lacrimal-, saliva-, sebaceous-, milk-glands   Inhibitors

Endocrinum: Blood cutting   Histamine
Internal secretoric balance/
stress-susceptibility   (Interferon)

Flora of the gastrointestinal tract and
of the body openings

Internal corporal barriers:
Blood-liquor-barrier, blood-thymus-
barrier, fasciae, serous linings,
tendon sheaths

Figure 3. Main resistance mechanisms (modified from Rolle and Mayr 1993).

organism triggers off a localized and limited defence. Subsequently, phagocytosis and corresponding serum factors are included into the host defence. Phago-cytosis may be triggered off by opsonin and thus this defence mechanism be-comes specific to certain degree (Fig. 3). The interferon system may also be considered as an unspecific antiviral mechanism. Produced as a response to a viral infection by the cell, it subsequently blocks the intracellular proliferation of the virus. Interferon is even thought to possess viral interference (resistance of an organism infected with the virus against superinfection caused by another un-related virus). In the case of a general infection, these mechanisms of unspecific host defence are gradually replaced by specific defence mechanisms (immunity).

The immune cell may generally be considered as the substratum of cellular immunity and the antibody may be considered as the carrier of the action of humoral immunity. Though different in their functions, both systems have a common source. Immunity is the most efficient and most developed host defence mechanism which nature, during phylogenesis, has developed for humans and animals against endo- and exogene pathogens.

2.1. Resistance

Resistance is the totality of defence which the organism is able to present against an antigen before an immune response has been developed. It enables the macro-organism to suppress the clinical manifestation caused by the invading micro-organisms or to mitigate them and eventually to stimulate and enhance the development of specific host defence. Resistance is not specific to any antigen; it is the sum of independent yet correlated reactions which are genetically and constitutionally determined. Resistance has to be divided into **absolute** and **relative** resistance. The most important mechanisms of resistance are presented in Fig. 3.

2.1.1. Absolute Resistance

Absolute resistance is the complete insusceptibility of an organism to a specific pathogen. In this case, the unspecific resistance mechanisms are so effective that no confrontation with the system of specific immunity is necessary. Therefore, no immune response which can be serologically demonstrated will appear. As an example, equines are completely unsusceptible to the FMD virus. Absolute resistance is usually specific to any given species.

2.1.2. Relative Resistance

In contrast to absolute resistance, there is relative resistance which is composed of factors of **passive** and **active resistance**.

2.1.2.1. Passive Resistance

The traits of passive resistance are basically components of the constitution of the animal. Conditional factors may influence the genetically determined qualities of an animal's constitution secondarily. All physiological and anatomical properties of the animal are constitutional factors which are relevant for the resistance complex, especially those which may become important in connection with the adaptation to the tropical climate. The adaptation of cattle, e.g. *Bos taurus indicus*, to the hot humid climate of the tropics is enhanced through specific properties, such as

– the skin (surface, thickness, colour, pigmentation, localization and structure of sweat glands);
– the coat (length, density, colour);
– the metabolism (basal metabolism, capacity for digestion of crude fibre, capacity for water retention, tolerance to salt, tolerance to heat, upper limit of level of comfort, capacity for energy reserves);
– the anatomical and physiological characteristics of the musculo-skeletal and circulatory systems, which, for example, have an important influence on the capacity of the animal to work under tropical conditions;
– the relatively low level of production.

Variations in passive resistance are obvious between the breeds of the same animal species under comparable ecological and production conditions. As an example, West African Sanga Cebus are highly susceptible to an infection with *Trypanosoma* while West African taurine N'Dama cattle do not present serious clinical symptoms in spite of high infection rates. Susceptibility to CBPP is exactly the opposite: *Bos taurus indicus* cattle are much more resistant to this disease than N'Damas (II/3.3.1.1).

Passive resistance is also important in pre-infectious processes on the skin and the mucous membranes, which possess mechanisms which prevent the multiplication of pathogens. Micro-organisms are washed out or off mechanically through secretion and excretion; ciliated epithelium, as well as coughing and sneezing also eliminate pathogens mechanically. The composition of the flora on the mucosal membranes and in the rumen may be determined genetically (Cebus) and contribute to enhancing the resistance of the organism in a tropical environment (enhanced capacity to digest crude fibre, catabolism of the poisonous content of plants). Many secretions of the organism, such as milk, saliva, nasal mucous, lacrima, vaginal secretion along with the secretion of the respiratory and intestinal tracts contain lysozyme which has a bactericidal action.

2.1.2.2. Active Resistance

Active resistance is complex in its nature. The most important component of this mechanism are the **cellular resistance factors**. These are phagocyting cells (micro-

and macrophages) which function in a chain reaction in the initial phase of host defence, but have to be assisted and stimulated through the **humoral resistance factors**.

• *Cellular Resistance Factors*

Macrophages have a single rounded nucleus and are avidly phagocytic (mononuclear phagocytes): they are capable of sustained, repeated phagocytic activity. They also process and present an antigen in preparation for the immune response; they release soluble mediators which amplify the immune response. They control inflammation; they contribute directly to the repair of tissue damage by removing dead, dying and damaged tissue, and they assist in the healing process. They travel freely within the blood as monocytes, fixed to the tissue within the connective tissue as histiocytes, lining the sinusoid of the liver as Kupffer cells, in the brain as microglia, and in the lung as alveolar-macrophages. Furthermore, they appear in the spleen, in bone marrow, in the lymph nodes, and within the sinus epithelium where they extend their cytoplasmic processes into the lumen.

The cells of the mononuclear phagocytic system arise from the bone marrow stem cells (monoblasts) which appear as monocytes in the blood stream and either remain there or develop into tissue macrophages after entering the tissue (Fig. 4). Their structure is very varied and depends on the tissue in which they are found; mostly they are round, with a diameter of 14–20 μm and have a single round, bean-shaped or indented nucleus; in glia cells it is rod-shaped. According to their function within the cellular system of host defence, they change their shape, enlarge their cells or increase their lysosomes. An epithelioid structure may appear (epithelioid cells). Beside their function in eliminating foreign particles and dead cell remnants, macrophages are able to synthesize some components of complement. They may also produce monokines which regulate the immune response and may also protect cells which are sensitive to antigens against a higher antigen dose.

The most important function of macrophages is **phagocytosis** which, when schematized, develops as follows:

- chemotactically the phagocyte is attracted to the particle which has to be eliminated;
- adhesion of the phagocyte to the particle;
- the phagocyte flows around the particle and ingests it;
- phagocytic vacuoles (phagosomes) appear within the phagocyte through invagination of the outer cell membrane. Subsequently the lysosome fuses with the phagosome and forms the phagolysosome whereby the enzyme content of the lysosome is emptied into the phagosome (endocytosis). The arsenal of these bactericidal mechanisms (lysozyme, proteolytic enzymes, ribonuclease and phospholipase) digests the particle/antigen. The phago-lysosome is where most of the bactericidal and all metabolic activities of the phagocyte take place. There the pathogen is killed off and broken down into smaller molecular components;

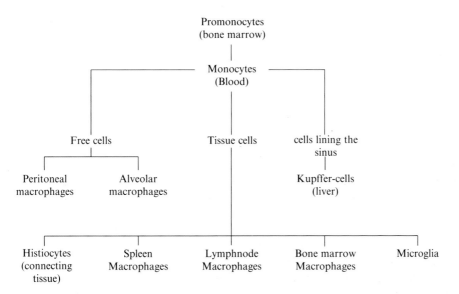

Figure 4. Origin and distribution of macrophages (Tizard 1992).

- the phagolysosome coalesces with the cell wall and excretes its content out of the cell (exocytosis). Through this, lysed components of the particle are provided to the immune competent cells as antigen for the production of specific antibodies.

A defence mechanism which is similar to phagocytosis is **pinocytosis** or **viropexis**. It is the destruction of a virus through the cellular system. Again, this is an active invagination of the cell wall, and not only of macrophages, whereby the virus is invaginated and enclosed within a vacuole of the cell plasma. Together with the uncoating of the virus, the membrane of the vesicle and subsequently the coat of the virus are disintegrated and the virus nucleic-acid is set free.

Tissue macrophages are rather long-living, especially if they are able to break down the ingested particles through lysosomal enzymes. Macrophages may also carry foreign particles to the lung or into the intestine where they are excreted. If the macrophages are destroyed by the particles which they are carrying, large amounts of lysosomal enzymes are set free which leads to inflammation and formation of granulative tissue. Hydrophobic bacteria (*Mycobacterium bovis*) are phagocyted spontaneously which is not possible with hydrophilous; antibodies and complement have to enclose such pathogens and make them hydrophobic, thus preparing them for phagocytosis. Opsonin may also enhance the ingestion of particles through macrophages. The mixture of active enzymes within the phagolysosomes is highly bactericidal for Gram-positive pathogens; Gram-negative pathogens have a longer survival rate. Some pathogens like *Brucella abortus* and *Listeria monocytogenes* are resistant and may even multiply within the vacuole. Different pathogens have developed different strategies for intracellular parasitism. Some which are able to multiply intracellularly are able to hinder the fusion of the lysosome with the phagosome. The intracellular deposition of the phagocyted pathogen may not only protect it against a host defence but also against antimicrobially active chemotherapeutics and antibiotics, since only certain substances may penetrate into the phagocytes when in an effective form and concentration. This is the reason why pathogens in particular which multiply intracellularly may persist for a long time in spite of specifically targeted chemotherapy. Attempts have been made to manipulate the intracellular defence against pathogens. Some substances may stabilize the membrane of the lysosome and hinder the break down of the phagocyted pathogen (concanavalin A, suramin, colchicin, mycobacterial sulfatides). Pathogens which are completely destroyed through the mechanism of phagocytosis do not produce any antigenic stimulus for the immune system (absolute resistance).

The activity of phagocytosis depends on endogenic and exogenic influences. Oestrogen has, for example, a stimulating effect, whereas corticosteroids have a depressive one. It has been possible to stimulate the activity of phagocytosis experimentally through bacterial-lipopolysaccharides as well as through a number of organic and inorganic substances. Lysozyme may also act upon the phagocytic mechanism.

Polymorphic nuclear neutrophil granulocytes (neutrophils) are formed in the bone marrow, from where they migrate to the blood stream and later move into the tissues. Their total life span is only a few days. They constitute about 60–75% of leukocytes in most mammals, but only about 20–30% in ruminants. Like phagocytes, neutrophils may phagocytize and enzymatically destroy foreign material following chemotactical stimulation and opsonization. The nucleus of a neutrophil is unsegmented and is horseshoe- or rod-shaped. In the aging neutrophil, the nucleus becomes segmented. A mature neutrophil contains two populations of granula: primary or azurophil, and secondary or specific granula. Both types differ in their place and time of origin, in the course of their maturation, in shape, size and enzyme content.

The azurophil granula are produced within the promyelocyte stage, and the specific granula in the myelocyte and metamyelocyte stages. The azurophil granula contain

- lysosomal enzymes with an optimum acid pH activity;
- myeloperoxidase;
- neutral proteasis;
- lysozyme.

Granulocytes in birds do not contain myeloperoxidase.

Most components of the specific granula are lysozyme and lactoferrin. The granula in rabbits, rats, guinea-pigs, horses and cattle contain additionally alkaline phosphatase.

The mature neutrophil is highly specialized and only has a short lifetime. Its biological half-life is less than a day. With the enzymes contained in the granula, the mature neutrophil may kill and break down assimilated foreign material.

Since neutrophils reach any site of infection through the blood stream very quickly, they are activated more quickly than the lower macrophages. Unlike macrophages, they perish during phagocytosis. Their bactericidal action relies on a specific feature of their metabolism. They gain energy through stimulation of the hexose-monophosphate-cycle, whereby NDAP is produced, setting free peroxides which, in connection with halogens and myeloperoxidase, are highly bactericidal. Neutrophils are neither able to eliminate cell remnants, nor can they prepare antigens for a targeted immune response. Adult cattle and calves have a higher phagocytose activity after drinking and can eliminate invading pathogens more quickly. This is caused by a higher magnesium content in the blood serum; through increased magnesium content within the colostrum, the magnesium level of the blood serum of calves may be increased, and through this the resistance level may be enhanced (Einfeld 1975, Rolle and Mayr 1993, Tizard 1992).

- *Humoral Resistance Factors*

Humoral resistance factors are a heterogeneous mixture of substances which mostly act in a bactericidal manner, or indirectly enhance the immune response. Their amount and activity within the organism is genetically determined. They do not appear in connection with an infection; they are normal products of the metabolism (fatty acids, monocarbon acids, porphyrines, polylysin- and poly-arginine-peptides). Protamines of the sperm (spermines, spermidines) and the phagocytin of the phagocytes are also humoral resistance factors. The leukines and plakines within inflamed tissue (as well as peptides, amines and histones which are set free during cell destruction) are products of specific reactions.

The bactericidal-acting enzyme lysozyme, which is produced by phagocytes and macrophages, is of special importance for the main thematic topic of this book, as well as the serum factors complement, opsonin and properdin.

Lysozyme is the product of cellular resistance. It is produced by phagocytes through degranulation and is also released from activated macrophages. Lysozyme attacks the inner layer of the bacterial cell wall which lies directly on the cytoplasmic membrane. This layer is common in Gram-positive and Gram-negative bacteria. It guarantees the mechanical strength of the cell wall and is composed of the same polymer in all bacteria: peptido-glycane (mureine). The cell wall may contain polysaccharides, tichon-acid, lipids and proteins. Apart from the mureine, Gram-negative bacteria have an outer membrane which is composed mostly of polysaccharides and proteins. They have physiological functions and are the carrier of the antigenic properties. The basic structure of the mureine consists of a polysaccharide which is composed alternately of two β-(1–4)-glycosidal connected monosaccharides: N-acetylglucosamine (NAG) and N-acetyl-muramin-acid (NAM).

The lytical function of lysozyme consists of breaking down the linkage between NAG and NAM. Thus lower molecular soluble fragments appear, whereby the bacteria lose their protection and deliver their content into the sur-

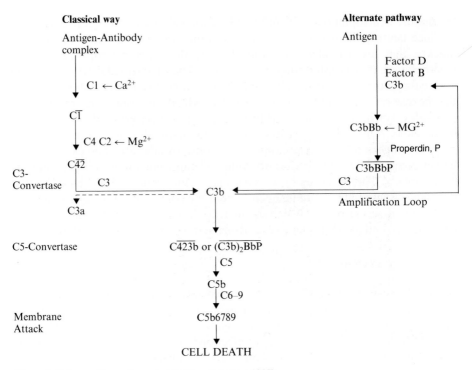

Figure 5. Scheme of complement activation (Margan 1987).

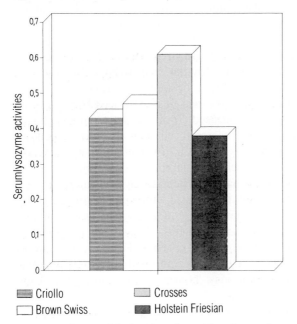

Figure 6. Influence of the breed on the activity of serum lysozyme of cattle which have been kept in the Mantaro Valley/Peru under similar conditions (Einfeld 1975).

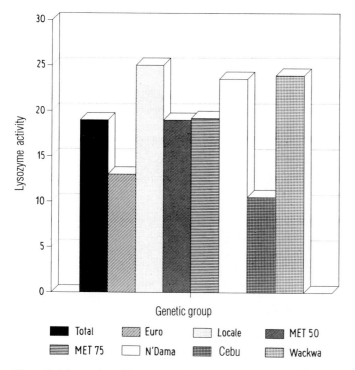

Figure 7. Mean value of lysozyme activity of European breeds, African autochthonous breeds and their cross breeds (Greiling 1979).

rounding substratum when their cytoplasmic membrane (lysis) ruptures. In contrast to penicillin for example, lysozyme is not only able to kill living bacterial cells, but also to destroy dead ones.

Different species of bacteria are varyingly sensitive to lysozyme. If lysozyme alone is allowed to act upon bacteria, it is usually Gram-positive bacteria which are attacked. An exception is *Staphylococcus aureus* which is resistant to lysozyme. In Gram-negative bacteria the lipopolysaccharide layer hinders the action of lysozyme.

Through interaction of lysozyme and the antibody-complement-system, Gram-negative bacteria may also be lysed. While the antibody-complement-system attacks the lipopolysaccharide layer, lysozyme breaks down the mureine structure.

The **complement system** has a central function within the development of an infection. It is a complex system of plasma proteins which together with immune globulins and cellular components guarantee the host defence against invading pathogens and foreign cells. The simple connection of antigen and antibody alone would be an unsuccessful procedure.

Complement can be activated in two ways: either specifically in the **classical way** through the antigen-antibody-complex, or unspecifically in the **alternate pathway** but only through the presence of certain antigens.

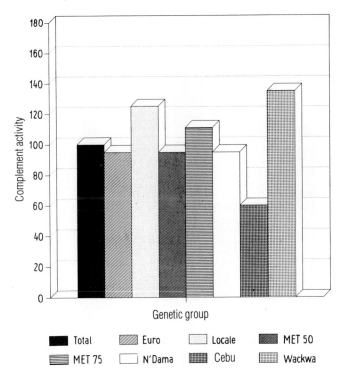

Figure 8. Mean value of complement activity of European breeds, African autochthonous cattle and their cross breeds (Greiling 1979).

The **classical** complement system is composed of 11 plasma proteins which exist in the blood serum in an inactive form and may be divided into 3 functional groups: the unit of recognition C_1 (C_1q, C_1r and C_1s), the unit of activation C_2, C_4, C_3 and the membrane attack complex C_5, C_6, C_7, C_8 and C_9.

Through the antigen-antibody-complex in which IgM or IgG collaborate, complement is activated in the classical way (Fig. 5).

During complement activation in the classical way, several peptides are produced of which some have biological properties:

- anaphylatoxins, C_3a, C_4a, C_5a, have anaphylatoxic properties; they may set histamines free from mast cells, basophils and thrombocytes, and cause inflammation and contraction of the smooth musculature. Through these mediators, vasopermeability is increased and local oedemas appear. Furthermore, thrombocytes aggregate and granulocytes produce lysosomal enzymes (lysozyme). Lysozyme enhances the action of complement;
- kinins, the product which results from splitting C_2 and C_2b has a kinin-like activity which directly increases the permeability of the wall of the blood capillaries;
- chemotaxis, the peptides C_3a, C_5a and the C_5b, C_6, C_7 complex act chemo-

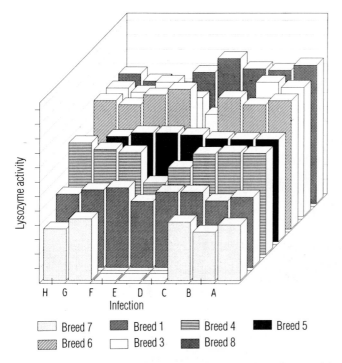

Figure 9. Influence of breed and challenge of infection on the activity of lysozyme of European breeds, African autochthonous cattle and their cross breeds (Greiling 1979).

tactically on neutrophils, eosinophils and macrophages. They are drawn to the site of complement activation, where they take up the pathogen and digest it;

– immune adherence and opsonization: through linkage of C_3b and C_4b on the surface of a pathogen, it is easier to phagocytize the respective antigen. This procedure is called opsonization. Phagocytes have special receptors for connecting to the membrane-bonded complement molecules. Thus the adherence of the antigen to the phagocyte is assured and the subsequent phagocytosis made easier.

Control mechanisms regulate the activation of the complement system and protect it against excessive complement consumption.

The **alternate pathway** of complement activation is a natural system of resistance in the organism for recognizing and eliminating foreign material. It belongs to the humoral resistance mechanism within the non-immune host. The activation of the alternate pathway can be triggered through certain polysaccharides in the cell wall of an antigen as well as through bacterial endotoxin and zymosan. An activation is also possible through aggregated immune globulins which are unable to bind to C_1q such as human IgA and guinea-pig IgG1 (Fig. 5). For the

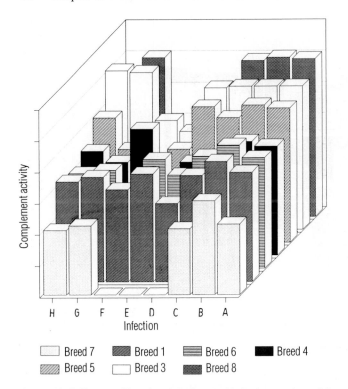

Figure 10. Influence of breed and challenge of infection on the activity of complement of European breeds. African autochthonous cattle and their cross breeds (Greiling 1979).

activation of the alternate pathway, the complement components C_1, C_2 and C_4 are unimportant. They are replaced by factor B, factor D and properdin. The real process of activation of the alternate pathway starts with the bonding of C_3b to specific receptors on the antigen membrane. Subsequently, factor B is added and together with C_3b and magnesium ions forms a bi-molecular C_3bMgB-complex. The further development of the cascade of C_3-activation is illustrated in Fig. 5 (Margan 1987).

Opsonin plays an important part in the first phase of phagocytosis. Opsonization is a procedure by which the ingested particle is sensitized for the phagocytosis by bonding with the proteins of the host organism. Pneumococci for example can only be phagocyted through neutrophils after opsonization. Antibodies and complement assist and complement each other. If the C_3b is bonded to a bacterium it will be opsonized; C_5 may also assist in opsonization.

Properdin is a bactericidal substance which only becomes active in collaboration with the complement system and in the presence of Mg^{++} ions. Two further serum factors, factor A and factor B, are regarded, together with properdin, as the **properdin system**. The properdin system, also called "alternate pathway" or "bypass system", is able to carry out a number of biological

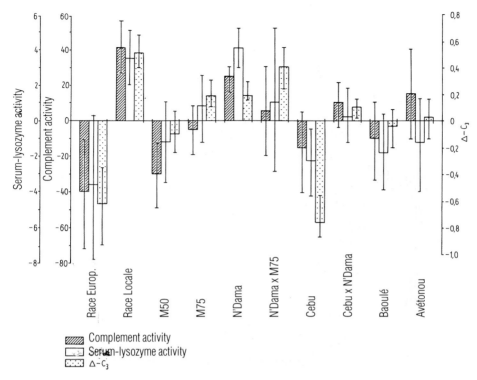

Figure 11. Complement .(C$_9$), serum-lysozyme activity and C$_3$-level of European breeds, African autochthonous cattle and their cross breeds (Lampe-Schneider 1984).

activities, such as bactericidal action, neutralization of viruses and haemolysis (see also alternate pathway of complement activation).

2.1.3. Genetical Determination of Factors of Active Resistance

The activity of lysozyme and complement (C$_3$, C$_9$) is easy to determine. Appropriate methods for analysing large numbers of samples have been developed. Comparative testing of autochthonous and imported exotic breeds in a tropical environment have shown that the amount and activity of resistance factors is specific to any given breed. Analysis of cattle kept under similar ecological conditions in the Andes of Perú (Mantaro Valley) have shown parameters of activity of serumlysozyme as presented in Fig. 6 (Einfeld 1975). A research project in collaboration with CREAT[1] using

---

[1] Centre de Recherche d'Elevage, Avétonou, Togo, GTZ PN 80.2170.1.

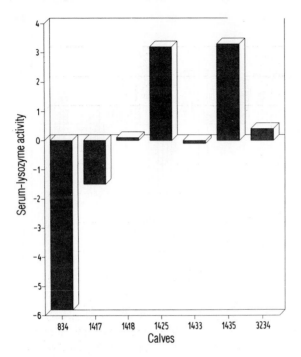

Figure 12. Serumlysozyme activity of calves from different sires (Morkramer 1981).

- German Yellow and Brown Swiss cattle (Euro);
- local nondescript taurine Shorthorn cattle (Locale);
- N'Dama, which are particularly resistant to trypanosomosis (N'Dama);
- Métisse 50, F_1 crossbreds, crossed between European breeds and local cattle (M 50);
- Métisse 75, a cross between the Métisse-50-cows with N'Dama pure-bred sires (M 75);
- Avétonou, inter-se-cross of the M-75-crossbreds

has had the results presented in Figs. 7 and 8. In Figs. 9 and 10, three-dimensional histograms show that these breed-typical parameters are not altered in their tendency through the influence of typical infections prevalent in the local environment of the test animals. The histograms point out that European breeds, German Yellow and Brown Swiss cannot compete as far as their resistance is concerned either with the autochthonous local cattle or N'Damas, while if cross-bred, they show a proportional intermediate position (Greiling 1979). Through the inclusion of C_3 into the research programme we were able to confirm that humoral resistance depends on the breed or rather the percentage of exotic genes within the breed (Fig. 11, Lampe-Schneider 1984). The autochthonous breeds are far superior to the exotic breeds and those crossed with local breeds where their resistance factors, complement, lysozyme and C_3 are concerned. The crosses between autochthonous

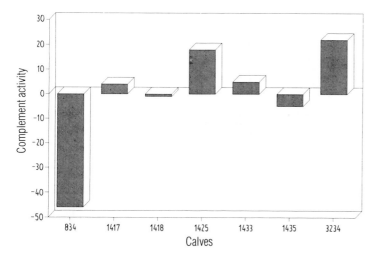

Figure 13. Complement activity of calves from different sires (Morkramer 1981).

West African cattle and European breeds show almost a mathematically proportional reduction in their resistance level in accordance with the percentage of their European genes. This is especially true for the Avétonou breed. This scientific experience has been confirmed through practical experience when these animals were exposed under natural conditions to *Cowdria ruminantium*. The genetical determination of the traits lysozyme and complement is additionally illustrated in Figs. 12 and 13. Here complement and lysozyme levels are presented from groups of the off-spring of different sires (calves of the age up to 180 days). With the exception of number 834, all sires are N'Damas. As the histograms show, the off-spring of the M-75 sire 834 are far inferior to the offspring of the N'Dama sires, where the traits lysozyme and complement are concerned. Within the offspring of the N'Dama sires, the calves of number 1425 and 3234 are obviously superior. The genetical determination of the activity of lysozyme and complement can clearly be recognized (Morkramer 1981). In Norway, research on half-siblings of cattle has shown that the sire has a highly significant influence on the heredity of the lysozyme activity. Because of the clearly bimodal distribution of serumlysozyme activity, it is postulated that this resistance factor is influenced from only one locus of high activity. Therefore, it can be concluded that the lysozyme activity is controlled through one major gene (Lie and Solbu 1983).

2.1.4. Interferon System

The production of interferon within the cell is stimulated by an infection from either a living or inactivated virus, and as well through stimulation by natural or synthetic nucleinic acids (especially RNA), bacterial toxins and other chemical substances. Since interferons are not targeted against a specific virus and thus

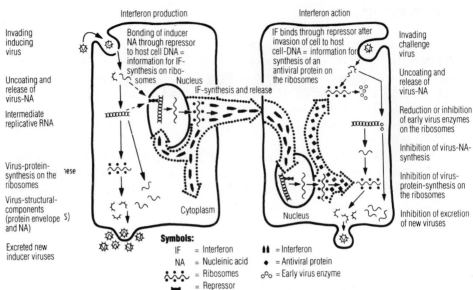

Figure 14. Interferon production and action (example of an RNA-inducing virus).

cannot be compared with antibodies, it seems to be correct to group them into the host defence mechanism of resistance. The capacity of the organism to produce interferons is probably constitutionally determined.

Interferons are proteins (glycoproteins, MW 20000–100000) with an unspecific antiviral action, which are already produced a few hours after infection or stimulation by the cells of the host and also lymphocytes with a so-called inductor. Because of their molecular weight and their physical properties, they are divided into different groups. They develop their best action within the homologous system, i.e. within the genus in which they have been produced. Since they are eliminated with the body secretions rather quickly, they can only continue their protective action if the induction is also continued, or repeated through parenteral injection. Some virus species are able to inhibit the interferon defence. Interferons enhance phagocytosis as well as cell division, but may on the other hand reduce the production of immunocompetent cells.

The antiviral action of interferons does not result, as is the case with antibodies, from the specific bonding of the virus. Instead, the replication of the virus is hindered because with the induction of the formation of interferon through a viral infection, an antiviral protein appears, which already blocks the translation of the virus-specific messenger-RNA before the next cell can be attacked by the virus and thus protects the cell against infection. The inhibition of a viral protein is already enough to hinder the synthesis of an infectious virus (Fig. 14).

2.2. Immunity

The basis for the **specific** host defence mechanism is the **antigen-antibody reaction**. It comes about through the cooperation of the different cell systems and humoral mechanisms which are last but not least components of the resistance system. The immune capacity of the organism has a highly significant correlation with its resistance potential. Figure 15 demonstrates the different components of the mechanisms of immunity and its dependence on humoral and cellular resistance.

Several text books give detailed explanations of general immunity mechanisms, which is not the case for resistance mechanisms. Only the details of immunity which have special relevance to the host defence of the organism within a tropical environment are described here.

2.2.1. Antigens

Antigens are substances which appear foreign to the immunological control system and which trigger the specific defence reaction of the immune system. To be effective, an antigen must fulfil certain physical and chemical requirements. Its molecules must have a sufficient size (generally >20000 MG), and be stiff and complex. If they can be metabolized quickly, there is only insufficient material available for the antigen stimulus. Inert substances cannot be broken down by the macrophage and are, therefore, not effective as antigens. Antigens can be materials from plant or animal tissues. Above all, they are parts of microorganisms, e.g. capsules, components of the cell wall, flagella, nucleus material, cytoplasma and toxins produced by microorganisms. According to their origin, they can be classified as tissue antigens, histocompatibility antigens, tumour antigens or antigenic-acting hormones and enzymes.

According to their chemical structure, antigens are proteins, e.g. serum protein, haemoglobin, enzymes, collagens, protozoal, bacterial, mycoplasma, rickettsial and viral proteins; or they are carbohydrates, e.g. dextran, components of blood groups, the somatic and the flagella antigens of bacteria.

Those substances which stimulate the organism to produce antibodies on the one hand and react specifically with antibodies on the other hand are recognized as **complete antigens**. Beside complete antigens with double action, lower molecular antigenic substances exist, which do not trigger an immunizing effect (production of antibodies). They connect, however, with specific antibodies. In contrast to complete antigens they are known as **half antigens** or **haptens**; by linking up with certain proteins which serve as carriers, they can become complete antigens. Individual protein molecules are not singular antigens. They have chemical groups on their surface, also called hapten groups or **determinants**, which give the antigen or hapten singularly or in groups the specific ability to link up with the pertinent antibody. Whether a substance is a complete antigen or just a hapten depends on their number and order on the antigen and from the

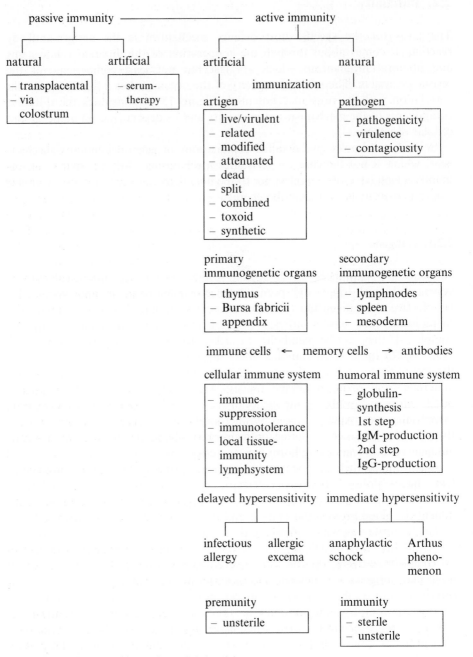

Figure 15. Mechanisms of immunity and their origin.

size of the complete molecule. Furthermore, antigens also possess the ability to stimulate the host organism to produce the immune competent cells of the resistance system: lymphocytes, monocytes and histiocytes.

Biological, chemical and physical elements influence the antibody production and with it increase or decrease the efficiency of the specific defence reactions of the organism. With the so-called **adjuvants**, the normal immune response of the organism can be enhanced. That can happen in such a way that adjuvants envelop the antigen (aluminium hydroxide) or so that it is broken down at a reduced rate. Through the formation of deposits (alum, aluminium phosphate, kaolin, tapioca and mineral oils), the transfer of the antigens via macrophages to the immune competent system is reduced. With water and oil and/or water-oil-water-emulsions, widely used as an adjuvant, a local inflammation is caused and the delivery of the antigen, which is found in the aqueous state, is delayed. Thereby the biological half-life of the antigen within the organism can be lengthened from days to weeks, which leads to a considerable extension of the duration of the immune response, above all through the stimulation of so-called memory cells. If cells from bacteria which have been killed off are added to the adjuvant (e.g. complete Freund adjuvants with mycobacteria), the sensitivity of the immune competent cells is stimulated still further and the immune reaction upon the antigen increases significantly.

2.2.2. Antibodies

The definition of antibodies is the opposite of the definition of the antigen. Antibodies are produced when antigens are introduced into an organism. Antibodies are **immune globulins**. Because of its specific affinity to the determinant group of the antigen, the antibody will be able to enter into a strong linkage with the antigen. This specifity of the immune globulin is caused by its antideterminants which are formed as reflected images of the structure of the determinant group of the antigen.

Normal antibodies are unspecific immune globulins which may exist in the blood serum and lymph fluid; they belong to the resistance system because of their unspecific anti-infectious action.

Antibodies are protein molecules composed partly of lipoids and carbo-hydrates. They are grouped into different gamma and beta fractions. Based on their molecular weight and structure they are divided into:

- **IgG** (immune gamma globulin G), MW 150000, S (sedimentation constant) 7. IgG is the major component of the gamma globulins and the most important antibody. It is synthesized in the spleen and the lymph nodes. It can be sub-divided into several species-specific subclasses, for example IgG1 and IgG2 in cattle and sheep, IgGa–c in horses, IgG1–4 in pigs and other animals. Because of its small size, it can easily leave the blood vessel and thus become active in immune reactions within the cavities of the organism and on its surfaces. It is able to opsonize, to agglutinate and precipitate antigens.

- **IgM** (immune gamma globulin M); MW 1000000, S 19. During immunization, IgM appears earlier than IgG; it has particularly agglutinating, phage-neutralizing, opsonizing, complement-fixing and haemolizing properties. Because of its size, IgM is restricted to the blood stream and has a special virus-neutralizing effect. It is synthesized in the spleen and lymph nodes.
- **IgA** (immune gamma globulin A), exists in saliva, colostrum, tear-liquid and mucosal secretion, as S 7 as a monomer and S 8 and S 18 as a polymer; it is the most important protective immune globulin within body secretions of the intestine, urogenital and respiratory tract, the udder and the eyes. IgA can neither activate complement nor opsonize, but can agglutinate specific antigens and neutralize virus. It is synthesized in the lymphoreticular tissue of the intestine.
- **IgD** (immune gamma globulin D), probably exists only in humans.
- **IgE** (immune gamma globulin E), MG 196000 S 8, appears only in slight concentrations in human and animal plasma. It is responsible for hypersensitivity and defence against parasitic invasions; it is synthesized in the intestine and the respiratory system.

Antibodies are produced by B-cells in the secondary lymph organs, the spleen, the lymph nodes, the Peyer's patches and other lymphoid tissue cells. B-cells as well as T-cells come from the bone marrow. After antigenic stimulation, the B-cells start to multiply in order to enlarge their cell clone. Certain antigens alone or together with T-cells and macrophages stimulate the B-cells. Where haptens only possess one determinant, the T-cell binds with the carrier substance itself. Generally, the **humoral antibodies** produced by T-cells, which are freely available within body liquids and are quickly effective, have to be differentiated from **cellular antibodies** which are produced by T-cells and remain bonded to them. The latter, if they are produced by cells bonded to an organ, are only effective at the point of origin. Cellular antibodies are important in tumour immunity, transplantation immunity and especially in those bacterial infections which mainly take place within the cell (tuberculosis, brucellosis, listeriosis, salmonellosis) as well as in certain viral infections. At the first contact with an antigen, the immune system reacts by producing activated T-cells and both mature and immature plasma cells. This first reaction or **primary immune response** is increased through further stimulation of the organism by the same antigen. Long living T- and B-memory cells are produced; they belong to the circulating pool of lymphocytes. Under renewed antigenic stimulus they multiply rather quickly, increasing their cell clone, becoming activated T-cells and mature plasma cells (memory cells) producing a higher level of either circulating humoral (B-cells) or cell-bonded, cellular (T-cells) antibodies. This development which initiates the **secondary immune response** is much faster and of higher intensity. Not only cell cooperation between macrophages, antigen and B-cells stimulates the immune response: macrophages may also ingest antigens, metabolize them intracellularly, bind them to the RNA of the cell and excrete them. This antigen-specific RNA allows antibody production within the B-cell over a certain time. Furthermore, the RNA produced within the macrophage may stimulate regulatory mechanisms which set free structural genes which produce the negative matrix of the antigen-specific RNA, multiplying it at the same time. These multiplied RNA molecules bind themselves to the ribosome and initiate the production of cell-specific and antigen-specific antibodies. T-cells also secrete substances which stimulate the synthesis of immune globulins through the macrophages both directly and indirectly.

An organism has different ways of developing antibodies. The most important production site of antibodies is the B-cell which is assisted by T-cells (helper cells) and macrophages.

The **development** of the **immune response** can be explained with two theories

- the instructive theory which postulates that the antigen determinant is channelled into the protein-synthesizing apparatus of the immune competent cell. Subsequently this pluripotent cell produces an antibody which fits precisely into the structure of the antigen;
- the selective theory which postulates that the DNA of the nucleus of the antigen encodes the primary sequence of an existing immune globulin using a messenger-RNA. According to the so-called clone-selection theory, cells with immune globulin receptors do exist *a priori* and are able to react in response to a specific single antigen determinant. If they come into contact with an antigen they proliferate and produce the required antibody. This theory is currently considered valid.

The **structure of the immune globulins** is Y-shaped, whereby the tail (heavy chain, H-chain, MW 50000) as well as the arms (light chain, L-chain, MW 25000) are doubled. The variable part of the light chain (V_L region), is built up of approx. 110 amino acids and consists of a terminal, also called a hypervariable region, to which the antigens are bound. The intensity of antigen antibody bonding depends on the affinity of the antibody to the antigen. It is increased during the course of the immune reaction. The constant region of the heavy chain (C_H region) is where the classic way of complement activation takes place (I/2.1.2.2).

Immune globulins are differentiated according to the reaction which they cause together with the respective antigen. An antibody which agglutinates bacteria is an agglutinine; the precipitation of soluble, colloidal and suspended antigens is caused by a precipitin; if cells are lysed, a lysine (cytolysin, haemolysin) becomes active. Antitoxins deactivate poisons of plant or animal origin as well as bacterial toxins. Neutralizing antibodies, for example, deactivate the action of a virus against the cell.

For diagnostic purposes, antibodies which circulate in body liquids can be determined. The **valency** is the number of antideterminant groups of the antibody available which can be bound with the antigen. Most are bivalent, i.e. they possess two reactive groups which can connect with antigens. They are mostly

IgG and monomer IgA. Since antigens may also be polyvalent, they may connect to several antibodies. Thus three-dimensional antigen-antibody complexes may appear which can be demonstrated through precipitation. Polyvalent antibodies (IgM and polymer IgA) contain up to 10 reactive groups. Therefore, they are able to bind themselves to several molecules of the same antigen which leads to the production of a massive immune complex and a significant serological reaction. Contrary to this, a special and important serological phenomenon exists, the so-called **incomplete** or **non-agglutinating antibodies**. They possess only one reactive area and can only connect to one antigen molecule. These weak, incomplete antigen-antibody complexes cannot be demonstrated using agglutination or precipitation. The explanation of this phenomenon lies in the two-phased course of the bonding of the antigen to the antibody. If the reaction partners are congruent, the first thing which takes place is chemical bonding between the reactive components. In a second step, which proceeds much more slowly, the reaction can be seen as a precipitation or flocculation. While bivalent antibodies react during both steps with their antigen and thus produce a visible reaction, the mono- or unovalent incomplete antibodies only react within the first step. Because of their particular bonding properties, the reaction between incomplete antibody and antigen within the second step is not recognizable. Incomplete antibodies are very important for the recognition of both the epizootiology and incidence of an infectious disease. On the one hand they may prevent the serological identification of carriers of pathogens because they might not be recognized with simple serological methods. On the other hand, they may disguise the presence of bi- or polyvalent antibodies since they block the normal reactions of precipitation and agglutination (hampering phenomenon). As indicators of an infection, incomplete antibodies may be determined with the blocking test according to Wiener and with other indirect serological reactions (Coombs test, indirect fluorescence, ELISA). This *in vitro* reaction is also influenced by components of the serological technique, such as composition of the substratum, relationship of antigen to antibody, pH, temperature and the affinity of the antibody to the homologous antigen. Infections which are especially chronic, and which lead to an immunological reaction primarily within the cellular area, usually only react with incomplete antibodies (brucellosis). We have been able to demonstrate that exotic cattle which are not adapted to the tropical environment, mostly produce incomplete antibodies (Einfeld 1975, Greiling 1979, Weiser 1995). This serological phenomenon has to be taken care of when culling animal herds of disease carriers in the tropics.

Lastly, the incidence of a disease may also be determined through the **cellular immune response**. This is especially true for such infections which do not produce serologically recognizable humoral antibodies (tuberculosis). Ordinary serological methods may only determine humoral antibodies. For the demonstration of the cell-bound immunity, *in vitro* and *in vivo* test methods have to be used which demonstrate the reaction of the lymphocytes (determination of the production of mediators and of direct cytotoxicity), or the reaction of the organism (determination of allergic reaction type III through intracutaneous testing).

Not only the level of the humoral antibody titre determines the degree of immunity: other protective mechanisms of the tissue are also active. This can be demonstrated when the antibody level settles down while the organism still remains specifically immune after a certain period of convalescence. The cellular immunity relies especially on a number of mechanisms, of which the development of cell hypersensitivity seems to be the most important reaction. A local tissue immunity develops which results in a hyperergic situation of cellular hypersensitivity type III. Under renewed contact with the antigen, a hyperergic inflammation develops which is very important for the defence against intracellular pathogens.

Polyclonal and **monoclonal antibodies**: To understand the hypothesis about the nature and the development of monoclonal antibodies, one has to follow the scientific dogma which Baron and Hartlaub (1987) have formulated as follows: each plasma cell synthesizes and excretes only a precisely defined type of antibody which is targeted against a particular epitope. An epitope is only part of an antigen. It is precisely the part which reacts with the hypervariable area of the V-region of the antibody. That means that an epitope is very small, often only forming the OH-, NO_2- or NH_2-group of a molecule; it is, however, mostly as big as 2–4 amino acids or sugar-units. Such an epitope-specific antibody belongs to only one IgG class and subclass, has only one affinity constant and only very particular biological properties, i.e. it may only activate complement or macrophages. At a defined time, therefore, different antibodies are produced by different plasma cells they may differ in one of those properties, even if they recognize the same epitope. Consequently, an antigen which possesses several epitopes has to activate several B-cells, whereby each B-cell fits into its own epitope. Each activated B-cell later on divides itself 5–10 times and differentiates into the plasma cell, whereby a true clone is finally developed out of each B-cell, which has originated out of a single mother-cell. All cells of such a clone excrete only a very particular antibody (monoclonal antibody). All cells which have been in contact with an antigen together produce a mixture of monoclonal antibodies, i.e. polyclonal antibodies.

During the development of immunity, a defined B-cell may be activated by several antigens if these are similar in their structure. Generally, however, most antigens will react with several B-cells and induce the production of polyclonal antibodies. The developing polyclonal mixture of antibodies contains antibodies which fit into the antigen either very well, well or only just. Subsequently, the original B-cells also multiply and produce their specific monoclonal antibodies. Amongst them are of course those which react less specifically. Consequently, the polyclonal antibody mixture is of rather low specifity. In general, B-cell clones with high affinity are preferentially activated, but all the others are included into the immune response. Since other immune specific cells mature out of the bone marrow at the same time, several B-cell clones develop, and from those emerge mutant cells with different stages of development, whereby young, proliferating, IgM-secreting and IgG-secreting daughter-cells may appear from each cell clone at the same time. The system is tremendously suitable for the natural immune defence system because better suited and more effective specifities which give better protection against contact with the pathogen are continuously selected through somatic mutation. Furthermore, a very efficient neutralizing effect appears through the bonding of all the antibodies which have been produced for the different epitopes of the antigen. Through the involvement of different IgG classes, different effector mechanisms may be additionally activated.

Until the mid seventies, antisera with polyclonal antibodies were used as diagnostic and therapeutic tools. Using such a mixture of antibodies is disadvantageous for diagnostic purposes because they are difficult to standardize. Often only 10% or even as little as 1% are targeted against the epitope which is to be diagnosed. The same is true for the therapeutic effect. In contrast, monoclonal antibodies have a precisely defined specifity and affinity constant since they only belong to the IgG class or subclass. At present, monoclonal antibodies can be produced *in vitro* – i.e. in fermenters – on hybridoma cells in large amounts.

2.2.3. Forms of Immunity

Figure 15 demonstrates the process of the development of immunity. It can be passive when the organism receives antibodies which are produced in another individual or it can be active when the organism itself reacts with an antigen and produces antibodies against it.

2.2.3.1. Passive Immunity

Passive antibodies may be acquired naturally or artificially. The foetus of a new-born animal receives passive antibodies via the placenta or colostrum. The diaplacental way depends on the histological structure of the placenta and the molecular size of the immune globulin. In humans, primates and rodents which possess a *Placenta haemochorialis*, IgG globulins penetrate the placenta, while IgM can not. In pigs and horses which have a *Placenta epitheliochorialis,* and ruminants which have a *Placenta syndesmochorialis,* IgG is not transmitted transplacentally. IgG and IgM antibodies can only be received by the new-born animal within the first 48 hours of life. The production of IgG in serum and the mammary gland reaches a maximum level at birth and is reduced drastically in a few days p.p.

In passive maternal colostral immunity, the protective effect appears immediately after the ingestion of antibodies. It is, nevertheless, only of short duration because the ingested antibodies are broken down catabolically and eliminated by the organism. Antibodies acquired with the colostrum last only a few weeks. Colostral immunity only exists in the case of infections against which the mother has become actively immunized. In autochthonous tropical cattle breeds, however, the passive immunity acquired via the colostrum may persist for several months.

The colostral IgG originates from the serum and is already secreted with the milk during the pre-colostral phase through hormonal regulation. At the time of calving, the IgG concentration in the colostrum is higher than in the serum. IgA and IgM originate only partly from the serum; they may also be produced directly within the mammary gland. The excretion of the immune globulins, especially of IgG, is reduced drastically a few days p.p. Contrary to other species, IgA does not play an important part as a component of colostral immuno-protection in cattle; this function is performed by IgG1. It is assumed that a gut-mammary-link exists which is able to transfer lymphocytes which are already sensibilized from the Peyer patches of the intestine into the mammary gland, where, as B-cells, they are able to produce IgG1 in large quantities. This mechanism is of considerable importance. Cows which are immunized against an intestinal infection may be additionally stimulated into producing better colostral antibodies through a parenteral vaccine application, and thus may be better able to protect their unborn calves. In all mammals, with the exception of ruminants, IgA ingested with the colostrum has a locally protecting function within the intestine of the new-born calf. IgA is more resistant to proteases of the metabolic system than IgG; it neutralizes viruses and acts bacterio-statically. Since it is bound to the epithelial cells of the intestine, it keeps pathogens away; basically it forms a protective layer within the intestine. In cattle this function is taken over by IgG1 to a certain degree (Bachmann et al. 1984).

For the protection of the calf during the first months of life, it is of utmost importance that the calf gets as much colostrum as possible immediately after birth. When intensifying systems of animal production in the tropics, care has to be taken that the strong relationship between mother and calf in autochthonous breeds is not disturbed. On the other hand, it has to be taken into account that autochthonous cattle very strongly protect their calves with passive antibodies. Active immunization of calves can, therefore, be carried out only at an age of 6–8 months, otherwise the antigen applied with the vaccine would be neutralized by the maternal antibodies. The immunity acquired via colostrum is supposed to protect the new-born calf against the contaminated environment until it is able to protect itself on its own by means of its own immune system.

It is now also known that not only humoral antibodies but also cellular resistance mechanisms are able to assist the colostral protection against infection of the new-born calves. Leukocytes, which are secreted from the mother with the colostrum, are not cast-off epithelial cells but rather, for the most part, fully functioning T-cells which are able to interfere with an antimicrobial and cyto-toxic action. Furthermore, these T-cells are able to produce immune globulins and interferons. Apparently colostral lymphocytes penetrate the intestinal wall of the new-born calf and invade its circulatory system. In addition, they may ex-crete cytokines in the intestine as well as generally within the system of the organism which again will stimulate the proliferation of the endogenous lymphocytes of the new-born calf. As true factors of resistance, such cells may additionally and very effectively support the host defence system on the surface of the intestinal system (Riedel-Caspari and Schmidt 1991).

Passive antibodies may also be transmitted artificially by injecting serum from donor animals which have been immunized against a specific infection. Since these specific immune globulins consist of proteins which are considered alien by the organism, they are destroyed very quickly by the auto-defence mechanism of the organism. They are considered to be antigens by the host. The advantage of artificial passive immunization is its fast effect, the disadvantage its short duration, and, when applied repeatedly, the danger of allergy. Since the species-specific globulin component of the foreign antibody reacts as an antigen and as an allergen, it leads to the production of allergic reacting antibodies. Severe reactions may already occur shortly after the first application of serum from another animal species if the injection is repeated (I/2.2.3.6).

2.2.3.2. Active Immunity

Under natural conditions, active immunity is the replacement, or rather the continuation of the immunoprotection acquired by the new-born calf through passive antibodies. After the application of an antigen, the organism is able to produce specific antibodies by itself. Since the production of antibodies takes some time after the initial antigen contact, the actively produced immunoprotection does not appear immediately after infection, but rather after a certain waiting period which may last some days or even weeks. On the other hand, this type of immunity lasts longer and the organism is able to react more quickly and strongly after the repetition of the application of the antigen. The cause lies in the slower rate of catabolism of the antibodies produced by the organism itself and in the development of the primary and secondary immune response. A **booster reaction** is the consequence of the stimulation of memory cells from T- and B-lymphocytes. If the vaccination is repeated about 3–6 weeks after the first vaccination, a significant increase of the antibody level is reached. At the same, the protective IgG is produced in increased amount and the affinity of the serum antibodies enhanced (Fig. 16).

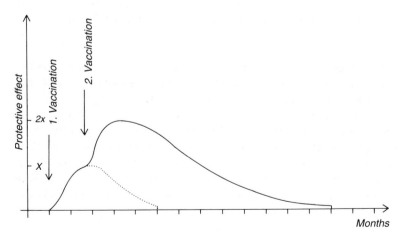

Figure 16. Course of the antibody level after booster vaccination.

The actively produced immune reaction may be acquired artificially or naturally. The organism may be stimulated artificially through the application of an
antigen to start the immunological processes in order to protect itself against an
eventual infection with the actively acquired antibodies.

The vaccination with an antigen is always a prophylactic measure with the
goal of stimulating the organism to produce an active immunity without any
pathogenic side-effect. The efficiency of the vaccination depends on

- the type and properties (virulence, pathogenicity) of the pathogen against
  which the vaccination is directed;
- the antigenic and immunizing properties of the vaccine;
- the antigen dose;
- the site and type of the application;
- the constitution, condition and disposition of the vaccinated organism.

The capacity of the organism to produce an immune response is clearly genetically determined by its resistance mechanisms (I/2.1). It is well known that different breeds under comparable conditions may produce a specific reaction to a
certain antigenic stimulus (Senft and Meyer 1977). This fact often is not taken
into consideration when vaccination campaigns are planned especially under the
conditions of animal production in the tropics often with very heterogeneous
types of animals.

Very similar conditions exist during natural infection. The organism tries to
eliminate the pathogen through specific defence mechanisms of its immune
system. If it is successful, a more or less lasting immunoprotection will remain
after the disease has been overcome. In this case, the actively acquired immunoprotection is also the result of the antigenic and immunizing effect of the pathogen or its toxin, as well as the type and number of invading agents, and the condition and disposition of the host.

2.2.3.3. The Importance of Active Sterile and Unsterile Immunity for Animal Husbandry in the Tropics

Contact infections caused by *Bacteria* and viruses, and soil-borne diseases caused
by *Bacillae* lead principally to a **sterile immunity** which is built up by the cellular
and humoral immune system of the organism. The sterile phase may appear after
an **unsterile** phase of indefinite length. Because of principal considerations, the
initial unsterile phase of immunity in contact and soil-borne diseases should by
no means be confused with the so-called **premunity** or unsterile immunity which
appears as consequence of an infection with vector-borne or parasitic pathogens.
Figures 17 and 18 demonstrate the principal differences between the unsterile
and sterile phase of contact and soil-borne diseases and the persisting unsterile
phase in vector-borne diseases and parasitic invasions (premunity).

Figure 17 demonstrates schematically that the increasing level of antibodies in
sterile immunity in the end leads to the disappearance of the antigen. When the

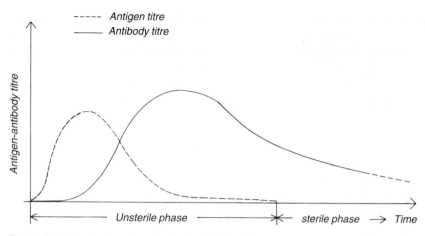

Figure 17. Immunity through contact and soil-borne diseases.

titre of the antigen descends to zero, the unsterile phase ends and the sterile one begins. During the entire unsterile phase, which may last for years with a rather low titre, the infected organism (carrier) may excrete pathogens (carrier state). If applying live vaccines, this situation has to be taken into account when examining certain epizootiological situations. A shortening of the unsterile phase by means of technical manipulations of the vaccine is not possible since the antigens are catabolized in the different organs during a determined length of time.

The biological half-life of antibodies of the IgG class is 5–6 days, and that of IgM antibodies 2–3 days. After reaching their peak of production, immune globulins are catabolized rather quickly at first, and later on much more slowly. If a renewed contact of the host with the same antigen takes place during the phase of catabolism of the antibodies, a faster and stronger antibody response occurs (booster effect). If a revaccination takes place right at the start of the catabolism of the antibody which was produced by the first vaccination, the level of immune gamma-globulins may be enhanced considerably and the catabolic rate may be prolonged considerably. The same effect may be reached with dead vaccines, but in order to obtain the same titre of antibodies as with live vaccines, the vaccination has to be repeated several times at shorter intervals. When repeating the application of dead vaccines several times, the increase of the antibody level appears step by step (II/2.2.3.2).

All vector-borne diseases caused by *Protozoa* and *Rickettsia* produce the unsterile **premunity** which is demonstrated in Fig. 18. This type of immunity almost never leads to a sterile phase after natural infection. The animal which has been infected with a vector-borne pathogen develops premunity, i.e. unsterile immunity. It remains, therefore, a lifelong carrier and a reservoir of infection for potential vectors. The curve of antibodies appears parallel to the curve of antigens whereby the respective titre of antigen determines the titre of antibodies and vice versa. When the level of antibodies is reduced for whatever reason, and with

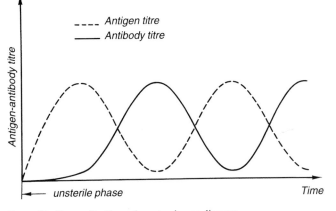

Figure 18. Premunity through vector-borne diseases.

that the immunological protection of the organism, the pathogen will increase its activity and consequently its titre. The increase of the antigen titre will then again enhance the production of antibodies which leads to the depression of the antigen titre and vice versa (Seifert 1971).

Through treatment with the dose of a drug high enough to eliminate the antigen, the so-called *sterilisatio magna* will appear (Fig. 19, point 1). After a short time (3–6 weeks), however, the antibody level will again decrease and eventually disappear, since no new antigenic stimulus exists for antibody production (point 2 in Fig. 19). Afterwards, the animal is fully susceptible to a new infection. The distance between point 1 and point 2 demonstrates the sterile phase of immunity which only lasts for a short time. Lately the possible existence of a sterile phase has been considered in theileriosis (II/1.2.2.2; McKeever and Morrison 1990).

In contrast to the definition of premunity as a level of enhanced resistance (paraimmunity) which is induced through chemostimulants (Rolle and Mayr 1993), the author has every intention of staying with the internationally well-known definition of premunity as being an unsterile immunity. Premunity as an unsterile immunity represents a dynamic balance between pathogen and host which requires the permanent presence of the pathogen in order to survive. If this balance becomes uneven towards one or the other side, either in favour of the

– **pathogen** because of decreased resistance of the host, i.e. through climatic stress, high production, hunger, thirst or activities like active immunization, an outbreak of disease can occur;

or in favour of the

– **host** through optimum keeping and feeding and an adapted production level or perhaps chemoprophylactic support of the host defence system, the immune system may increase its immune response and eliminate the pathogen. The antibodies will usually disappear completely from the organism after

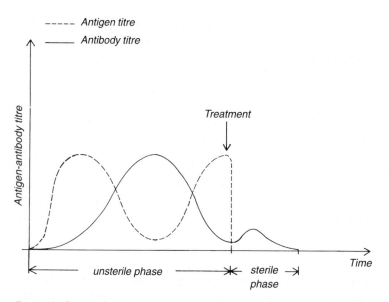

Figure 19. Course of premunity after treatment and sterilisatio magna.

a short phase of sterile immunity and the animal again will be fully suscept-ible to the infection. In this respect, application of chemoprophylactics may destroy the protective status of premunity.

Immune prophylaxis of vector-borne diseases, especially utilizing the booster effect, is rather problematic in connection with the phenomenon of premunity. Any vaccination with live protozoal or rickettsial antigens will artificially induce the status of premunity which follows the same principle as premunity caused by natural infection. This of course is very important if the local epizootiological situation has to be scrutinized. This is also true for antigens which have been attenuated and are less virulent. In this case, the interaction between antibody and antigen will be the same but on a lower level. Such an unsterile immunity does not prevent natural superinfection with virulent field strains. Through a superinfection with field strains, the load of infection will gradually increase which may lead to reduced production and finally, under stress, to the outbreak of disease. According to current information, clinical symptoms of field infections can only be prevented with an attenuated strain of *Theileria annulata* (II/1.2.2.2).

Vaccination with dead antigens in order to prevent vector-borne diseases can only produce temporary protection since dead antigens are eliminated rather quickly and thus cannot produce a longer lasting immunity. Through booster vaccination or regular application of the vaccine in short intervals, this sterile immunity may be extended. It will, however, not prevent field infection with live pathogens and will not prevent the onset of natural premunity which finally leads to the consequences mentioned above. It makes sense to apply dead vaccines under circumstances where only temporary protection is required in order to allow the onset of natural premunity similar to the colostral protection of the calf. It is a prerequisite, of course, that animals which are vaccinated in this way are constitutionally adapted to the location and are able to tolerate the gradual build-up of premunity. Because of its good adaptation and low level of production, the autochthonous animals will be able to compensate for the exogenous load and maintain the host-pathogen balance in most cases. The mechanism of premunity as a combination of resistance and active unsterile immunity may best be explained by using so-called **trypanotolerance** for example. The term trypanotolerance was coined during the forties. It is only used today with reservation because the immunological phenomenon it describes is in the sense of immune tolerance exactly the opposite. Therefore, there is a proposal to call it **trypanoresistance** but, it seems impossible to eliminate the technical term trypanotolerance because of its well established position.

Touré (1977) has two reasons for keeping this definition: firstly because it is well established and secondly because tolerance means something unstable which calls for attention. This explanation points to a special feature of trypanotolerance: it is a very important but very easily destroyable protective mechanism. When applied to cattle, it determines the capacity of a breed to produce and stay in good shape within an environment which is infected by glossines which are the vectors of pathogenic trypanosomes, while other breeds which do not possess this property would not be able to survive.

Omerod (1977) defines trypanotolerance as a genetically determined property of certain breeds or populations of cattle. It enables the animals to live and produce through the production of an enhanced immune response to infection with trypanosomes in areas where tsetse flies are endemic.

Investigations in the WAITR/NITR[1] have shown that N'Dama and Muturu cattle, when infected with trypanosomes, suffered just as much as "trypanointolerant" Cebu cattle. In the long run, however, they were able to produce an immune response which did not break down under stress. N'Damas were superior to Muturus, and cross-bred cattle (*Bos taurus typicus* × *Bos taurus indicus*) took an intermediate position (II/2.1.3).

In a search for genetic markers of resistance to trypanosomiasis, Angoni cattle in Zambia were studied. A whole range of blood group and biochemical polymorphisms were investigated as well as total body tick counts, estimates of trypanosome infection rates, and skin thickness. One of the main findings was that the frequency of blood group M was lower and that the skin was thicker in a high-challenge environment, and within animals both in the high and low-challenge environments, the appearance of trypanosomes in peripheral blood was correlated with the presence of blood group M and thin skin (Spooner 1993).

Apparently taurine cattle are as equally susceptible to trypanosomiasis as

---

[1] West African Institute for Trypanosomiasis Research, Vom, Nigeria.

Cebus, but are able to build up and maintain an unsterile immunity (premunity). The maintenance of the premunity on the other hand means that the host organism is continuously challenged by the pathogen. The resistance factors, which interfere in the course of the specific immune response of the organism, are thus continuously used up. We were able to show in a trial with rats that during an experimentally induced trypanosomiasis, the complement and lysozyme titre was reduced practically to zero. Therefore, it appears logical to look for an explanation for the phenomenon of trypanotolerance within this context. The results of our research presented under II/2.2.3.3 have shown that autochthonous cattle in the tropics possess a higher level of resistance than exotic animals kept within the same tropical environment. Therefore, it is understandable that such animals have better resources for maintaining a premunity against a pathogen, regardless of what it is. Thus the definition of trypanotolerance, which does not fit into medical terminology, can be formulated as the "special capacity of autochthonous breeds to develop and maintain a stable premunity against trypanosomal infection".

2.2.3.4. Immunosuppression

The phenomenon of immunosuppression has to be grouped into the cellular immune system. It is a specific blocking or suppression of the immune response to a certain antigen stimulus by means of immunological, physical and pharmacological procedures which are applied to several parts of the immune system. Through application of antilymphocyte-serum, the lymphatic system can be blocked selectively. In the course of this procedure, the lymphocytes are lysed with the assistance of complement and inhibited in their function through blockage of the membrane of the lymphocytes on the receptor zone. A similar effect may be triggered by ionizing radiation which disturbs especially the inductive phase of antibody production. Cytostatics and steroids also inhibit the antibody production. Steroids in particular suppress the ingestion and catabolism of antigens within the immune system.

2.2.3.5. Immunotolerance

A special form of immunity of the tissues of the organism is immunotolerance. Immunotolerance is a situation in which the specific immune response (production of antibodies or immune cells) to certain antigens is absent while the ability of the organism to react to other antigens remains active. The organism tolerates only a certain antigen which is therefore called **tolerogene**. Of course, the immunotolerance may be either innate or acquired. **Natural immunotolerance** guarantees that cells, tissues and proteins are protected against the immune system of the organism. This means, however, that the endogenous material must already have been in touch with the immunological active centres of the organism before birth. If the natural immunotolerance collapses, **autoaggression** or **autoimmune** diseases appear. This means that endogenous tissue is suddenly recognized as foreign by the immunological control system so that specific defence mechanisms are triggered off against the endogenous tissues. Reasons for this may be alterations of the antigenic structure of the endogenous cells or the release of cell material which has been encapsulated during the embryonic development; changes in the immunological defence system may also be the cause of a break down of the natural immunotolerance. Pathogens may also induce the immune system to target its defence system against endogenous cells. A well-known example of this is the appearance of autoimmune erythropenia during the course of anaplasmosis and babesiosis in cattle. First of all, infected erythrocytes are phagocyted after opsonization. Because antigenic material of the red blood cell is catabolized through the immune system, antibodies are subsequently produced against the material of erythrocytes and later on, non-infected red blood cells are eliminated, which leads to autoimmune erythropenia.

In **acquired immunotolerance**, a foreign antigen is tolerated after its absorption so that after renewed infection, no antigen-antibody reaction takes place. This type of immunotolerance is not only influenced by the degree of antigenicity of the antigen, but also by the degree of development of the immune system (Seifert 1971). A further cause for the development of immunotolerance with practical importance for animal production in the tropics is the practice of "over-vaccination". Often because of a wrong diagnosis of the disease, cattle are repeatedly vaccinated in short intervals. Consequently, the immune competent cells are already charged during their development with the pathogen and are no longer able to respond after maturation. The same effect may appear if the producer of the vaccine excessively increases the amount/dose of antigen according to the principle "a lot helps a lot". This is a phenomenon which can be observed especially in FMD vaccines in South America.

2.2.3.6. *Hypersensitivity*

Beside the normal (normergic) immune response, an **immunopathological reaction** may also develop in both the humoral and the cellular immune systems after the primary immune response as the result of a repeated antigenic stimulus. This divergent reaction of the immune system is called **allergy** or **hypersensitivity**. It is the result of a specific exchange between the partners of the immune reaction (antigen-antibody) which leads to pathological alterations in the cells or tissues of the organism. This mechanism is triggered by the intake of antigenic substances (**allergens**). These are either complete antigens or haptens and are differentiated according to the way they are absorbed, as inhalation, ingestion, injection, skin and resorption allergens for example. If immune cells are associated with the immunopathological reaction, an **allergy** of the **delayed reacting type** takes place. In this case the pathological alterations, i.e. on the skin within the hypersensitive organism, appear 24–48 hours after the injection of the antigen. An example for a local delayed reaction is the tuberculin-reaction. This type of allergy will be enhanced if the assimilation of the antigen occurs through the lymph system.

If immune globulins are involved within this immunological reaction an **immediate-acting allergy** develops which already sets in between a few minutes and a few hours after the secondary contact with the specific allergen. In this case, antibodies have to be transmitted passively, i.e. with serum or plasma, to the hypersensitive organism. The following immunopathological reaction takes place in three phases. During the first phase antibodies are produced which after renewed contact with the antigen react specifically with it. During the second unspecific phase an activation and/or release of so-called mediators results which are a product of the reaction of the first phase. They transform the specific immune reaction into a pathological phenomenon. During the third phase, this pathogenic immune phenomenon appears either as an anaphylactic reaction, in the form of the Arthus-phenomenon, as a serum disease, or as immune toxicity.

Contrary to this older division of allergy types, in German literature the types of hypersensitivity are at present classified according to their origin, as types I–IV (Tizard 1992). The problem of histocompatibility is explained through the characteristic histocompatibility antigens which each individual animal possesses. They are called the major histocompatibility antigens which are inherited through the activities of a linked set of genes known as the **major histocompatibility complex** (MHC).

The MHC contains three classes of genes. Class I genes code the proteins found on the surface of most nucleated cells, and class II code the proteins on the surface of lymphocytes and macrophages. Both class I and II proteins act as peptide receptors which control antigen presentation to lymphocytes. In this way, they regulate the immune responses. Class III genes code a variety of different molecules involved in the immune response. These include complement proteins, tumour necrosis factors, and heat-shock protein.

Type I hypersensitivity (**allergy** and **anaphylaxis**) is an inflammatory reaction which is caused primarily by IgE and leads to the production of pharmacological active substances, mostly histamines. Basically it comprises the allergic reactions of the immediate type or anaphylactic reactions. These reactions are mostly stimulated by antigens from helminths, pollen, and mosquito saliva, but also by lower molecular metabolites of culture mediums as it is contained in unpurified bacterial vaccines, and cell detritus in tissue culture vaccines as well. Adjuvants which are added to vaccines in order to enhance their immunogenicity favour IgE production. The tendency towards the production of allergenic IgE may be determined genetically. The allergic reaction mainly occurs when IgE reacts with mast cells in the connective tissue, but also with basophil leukocytes. This reaction occurs extraordinarily quickly after the formation of the immune complex. At the same time, granula from the interior of the cell are expelled into the surrounding substratum; channels are formed which open into the cytoplasma through which extracellular liquid can enter. Through contact with the extracellular liquid, the mast cell granula excrete vasoactive substances into the surrounding substratum. The cells may also synthesize other vasoactive substances after IgE and the antigen have fused on the surface of the mast cells, especially histamine, which effects blood vessels as well as the smooth muscles locally and generally. Either an acute systemic anaphylaxis or specific allergic reaction (erythema, oedema, urticaria) will result.

The immune reaction of the organism against erythrocytes is characterized as a **type II reaction**. The immune reaction caused by the surface antigen of red blood cells is utilized for the determination of blood groups and is also the cause of haemolytic diseases in new-born children. It has no connection with the main topic of this book and will therefore not be discussed any further.

Type III hypersensitivity is the term for pathological reactions which appear when immune complexes

- are deposited locally within the tissue (Arthus-reaction), when antigen is injected into tissue for which the specific precipitating antibodies are present; this leads to a localized inflammation a few hours later;
- get into the circulatory system in large amounts, i.e. if antigen is applied to a hyperimmune receptor. At the same time, the developed immune complexes are bound to the wall of the blood vessel, complement is activated, neutrophils accumulate, and vasculitis appears. Glomerunephritis may also appear as a consequence of sedimentation within the glomerula along with anaemia, granulocytosis and thrombocytopenia.

The tuberculin reaction is characteristic for the cell-mediated **type IV hypersensitivity**; it is an allergic reaction of the delayed type. It appears when tuberculin (purified bacterial protein, PPD-tuberculin = purified protein derivate) triggers off a T-cell-mediated specific allergic immune reaction. Circulating, antigen-sensitive T-cells make contact with the antigen, and react by dividing, differentiating, releasing lymphokines, and recruiting other lymphocytes. Symptoms of inflammation which appear during the course of the tuberculin reaction are the result of the release of skin-reactive factors as well as the excretion of lysosomal enzymes out of the macrophages (Tizard 1992).

The tubercles which appear during the natural infection with *Mycobacterium tuberculosis* are also the consequence of a type IV reaction. The same is true for the dermatitis or chancre following mosquito, tsetse fly or tick bites. Lower molecular components of the arthropod saliva which are released into the lesion caused by the bite react as haptens and bind themselves to the collagen of the skin. Subsequently, a type IV reaction with strong infiltration of mononuclear cells appears. If the application of the saliva is repeated, the type IV reaction turns into a type I reaction with its respective consequences. Trials to combat ticks with a vaccine produced from substances contained in the saliva of the insects are based on this immune mechanism (I/4.2.1.5).

3. Methods of Immuno- and Chemoprophylaxis

3.1. Immunoprophylaxis

As described under I/2.2.3 the immunization of an organism can occur actively or passively.

3.1.1. Passive Immunization

The natural colostral immunization of the calf is of a special importance amongst the passive methods of immunoprophylaxis within the extensive animal production of the tropics. In autochthonous cattle, premunity against vector-borne pathogens will develop without complications only under the colostral immunoprotection of the mother (I/2.2.3.3). A lot of the empirical experience which the farmers have about natural active immunization is based more or less unconsciously on the effect of colostral immunoprotection on the calf. As a reminder, it should be mentioned again that the maternal antibodies excreted with the milk can only penetrate the wall of the gut and be reabsorbed during the first 48 hours of the calf's life (I/2.2.3.1). Therefore it is important to allow the calf the chance to take milk from its mother as soon as possible after birth under natural conditions. This is the only way it will be able to get a maximum amount of antibodies. In autochthonous breeds, the colostral immunoprotection is much better developed than in exotic cattle which, because of their higher level of resistance, have a higher absolute and relative antibody level in their milk. Calves of autochthonous dams obtain a passive immunoprotection which may last more than 6 months while dams of exotic breeds may protect their calves at the most only 6 weeks against locally enzootic infections.

In connection with the animal health schemes for the different systems of animal production presented in part III, it should be pointed out here that the management of breeding, production and health must be integrated into the conditions of animal production in the tropics. If the mother is vaccinated shortly before calving while grazing conditions are good, she will be able to provide her calf with maximum passive antibody protection. As long as maternal antibodies are present, no active immunization of the calf against soil-borne and contact diseases should be allowed because the high level of circulating humoral passive antibodies would neutralize the antigen which has been applied. This is especially true for FMD and rinderpest, where early vaccination of the calves inevitably leads to the failure of the active immunization. Nevertheless, the maternal antibody production is essential for the natural or artificial premunization of calves. In a management system where the calving season is at the end of the rainy season, it is possible to allow the calves to be premunized under maternal antibody protection during the following dry season using a mild tick-invasion without any harmful side-effects.

The application of hyperimmune sera obtained from donor animals for a short-term prophylaxis, i.e. against anthrax, has become less important. Chemoprophylactics are cheaper, more effective and harmless (hypersensitivity I/2.2.3.6). Only the application of hyperimmune sera for the treatment of tetanus and snake-bite with antitoxins is still necessary. Viral infection in expensive animals (race-horses) will presumably be able to be treated with expensive monoclonal antibodies (horse influenza) in the future.

3.1.2. Active Immunization

In active immunization, an antigen is applied either naturally or artificially to the organism against which it is actively able to produce humoral and/or cellular antibodies itself. The development of the antibody titre depends very much on its constitution, i.e. its level of resistance, and also on the condition of the animal which is mainly influenced by the environment. It also depends very much on the production level of the animal, and also from its stage of pregnancy for example. Cows in early lactation are in a worse condition to produce an active immunity than non-lactating animals. During times of hunger and under conditions of climatic stress the animal only has incomplete resources available with which to produce active antibodies.

During natural or artificial immunization with live, and possibly fully virulent pathogens, the dose of the infection and the condition of the animal are factors decisive in the success of the vaccination. According to the type of antigen a sterile or unsterile immunity or premunity may be the consequence of the immunization. Simultaneous application of chemoprophylactics can prevent the development of the immunity through suppression of the antigen. By protecting the organism from the pathogenic effects of the antigen, a good immunizing effect can also be facilitated (I/2.2.3.2 and I/2.2.3.3).

An effective active immunity against several pathogens can only be obtained with a homologous antigen which is produced for a specific area. This is especially true for clostridial infections, pasteurellosis, and FMD as well. This particular problem is not taken into account in many places of the tropics. In other disease complexes, a good immunity can be obtained through the application of a related pathogen which is apathogenic for a specific species. Such antigens may produce a good challenge-resistant cross-immunity in some cases. Antigens applied in different vaccines for active immunizations are

- **live pathogens**, which may continue to multiply and persist during certain time within the organism. They will produce a high level of homologous antibodies following natural or artificial infection since they stimulate the secondary immune response before very long. To suppress the pathogenicity, either a very low infectious dose is utilized, or the animal is vaccinated while still protected by colostral antibodies or chemoprophylactic measures. Natural strains with low virulence or those which have been attenuated by passages in cultures or in other animal species may be applied without producing serious side reactions. After the application of a live antigen, the animal continues to be a carrier (premunity), depending of the type of the pathogen, for either a limited or unlimited period, thus being a reservoir of infection for other animals (I/2.2.3.3);
- **inactivated antigens** are produced through chemical or physical treatment of virulent pathogens. Along with that, the pathogenic component is damaged while the antigenic and immunogenic component has to be kept in tact as much as possible. This ought to prevent side reactions. When antigens are in- activated they ought not to multiply within the organism. Inactive antigens produce sterile immunity. Vaccines produced from inactivated pathogens mostly contain **adjuvants**. These are chemical additives which, through chemical or physical stimulation of the resistance system of the organism (macro-/microphages), lead to a better immune response, or prolong the re- sorption of the antigen in order to produce a longer lasting immunization and higher antibody titre (I/2.2.1);
- **split-antigens** are non-living parts or metabolic products of the pathogen which have been obtained through chemical or physical manipulation. They have an immunizing property and produce a more or less lasting sterile im- munity. Their effect may be enhanced through the application of adjuvants;
- **toxoids** are obtained by detoxifying (transformating into an insoluble state) exotoxins from bacteria. They are able to immunize the organism against the toxigenic effect of the pathogen. Such toxoids or anatoxins are applied not only where the infection does not take place at all (botulism), but also in in- fections which mainly produce a pathogenic effect through toxins (clostridial infections);
- **synthetic antigens** are mostly parts or epitopes of an antigen which have been produced after gene manipulation on a carrier bacterium (*E. coli*) or virus (*Vaccinia*). The adaptation of such antigens to the mutants and variety of antigens within the natural epizootiological development seems rather difficult. Synthetic vaccines have been tried to prevent FMD, rinderpest and rabies for example.

Efficiency and duration of the immunoprotection obtained through vaccination may be enhanced through several technical and management methods during production and application of the vaccine. When producing the vaccine, one of the most important prerequisites for a good effect is the specifity of the antigen. Homologous pathogens created against specific local pathogens which are used as antigens produce the best and longest lasting protection. For this a precise diagnosis of the disease is a prerequisite. Precise diagnosis is one of the most important problems of the immunological prophylaxis of infectious diseases in the tropics. Trying to overcome the problem of an inaccurate diagnosis using a "shot gun" effect usually fails. The best multiple vaccine is useless if it does not contain the specific antigen. Furthermore, with such vaccines, the immune system is charged with antigens which have nothing to do with the cause of the disease. An additional negative effect arises because multiple vaccines are supposed to contain each antigen component in as great an amount as the number of antigens which are contained in the vaccine. That means a triple-vaccine has to contain three times the amount (doses) of each single antigen in order for its immunizing effect per component to be as good as that of a monovalent vaccine with its single component. It is often obvious in practice that producers do not fulfil these requirements because of economic reasons. This is a further argument for producing homologous monovalent vaccines against the locally relevant pathogen or strain (Seifert et al. 1983).

The amount of antigen applied is directly proportional to the duration of the protective effect. It depends above all on the animal's tolerance to the antigen. High doses of antigen may cause allergic reactions because of lower molecular components which originate in the culture medium. Attenuated or modified vaccines may still be infectious or toxic and even lead to immunotolerance (I/2.2.3.5). The volume of the antigen can also become a limiting factor as far as production costs and shipment are concerned. Through the concentration of the antigen by means of cross-flow filtration, a better efficiency of the antigen and a lower volume which is more easily transportable and applicable can be reached (Seifert et al. 1983).

The application of adjuvants chemically and physically stimulates greater activity in the unspecific resistance system of the organism. The production of cellular antibodies is especially enhanced this way. At the same time, by choosing the appropriate chemical (e.g. aluminium hydroxide), a more delayed resorption of the antigen is achieved and consequently a longer lasting production of antibodies. Adjuvants may assist in the attenuation of antigens and guarantee the innocuity of the vaccine (I/2.2.1).

The efficiency of the vaccine is influenced the way in which it is applied. Some important influencing factors on the production systems with an insufficient veterinary service in the tropics should be pointed out. If the vaccine is applied at a place where it is easiest to do so during blanket vaccination, e.g. into the muscles of the neck, not only may abscesses be produced through contamination or by the adjuvant which has been added to the vaccine, the protective effect will also be rather short because the antigen will be reabsorbed i.m. comparatively quickly. Generally, therefore, attention should be paid to the subcutaneous application of the vaccine. The best place for this in cattle is the dewlap since the vaccine deposit remains there for a longer time. It is well tolerated and will not damage the quality of the carcass and guarantees furthermore a very slow re-sorption and thus an excellent immunizing effect. Nevertheless, where the ap-plication is made also depends on the tolerance of the respective animal species to the antigen preparation. Oil-water or oil-water-oil emulsions are not tolerated well subcutaneously if they are not injected under almost aseptical conditions. In camels, the application of a *Clostridium perfringens* toxoid which was prepared with addition of an oily adjuvant (MONTANIDE) led to a severe inflammatory reaction at the point where the injection was made; it was tolerated without any side reaction by sheep. If the oil-adjuvant-FMD-vaccine used at present in South America is also applied within the "open chute" (manga abierta), it leads to severe local side reactions. In contrast, the i.m. application is tolerated much better under such conditions although this may damage the quality of the carcass, an effect which prevents many of the animal owners from vaccinating at all. The best place for vaccine application is doubtless the outer skin itself. If a vaccine is injected intracutaneously into the fold of skin at the base of the tail using a suspension of anthrax spores, for example, a relatively pathogenic vac-cine can be used because the resorption takes a long time. On the other hand, using a continuous and slow stimulation of the immune system a booster effect which eventually may provide life-long immunity is created. Highly concentrated toxoid vaccines for protection against clostridial infections can either be applied with a tuberculin needle intracutaneously or with a needleless pressure injector under high pressure in a dose of up to 0.1 mL intracutaneously. The needleless "dermojet" is safe and simple to use and guarantees a vaccine deposit which is reabsorbed gradually. The iatrogenic transmission of pathogens which may cause vector-borne diseases is prevented by using needleless injections (Schaper 1991, Seifert et al. 1983).

Through repetition of the injection at the appropriate time, a booster effect can be achieved which may eventually produce an antibody titre which lasts an entire lifetime (tetanus vaccination). If, as demonstrated in Fig. 21, a booster vaccination is applied at the moment where the antibody level from the first vaccination has reached its peak, a maximum number of memory cells will meet with the antigen (I/2.2.2). Since they are still sensitive from the first vaccination, they react with a titre which is about twice as big as the peak reached after the first vaccine application. The result of this is a protracted reduction of the titre and thus a longer-lasting immunity. If, for example, a human is vaccinated with tetanus toxoid and the vaccination is boosted after 1 month and then again 1 year later, immunity against tetanus toxoid can be achieved for a period of up to 10 years.

A further point of view regarding the management of vaccine application is the choice of the time of vaccination. If the vaccination is carried out, when

- the production of the animal has reached its minimum,
- climatic conditions are favourable,
- other health problems are prevented,

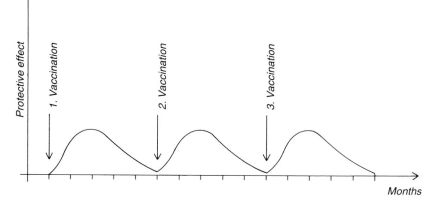

Figure 20. Course of the antibody titre after periodical vaccination during one year.

one can assume that maximum immunity and a long-lasting immunoprotection will be achieved. Within the production systems of the tropics with their very different structures and changing environmental stress on the animal (rainy – dry season) it does not make any sense to establish vaccination schemes along the rules of the industrialized countries of the temperate climate. If in Brazil, for example, it is decreed by the government that FMD vaccinations are to be carried out every 4 months because the vaccine only provides 4 months of protection under experimental conditions, the vaccinations may occur at a time when environmental and technical production factors are putting a lot of stress on the animals. Any responsible ranch manager will refuse to carry out an FMD vaccination during the peak of the calving season and try to prevent the vaccination employing local customs and methods to do so. Vaccination in a highly productive dairy herd may lead to considerable losses in productivity and managers will therefore try to avoid it. If on the other hand the vaccination is adapted into the normal course of the management of the farm, is, furthermore, adapted to the environmental conditions of the region, and is carried out at an optimum moment as a booster vaccination, a longer-lasting effect might be achieved with only 2 instead of 3 vaccinations (Figs. 20 and 21, Seifert 1978).

The prevention of vector-borne diseases caused by *Protozoa* and *Rickettsia* through vaccination leads to a special problem which depends on the phenomenon of premunity, typical for this disease complex. As described under I/2.2.3.3, the premunity only lasts as long as a living antigen is present in the organism. If the antigen is eliminated through the defence mechanisms of the host, or through chemotherapeutic intervention, the antibodies which can be serologically detected disappear after waiting periods of varying lengths. It has been shown that animals are fully susceptible to a natural or artificial infection afterwards. Using this knowledge, an active immunity to these pathogens can only be acquired either through the application of a small dose of the virulent

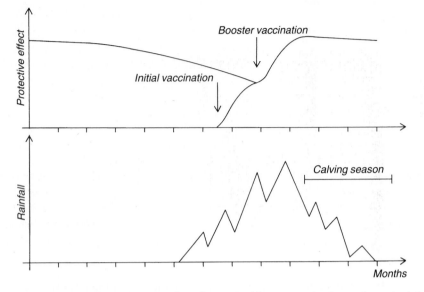

Figure 21. Course of the antibody titre after targeted booster vaccination at the optimal time of the year.

pathogens while simultaneously administering chemotherapeutics to protect the host against the pathogenic effect of the pathogen, or the application of attenuated strains. Examples of this are the artificial premunization with *Theileria parva* in combination with oxytetracycline and premunization against *T. annulata* by administering an attenuated strain (II/1.2.2). The aim here is to induce the premunity artificially whereby only slight or even no clinical side reactions should appear. As experience has shown in the case of babesiosis and anaplasmosis, premunity of this kind is a form of immunity which gives protection against the clinical symptoms of a field infection in very varying ways. Protection, as it is obtained through the sterile immunity known from soil-borne and contact diseases, can never be reached. The most important problem of the artificial premunization is however its inability to provide protection against a superinfection with a field pathogen. The field infection may take an asymptomatic course but will have the consequences of premunity which have been discussed here several times already, meaning that the animal

- may become acutely sick under stress;
- has to support a reduction in productivity which depends on its genotype;
- remains as a carrier of the pathogen;
- because of its carrier state will prevent carrying out measures for the eradication of the disease, for example through serological recognition of carriers which is an eradication scheme which has been used successfully in the control of anaplasmosis (Seifert 1971).

Beside immunization with live antigens, attempts have been made to obtain sterile immunity against vector-borne diseases through the application of dead antigens. Sterile immunity will be achieved through a booster vaccination which may last several months, and allows premunization with minimal clinical symptoms through the natural immunization with live pathogens which follows. Nevertheless, such premunity will encounter the same problems as have been described above. It has been possible lately to multiply *Babesia* and *Theileria* in tissue cultures and isolate soluble antigens of the pathogens. Those antigens are able in particular to stimulate the cellular immune response. In principle, these modern vaccines do not change the basic principal that the unsterile premunity caused by the field pathogen will sooner or later take over and continue the protection against clinical symptoms during the course of sterile immunity. This will be accompanied by the problems described above. The same should be true when intending to vaccinate humans against *Plasmodium falciparum* with such vaccines. The special complication of autoimmune anaemia in the course of such a latent infection is described under II/1.2.2 and II/1.3.4.2.

3.2. Chemoprophylaxis

Chemoprophylaxis is a single or repeated application of chemotherapeutic drugs for the prevention or blockage of infectious diseases. Rolle and Mayr (1993) describe **metaphylaxis** as being the preventive chemotherapeutic treatment of animals which are supposedly infected. This is an indication which becomes relevant under tropical animal production conditions if, for example, the first cases of anthrax appear in a herd and the supposition is that more animals will fall ill. The following may be used as chemoprophylactics/metaphylactics:

- pure chemical substances (sulphonamides) or
- antibiotics.

They may either

- inhibit the development of the pathogen and thus allow the system of host defence to destroy it, or
- the pathogens are directly irreversibly damaged, whereby the cooperation of the host defence is not necessarily required at all.

The application of such compounds which are specific for the pathogen is prerequisite for a well aimed chemoprophylaxis. Through the combination of suitable compounds, a broader spectrum, especially of tropical pathogens, may be registered. It has to be taken into account that the compound has not only to be effective against the pathogen, but also that it has to be harmless to the organism of the animal, and that its metabolites, which may appear within the products of the animal, do not present any danger to the human consumer.

Chemoprophylaxis can be applied within tropical production systems when animals are in danger, because

- they have to be exposed to an infection because of management rèasons, for example during transportation (pasteurellosis), lack of fodder resources during the dry season (soil-borne diseases), and also when it is unavoidable to graze them in an area with high vector activity;
- an enzootic has appeared in the herd, e.g. anthrax or vector-borne diseases, which may be controlled through metaphylaxis. In this case chemoprophylactics are more economic and more efficient than immune sera; furthermore they are easier to be obtained.

The problems involved in chemoprophylactic disease prevention may be best explained using the example of trypanosomiasis.

For the **chemoprophylaxis of trypanosomiasis** a number of compounds is available which differ in their principles of action. Here there is a direct connection between the dosage and the way it acts. If for example diamidin is applied in a low dosage, a short suppression of the pathogen may be obtained. The host defence mechanism also gets a chance to stabilize the premunity and prevent clinical symptoms. In contrast, a *sterilisatio magna* is obtained with a high dose of diamidin which leads to a complete if short-term recovery of the animal. The result, however, is the complete disappearance of the antibodies (I/2.2.3.3) and the animal will again be susceptible to a field infection. Other drugs, like naphthalenes, phenanthridines and chinolines, have a low solubility and consequently are slowly reabsorbed. Therefore, long intervals between treatment become possible with these compounds during which the drug has a suppressive action on the pathogen. The advantage of long intervals of application, however, contains the problem of development of resistance because with the gradually reduced blood level of the drug, a population of pathogens is selected which is resistant (Fig. 22).

Taking into account the special mechanisms of immunity in trypanosomiasis, babesiosis, theileriosis and anaplasmosis, one has to pay attention to the scientifically proven facts that depending on the applied drug, chemoprophylaxis

- can be used to stabilize the premunity;
- should be repeated before the chemotherapeutically suppressed parasitaemia gets a chance to recover;
- is dependent in its duration of efficiency òn the rate of excretion of the respective compound and on the resistance of the specific pathogen against it and
- is limited in its efficiency by the toxicity of the respective drug (therapeutic range).

Chemoprophylactics can be administered orally, subcutaneously, intramuscularly and dermally. For the production systems in the tropics, the oral application

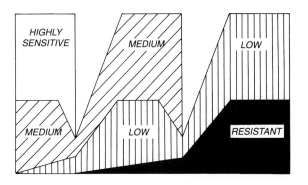

Figure 22. Selection of a resistant population of pathogens.

is the most suitable. Compounds can be combined with the salt lick or supplement feeding. Feed-blocks can be distributed on the range and thus allow continuous application of the compound to the grazing animals. The important thing for the oral application is that it is provided continuously, or at least in calculable intervals. At present, dermal application is the most promising method. Compounds can be dispensed with the "spot-on" method, or regularly and continuously with the "back-rubber" when the animals either walk to their watering place or salt-lick (I/4.3.2.2). The appropriate formulation of the compound which has to allow dermal reabsorption is prerequisite. By adding DMSO (dimethylsulfoxide), the potential for reabsorption can be enhanced. The iatrogenic application should be avoided in tropical production systems since it allows the transmission of pathogens which cause vector-borne diseases. Chemoprophylaxis can assist in carrying out animal production in the tropical environment under rather unfavourable conditions with a high pressure of infection and vectors of disease. But the recommended withdrawal time must be complied with in order to prevent contaminated animal products from being marketed.

References

Bachmann, P. A., W. Eichhorn and R. G. Hess (1984): Active immunization of the dam: Passive protection of the newborn. Anim. Res. Dev. **20**, 23–51.

Baron, D. und U. Hartlaub (1987): Humane monoklonale Antikörper. Fischer, Stuttgart.

Einfeld, H. (1975): Untersuchungen über Zusammenhänge zwischen Faktoren von Haltung und Zucht auf der einen sowie Resistenz und Immunität auf der anderen Seite bei Rindern an marginalen Standorten, dargestellt am Beispiel der Brucellose im Mantarotal der peruanischen Zentralkordillere. Diss. FB Agrarwiss., Göttingen.

Greiling, J. (1979): Untersuchungen über rassebedingte Unterschiede der Aktivität von Komplement und Lysozym im Blutserum autochthoner und exotischer Rinder in Togo unter Berücksichtigung von Umweltfaktoren. Diss. FB Agrarwiss., Göttingen.

Lampe-Schneider, G. (1984): Untersuchung zur Bedeutung der dritten Komplement-Komponente (C$_3$) als Parameter der Resistenz bei autochthonen und exotischen Rindern in Westafrika. Göttinger Beitr. Land. Forstwirt. Trop. Subtr. 8.

Lie, O. and H. Solbu (1983): Evidence for a major gene regulating serum lysozyme activity in cattle. Z. Tierzücht. Züchtbiol. **100**, 134–138.

Margan, U. (1987): Vergleichende Untersuchungen zur Bedeutung der alternativen Komplement-aktivierung bei Rindern und Kamelen. Göttinger Beitr. Land. Forstwirt. Trop. Subtr. 33.

McKeever, D. J. and W. I. Morrison (1990): *Theileria parva*: the nature of the immune response and its significance for immunoprophylaxis. Rev. sci. tech. Off. int. Epiz. **9**, 405–421.

Morkramer, G. (1981): Biometrische Auswertung rassetypischer Parameter von Serumlysozym und Komplement bei autochthonen und gekreuzten Rindern in Togo. Diss. FB Agrarwiss., Göttingen.

Omerod, W. E. (1977): cited by Lampe-Schneider 1984.

Riedel-Caspari, G. and F. W. Schmidt (1991): The influence of colostral leukocytes on the immune system of the neonatal calf. I. Effects on lymphocyte responses. Dtsch. tierärztl. Wschr. **98**, 77–116.

Rolle, M. und A. Mayr (1993): Medizinische Mikrobiologie, Infektions- und Seuchenlehre. 6. Aufl. Enke, Stuttgart

Schaper, R. (1991): Methodische Untersuchungen zur Produktions- und Wirksamkeitskontrolle von Rauschbrand. Diss. FB Math. Nat., Göttingen.

Seifert, H. S. H. (1971): Die Anaplasmose. Schaper, Hannover.

Seifert, H. S. H. (1978): Hygiene management on extensive and intensive cattle units in the Tropics and Subtropics. Anim. Res. Dev. **7**, 49–102.

Seifert, H. S. H., H. Böhnel und A. Ranaivoson (1983): Verhütung von Anaerobeninfektionen bei Wiederkäuern in Madagaskar durch intradermale Applikation von ultrafiltrierten Toxoiden standortspezifischer Clostridia. Dtsch. tierärztl. Wschr. **90**, 274–279.

Senft, B. und F. Meyer (1977): Untersuchungen über die immunologische Reaktionsfähigkeit von Kälbern. Züchtkde. **49**, 193–201.

Spooner, R. L. (1993): Genetics of disease resistance in domestic animals. In: Resistance or tolerance of animals to disease and Veterinary epidemiology and diagnostic methods (eds: Uilenberg, G. and R. Hamers), CIRAD – EMVT, 3–5.

Tizard, I. (1992): Veterinary Immunology – An Introduction. 4th ed. Saunders Company, Philadelphia.

Touré, S. M. (1977): La trypanotolérance – revue de connaissance. Rev. Elev. Méd. vét. Pays trop. **30**, 157.

Weiser, A. (1995): Analyse der Tierhygiene-Situation in mobilen pastoralen Tierhaltungssystemen in der Butana/Nordost-Sudan. Göttinger Beitr. Land. Forstwirt. Trop. Subtrop. **101**.

4. Biology and eradication of Vectors of Animal Diseases in the Tropics

Vector-borne diseases are the most difficult ones to combat in the tropics, but the disease complex itself is the most important one economically in the region. The vector is the indispensable epizootiological link between carrier and susceptible organism where the appearance of the infection is concerned. Because of the extraordinary importance of vector-borne diseases for animal production in the tropics, the biology and eradication of tropical animal disease vectors are discussed in a special chapter. The integration of the eradication schemes of vectors into the production systems of the tropics is discussed in Part III.

Vectors of vector-borne diseases in domestic animals in the tropics are mainly arthropods. Those genuses within the biological system which are important as disease vectors are compiled in Table 1; their importance as disease vectors is presented in Table 2.

The processes for transmitting protozoa, rickettsia and viruses through arthropods are diverse. They depend on the behaviour of the pathogen within the system of the vector on the one hand, and on the alternating ways of transmission on the other. Two ways of **pathogen-vector-relationship** have to be distinguished:

- the **acyclical system**, with which the pathogen is not multiplied within the vector, i.e. both partners are independent of and neutral to each other. The pathogen is carried either externally or internally by the vector and survives within or outside and attached to the vector only for a short time;
- the **cyclical system**, with which the pathogen is multiplied within the vector and may even develop into a further stage which may be indispensable for creating infection. It may even have a parasitic effect on the vector and kill it.

There are four alternatives for the **transmission mode** of the pathogen from the vector to the susceptible host:

- **mechanical or tactile**, with which the pathogen is attached externally to the vector and is transmitted simply through touching;
- **alimentary**, with which the ingestion as well as the transmission of the pathogen is carried out through biting, stinging or licking;

Table 1. Important arthropods as disease vectors in the biological system

Phylum	Subphylum	Class	Subclass	Order	Suborder	Family	Genus
Arthropoda	Chelicerata	Arachnida	Acarina	Metastigmata		Argasidae	Argas, Ornithodorus, Otobius
						Ixodidae	Amblyomma, Aponomma, Boophilus, Dermacentor, Haemaphysalis, Hyalomma, Ixodes, Rhipicephalus
				Mesostigmata		Dermanyssidae	Dermanyssus
	Tracheata (Antennata)	Insecta (Hexapoda)	Apterygota	Hemiptera	Heteroptera	Reduviidae	Triatoma
			Pterygota	Diptera	Nematocera	Phlebotomidae	Phlebotomus
						Culicidae	Anopheles, Aedes, Culex
						Ceratopogonidae	Culicoides
						Simuliidae	Simulium
					Brachycera	Tabanidae	Tabanus, Haematopota, Chrysops
					Cyclorrhapha	Muscidae	Stomoxys, Haematobia, Musca
						Glossinidae	Glossina
						Calliphoridae	Calliroga
				Siphonaptera			Pulex, Xenopsylla, Ctenocephalides, Tunga etc.
				Phthiraptera	Anoplura	Haematopinidae	Haematopinus
					Mallophaga		Melophagus

Table 2. Vector-borne diseases and their vectors

Disease	Ixodidae	Tabanidae	Muscidae	Hypodermatidae	Culicidae	Ceratopogonidae	Simuliidae	Phthiraptera	Rhynchota	Siphonaptera
Nagana		○	○							
Surra		○	○		○					
Mal de Caderas			○							
Chagas disease									○	
Texas fever	○									
Redwater	○									
East Coast fever	○									
Corridor disease	○									
Mediterranean coast fever	○									
Benign bovine theileriosis	○									
Q-fever	○									
Heartwater	○									
Haemobartonellosis	○									
Eperythrozoonosis	○						○	○		
Anaplasmosis	○	○	○	○	○	○				○
Rift Valley fever					○					
Three day sickness						○				
Lumpy skin disease					○					

- **alimentary-excretoric**, with which excretions contaminated with the pathogen (coxal liquid) are deposited while feeding on the host;
- **germinative or transovarial**, with which the relationship between pathogen and vector is cyclical or acyclical, but so close that the pathogen is transmitted by the female vector through its ovaries; a congenital infection takes place (Mitscherlich and Wagener 1970, Seifert 1971).

4.1. The Biology of Vectors

Vectors of vector-borne diseases are usually also ectoparasites which are parasitic externally on humans, animals or plants, feeding on blood, parts of the skin, hair, feathers or plant juices in order to preserve their species. Parasites appear mostly in large numbers and damage their host through the withdrawal of nutrients, by damaging tissues or through excretes (toxins), which they produce while feeding on the host. Furthermore, those ectoparasites which are vectors are transmitters of pathogens.

4.1.1. Ixodidae (Ticks)

Ticks belong to the arthropod phylum and to the class of *Arachnidae*, subclass *Ixodidae*. For domestic animals they are the most dangerous and most widespread of ectoparasites and disease vectors.

Ticks occur in the temperate as well as in the tropical and subtropical regions of the world. They adversely affect animal health especially in the tropics. Ticks and tick-borne diseases constitute a major constraint to livestock production in tropical and sub-tropical areas (Bram 1975). Total losses have been estimated in the range of US$ 7 billion annually, with 80% of the world cattle population of approximately 1.214 million at risk from ticks and tick-borne diseases (McCosker 1975, 1979). Ixodid ticks (hard ticks) are important in veterinary medicine, primarily as vectors of a wide spectrum of pathogenic microorganisms, such as protozoa, rickettsia, spirochaetes and viruses. In addition, they may also cause direct damage such as reduction of quality of hides, reduction in live weight gain, anaemia, toxaemia, and paralysis (Jongejan 1990).

The various tick species are numerous in shape, size and way of life. Teneral ticks almost have the shape of a small flat bean; when engorged, the females may increase their weight many times over and appear round and plump. Ticks have a tough, leather-like skin and live from the blood of the host animal which they attack in large numbers.

In contrast to insects, ticks do not bite. Instead they dig a small pool with their proboscis which gradually fills with blood which they then suck up (pool feeder). This can take up to several days. Because of its elastic cuticula the parasite may increase its volume considerably.

Ticks are obliged to be parasites because they are unable to feed on anything but a living being. The host will lose important nutrients under a heavy attack by ticks. Because of the loss of blood, anaemia and general weakening may appear which leads to a reduction of milk production, daily weight gain and a reduction of host resistance against other diseases. Because of the many lesions caused through feeding by the ticks, the quality of the leather and even the quality of the carcass may be damaged; these lesions may become the location for skin infections (dermatophilosis, II/3.3.2.2). Enzymes which are secreted during feeding in order to inhibit the coagulation of the blood of the host may cause severe diseases because of an allergic reaction (tick toxicosis, sweating sickness). The most important damage which is caused by ticks in animals and at the same time for the owner of the animal is the transmission of pathogens of vector-borne diseases.

Two families belong to the subclass *Ixodidae*: the *Argasidae* and the *Ixodidae* which differ in their morphology and parasitism. Both families have in common the development from the egg to the six-legged larva, to the eight-legged nymph and then after renewed moulting, to the imago. In order to localize the host, both families of ticks possess a number of chemoreceptors which are placed along the fore legs, above all in a cavity, called the Haller organ.

The families may be divided according to the following characteristics (modified according to Mehlhorn and Piekarski 1989):

	Ixodidae	*Argasidae*
Cuticula:	Tough	Leathery
Shield:	Covers the entire back of the male, and a part of the back of the female, nymphs and larvae. It is extended during engorgement.	Non existent
Capitulum:	Protrudes over the cranial edge of the tick	Only dorsally visible in larvae
Stigmata:	Situated behind coxes IV	Situated on coxes III
Eyes:	Made up of a cuticular lens and sensory cells, placed at the edge of the shield (except *Ixodes* and *Haemphysalis*)	Mostly non existent
Nymphal stage;	1	Mostly 2, in some species up to 8
Mating:	Once; males die afterwards	Several times
Oviposition:	Females die afterwards	Repeatedly after mating following engorgement
Egg number:	*Amblyomma* 15000 *Boophilus* 20000 *Dermacentor* 6000 *Ixodes* 3000	Several hundred
Habitat:	Outside	Stable, nests, etc.
Hosts:	1–3	Several

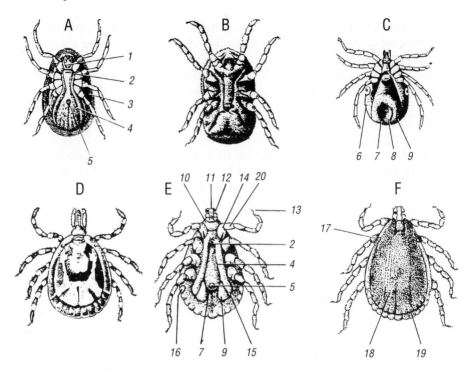

Figure 23. Morphological diagnostic features of *Argasidae* and *Ixodidae* (Mitscherlich and Wagener 1970).

A   *Argas reflexus* ♀, ventral   B   *Otobius megnini,* nymph, ventral
C   *Ixodes pilosus* ♀, ventral   D   *Amblyomma variegatum* ♂, dorsal
E   *Rhipicephalus capensis* ♂, ventral   F   *Rhipicephalus capensis* ♂, dorsal

1 Capitulum; *2* Genitalporus; *3* Supracoxal fold; *4* Genital groove; *5* Anus; *6* Lateral shield; *7* Anal groove; *8* Anal shield; *9* Adanal shield; *10* Basis capituli; *11* Hypostome; *12* Palpae; *13* Glaws; *14* Coxa I; *15* Accessory adanal shield; *16* Spiraculum; *17* Eye; *18* Scutum; *19* Clasps; *20* Trochanter

The morphological differences between *Ixodidae* and *Argasidae* are also described in Fig. 23.

Argasidae are wandering ticks which only remain on their host while feeding. During their development they attack several hosts, and so are temporary parasites. *Argasidae* prefer stables, cracks in walls and similar hiding places where the female can deposit its eggs. After about 3 weeks, the six-legged larvae emerge out of the eggs which then remain on a host for 5–10 days. They gorge themselves on blood, drop off the host and look for a hiding place where the eight-legged nymph emerges after several weeks. The nymphs go through two stages and attack a host once in each phase for about ½–1½ hours in order to engorge themselves. Between both phases there can be an interval of several months. After moulting, the adult (imago) tick emerges out of the nymph.

The entire cycle of development of *Argasidae* from egg to imago lasts about 9 months; their lifetime may last several years since they survive for years without

Table 3. Ixodidae important as vectors (listed according to their importance as vectors of pathogens[a])

One-host ticks	*Boophilus annulatus*
	Boophilus decoloratus
	Boophilus microplus
Two-host ticks	*Hyalomma detritum*
	Hyalomma turanicum
	Rhipicephalus bursa
	Rhipicephalus evertsi
Three-host ticks	*Amblyomma gemma*
	Amblyomma hebraeum
	Amblyomma pomposum
	Amblyomma variegatum
	Dermacentor spp.
	Haemaphysalis bispinosa
	Haemaphysalis punctata
	Hyalomma dromedarii
	Hyalomma excavatum
	Hyalomma marginatum
	Hyalomma truncatum
	Ixodes ricinus
	Ixodes dammini
	Rhipicephalus appendiculatus
	Rhipicephalus simus

[a] Those pathogens which are transmitted by ticks are described and compiled in Part II of this treatise.

feeding. Vectors of vector-borne cattle diseases from the *Argasidae* family are the genus *Ornithodorus (Coxiella burnetii)* and the species *Argas persicus (Anaplasma marginale)*.

Ixodidae (hard ticks) are the disease vectors which are most important for animal production in the tropics; those species which are important vectors of vector-borne diseases are compiled in Table 3. *Ixodidae* are mostly field parasites. If the management of the pasture and the stocking rate are increased, the chances of the tick of finding a host are increased as well. Especially in the case of the tactile transfer of diseases through ticks, a new host may be infected rather quickly if the pathogens which are transmitted have a low resistance outside the host organism.

The family of *Ixodidae* is very varied in shape and species. As the scientific nomenclature indicates (ixos gr. = fly glue, ixodes = glued), they are "sticking" ticks and stationary parasites. They cling to their host with their oral apparatus and only engorge themselves once during each stage of development.

The adult female remains about one week on the host where it soaks up considerable amounts of blood and copulates with the male while feeding. Afterwards, it drops off and deposits up to 20000 eggs under the protective cover of grasses, stones or dry leaves. After that, it dies.

Under suitable climatic conditions (mostly hot and humid), the eggs open after a while and the six-legged larvae emerge. They wait on the tops of plants for a host animal to pass by. They localize their target by means of their chemoreceptors (Haller organ) which are on the upper side of the tarsus and, by waving their fore legs in the air, they manage to cling on to the host animal. They are especially attracted by dark, moving objects; a long coat makes it easier for the ixodid tick to attach itself to the host. Experience in South America has shown that short-haired and bright coloured Charolais-Cebu-crossbreds have less of a tick problem than Aberdeen-Angus cattle with a long, dark coat. Many larvae which do not find a host die. The preservation of the species is only guaranteed because *Ixodidae* produce enormous numbers of eggs. As soon as they have found a host, the larvae seek out places on the animal where they are protected and have favourable conditions for their development, e.g. the inner-side of the fore and hind legs, and around the udder and scrotum, anus and vulva. They prefer to bite into thin parts of the skin. At this point, the saliva is used not only to inhibit blood coagulation, but also as a local anaesthetic. In a course of about 6 days they feed without interruption on blood after which they moult. This is the stage when the eight-legged nymphs, which are sexless, emerge. As soon as the nymphs hatch out on the ground or on the host depending on the species, the act of feeding is repeated and after moulting again, the eight-legged imagines which are sexually differentiated emerge. It is usually only the female imagines which feed on blood and only half of their bodies, which is covered with the shield, expands to the size of a bean or even a pigeon's egg. During feeding, they copulate with the males which scarcely feed on blood and spend their time creeping around on the skin of the host.

Ixodidae are one-, two- and three-host ticks. According to the species, the development is either completed on one, two or three hosts. The knowledge of the behaviour of the tick on the host animal is prerequisite for any measure of control and eradication.

In **one-host ticks**, the larvae which emerge from the eggs 3–4 weeks after deposition at the earliest attach themselves to a host animal where they complete their entire development. On the host they develop from nymph and imago and then copulate. Afterwards, they drop off and deposit their eggs on the ground. The entire development cycle takes mostly 19–21 days as a rule, with a minimum of 15, and a maximum 40 days, each stage taking 1 week (Fig. 24).

The **two-host tick** attaches itself as a larva to a host, feeds on blood and develops into the nymph stage. After a maximum of 14 days, it drops off on to the ground where it reaches the imago stage in 20–30 days time (Fig. 25). Male as well as female ticks then look for another host, feed on blood and copulate. After a further 6–11 days, the female drops to the ground and deposits its eggs. The entire cycle from the time the larva emerges from the egg until the engorged female deposits the eggs depends on the time the nymph needs on the ground to find a new host. According to the species the nymph may survive on the ground for several weeks.

The **three-host tick** looks for a new host during each stage of development in

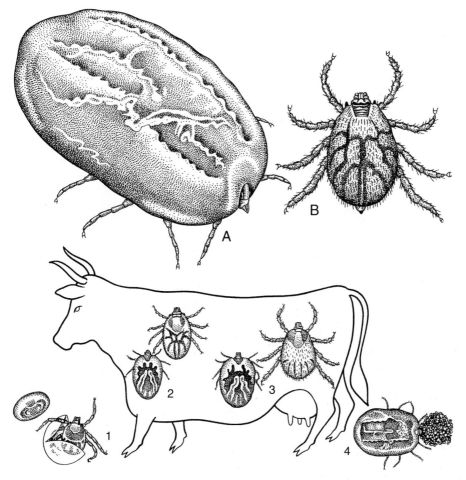

Figure 24. Development of one-host ticks (*Boophilus* spp.; Wellcome 1976).

A Fully engorged female tick; B Male tick.

1 Larvae hatch from the egg; 7–10 days later the larvae climb onto vegetation and search for a host. 2 The larvae find a host and feed on blood for 3–5 days. They moult afterwards, and the hatched nymphs feed on blood for 3–6 days. 3 The fully engorged nymphae enter a stage of moulting which lasts 2 days, and from which develop sexually differentiated males and females. During mating, the female feeds on blood for 4–5 days and finishes its residence on the host with a large blood meal. 4 About 30 days after the larva has found a host, the engorged female drops to the soil and deposits about 20000 eggs in a humid hiding place.

order to feed. The larva emerges from the egg on the ground, looks for a host, feeds on it for 3–7 days, drops off and moults after 3–4 weeks on the ground. The nymph attacks a second host in order to feed on it for 3–7 days, leaves it and develops into an imago on the ground after 2–8 weeks. After that, the adult tick looks for a third host to feed on and for copulation which takes 1–3 weeks. Finally it drops off and completes the cycle with oviposition on the soil. Because

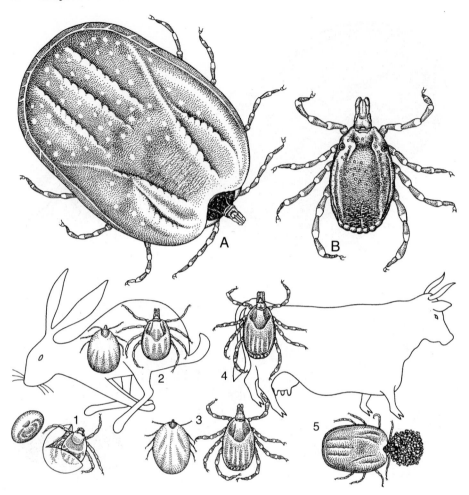

Figure 25. Development of two-host ticks (*Hyalomma*; Wellcome 1976).

A Fully engorged female tick; B Male tick.

1 The hatched larvae search for a host from small mammals or birds. 2 The larvae feed on blood and hatch to nymphae. Subsequently the nymphae take their blood meal. 3 The fully engorged nymphs drop to the soil, the sexually differentiated males and females hatch and find a new host; they regularly choose large animals. 4 The imagines mostly stick to the region around the anus and vulva, copulate and the female takes a large blood meal. 5 The fully engorged females drop to the soil and deposit up to 10000 eggs.

of the different amounts of time in each stage spent on the ground, the entire development cycle may last up to 1 year (Fig. 26).

Ticks are found on all continents and are bound to certain climates as far as the requirements for temperature, humidity, sun-radiation and shade of each species are concerned. Furthermore, each species requires specific environmental

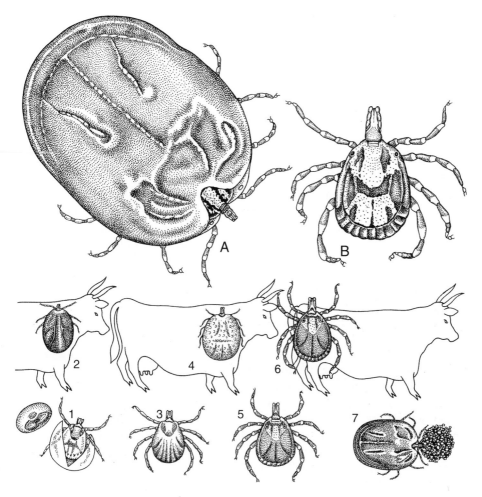

Figure 26. Development of three-host ticks (*Amblyomma* spp.; Wellcome 1976).

A Fully gorged female tick; B Male tick.
1 Larvae hatch and climb on the vegetation in order to find a host. 2 On the host the larvae search for mostly hairy parts on the head or the trunk and feed on blood. 3 The fully gorged larvae drop to the soil and hatch to nymphae. 4 The nymphae search for a host and feed on blood on similar parts as the larvae. 5 The fully gorged nymphae drop to the soil; males and females hatch. 6 The imagines find a host and search for sites on the underbelly, scrotum, udder or around the anus or vagina. After copulation, the females gorge on blood. 7 About 10 days after finding its host, the female drops to the soil and deposits up to 20.000 eggs in a protected humid place.

conditions for its habitat. The respective species only have a chance for survival when these prerequisites are fulfilled.

An identification key of the ixodides species which are important as disease vectors is presented by Mitscherlich and Wagener (1970) and Seifert (1992).

4.1.2. *Diptera* **(Flies)**

Important disease vectors for animals in the tropics are also the families of *Tabanidae* and *Muscidae* which belong to the subgenus *Brachycera* and *Cyclorrhapha*. *Hippoboscidae* and *Calliphoridae* are only indirectly important as disease vectors.

4.1.2.1. Tabanidae (Horse flies, Gadflies)

Many species belong to the family *Tabanidae*; the genuses *Tabanus*, Chrysops and *Haematopota* are important vectors. *Tabanidae* are sturdy insects, up to 30 mm long, having thick hind quarters, over which its broad wings protrude when the insect is in a resting position. The strong broad head has two large compound eyes which are coloured and iridescent. In several species, an additional brow-eye may exist. The feeler and the proboscis are sturdy. Because of their sturdy bodies, tabanides are good flyers with a lot of stamina. The males live only on parts of plants while the females are amongst the worst blood-sucking parasites known. The bite of these vectors is very painful, the wound may bleed a lot and easily get infected. A blister (chancre) appears accompanied by severe itching. High activity of tabanides may make cattle restless meaning that rumination and feed utilization will be reduced. For their development, tabanides need humid muddy soil or stagnant water ponds with plant vegetation where the female deposits 100–1000 eggs. After a week, the tiny larvae emerge, living under the surface of the water or within the mud where they feed on the larvae of other insects. After moulting at least 7 times, the pupa appears. Moulting takes place for the last time on the surface of the water, and the change into a chrysalis takes place on land. During the dry season, the tabanides emerge out of the pupa. The development cycle of the tabanides lasts about a year. It depends on the temperature and may be interrupted by a resting period during the colder season of the year. Tabanides are vectors of anaplasmosis and trypanosomiasis. While feeding on an infected animal, the pathogens are taken in through the proboscis and transmitted to a healthy animal the next time they feed; they do not penetrate further the organism of the tabanide. A second host has to be found soon since the pathogens only survive a short time within the hypopharynx. The mode of transmission is exclusively acyclical and alimentary.

4.1.2.2. Glossinidae (Tsetse flies)

Glossines are yellow-brownish insects, 6–13.5 mm long. They transmit Nagana between domestic animals and sleeping sickness between humans. The position of their wings while resting is characteristic: one wing covers the other like a pair of scissors. The typical proboscis with lateral palpae protrudes over the head looking like a protruding tongue (in German tsetse flies are also called "tongue" flies) (Fig. 27).

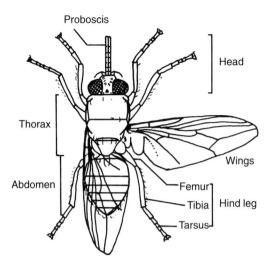

Figure 27. Morphological features of glossines.

The family *Glossinidae* is divided into 22 species and 33 subspecies, they can be summarized into 3 groups (Table 4).

While the larger species of the *G. morsitans* group (9–14 mm long) mostly live within the open savannah, the *G. palpalis* group is limited to living along the rivers. The environment of the flies of the *Fusca* group is the tropical rain-forest. The different species may be recognized by their markings and their size (Figs. 28 a–e).

Table 4. Groups of tsetse flies – species/subspecies and their habitat

Species/subspecies	Habitat	Mean temper-ature (°C)	Humid-ity (%)	Feeding time (h)	Vector
• *Glossina morsitans* group **Savannah**					
G. austeni	Savannah, open forest,	27	40–60	9–11	T.b. brucei
G. m. morsitans	independent of open water			15–17	T. congolense
G. m. submorsitans	up to 1600 m				T. simiae
G. m. orientalis					T. vivax
G. longipalpis	Coast and inland savannah, fields in the rainforest	24–30	50–80	9–10 15–16 (at night)	Nagana
G. pallidipes	Open and dense forest of the savannah up to 1860 m, also dry savannah			7–8 17–18 (at moon-light)	T. congolense
G. swynnertoni	Acacia savannah as a rule, excluding, G.m. morsitans			9–17	T.b. brucei

Table 4. Continued

Species/subspecies	Habitat	Mean temperature (°C)	Humidity (%)	Feeding time (h)	Vector
• **Glossina palpalis group** **Transitional forest**					
G. palpalis palpalis G. p. fuscipes G. p. gambiensis G. fuscipes martinii G. fuscipes quanzensis G. caliginea	Rain- and gallery forests up to 1800 m	28	70–80	8–11 15–18	T.b. brucei T.b. gambiense T. vivax
G. pallicera G. p. newsteadi	Dense, humid forest close to water			16–18	
G. tachinoides	Forest and bush at the edge of the savannah	28	60–70	8–11 15–18	T.b. brucei T. vivax
• **Glossina fusca group** **Rainforest**					
G. brevipalpis	Adaptive from gallery forests to scattered brush				T. congolense T. simiae T. suis
G. fusca G. fusca congolensis	Rain- and gallery forests at least 200 m wide and up to 1500 m	20–30	80–90	19–23	T. simiae T. congolense
G. fuscipleuris	Dense riverbank vegetation up to 1800 m		80–90		Nagana
G. haningtoni	Shady forest close to rivers and creeks			16–18	
G. longipennis	Thorn brush vegetation		40–60	16–18	
G. medicorum G. nashi	Riverbank thicket up to 2 km in distance from the river				
G. nigrofusca	At the edge of rainforests and the transition to the savannah				
G. schwetzi	Less dense gallery forests			17–18 (at night)	
G. severini					
G. tabaniformis	Rain forest			17–18 (at moonlight)	

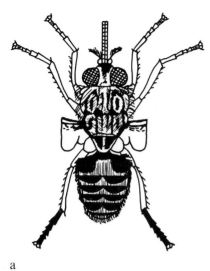

a

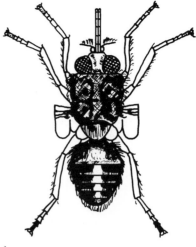

b

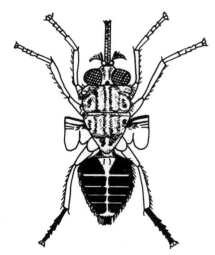

c

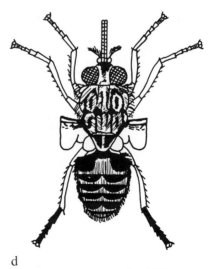

d

Figures 28a–d.

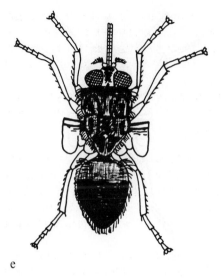

e

Figures 28a–e. Body markings of glossines. a *G. pallipides*, b *G. morsitans*, c *G. palpalis*, d *G. tachinoides*. e *G. fusca*.

Figure 29. The big African river systems

The most important vector of bovine trypanosomiasis is the *G. morsitans* group. It transmits *T. vivax, T. congolense* and *T. b. brucei* which causes Nagana. *T. b. gambiense* and *T. b. rhodesiense*, which cause sleeping sickness are mostly transmitted by flies of the *T. palpalis* group, but also by species of the *G. fusca* group.

The transmission of trypanosomes is very much correlated with the preference of the respective tsetse species for a particular host. While *G. morsitans* mostly feeds on *Bovidae, G. palpalis* prefers reptiles. It may also feed on primates, *Bovidae, Suidae* and birds.

The area where Nagana is enzootic coincides almost completely with the habitat of the glossines. The tsetse region, about 10 million km^2 in size, is located on both sides of the Equator in West and Central Africa and as far as East Africa. It is the humid region of the African continent in the area of the large river systems located generally between the latitudes of 14° north and 29° south (Fig. 29).

Glossines are not distributed evenly since there are also areas which do not have an environment which is suitable for their physiological requirements within the tsetse belt. Each species is bound to a specific ecology; it reflects the arrangement of a particular plant association (Table 4). Temperature and humidity as well as the vegetation are limiting factors which are especially important for the pupal period of the vector. Optimum temperatures are between 20 and 28 °C, with a relative humidity between 50 and 80% and an annual rain-fall between 635 and 1524 mm. The lower temperature limit is 16–18 °C, because at lower temperatures the fat reserves are used up; at higher temperatures they are metabolized too quickly. Equally, highlands at over 2000 m, dry and desert-like regions, and cultured areas where no grazing exists are unsuitable habitats for tsetse flies. Highly populated areas also tend to split the tsetse populations up into smaller ones.

The size of the area of distribution of a fly population depends on the season of the year. During the late dry season, when most parts of the savannah are defoliated and the climatic factors are drastically altered, the glossines retreat into areas which also provide shade during this season and guarantee the environment they require. These are above all the wooded borders of rivers as well as groups of evergreen trees where the climatic contrasts are milder. The tsetse flies live and reproduce the whole year round either within these retreats or in their permanent habitat. With the onset of the rainy season in the surrounding areas, the foliage of the trees re-emerges and with that a favourable habitat for the glossines. These areas are populated as secondary habitats until the next dry season when the tsetse fly is pushed back into its permanent area of distribution. The fly belt which is specific for each species is made up of the permanent and secondary habitats together with those areas which are invaded when the flies are hunting for their blood meal.

The extension of a fly belt through the invasion of an area which has so far been free of tsetse flies is only possible during years with an especially favourable climate, provided the flies have not been carried into the area. The flies multiply much more than normal under those conditions. Over-population occurs, and the flies may emigrate and pass through areas which would normally be unsuitable to cross. During such favourable years, areas may be reached which are suitable as new permanent retreats. During unfavourable years, however, such habitats can be eliminated.

The tsetse fly may also reach new permanent habitats through the help of a carrier. The spreading of the flies along stock-routes is well known and especially problematic. The flies follow the animals. Flies may also be carried with vehicles, both on land and on water. Contrary to the earlier opinion that glossines will be ousted from their habitat through human settlement, it has been found out that the tsetse flies are even able to adapt to a cultivated environment. Kaminski (1983) found breeding places of *G. palpalis* in oil palm plantations in Liberia.

The ***Glossina morsitans* group** of the savannah flies is the most important for animal production within the tsetse belt. The group is usually distributed in *Brachystegia* forests in the dry savannah in East and Central Africa, and in *Isoberlinadoka* forests in West Africa. During the rainy season it also invades the northern Sudan and Guinea zones, while during the dry season it retreats to the humid areas along the rivers as well as into denser vegetation. This seasonal rhythm is almost non-existent within the southern Guinea zone so that *G. morsitans* becomes a year-long hazard for cattle which graze during the dry season in this area. As already mentioned, as the most important vector for trypanosomes pathogenic for cattle, *G. morsitans* is the most dangerous of flies and also the most difficult to eradicate, especially since it feeds a lot on wild game which often is heavily infected.

The ***Glossina palpalis* group**, also known as the riverine or gallery forest fly, is distributed along the rivers which flow into the Atlantic or Nile. It is mostly endemic to the West African forests, in the northern part up to the Sudan zone, and in the south down as far as the coastal areas. Within the tropical rain forests, *G. palpalis* is concentrated in the humid areas along the rivers. In northern Nigeria, *G. palpalis* and *G. tachinoides* occupy up to 75% of the country. When the cattle have to enter the humid areas and reach the rivers during the dry season, these vectors become important vectors of Nagana. They have, however, a special importance as vectors of sleeping-sickness.

The ***Glossina fusca* group**, or rain-forest fly, is divided into a large number of sub-species. It is endemic to the tropical rain-forests of the large rivers and along the Guinea coast. Two sub-species, *G. brevipalpis* and *G. longipennis,* are not necessarily dependent upon the rain-forest. Rather little is known about the importance of this group of vectors because, amongst other things, almost no animal production takes place in the rain-forest (Table 4).

In general it is true that no species is able to live on the treeless steppes. Some species may enter this area for hunting, especially the *G. morsitans* group. They always have, however, to get back to relatively near forests or groups of trees in order to get protection against radiation from the sun. Their pupae are unable to complete their development on open ground. They also die because of the action of the sun even if deposited some centimetres into the soil.

Light-sensory and smelling organs assist the tsetse to recognize its host from a distance. It has a special preference for dark objects. This piece of knowledge has been used for the construction of fly-traps (I/4.2.1.2). Flies mostly settle on the side of their host which is in the shade. The body-temperature of the host is also important. Recent studies have shown that hair-shaped sencillae are present on the front side of the wings of *G. f. fuscipes*. The external appearance of these hairs suggest that they have a gustatory chemoreceptor function probably in association with the mechanoreceptor. The gustatory cells are stimulated by different substances (tsetse flies excrements, sex pheromones) (Deportes et al. 1994).

Before starting to feed, the tsetse probes the skin of the host with its proboscis. It is, like ticks, a pool feeder and cuts the capillaries with its rasp-like oral apparatus in order to produce small haematomas which it sucks out. Before and during sucking, heparin-like saliva is secreted from the labial glands in order to prevent blood clogging within the hypopharynx, oesophagus and the first part of the gut. The fly feeds every 2–3 days, the intervals varying according to the season. During the dry season, the fly may feed daily, while during the rainy season the fly may fast up to 10 days. The amount of blood taken during one meal can be greater than the body weight of the fly, liquid and organic salts already getting excreted during feeding. This primary excretion takes place by means of the Malpighian vessels. The secondary excretion of nitrogen containing metabolites occurs many hours later (Vogler 1976).

As far as their **reproduction** is concerned, glossines have a special place amongst insects. They give birth to living larvae and are extremely careful with their brood. One single larva is fed at a time with "milk glands" in the uterus each time and is developed until the third larva is born after 10–12 days. After hatching in loose organic material in the shade of stones, groups of trees, fallen trees, tree trunks or similar material, the larva immediately hides in the soil. The flies hatch after about 43 days. Copulation usually only takes place once after the 3rd or 5th day of life of the female, frequently after the first feeding period. The males copulate with the female when they are 7 days old. 9–10 days after hatching, the first mature egg is fertilized within the uterus with the spermatozoa which are deposited within the spermatotheque during the entire reproductive life of the female. The female lives about 3 months and during this time only gives birth to 7–12 larvae. From this point of view, the prospects of eradicating glossines are good because of their low fertility rate. The larva enters into the pupal state within 1–2 hours after birth. The puparium which covers the larva has a different shape within the different species. The pupal stage lasts 2–13 weeks. It depends on temperature and humidity. The conditions which have to be appropriate for each species guarantee a minimal consumption of the fat reserves. Unfavourable conditions lead to a premature consumption of the fat reserves which may hinder the hatching of the imago. The resistance of the species to environmental conditions is also an indication of the type of habitat of the respective species. Temperature is more important for the development of the pupa than for the imago. Like all holometabolic insects, the tsetse is able to walk immediately after hatching and to fly within a few hours.

Both *Glossina* sexes only feed on the blood of vertebrates. Since several host animals which exist in the tsetse belt are infected with trypanosomes, the infection of the tsetse fly takes place while it is feeding (Vogler 1976).

4.1.2.3. Muscidae (House and Stable flies)

There are species belonging to the family *Muscidae* the adults of which possess a licking (*labellae*) or biting apparatus. They feed either on decaying organic matter or blood. Flies which only feed by licking can also become vectors of pathogens through passive acyclical transmission, i.e. between animals with bleeding wounds (myiasis).

Musca domestica, the big house fly, has as a distinguishing feature within the hyaline wing which is a central vessel (4th) which becomes bent towards the tip. The fly is 6–9 mm long, dark grey, partially yellowish in colour, and has dark stripes on the thorax. It deposits 30–130 eggs all at once which are up to 1 mm long, creme coloured and slightly curved. The female lives only 3–5 weeks during which it deposits its eggs 2–5 times in the dung in stables on humid straw and the humid remains of feed and organic refuse. The first characteristic maggot-like, worm-shaped larvae develop out of the egg, 1–2 mm long, pointed at one end, whitish, transparent, and they feed on decaying organic matter. They may also nourish themselves on blood (myiasis) and develop on to the second and third stage. The third-stage larva leaves the brood substratum and bores itself about 50–65 cm deep into the floor of the stable in loose dung of the pen in order to start its pupation. The barrel-shaped brown or brownish-black wart-like pupa, which is up to 8 mm long, hatches out into the imago by expelling a circular lid. Depending on the surrounding temperature, the development takes 8–50 days, and on average 14–21 days. At temperatures of below 15 or above 40 °C, the mortality rate of eggs and larvae is increased. The same is true for dryness. Within a dung heap, the larvae migrate to the edges since the temperatures are too high in the centre. Larvae and pupae survive the dry season protected in humid dung or similar organic matter. The imagines have a scope of activity of 1–3 km, or even up to a maximum of 40 km. In the tropics, 9–12 generations of flies may develop in 1 year. Within intensive poultry and pig production systems, 15000–20000 larvae may be found for every kilo of dung. House flies become vectors of pathogens in intensive production systems (feed-lots) if the animals are bleeding because of horn or *Cochliomyia hominivorax* maggot wounds. The pathogen can be extracted from the wound by licking and be delivered to another animal the same way (Mehlhorn and Piekarski 1989).

Stomoxys calcitrans (stable fly) is a vector which occurs worldwide and is especially common in the tropics. The insect is about 6 mm long and has a long horizontally protruding pointed proboscis. The body is grey and marked with 4 dark stripes. In the resting position, the wings are held in a characteristically spread out manner. Both sexes feed on blood and are not only vectors of trypanosomiasis and anaplasmosis, but also make animals restless and feeble if they appear in large numbers. The female stomoxydines live about 70 days and deposit hundreds of eggs in decaying material, but best of all in dung with a high straw content. The larvae hatch after a short time and feed on decaying organic matter. After moulting, they enter the pupal stage in about 2–3 weeks. After a further 9 days, the imagines hatch out of the pupae (Mehlhorn and Piekarski 1989).

Haematobia (Syphona) irritans (horn fly, span. mosca de paletilla) is the most important vector of the genus *Haematobia*. The sting of the small fly which is only 4 mm long is very painful. If cattle are attacked by a large number of flies, they suffer not only a remarkable loss of blood but also become very restless. When feeding on the animal, the fly rests in a characteristic position with wings spread out which gives it a delta-shaped look. The vectors prefer to sit around the area of the shoulder (paletilla) of the host. Bulls seem to be preferred apparently because of the attraction of their specific pheromones. During vaccination, mating and at other times when animals are concentrated, the flies change their host in large numbers which leads to a massive acyclical alimentary transmission of, for example, anaplasmosis. Trypanosomes can also be transmitted by the horn fly.

The horn fly does not only develop in the hot humid climate of the tropics: in the Andes, anaplasmosis has been transmitted in altitudes of up to 3500 m. The fly remains for a long time with the same animal, or at least with the same group of animals, and lays its eggs into the freshly deposited dung of the animal. A dung heap provides ideal conditions, and in a favourable environment, the larvae may already hatch after 20 hours. 4 days later the larvae have matured to pupae, and the adult flies emerge one week later. In the tropics, the outer layer of the dung heap dries rather quickly and forms a crust which maintains the humidity required for the development of the fly. This means that the fly has a habitat which is closely connected to its host. The horn fly finds especially good conditions where dung beetles (*Geotropinae*) are not common (Seifert 1971).

4.1.2.4. Hippoboscidae (Louse flies)

Hippoboscidae are louse-like, small, stocky flies which, like lice, hide in the coat of the host. With large claws at the end of their sturdy legs they attach themselves to their host. The wings are reduced to stumps, and only the horse louse fly, *Hippobosca equina*, has wings about 8 mm long. *Melophagus ovinus* which is also an important vector of bluetongue is about 5 mm long and often is wrongly considered to be a tick and is therefore called "sheep tick". *M. ovinus* completes its cycle on the host and can move to other animals through direct contact between sheep. Not only is it a vector of pathogens, it may also cause painful itching and bacterial secondary infections with its bite. Infestation with *M. ovinus* may reduce the quality of the wool considerably. The females lay 10–15 third larvae individually during their lives which lasts 4–7 months. Similar to tsetse flies, the larvae turn in a short time into barrel-shaped pupae 3 mm long which are stuck to the wool. After 20–23 days, the adults hatch which, after feeding on blood, will copulate after 3–4 more days (Mehlhorn and Piekarski 1989).

4.1.2.5. Calliphoridae, Cochliomyia hominivorax (Screw worm)

Contrary to other flies whose larvae **facultatively** cause myiasis on mostly dead tissues, the larva of *C. hominivorax* is an obligate parasite which needs living tissue for its development. In connection with vector-borne diseases, the infestation of the screw worm is of special importance since wounds which are infested with a countless number of larvae and are permanently bleeding are an ideal source of infection for licking insects. These are vectors for the tactile, acyclical mode of transmission. Anaplasmosis, for example, can be transmitted especially in feed-lots (Seifert 1971). Though the fly is not a vector itself, it is described here because of the importance of its larva in connection with disease transmission.

Until now, *C. hominivorax* had only been endemic to the humid tropical parts of America, but recently an invasion of North Africa (Libya) has been reported.

The pregnant fly seeks out wounds or small lesions – tick bites are enough – when groping around for a suitable site for oviposition. They also invade natural openings of the body like vulva, eyes, ears and nostrils. In about 15 minutes up to 200 eggs are deposited as a flat shingle-like mass at the edge of a wound or body opening. By sweeping with the hind quarters, the eggs are positioned in one direction around the same wound, often in several groups. The flies do not mind depositing their eggs where other eggs already have been placed. Normally such batches are deposited 3–4 times; sometimes up to 33 ovipositions can take place. 11–24 hours after oviposition, the larvae hatch and immediately start feeding, sinking their entire body into the tissue. Only the rear end appears from the wound. A screw worm wound looks like a raspberry: only the blood-covered hind quarters of the larvae can be seen in the tissue. In this position, the larvae grow and develop into the third larva stage, producing secretions which favour secondary infections. Because of the growing and developing larvae, the wound becomes enlarged, and it is also possible because of the smell of the wound that more flies are attracted and deposit the eggs at the edge of the wound. The vulva of a calf which is born during the rainy season can turn into a bloody wound in a short time. Other licking flies feed on these wounds and thus also transmit acyclical and tactile vector-borne diseases. Depending on the temperature, the larvae fall to the ground after 4–8 days and dig into the earth. Where no soil cover exists, they dig more deeply into the earth than where it is covered with vegetation. Within 24 hours, the larva turns into a dark-reddish/coffee-brown pupa which rests from 7 days up to 2 months depending on the temperature. The adults hatch in the early morning hours, and after 5–20 minutes are able to fly. They disperse quickly over wide areas. Low brush vegetation is a favourable habitat, whereas dense forests are avoided. On the third day after hatching the females are fertilized once; the males are polygamous. 4 days later, the oviposition starts. To find both a suitable host and place for oviposition, the fly may eventually cover hundreds of kilometres. The males often remain in a waiting position from which they attack potential hosts. The flies rest at night on leafless branches 1–1.5 m above the ground, mostly close to rivers or creeks where the humidity is high. They live for about 2–3 weeks.

The screw worm is an opportunistic parasite which because of its longevity, independence and enormous reproductivity, as well because of its capacity to cover large distances, is able to adapt to a large number of available habitat conditions (FAO 1990).

Hypodermatidae are not vectors of diseases in the tropics and are thus not discussed here.

4.1.3. *Nematocera* (Midges and Gnats)

In the zoological system *Nematocera* are a sub-order of the *Diptera*. They have a slim delicate body and strikingly long, thin legs. The thread-like finely structured antennae (at least 6) of the male are very fury. The larvae have a head capsule with strong mandibula. The development cycle of the larvae depends on the habitat to which the respective species is adapted; many live in stagnant water or on decaying plant material, and only a few prefer living plants. The pupae are of the so-called free type, i.e. the legs are seen clearly through the transparent cover of the pupa. Gnats from the families *Culicidae*, *Ceratopogonidae* and *Simuliidae* are important vectors of vector-borne animal diseases in the tropics. The transmission is mostly acyclical-tactile or alimentary-excretoric. *Culicidae* are supposed to transmit arboviruses transovarially.

4.1.3.1. *Culicidae (Midges)*

From the family *Culicidae*, the genera *Aedes* (surra, anaplasmosis and Rift Valley fever), *Anopheles* (anaplasmosis), *Culex* (anaplasmosis), *Culiseta* (anaplasmosis) and *Eretmapodites* (Rift Valley fever) are vectors of diseases in domestic animals. Only the females feed on blood: they seek out a host every 3–4 days, mostly during twilight and at night. The site of the bite is painful. It itches and swells rapidly. The males live on plant material. According to the species, the females deposit 40–400 eggs either individually (*Aedes, Anopheles*), or stuck together like small ships (*Culex*). The eggs are deposited onto water plants or other swimming material under the surface of the water. The development from egg to pupa takes place in the water. Depending on the temperature, the eyeless larvae hatch between 12 hours and 2 days later. They take air into their tracheas which comes through a respiratory opening at the rear end of the larva which hangs just under the surface of the water. The position in which the larva hangs under the surface is specific to each species; *Culex* and *Aedes* hang at an acute angle to the water surface, while the larvae of *Anopheles* are positioned parallel to it. The larvae are vegetarian, moult 4 times and pupate under the water surface. There the pupa is able to move and thus able to obtain oxygen. Depending on the temperature the entire cycle lasts 10–21 days.

4.1.3.2. Ceratopogonidae (Biting midges)

Ceratopognidae are important vectors of arboviroses in the tropics (African horsesickness, bluetongue). They are small gnats, only 0.5/1.0–4.0 mm long, which cross their wings very precisely when at rest. Only the females suck blood during late evening. Where they appear in masses, they may cause unrest in animal herds because of the painful burning bites they cause around the eyes and on the abdomen of the animals. The larvae live on organic material in the mud of rivers, creeks and ponds, as well as in brackish and salt water, or the leaf axis of tropical plants where the pupae hatch. Depending on the temperature, the development from egg to imago takes 1–2 weeks.

4.1.3.3. Simuliidae (Blackflies)

Simuliidae are vectors of anaplasmosis and of filariosis; they are only 3 mm long and live along fast running, oxygen-rich creeks or rivers. Only the blood-sucking females which have a painful bite become a plague for cattle when they appear in masses. *Simuliidae* are pool feeders which cut the skin with their saw-like proboscis and suck up the blood. A hypersensitivity type I may be caused by the secretion which is applied during feeding. Up to 20000 blackflies may infest a cow.

The females copulate immediately after hatching and need only one blood meal for oviposition, which takes place 4–5 days later. They deposit about 250 eggs on plants or stones in oxygen-rich, fast flowing waters. The larvae hatch after only 4 hours and attach themselves to plants or other material, pupating after moulting 5 times. About 4 days after pupation, especially on humid warm days, the imagines hatch and immediately attack grazing animals (Mehlhorn and Piekarski 1989).

4.1.4. Rhynchota (Bugs)

From the genus *Rhynchota* the family *Reduviidae* is important as a vector of the Chagas disease which is caused by *T. cruzi*. They are beetle-like ectoparasites with a half horny, half cutaneous pair of wings. Bugs hide during the day in cracks of the wall of shacks and sheds and suck blood at night; they can live up to one year without feeding. The female deposits its eggs in hiding places in buildings. After 3 weeks, the larvae hatch out and develop into the adult after going through 5 nymphal stages. The metamorphosis is incomplete since the larvae are similar to the imagines. The entire development cycle lasts at least 6 weeks depending on the temperature. *T. cruzi* is transmitted in the cyclical alimentary-excretoric manner in such a way that the pathogen is ingested with the proboscis during feeding and excreted with the faeces which contains the metacyclical trypanosomes. The bug massages the pathogen into the site of the bite using its hind quarters and thereby injects the system of the host.

4.1.5. *Siphonaptera* (Fleas)

The sand flea (*Tunga penetrans*) can transmit anaplasmosis. The rodent flea *Xenopsylla cheopsis* is well known as a transmitter of the plague. Bacterial infections (tularaemia, pseudotuberculosis, swine erysipelas, listeriosis, brucellosis) and rickettsiosis are also transmitted by fleas. Fleas have no wings, are flat, approx. 2–3 mm long, brown, smooth, hard-skinned and have strong jumping legs. They are able to jump a distance of 30 cm and a height of 10 cm and, as temporary parasites, only look for a host for feeding. Both sexes live on blood and may survive many months without feeding; if they have fasted, they will accept blood of almost every unspecific host. The proboscis of the adult has two channels; blood is ingested through one, and saliva is excreted into the biting channel through the other in order to prevent blood clotting which leads to the local symptoms at the site of the sting. Fleas can be recognized because of their typical scale-like segments. *T. penetrans* has no neck crest; the male is 0.5–0.7 mm long, and the female 0.5–6.0 mm. Fleas normally deposit their eggs (10–1000) on the ground. After 5 days the larva hatches and develops into the pupa after moulting twice in about 2–3 weeks; the imago hatches 1–2 weeks later. The female sand flea bores into the skin. In the case of cattle, they prefer to look for non-lactating cows or heifers. They enter the teat canal of the udder and invade the cistern. At first they develop there into a ball, copulate with the male and deposit their eggs which later on are set free into the open with the pus of the emerging abscess. Sand fleas, therefore, are not only vectors of diseases but may also cause serious damage through mastitis (Mehlhorn and Piekarski 1989, Seifert 1971).

4.1.6. *Desmodontidae* (Common Vampire)

Desmodontidae are vectors of diseases because they only feed on the blood of warm-blooded animals. Because of their special importance as disease vectors in semi-arid and hot humid areas of South America east of the Andes, in Central America and in southern Mexico, they have to be discussed here. Of the 3 known species *Desmodus rotundus, Diaemus youngi* and *Diphylla ecaudata, D. rotundus* is of special importance and occurs widely in the grazing grounds of Latin America.

D. *rotundus* is a medium sized grey-brown coloured tailless bat, 10–12 cm long which can easily be recognized because of its nose which is similar to the snout of a pig. The medium sized ears do not have a tragus. Contrary to other bats, the vampires have large razor-sharp upper central incisors; the lower incisors are set apart to allow the tongue to suck up the blood of the host.

Vampires live in colonies in caves, under rocks and also in hollow trees or in abandoned buildings. They find their habitat especially in the semi-arid savannah, the typical grazing ground for cattle in Latin America.

The vampires attack their host only during complete darkness and they find it by means of ultrasound (above 100 kHz). Dogs, which are able to hear ultrasound wake up when a vampire appears and run away. As soon as a vampire has found its prey it walks on its feet and thumb pads over the body until it finds a hairless site which will remain undisturbed by any defence movement of the host (tail). In cattle, preferred sites are the thorax, the sides of the neck, the withers and the skin on both sides of the croup. With its sharp incisors, the vampire cuts a deep wound out of which the blood may ooze for hours, since the vampire excretes saliva during sucking which contains enzymes which prevent coagulation. The vampire sucks up the blood, forming the tongue and lower lip into a tube. It feeds as long until it is extremely full and almost unable to fly. A vampire may take up more than the double of its general body weight; it needs 25 L blood per year. Because of their close social relationship, fully engorged vampires share their blood meal with weaker or young animals which did not get a chance to feed on a host. This is done by belching up blood out of the stomach and delivering it into the mouth of the companion. Thus, pathogens also ingested with the original blood meal may be distributed within the colony of vampires, and the infectious pressure of the colony increased in this way. The floor of the vampire caves is covered with the tar-like stinking excrement of the animals. This is an important hint to recognize the caves of vampires in order to eventually eradicate the vectors (I/4.3). During feeding, the vampires may ingest pathogens of vector-borne diseases and excrete them at the same time. *Protozoa* and *Rickettsia* are excreted in the acyclic-excretoric mode, and the rabies virus in the cyclic-excretoric way. When transmitting rabies, the vampire gets sick itself, loses its natural shyness and thus becomes a more aggressive vector. Through the enormous amount of blood which the vampire ingests, and simultaneously the large volume of saliva which it will inject a large amount of pathogens, is ingested and transmitted. This mechanism is further enhanced when the vampire attacks the same prey on several nights and feeds on the same wound.

4.2. Methods for Vector Control in the Tropics

Methods for vector control can be divided into two groups of measures:
- in the habitat of the animal when the vector is either at rest or developing and
- on the target animal while the vector is feeding, developing or resting.

4.2.1. Control Methods in the Habitat of the Animal

Methods for vector control in the habitat of the animal aim to destroy the habitat of the vector, or to alter it in such a way that its resting stage is disturbed or its development prevented. Management measures may also make the contact between vector and host more difficult and prevent the vector from feeding.

4.2.1.1. Measures of Herd and Pasture Management

Pasture management measures in the animal's surroundings may help to control disease vectors. They are hardly likely to be permanently successful on their own, however. Together with chemical methods, they may contribute to the eradication of the disease vector. Methods of pasture management have to change the natural habitat of the vector in such a way that its chances of survival are reduced.

A reduction of the tick population may be reached by combating the tick stages which rest or develop on the soil. Depending on the species of ticks

– the pasture can be burned;
– the pasture can be ploughed regularly;
– rotational grazing can be carried out;
– zero-grazing and production of fodder plants can be used as an alternative.

Burning of the pasture ought be carried out at the end of the dry season because the larvae have already hatched by this time. It is, however, scarcely possible to destroy the tick brood completely. Enough hiding places remain on the ground where the larvae and eggs could survive a fire.

Regular **ploughing** of the pasture is often successful, especially with two- and three-host ticks, because the microclimate which the tick stages which live on the ground require is destroyed. The cultivation of fodder plants, like clover and alfalfa for example, will take away the natural habitat of the tick. *Stylosanthes* spp. is known to reduce the tick infestation of the pasture since certain substances which it contains have an acaricide action (Skerman et al. 1988).

Rotating pasture management tries to starve out the ticks. By keeping the animals for only a limited time on the pasture, the ticks' host is removed. The success of this scheme depends upon working uncompromisingly: no host and no wild game can be allowed to enter the pasture therefore this concept necessitates fencing. The pasture has to be kept host-free for a period as long as the longest lifetime of a tick stage. In order to starve out larvae and nymphs of two- and three-host ticks, the animals have to be kept away from the pasture for 6–7 months; to obtain the same success with adult ticks, the closure of the pasture has to last 14–15 months. Rotating pasture management, therefore, requires enough pasture reserves. The rest period for the pasture must occur during the hot dry months of the year, since the tick stages have little chances for survival then. In areas with strong wind, rotating pasture management of neighbouring paddocks may result in failure since tick larvae may become wind-borne and be carried over rather long distances.

As a last resort, **zero-grazing** and feeding with cut fodder may be introduced during the entire year (III). In pens (feed-lot) with a high stocking rate, tick infestations can scarcely occur if certain prerequisites are taken into account. In order to prevent ticks from getting into the premises with the fodder, it can be decontaminated by preparing silage (III/4 and III/5).

The control of **midges and flies** in the surroundings requires more financial funding as is the case with ticks. The destruction of breeding places of these vectors can be achieved by either draining swamps or humid places on the grazing land, or regulating creeks and springs in order to prevent them from producing a favourable environment for ticks on the pasture. Watering places for animals must be built above the ground and paved so that mud cannot form around them; humid pastures have to be drained. Where artificial irrigation is used, the attempt should be made to introduce intermittent irrigation in order to keep the pasture dry for at least 3–4 weeks. Thus the development stages of the vectors are not able to survive, and pupae also die. In order to be successful, the system of irrigation has to be built in such a way that the water can run off completely after irrigation and that no puddles of water remain (profile and gradient of the irrigation channel). Regular cultivation of the pasture may not only reduce weeds but also impede the development of *Haematobia irritans* by way of destroying cow-pats.

Animal stables have to be constructed in such a way that they stand square to the direction of the wind. They have to be kept clean, and the doors and windows may have to be screened. Regular dung disposal and appropriate management within the intensive production system is a prerequisite. If open pens are constructed and stocked appropriately, larvae are unable to find a suitable habitat (III/5). The dung heap must be kept moist, and perhaps it ought to be sprinkled in order to maintain the internal temperature at around 70 °C through the process of fermentation, thus leading to the destruction of the fly larvae. The premises should be cleared of any surrounding dung or bushes in order to deprive the insects of places for hiding and breeding (Seifert 1971).

Special management is required for the control of **tsetse flies**. The removal of resting and breeding places used to be considered to be important. By removing the habitat, however, only a displacement of the flies is achieved, and not their eradication. A total clearing of the trees for this purpose would be required, since tsetse flies are unable to exist on open ground without trees and bushes which provide shade. The glossines will avoid flying over an open area 2–3 km wide. Intensive production systems may be protected by establishing a treeless protective belt around the stables. Tsetse flies normally need a resting place close to the animals which they attack (Fig. 30).

In spite of the total removal of their habitat, glossines may be carried with wild game, cattle or vehicles over large distances (follow-up effect). Today people are conscious of protecting natural resources and the environment. Large-scale clearing of forests is therefore avoided. Partial clearing, however, may be carried out. To this end, the lower branches of trees and undergrowth are removed, especially along the rivers. This allows the hot wind which blows from the savannah to penetrate the forest and change the environment in such a way that the habitat of the glossines is destroyed.

Removal of game and uncontrolled domestic animals takes away the alternative source of feed and infection of the vector. It is well known that game is an important reservoir of infection for cattle. Pigs are a favoured host for tsetse flies.

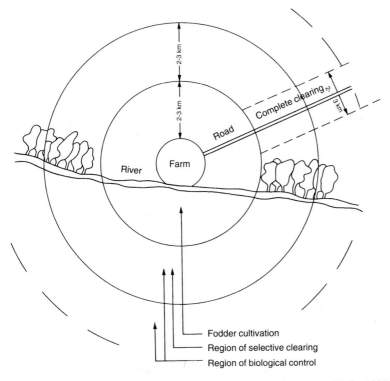

Figure 30. Tsetse barrier around an intensive animal production system (Holler 1986).

An increase in the pig population within the tsetse belt automatically leads to a higher incidence of tsetse flies. Game has a higher natural resistance to vector-borne diseases and thus is extensively protected against this disease complex. Game, however, often remains latently infected and is thus a permanent carrier and source of infection for domestic animals. Game should therefore be kept as far away as possible from domestic animals. It can be best achieved by constructing fences which cannot be penetrated by small game if intensive production systems (dairy- or feed-lot) are to be protected. Not only large game, but also small game and rodents are alternative hosts for *G. morsitans*.

4.2.1.2. Fly Traps

For centuries, fly traps have been used to control tsetse flies. The construction is based on the knowledge that tsetse flies are attracted by the contrast between light and shade and seek shade or dark spots. They choose resting places as well as the sites where they feed (on the under belly) in the shade. Consequently, traps have been constructed in such a way that the entrance is placed on the underside and made out of dark-coloured material. The principle of the classical Harris trap has been developed by Challier and Lavaissière into an efficient catching system which, with minor changes, is still used today. It is a box in the shape of a pyramid or double pyramid of about 1 m in length and 50 cm in diameter at its widest point. It is covered with gauze made of synthetic material and has an entrance at the lower pointed end of the trap. On the lower side of the trap, the clear cover is underlayed with dark cloth in order to attract the flies. The glossines, flying to the opening, enter the trap and when trying to get back to the light, reach the upper part and enter a kind of creel, where they remain trapped. Nowadays, the lower part of the trap is covered with blue gauze and only the entrance is underlayed with black (Fig. 31). The tsetse fly is attracted from a distance by the blue colour and then, when approaching, by the black colour. Several variations of the construction of the classical Challier-trap have been developed, for instance by Lavaissière, Flint, Mérot, Gouteux and Noireau. They all follow the same principle but are different in design (one or two-cone, or only square-cage).

Gouteux (1991) has described the properties and manufacturing of a bipyramid trap which is easy to handle, set up and dismantle for transportation and does not need any special preparation such as insecticidal impregnation. Its cost price is about 30 FF and it is supposed to be used by semi-nomadic cattle owners. Gouteux et al. (1991) claim that this dry capture device used for catching and killing of flies is not only cheaper, but also 3–4 times more efficient than any other biconic, pyramidal or monoconic traps already known or designed.

Generally the traps are often combined with odorous substances which are attractive for tsetse flies. Glossines are attracted by pheromones which are

Figure 31. Challier trap in Ethiopia (the lower part is blue, the entrance underlayed in black.

produced by animals. Pigs are the animal species which is most attractive for tsetse flies. Consequently, the efficiency of the trap can be enhanced if a bottle with cattle or pig urine is placed at its base. The urine will attract flies for weeks. Other enticing substances may also be applied, such as metacresol and octenol and others, (see also I/4.2.1.3 targets). Amsler et al. (1994) found that metacresol increased the catches 1.5 fold and metacresol/octenol (3/1) 2.5 fold in comparison with the control trap with no attractant at all.

Along cattle trails where people regularly walk to a water supply or to the market, traps may reduce the tsetse population considerably. The traps are usually used, however, for surveys which observe the tsetse infestation before and after control measures. The efficiency of the fly-control-programme can be determined especially in a stationary system with targets.

Musca domestica larvae can easily be caught at the edge of a dung heap if it is well packed and covered and surrounded by a small ditch filled with water. As soon as the third larvae leave the dung, they fall into the water where they can be removed regularly.

4.2.1.3. Targets

Shortly after the turn of the century, Maldonado had already begun to catch *G. palpalis* on the island Principe, using a target. Farm workers carried dark cloths on their backs which were covered with glue to which the flies became stuck having been attracted by the human smell. Nowadays, targets for tsetse control are pieces of cloth or boards made of wood, cardboard or plastic of about 1.20 × 0.60 m, the outer third of which are coloured blue and the central third black. They may also only be fixed to a pole on one side and thus be able to rotate in the wind like a flag (Dehoux 1993). They are placed on appropriate sites at determined intervals. Again, the division of the colour is based upon the empirical knowledge that tsetse flies are attracted at a distance by the blue and by the black colour when approaching. In addition, 50 mL bottles with acetone, plastic bags/containers with octenol or phenol, or bottles with pig or cattle urine are placed at the foot of the target as a lure. According to recent findings, it has been proven that M-cresol and 1-octen-3-ol in a mixture of 3:1, placed in a small plastic container ("Diffuseur"), have a longer-lasting effect than animal urine (Filledier and Mérot 1989). In order to kill the flies which settle on the target, it is sprayed with the pyrethroid (deltamethrin 0.1%) at regular intervals (8 months). 4 targets per km² located 1 km apart is enough to reduce the fly infestation considerably. The targets may also be placed along creeks and rivers, perhaps together with traps in order to catch riverine flies.

Recently, so-called tyre traps ("piège-pneu") have been introduced (Dehoux 1993). These are pieces of used car tyres (one third) which are suspended from trees between 0.1 m and 1.0 m above the ground. They are suspended in such a way that the outer side of the tyre points upwards. They are also painted partly blue at the centre and a piece of cotton wool is fixed to the underside of the tyre which is soaked in 5% deltamethrin in order to obtain approx. 100 mg of active substance per trap. Because the cotton is protected by the tyre against rain and sun, it does not need to be replaced or reimpregnated.

In Zambia, attempts have been made to reduce tsetse areas or prevent their expansion by placing targets in several groupings with a row of traps behind them in order to control the effect of the targets. This rather simple and efficient system seems to be suitable for pushing back the tsetse fly within the African tsetse belt. The problem is that the peasants are so far not willing to contribute to the maintenance of the control system.

4.2.1.4. Artificial Breeding Places

It is possible to control tsetse flies by providing suitable places for the deposition of their larvae. In order to achieve this, some form of shade ought to be made under which the soil is loosened in order to produce a suitable habitat for the deposition of the larvae and its pupation. Afterwards, the pupae can either be destroyed, or the shade removed which leads to the destruction of the pupae.

Simuliidae can also be controlled efficiently in artificial breeding places. To do this, creeks or rivers in which blackflies are endemic have to be dammed up. Since the natural rapids (stones, rocks) which provide the habitat for the development of the larvae become flooded, the cycle of the blackfly is interrupted. The blackfly is then forced to deposit its eggs on dams which are built out of concrete, and then they can be removed by regularly cleaning the dam.

4.2.1.5. Biological Measures

For years, vector control with chemical compounds has been carried out without taking into account either the effect of the chemical on the animal, on the consumer of the animal product or on the environment. New pesticides have been developed mainly because of the emerging resistance without thinking about possible environmental hazards. In the following, biological methods for vector control are described which will be unlikely to be able to replace pesticides; they may however contribute to the reduction of their application.

Vectors may be controlled biologically within the environment of humans and domestic animals by

– increasing and releasing predators, parasites and arthropodal pathogens or the metabolic products of those pathogens;
– interfering with the physiology of the vector.

The critical situation of vector control through development of resistance and its unavoidable side-effects makes the application of biological alternatives more and more important.

* *Vector Control through Predators, Parasites and Pathogens of Arthropods*

Methods which are available at present try to destroy the larval stage of the vector within its habitat using predators, parasites or pathogens which normally live in this environment. Almost all midge and gnat larvae can be controlled by increasing the population of **fish** within the water, e.g. *Tilapia* in rice fields.

Some species of **arthropods** prefer to hunt tsetse flies. *Asilidae* are flies which attack the fully engorged tsetse fly which has difficulty flying, or descend on them while they are just landing or already resting. Dragonflies and spiders are also predators of adult glossines. For some beetles and ants, the pupae of glossines are a potential prey. None of these predators has so far been multiplied artificially in order to carry out organized tsetse control.

Parasitic **nematodes** (*Romanomeris culicivorax*) have been used in the USA to control *Anopheles* larvae. Depending on the type of habitat, it has been possible to control 50–95% of the larvae by applying 10^3 nematodes/m^2 of water. The product SKEETER DOOM (FAIRFAX) contains 10^5 *R. culicivorax* larvae in suspension per package which is enough to control mosquitoes on an area of water of 100 m^2. So far, *Culex tarsalis, C. pipiens, Culiseta* spp., *Anopheles* spp., *Psorophora* spp. and *Aedes* spp. have been eradicated successfully. In Central Africa, *R. culicivorax* has been used against *Simuliidae*.

Protozoa are also diptera pathogens and can be set free in order to control these vectors. Microsporidia of the species *Parathelohania legeri* and *Ambylospora opacita* are parasites for mosquitoes and are transmitted transovarially onto the next generation. The same is true for *Pleistophora (Vavreia) culicis* which attacks *Culex pipiens* and *Nosema stegomyiae* (*algerae*) which attacks *Anopheles gamiae*. The infection of the eggs takes place in the ovaries causing a reduction in the fertility of the insect and it also inhibits the production of the plasmodia ookinetes which prevents the *Anopheles* from transmitting the *Plasmodia*. Under laboratory conditions, *Microsporidia* can be produced in large amounts on *Lepidoptera*. *Caudospora simulii* is pathogenic for *Simuliidae*. *Triatomae*, the vectors of Chagas disease, are not only invaded by *Trypanosoma cruzi*, which they transmit, but also by *Blastocrithidia triatomae*. This flagellate delays the development of the larvae of the triatomes and increases the mortality rate. Attempts have been made to multiply the pathogen *in vitro* and release it into the hiding places of the triatomes (Schaub 1986, Weiser 1987).

Bacillus thuringiensis, serovar H-14 or var. *israelensis*, is the best known and most successful pathogen from the group of *Bacteria* and *Fungi* which is being applied at present under practical conditions to control midges and gnats, especially *Simuliidae* and *Culicidae* (*Anopheles*). Spores and parasporal crystals of the bacillus are set free which on the one hand leads to the multiplication of the pathogen within the biotope of the larvae, but on the other hand, they also penetrate into the abdomen of the larvae where they cause a deadly sepsis. The toxic metabolites of *B. thuringiensis* stop the larvae from feeding and damage the wall of the gut. Meanwhile, β-exotoxin of *B. thuringiensis* is also produced in fermenters on an industrial scale and can be applied as an insecticide in breeding places (dung, mud). The pathogenicity of *B. thuringiensis* is targeted exclusively towards diptera larvae. No side-effect is known in other living beings. As a further larvicide, bacillus *B. sphaericus* 2362 is applied to control *Simuliidae* in particular. It has been obtained from a combination of *Simuliidae*. It can be released within and around the larvae of vectors and controls only target-larvae while otherwise living without any side-effects upon non-target organisms or dead material. It has been possible to isolate the strain BT-14 from Vietnamese silk worms which is completely harmless to silk worms, but it destroys *Simuliidae* (Weiser 1987).

Fungi as natural pathogens for *Simuliidae* (*Coelomomycidium simulii*) can also be used to control this species and other *Culicidae*. The same is true for *Phyco-mycetae* (*Entomophthora destruens*), which are produced industrially in ferment-ers and distributed in the hiding places of *Culex pipiens* and *Culiseta annulata* in their adult stages. Amongst the *Deuteromycetae*, *Tolypocladium cylindrosporum*, *T. inflatum*, *T. niveum*, *T. geodes*, *T. tundrense* and *T. terricola* are natural patho-gens for mosquitoes. The insecticidal activity is due to secondary metabolites which are produced by fungi, such as tolypin, a polypeptide, and cyclosporine A, a depsipeptide, which is known as an immunosuppressive drug (Weiser 1987).

R. chironomi belonging to the genus *Rickettsiacae* is pathogenic for gnats and midges.

Pathogenic **viruses** have also been identified as pathogens for mosquitoes, a so-called MIV (mosquito iridescent virus) for *Aedes* spp. as well as another MIV and a CP-virus (cytoplasmic polyhedrosis) for *Simuliidae* (Weiser 1987).

● *Vector Control through Manipulation of the Physiology of the Vector*

Flies, gnats and midges may be controlled by interfering with the hormonal physiology and the processes of development during metamorphosis. Methods which are used for this purpose are:

- application of insect growth regulators (IGRs)
- application of juvenile hormone analogues or mimics;
- chitin synthesis inhibitors;
- others;
- hybridization;
- the sterile male technique;
- genetic manipulation;
- chemosterilization;
- the use of attractants;
- vaccines which contain antigens from saliva or tissue from the vector which affect the arthropod through an allergic reaction at the site of the bite.

Insect growth regulators (IGRs) represent a relatively new category of insect con-trol agents aimed mainly at covering the need for safer compounds and over-coming the development of resistance to classical insecticides. IGRs are characterized by the fact that they do not necessarily kill the target pests directly, but interfere in some way with their processes of growth and development. They are thus likely to address target organs other than the central nervous system, exhibit a more selective activity, and show an improved safety profile. They act principally on embryonic larval and nymphal development by interfering with metamorphosis and reproduction. IGRs will require more time to reduce insect populations than conventional insecticides and will sometimes have to be used in combination with adulticides to achieve an immediate knock-down effect. The IGRs include various chemical classes with different modes of action and can be divided into three categories: juvenile hormone analogues or mimics, chitin synthesis inhibitors and others (Graf 1993).

Juvenile hormone analogues (JHAs) are synthetic compounds which are similar to the natural moulting or juvenile hormones and are able to prevent the adult gnats or midges from hatching out of the pupa. The larvae and pupae are not damaged externally; the imago dies because it remains connected to the membrane of the pupa during hatching. The JHAs methoprene and hydroprene are the major representative of this class of compounds. Methoprene is applied to the water in which gnats and midges develop in microcapsules which have a slow release formulation (ALTOSID). The compound which is slowly released is ingested by the arthropods with the feed. Methoprene has found a major application in the control of *Haematobia* spp. through feed application or sustained-release bolus in cattle (INHIBITOR). Since methoprene cannot cope with sunlight, it has to be applied repeatedly because the population of gnats and midges does not develop simultaneously. By feeding methoprene to cattle (0.7 mg/kg b.w./day) *Haematobia irritans* and *Stomoxys calcitrans* can be controlled (feed-through). The compound, which is excreted with the faeces prevents the development of the flies. The product can be administered by adding it to the drinking water; it is safe for cattle. Beside the horn fly, the face fly (*Musca autumnalis*) can also be efficiently controlled. If the product is added to poultry feed (10 g/ton), the stables remain free of flies for 1 week (Hoffmann 1987, Holler 1986). Topical application of methoprene also in combination with pyrethroids is used to control ectoparasites on pets. Hydroprene is applied for pest control in agriculture.

Chitin synthesis inhibitors act either on the membrane-bound chitin synthetase or on the polymerization of the chitin cuticula. Diflubenzuron, an acyclical carbamide which has a similar effect to a juvenoid, inhibits the development of chitin and can be given to cattle in salt licks. The product marketed as DIMILIN or VIGILANTE as sustained-release bolus prevents the formation of the cuticula and thus moulting at all stages of gnats, midges and flies since the larvae become unable to hatch out of the old cuticula. Because of its persistence, the compound can be applied at long intervals. It may also be used with the feed-through (DUPHACID) method, in order to control horn fly and stomoxydines, or be applied topically to the manure for this purpose. In New Zealand, ZENITH is used to control blowfly and lice in sheep. Several other compounds are applied mostly for the control of ectoparasites in pets or flies in livestock housing (Graf

1993, Holler 1986). For tick control in cattle, the compound fluazuron has been marketed as ACATAK pour-on tick development inhibitor. It has the typical mode of action of a chitin synthesis inhibitor, meaning that it does not directly affect the different tick stage but rather interferes with moulting and hatching, this activity being especially well suited to the control of the one-host tick *Boophilus*. *Boophilus* larvae feeding on treated cattle will not moult to nymphae, ṇymphae will not moult to adults, and engorging females, also imbibing their normal blood meal, will produce eggs from which no larvae will hatch. Systematic treatment of all cattle from a region will lead to a rapid reduction of the *Boophilus* population to a harmless level and the presence of a few larvae will permit the maintenance of a certain level of host resistance to ticks. This mode of action is especially suitable for strategic control programmes. The control of multi-host ticks is more difficult to achieve since the immature stage usually feeds on wild animals (Graf 1993).

Other IGRs interfere with moulting and pupation without acting directly on chitin synthesis, cyromazine (VETRAZIN) at present being the only representative. The mode of action of cyromazine is different from chitin synthesis inhibitors. Another difference concerns the spectrum of activity, while chitin synthesis inhibitors usually act against a wide range of insects, cyromazine shows a high specificity for dipteran larvae and, at least at the concentration used in fly control, is almost harmless against larvae of most other insect species. VETRAZIN is recommended for the control of blowfly larvae and also for topical application on manure (NEPOREX) or as feed-through in poultry (LARVADEX). Currently a new molecule code name CGA 183893, a pyrimidine derivative is tested by CIBA, Switzerland, which apparently has a similar mode of action to cyromazine, but possesses an *in vitro* activity ten times higher than diflubenzuron and cyromazine (Graf 1993).

Hybridization, the goal of this technique is to obtain a sterile offspring by pairing related species. This method has been tried experimentally in the control of tsetse flies. For this purpose, alien species have been introduced to a particular tsetse-infested area. Since different *Glossina* species copulate with each other, crossbreeds appear which are mostly sterile and are completely unable to survive. The method has not been successful under field conditions since the introduced insects were unable to adapt to the new environment.

With the **sterile male technique**, or sterile insect technique/SIT, a large number of sterile males is introduced in a ratio of about 6:1 into a population. The sterile males will thus have a greater chance than the wild males of copulating with the females. Since the females can only mate once (I/4.1.2.2), they are unable to produce offspring.

In order to obtain sterile males, the gonads of artificially produced males are radiated in such a way that the *Potentia coeundi* is preserved while the *P. fecundandi* is destroyed.

Radioactive cobalt or caesium are used for the radiation, the dose of radiation being adapted to the resistance of each species. The dose has to be such that about 90% of the males become sterile; if it is increased, their survival rate in the wild is endangered. As a rule, sterile tsetse males may survive for 6 days, while the wild ones can live for 3 weeks. The radiation is applied shortly before hatching. Since male tsetse flies hatch 2 days before the females, they can easily be separated in this way. In laboratory colonies of flies, cross-breeding with wild strains has to be carried out regularly in order to maintain the genetic variability and through this the vitality of the flies.

The SIT has been applied very successfully to control the monogamous screw worm fly *Cochliomyia hominivorax*. This has been rather easy because these flies deposit their eggs on freshly ground meat where their development takes place. For this purpose, wire mesh is spread out in a suitable shack on which the meat is placed. The larvae which develop out of the pupae, fall onto fine sand which has been spread underneath the wire mesh. The pupae are collected from the sand for sterilization. With this technique, the FAO has also been able to control the *C. hominivorax* invasion which has appeared recently in Libya.

The control of tsetse flies by means of the SIT is much more difficult and expensive. Tsetse flies can only be fed with fresh blood. At first flies were fed on laboratory or domestic animals which is a very expensive method. In the meantime, the membrane technique has been developed. It allows the breeding of flies on conserved blood. For this purpose, pig blood is spread on a grooved glass plate which is kept at body temperature on a plate warmer and covered with a silicon membrane. The fly cages which are covered with synthetic gauze are placed on the membrane, and the flies suck through the silicon membrane just as they would through the skin of an animal (Fig. 32). This apparently simple technique, however, has a lot of problems. The flies may become infected with bacterial or fungal infections; the maintenance of the micro-climate which the flies require is cost-intensive; the process of fly production and subsequent sterilization requires a lot of well educated technical personnel. In practice it has only been possible so far to control isolated tsetse pockets with the SIT: in an area of

Figure 32. Feeding tsetse flies on silicone membranes (photo: J. Greiling).

3000 km² in Burkina Faso, and of 1500 km² in Nigeria. Economically it is not feasible to apply the SIT to a larger extent. Reinfestations could be controlled if enough sterile males of the relevant species could be provided in time (IAEA 1990).

Genetic manipulation, i.e. radiation of adult male gnats and midges gives rise to a translocation of parts of the chromosomes. If males which have been manipulated in such a way copulate with the female, translocated heterozygotes appear. If males with such a genetic defect are introduced into a population of gnats and midges, they transmit this defect to their offspring. Since they are semi-sterile but still fully functional and even more eager to copulate, a gradual reduction of the population will be obtained.

For **chemosterilization**, chemical compounds (chemosterilants) are used to reduce or destroy the fertility of the insects. These substances prevent the production of spermatozoa and ova or destroy them immediately after production; TEPA (triethylenephosphoramide), HEMPA (hexamethylphosphoramide) and TETRAMINE (triethylenemelamine/TEM) are used at present. So far no methods are available which can be applied under practical conditions. There are still also questions regarding the metabolism and the environmental effect of these compounds.

Pheromones (attractants) are natural chemical signals used by the partners for copulation to trace each other as well as to find the host for sources of blood. They can be used as insectistatics if the structure of the active substance and action of each insect species are known. The morphology of the antenna as well as the level of reaction threshold of the pheromones in the insect as well as its pattern of orientation and behaviour have to be analysed. The application of this method is well developed for the control of pests in forests, fruit and horticulture.

Insects cannot be eliminated using lures, but their population may be reduced until they no longer present a hazard. Using pheromones, the arthropods may be stimulated to copulate with a dummy which has been soaked with the female pheromone. In order to control *Lymantria dispar*, a forest pest, the relevant pheromone is spread in the forest so that the males become confused and are unable to find the females. The application of pheromones to make traps more attractive has already been described.

Vaccines have been produced which cause allergic reactions in the skin of cattle and thus interrupt the metamorphosis of the vector on the host. They are still in an experimental state of development, but the results are promising.

It is well known that Cebu cattle have a lower tick-count than taurine breeds after repeated infestation with *Boophilus* ticks. It has been shown that *Bos indicus*, including Sanga cattle, carry significantly less one-, two- and three-host adult ticks than exotic *Bos taurus* cattle, while *Bos taurus* × *Bos indicus* crossbreeds carry middling numbers of ticks, particularly in summer, when adult tick activity is increased. Comparison of the tick resistance of indigenous Mashona and crossbred Africaander × Sussex following natural tick infestation indicated that significantly more *B. decoloratus* were found on the crossbred than the indigenous cattle (Fivaz et al. 1992). It is a question whether this defence mechanism against tick infestation belongs to the genetically determined resistance mechanism or acquired immunity. Certainly the prerequisite for developing an immunity against tick infestation depends on the resistance potential of the animal (I/2.1.2.1). Immunity to ixodid ticks has for many years been recognized as a possible biological control method (Trager 1939). This defence mechanism, acquired after repeated infestations by ticks is immunologically mediated by antibody and cutaneous hypersensitivity reactions (Allen 1989, Willadsen 1980). It is expressed as a reduction in the number of ticks which attach to the host, reduced engorgement weights, reduction of viability and capacity of moulting and reduced egg and larval production resulting in a significantly reduced tick population (Willadsen 1980). Two practical applications of the immunity to tick infestation appear to be feasible:

– selection of cattle with an efficient immune response to ticks and,
– artificial induction of immunity by vaccination against ticks.

Biological control of ticks by selecting for tick-resistant cattle to control populations of the one-host tick *Boophilus microplus* has become general practice in Australia (Seifert 1984). However, it is estimated that despite the use of tick-resistant cattle in this part of the world, there is still an annual cost of A\$ 100–150 million due to production losses caused by ticks and the costs of their control (Willadsen and Kemp 1988).

Biological control of ticks by using anti-tick vaccines would further reduce the reliance on acaricides and also permit more widespread exploitation of tick-susceptible cattle. In the development of vaccines against ixodid ticks, one may roughly discern two approaches.

The first approach is to mimic the naturally occurring immunity to *ixodid* tick infestation. The successful exploitation of this immune mechanism depends on the identification of tick salivary gland components that are involved in eliciting a protective immune response and using them for vaccine. Partial characterization of salivary gland extract has been reported for *Dermacentor andersoni* (Gordon and Allen 1987), *Hyalomma a. anatolicum* (Gill et al. 1986) and in *R. appendiculatus* (Shapiro et al. 1986), whereas studies on *A. variegatum* have not been conducted.

In laboratory animals (rabbits, guinea-pigs) it has been possible to demonstrate that an infestation with *Dermacentor andersoni* larvae produces an immunity which prevents reinfestation. Under a renewed invasion, ticks ingest smaller amounts of blood, their time of development is increased, the moulting of larvae and nymphae is disturbed and the number and vitality of eggs reduced. Under extreme conditions, the ticks even die on the host. This phenomenon can be explained through the antibodies which have been produced in the host and which disturb the development of the tick. While feeding, the tick injects a number of antigens together with the saliva which produce these antibodies. These antibodies can also be passed on passively on to other laboratory animals which have been exposed to a tick infestation producing a similar effect. It has been shown that the antibody titre is increased after the third and fourth tick infestation. IgG, complement, the latter mostly activated on the alternate pathway (C_3) as well as a type IV hypersensitivity (cutaneous basophil hypersensitivity) are above all responsible for this defence mechanism. Ticks on plury-infested rabbits ingest more C_3 and antibodies against saliva than on animals which have been infested for the first time. The globulins of the host apparently disturb the development of the tick. Hypersensitivity type I is also important for defence against ticks. When rabbits are reinfested with *Ixodes ricinus*, mast cells become degranulated and consequently histamines are released which cause an enhanced skin reaction, as well as oedema, interruption of the blood supply for the tick and the direct poisoning of the vector at the site of the bite.

A vaccine prepared from extracts from the saliva glands of *Boophilus microplus* has been used in Australia to protect cattle against infestation by this vector. A reduction up to 70% of tick infestation has been reached without any allergic reactions in the vaccinated animals. Histological changes in the midgut epithelium have been found in affected ticks from the vaccinated animals. With 3.3 mg antigen material collected from 1368 g of adult ticks, the relevant antigenic fraction was isolated. The protein has been sequenced and recombined with *E. coli* (Cobon et al. 1988, Kemp et al. 1989, Willadsen 1988, Willadsen et al. 1988; all cited by Aeschlimann et al. 1990).

Moreover, tick attachment cement has also been reported as an important source of immunity-inducing antigens in *Amblyomma americanum* (Brown et al. 1984) and *R. appendiculatus* (Shapiro et al. 1987). Vaccination of laboratory animals with 20 kDa and 94 kDa cement antigens induced a certain degree of immunity to subsequent tick feeding, but not as much as did repeated tick infestation (Brown and Askenase 1986, Shapiro et al. 1987). The advantage of this approach may be that the response, once induced, could be continually boosted by natural exposure of livestock to ticks in the field (Shapiro et al. 1989). However, isolation of immunorelevant antigens from salivary glands or salivary gland secretory products in sufficient quantities for vaccination purposes may be the limitation for the development of such anti-tick vaccines. To overcome this constraint, recombinant DNA techniques have recently been introduced in this field of research. For instance, cDNA libraries have been constructed from mRNA of salivary glands of *A. americanum* and screened for expression of fusion proteins using polyvalent antiserum from rabbits hyperimmunized against tick salivary glands (Needham et al. 1989).

The second vaccination approach makes use of tick gut antigens, rather than salivary antigens, as the targets for the immune response. That this approach works in principle was shown by Allen and Humphreys (1979), who succeeded in inducing immunity in guinea-pigs and cattle against *D. andersoni* using extracts of midgut and reproductive organs of adult ticks. Kemp et al. (1986) succeeded in immunizing cattle against *B. microplus* with extracts derived from internal organs of adult female ticks. They showed that the immunity was not mediated by a hypersensitivity reaction and that vaccination caused death of adults rather than larvae and that many of these ticks had gut damage (Agbede and Kemp 1986). It was concluded that these antigens normally located on the plasma membrane of tick gut cells used in this type of vaccination were concealed from the host immune response. The immunity induced with these antigens appear to be based on specific action of hosts' immunoglobulins alone, or with the aid of complement resulting in damage of the tick's gut during feeding (Kemp et al. 1989, Wong and Opdebeeck 1990).

The methods for immunizing cattle against *B. microplus* have been improved by fractionation and semi-purification of crude tick material (Opdebeeck et al. 1988 and 1989). One membrane-bound tick gut glycoprotein has been isolated from *B. microplus* and the gene coding for this protein has been cloned in *E. coli*. The recombinant protein has recently been shown to be capable of inducing a substantial degree of protection of cattle against infestation with *B. microplus* (Rand et al. 1989). In Australia a vaccine named TickGARD is now on the market which is prepared from antigens of tick gut multiplied with genetically engineered *E. coli*. It acts essentially by damaging the tick gut, thereby reducing tick fertility. TickGARD has been evaluated in extensive field trials on beef and dairy properties during the last 5 years and has demonstrated its safety and efficacy as part of an integrated tick control programme. The product contains 25 µg/mL Bm86 (synthetic tick gut antigen) in water in oil emulsion. Cattle vaccinated with tick gut produce specific antibodies against the gut of the cattle tick. When a cattle tick feeds on a vaccinated animal, it will ingest these antibodies which bind to and damage the gut of the tick. The most significant result of this is a reduction of up to 70% in the fertility of the ticks. Damaged ticks lay fewer eggs, fewer of these eggs hatch successfully and fewer of those larvae which do hatch are viable. TickGARD is aimed at reducing the tick population in the paddock and not controlling ticks on the individual animal. Use of the vaccine every 6–10 weeks during the tick season will maintain continuous effective antibody levels which will reduce the fertility of ticks on cattle throughout the season (Hoechst product information). In Cuba a vaccine has been prepared following the same principle which is called GAVAC (vacuna contra garrapatas).

Antigens have also been isolated from glossine saliva which produces an immune response in the host organism. Glossines which feed repeatedly on the same animal become severely affected. By injecting antigens which were prepared from saliva glands, breast muscle, brain and nerve ganglia as well as uterus protein and bacteria from the midgut, antibodies have been produced in rabbits which reduce the rate of reproduction of tsetse flies which feed on the immunized animals. The saliva which is injected when feeding tsetse flies regularly on laboratory animals produces antibodies which cause an increased mortality rate and reduces the feeding time in the flies. This is also a management problem in fly colonies which are kept to produce sterile males for the SIT. This defence mechanism is a hypersensitivity type I and II. Furthermore, a so-called killing factor has been reported which also is induced through repeated blood sucking. It does not correlate with the allergic defence reaction (Matha and Weiser 1988).

An allergic host response has also been proven in other haematophagous insects (*Anopheles stephensi, Aedes aegypti, Stomoxys calcitrans, Rhodnius prolixus*) (Matha and Weiser 1988).

4.2.1.6. Application of Chemical Compounds

In chapter I/4.3.1 (Tables 7 and 8), compounds are described which are not only applied directly to the animal but also to the environment in order to destroy vectors of vector-borne diseases. They are discussed here regarding their action on warm-blooded animals, their metabolism and their importance for human health. In addition, the formulation is explained with respect to persistence and efficiency.

In principal, vectors can be affected with pesticides as adults in the habitat of the host animal in places where they are at rest, or in their development stages in places where they hide as well as within the environment where they hunt for blood. The strategy of control is twofold:

– more or less regularly repeated treatment of the environment with non-persistent, rapidly hydrolizing and low concentrated compounds which should not be harmful to the environment (knock-down method);
– selective and well-targeted application of pesticides with long persistence in high concentration to the resting sites of the vectors (tsetse flies) (residual application).

Pesticides can be used in the form of wettable powders (W.P.), emulsion, dust, oily solution and emulsifiable concentrates. Emulsions penetrate the substratum more quickly than solid particles from suspensions and adhere well to the vegetation. Application of the compounds on the ground with hand sprayers can be well targeted, comparatively friendly to the environment and inexpensive.

For the treatment of large areas, fixed-wing aircraft or helicopters are used. From the **plane**, the pesticide is applied in the form of an aerosol as a driftspray. Because of its fine distribution and minimal particle diameter (5–50 μm), an aerosol penetrates obstacles like the leaves of bushes and trees. This minimizes the required amount and cost of the product as well as damage to the environment. Application with planes has to be carried out in windless conditions. It is more difficult to use planes in hill country and the result may also be reduced. The method is mainly applied with a low pesticide concentration for a knock-down effect.

Application with **helicopters** is more expensive in comparison; only 40 ha/hour instead of 1500 ha/hour with fixed-wing aircraft can be covered. On the other hand, the pesticide can be applied directly to the resting and breeding sites of the vectors since the helicopter flies only 1–2 m above the tree-level. Because of the small size of the droplets, aerosols get pushed into the leaf-cover of the vegetation with the rotor wash. Therefore, the helicopter is used for the application of pesticides with the ULV-technique (ultra low volume) with droplets on average of 120 μm in size. For this purpose, high concentrations of the compound are used (25–35%). The rotor wash pushes the drops through the leaves right into the lower parts of the tree-cover and also moistens the leaves underneath. To obtain this effect, the helicopter should not fly faster than 35–45 km/hour. There is no alternative to the helicopter in hill country. Because it is able to apply highly concentrated compounds accurately, the helicopter only has to fly over an area once. In contrast, the plane has to fly over the same area 3–4 times in order to obtain the same effect. The higher cost of the helicopter in comparison to the plane can be compensated for through the economic and selective application – often only 10% of the area has to be sprayed.

Because of economic reasons, **chlorinated hydrocarbons** are still used against the tsetse fly in Africa. They are favoured because they have a long-lasting effect due to their persistence. This leads to considerable damage to the environment and contamination of the food chain for birds and warm-blooded animals. It can also lead to the extermination of entire species. **Organic phosphoric acid esters** (OPAE) and **carbamates** are much better tolerated by the environment since they become catabolized rather quickly. They have not, however, been able to replace chlorinated hydrocarbons because of their short persistence. Nowadays, **pyrethroids** which are also very effective against tsetse flies and only toxic for fish are a good alternative which the environment can better tolerate. The problem is the high price of the available compounds.

In **choosing** the **appropriate pesticide** for application within a particular environment in order to control vectors of tropical diseases, the following questions must be asked:

– if concentrations of above or below 1 ppb (parts per billion) appear in the water;
– if there is more or less than 1 ng/m^3 in the air;
– if the tropospheric half-life level is more or less than 10 days;
– if the bioaccumulation potential (log P_{ow}) is above or below 3;
– if the acute toxicity in water (LC_{50}) is above or below 1 mm/l;
– if the acute oral toxicity for mammals (DL_{50}) is below or above 25 mL/kg b.w.;
– if data about mutagenic or cancerogenic properties are available;
– if the compound is prohibited in any country;
– if the proposed dose and method of application have any known negative effects on man;
– if any effect on the biosphere has to be expected (Müller 1991).

If pesticides are applied to the environment it also has to be taken into account that no compound reacts only with a specific vector. Non-target organisms are always affected as well, and the compound may remain for some time in game and fish depending on the compound applied. Furthermore, the persistence of pesticides within the environment and their effect on man and animal depends on the method and frequency of application, the formulation of the compound, the kind of landscape to which the pesticide has been applied, and the composition of the feed of game and fishes (Müller 1991).

Each species of vector requires a specific type of application and specific kind of pesticide because of its biology and habitat. Therefore, the appropriate methods and compounds for the most important groups of vectors are described in the following section.

- *Tick Control by Applying Acaricides into the Surroundings of the Animal*

Argasidae are temporary parasites which only look for a host at night and hide during the day in the surroundings of their host animal. Therefore they can be controlled by applying pesticides to the tick's hiding places (joints and cracks in the stable). Since only comparatively small areas have to be treated, chlorinated hydrocarbons (CHC) can still be utilized (2% LINDANE, 2,5% dieldrin). OPAEs (MALATHION 3–5%), carbamates (carbaryl 5%) and pyrethroids are more suitable. When utilizing OPAEs, it has to be taken into account that they hydrolyse quickly on alkaline ground (chalked walls) and thus are rendered ineffective. Using CO_2 traps (pieces of dry ice) which are wrapped in cloth soaked with acaricides, *Argasidae* can be enticed to leave their hiding places and are killed on their way to the CO_2 source with a suitable pesticide (LINDANE).

Before the environmental hazards of such measures were fully recognized, trials were carried out in South America and Australia to control *Ixodidae* on pasture land through the application of chemicals. CHCs as well as OPAEs mixed with fertilizer were spread on the fields. At present their is neither a known method nor a known compound which, because of the environmental hazard they present and the residual effect they have on the animal or its products, can be applied to the pasture for the control of ixodides.

In order to take into account the particular biology for each family of *Diptera*, specific methods of control have to be applied. *Culicidae, Cerato-pogonidae* and *Simuliidae* from the family *Nematocera* can be eradicated with similar schemes. The same is true for *Tabanidae* and *Muscidae*, while *Glossinae* require specific methods.

Nematocera can be controlled in their adult form where they rest, or in forests and bushes around the pastures. Endosulfan and dieldrin from the CHC group are still utilized. From the phenylcarbamate group, propoxur (BAYGON), which has a knock-down effect on *Nematocera*, is suitable. The compounds may be applied as a ULV-formulation on the vegetation. Mosquitoes can also be attacked during the hours when they are in flight. Because of the high frequency at which their wings flap, they moisten themselves with the insecticide. The developmental stages can be reached in ponds, lakes, moist mud, irrigation ditches, and the edges of creeks and rivers (blackflies) where the insects deposit their eggs on water vegetation. Since larvae and pupae need oxygen which they collect from the water surface, the development stages may be controlled by treating the surface of the water. However, care must be taken not to endanger the fish populations. Dieldrin and thiodan are used for this purpose. Heavy granulates, for example chlorpyriphos, wrapped in polyethylene, sink to the bottom of the water where they release the active ingredient gradually and under controlled conditions. After only 72 hours following the application of DURSBAN 10 R (chlorpyriphos), almost 100% of the existing larvae have been killed. This means that a maximum of only 0.6–2.2 ppb (parts per billion) of the compound is necessary, whereas treatment of the water surface requires the concentration to be in the region of ppm (parts per million) in order to be effective. Because of the long-lasting effect of this type of application, water infested with larvae only has to be treated 2–3 times during the rainy season.

An elegant method for interrupting the provision of oxygen to the development stages of *Nematocera* without damaging fish or introducing pesticides into the food chain for birds and warm-blooded animals is the application of surface films to the water. Alcohols and long-chained fatty acids can be applied with high pressure pumps (3–5 bar) to the surface of the water. The tension of the water surface is about 74 dyn/cm. The pressure for spreading the surface film is based on the difference between the surface tension of the water and that of the film. The film pressure has to be at least of 45 dyn/cm in order to be efficient, with a difference of 29 dyn/cm. This method is a purely physical phenomenon and does not have to produce a pharmacological effect.

The development stages of *Aedes, Anopheles* and *Culex* can also be destroyed with lecithin films which are applied to the water surface. Soyalecithin can also be used which contains more phosphorlipids and triglycerins than lecithin. Dissolved in paraffin, 0.3–1 L/ha covers the water surface for 10 hours. Egg-yolk and brain lipids also produce enough tension to be able to cover the water surface evenly.

The control of **Muscidae** within the environment has to be targeted at the adult insects and the development stages. In intensive animal production, application of minimal concentrations can be carried out without any hazard either to the environment or to animals if it is targeted directly at the places where the flies rest and breed. The most suitable compounds are OPAE-compounds with a short biological half-life, e.g. trichlorphon 0.75%, DDVP (dimethyl-dichlorvenyl-phosphate) 0.25%, tetrachlorvinphos 1% or carbamates 0.25% and pyrethroids (decamethrin, permethrin and cypermethrin), which are harmless to vertebrates. Through combination or regular change of the compound, the development of resistance can be prevented. This is also possible when compounds are used which have a distinct knock-down effect (trichlorphon, DDVP). Pyrethroids may also produce resistance. Before the compounds are applied, the places where the flies breed have to be localized – pupae being easily to be found in the dung – in order to apply the insecticides on target. Feed-troughs and feed should never be treated. Application of larvicides to the dung may become an environmental hazard; furthermore, larvae are more resistant to insecticides than imagines. The traditional larvicides which have been applied to dung are cresol at 4.5 L/ton, boric acid at 1–1.5 kg, sodium borate at 2.3 kg, paradichlorbenzene at 0.7 kg and sulphur carbamide at 28 g/0.7 m^3. The compounds can also be applied by feeding them to the animals (feed-through). Best suited for this purpose are the OPAEs azamethiphos, xanthene erythrosin B and rhodamine as well as lactone iver-mectine. The hatching rate of the flies can be reduced to almost zero through these compounds which in different ways effect the development of the flies. In calf sheds, larvicidal action has been achieved through feeding with dichlor-phenyloxadiazol, and in poultry and pigs with a derivative of azidotriazine cyromazine. The application of insecticides in sheds should not replace the ap-propriate construction of the premises, efficient management, cleanliness (regular dung removal) and hygiene (III/4 and III/5) (Behrens et al. 1987, Hoffmann 1987, Seifert 1971).

Glossines can be controlled on their resting sites with long-lasting insecticides. The resting sites preferred by glossines depend on the tsetse species and on the type of vegetation. Therefore, the most efficient method for controlling glossines is by operating ground spray teams which are able to spray the glossines' resting sites selectively without causing environmental hazards. *G. palpalis* can be ef-ficiently controlled if tree trunks and the underside of the leaves of trees which form shade are sprayed up to a height of 1.50–2 m. *G. morsitans* can be reached if the underside of the tree canopies is sprayed. At twilight, the flies leave their resting places and move upward where they settle on the underside of the can-opies in order to find greater humidity. Therefore, the fly can best be controlled with a helicopter. For tsetse control with helicopters, *Isoberlina* forests are only treated at the edges. It is enough to spray in a zigzag fashion in a band of about 200 m wide in larger areas. Thus the amount applied, the cost and the ecological consequences are minimized. Mixed forests at the foot of hills and areas which are flooded during the rainy season have to be sprayed completely. Undefinable mixed forests on the southern Guinea savannah are sprayed in a grid work pat-tern at intervals of 150–200 m. It is sufficient to apply a veil of spray which penetrates between 1 and several km into the fly-belt, especially along stock routes, as well as along bushways and roads. In this way, the degree of selectivity of application of the insecticide can be varied according to the properties of the tsetse species within a particular vegetation. Only 10% of the area has to be sprayed in the northern Guinea savannah, where *G. morsitans* is predominant. Along the gallery forests, where *P. palpalis* is endemic, only 5% need be treated.

The eradication of the tsetse flies in their resting places is only possible with the application of persistent compounds, i.e. through **residue application**, because the pupal stage of the fly cannot be eradicated simply with insecticides. New flies are continuously released from the pupae hidden in the soil. Long-acting insecticides are used so that they only have to be applied once every few years for economical and ecological reasons. With this method, the insecticidal action of the compound has to persist during the entire dry season on the vegetation in order to destroy flies which also hatch at a later time. This requirement is best fulfilled by the CHC compounds endosulfan and dieldrin as well as by pyre-throids. OPAE products and carbamates are relatively inefficient against tsetse flies and hydrolyse rapidly on the vegetation. Endosulfan is sprayed as a 25% solution formulated as ULV with the helicopter. It becomes catabolized faster than dieldrin and produces only little residue in the soil, as well as in plants and animals. From the pyrethroid group, permethrin and deltamethrin have been used the most. The insecticidal effect of permethrin against tsetse flies is 5 times that of dieldrin and is of the same persistency. The low toxicity for mammals is, however, unfavourably balanced out by its high toxicity for fish. Pyrethroids are also much more expensive than compounds of the CHC group.

Another alternative for controlling tsetse flies is the application of compounds with a **knock-down effect**. Products with a low concentration (15 g endosulfan/ha) are sprayed in short intervals. To be effective, this treatment has to be repeated every 3 weeks in order to prevent the tsetse-females from laying eggs. The application is carried out at the end of the dry season when the climatic pressure on the glossines is most intensive. An insecticide-aerosol is applied either shortly before dawn or 2 hours after sunset. Because of the meteorological inversion which takes place at night, the aerosol remains as a slowly drifting fog around the tree tops for a rather long time and thus comes into direct contact with the flies. The only flies which are destroyed are those which are affected when the application is carried out. Therefore, spraying has to be repeated 6 times within a time span of 3–4 months in order to catch flies which hatch later.

Using both strategies, areas which have already been cleared, have to be secured against reinfestation with persistent insecticides or with clearing and cultivation. It is generally assumed that each cleared area only needs to be treated once and will remain free of tsetse infestation later on.

The methods and strategies for tsetse control with chemical compounds are compiled in Table 5. In contrast to other insects, glossines are relatively sensitive to the compounds which are mentioned here. With the knock-down method, low concentrations which are well tolerated by the environment can be used. For

Table 5. Insecticides used for tsetse control

Type of application	Insecticide	Active substance/ha	Remarks
Residuum			Selective, also under difficult ground conditions
Ground spray	Dieldrin	2 kg	Mostly applied for barriers
Helicopter	Dieldrin	approx. 0.9 kg	Selectively applicable, but very expensive
	Endosulfan	approx. 0.9 kg	
	Deltamethrin	1–230 g	
Knock-down			Only applicable above flat ground with not too dense a vegetation
Aeroplane (4–6 applications)	Endosulfan	9–24 g	Most used compound, good residual effect
	Deltamethrin	0.1–0.3 g	Relatively short residual effect, expensive

residue application, high concentrations have to be applied in order to guarantee that lethal concentrations still remain on the vegetation at the end of the dry season. After application of endosulfan (1 kg/ha), almost no residuals could be detected by means of gaschromatographic analysis in spite of the fact that dead birds and mammals were found. Dieldrin is much longer-lasting and has been found still evident in birds which have fed on insects as well as in the environment after 3 years. In contrast, the knock-down application has lesser consequences for the ecosystem and may only endanger the fish populations.

So far, most of the large scale programmes for controlling tsetse flies in Africa have failed. The success which has been reached after years of intensive work has mostly disappeared again in a short time. There are only a few examples of tsetse flies being permanently eradicated in a local limited area, as was possible on Principe back at the turn of the century. With the exception of Zimbabwe (1986–1988), there is no example of this in the tsetse-belt (Müller 1991).

4.2.2. Vector Control on the Animal[1]

Vectors may be destroyed on the host animal while feeding, or during development. This depends on their mode of development and whether they feed only temporarily or permanently on blood. The compounds which are applied must have a toxicity level which is harmless to the animal while being efficient against the vector. Furthermore, it has to be made certain that no metabolites from the compound enters the human food chain. Last but not least, frequent attempts are made in the tropics to increase production by using highly productive exotic breeds which inevitably requires the indiscriminate application of pesticides for vector control. The FAO/WHO, Codex Alimentarius Commission, has determined **limits for residues** in order to protect the health of the target animal as well as the health of the consumers of animal products. These limits only determine the maximum residue limit in animal products and do not set any withdrawal time during which these products may not be used for human consumption. The instructions for the utilization of the compounds are also left up to the producer. There is a distinction between

- **no effect level (NOEL)**: the dose of a compound which does not show any effect under experimental conditions in the most sensitive animal species;
- **acceptable daily intake (ADI)**: the level of a compound which can be ingested by humans over a long time, it is determined by multiplying NOEL × SF (security factor);
- **maximum residue limit (MRL)**: the maximum amount of residue (mg/kg) which can be tolerated within an animal product; the standard data are ob-

[1] I am grateful to Dr. Friederike Thullner from Warrc/FAO World Acaricide Resistance Reference Centre – Federal Institute for Health Protection of Consumers and Veterinary Medicine, Berlin, Germany for revising this chapter and providing me with many valuable references.

tained through experimental studies under field conditions and calculation; it is the basis for the determination of the withdrawal time: time which has to expire after the withdrawal of the product from the animal until the animal product can be consumed;

– **extraneous residue limit (ERL)**: substances are registered which are not pesticides but may appear in animal products through "normal" environmental contamination.

The MRL levels for milk and beef for some of the pesticides which have been used so far for vector control are summarized in Table 6. The MRL for milk is with a fat content of 4% in raw milk. These levels must therefore be halved in skim-milk products (Holler 1986).

When using a pesticide on an animal, there is always a danger of **poisoning**. It depends on the biochemical mechanisms of the action of the compound, its dose and the group of compounds to which it belongs, and also its influence on

– the vector,
– the target animal, and
– the consumer of the animal product

whether application of a pesticide leads to poisoning in humans and animals.

The chemical is supposed to destroy the vector for as long a time as possible without developing a resistance in, or poisoning, the treated target animal. Furthermore, no, almost no or only a minimum of residues ought to remain in the animal product. The choice of the compound, its concentration (dose) and formulation depends on the

– LD_{50} for the mammal: lethal dose for 50% of treated target animals;
– TD_{50} for the mammal: toxic dose, which causes symptoms of poisoning in 50% of treated target animals;

Table 6. Maximum residue limit for milk and beef (Holler 1986)

Compound/ Generic name	Milk (4% fat)/ mg/kg	Body fat of carcass/ mg/kg
Aldrin	0.006 ERL[a]	0.2 ERL
Dieldrin	0.006 ERL	0.2 ERL
Chlorphenvinphos	0.008	0.2
Coumaphos	0.02	1.0
Crufomate	0.05	1.0
DDT	0.05 ERL	5.0 ERL
Dichlorvos	0.02	0.05
Ethion	0.02	2.5
Fenchlorphos	0.08	10.0
Lindane	0.01	2.0

[a] ERL: extraneous residue limit.

– LD_{50} for the vector: lethal dose for 50% of vectors;
– ADI for the human consumer.

Theoretically it would be the best way to prevent the development of resistance to a pesticide within a vector population, if the dose of the compound always is kept above the LD_{50} for the vector, but below the TD_{50} for the target animal. The broader the therapeutic width of the compound, the safer its application.

For two groups of compounds, the symptoms of poisoning which may occur together with their application can clearly be defined:

– Chlorinated hydrocarbons (CHC) cause:
  • loss of appetite, reduction of productivity and emaciation;
  • nervous hypersensitivity, restlessness, ataxia, muscular tremors, convulsions;
  • salivation and diarrhoea;
  • death because of respiratory paralysis.
  The symptoms are usually the consequence of cumulative poisoning and are untreatable for the most part. The course of the disease is protracted and always leads to death, since as soon as the symptoms appear, a critical amount of the particular drug has already been accumulated in the organism, especially within the fatty tissue.
– Organo-phosphoric acid compounds (OPAE) and carbamates exert their neurotoxic effects by inhibiting the enzyme acetylcholinesterase at the endplates of the nerves thereby prolonging the resting time of acetylcholine at cholinergic synapses and producing hyperexcitation of cholinergic pathways, the onset of the symptoms being mostly abrupt in OPAEs while more protracted in carbamates. Characteristic symptoms are:
  • agitation, muscular tremors, muscular weakness and contraction of the pupils;
  • salivation, frequent urination and diarrhoea;
  • death through respiratory paralysis with symptoms of asphyxia and tonic-clonic tremors.

Because of the pharmacodynamics of this group of compounds, symptoms mostly appear peracutely immediately after application. If the dose has only been sublethal, recovery is quick. Cholinesterase-blockage which is caused by compounds of the OPAE and carbamate groups can be easily treated with atropine, sometimes even in cases which are seemingly hopeless.

After dermal or oral application, or accidental oral intake, pesticides are **reabsorbed** through the skin or the mucosa of the intestine. The blood distributes the compound and either eliminates it through milk, faeces and urine, or uses it for biotransformation and deposition within the tissue. This process takes the following course:

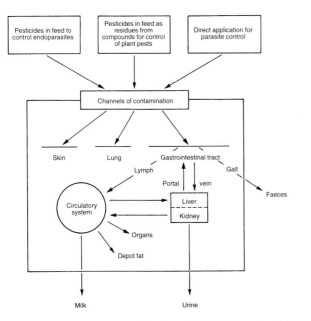

Figure 33. Reabsorption, distribution and elimination of pesticides (Holler 1986).

- at first the concentration of the compound within the organism is rapidly increased;
- during repeated ingestion, a plateau-concentration is built up;
- after the end of the application, the concentration of the compound is reduced depending on its biological half-life (Fig. 33).

The deposition of pesticides within tissue, organs and excretions *per vias naturales* (urine, faeces, milk, sweat, saliva) depends on the group of compounds and the type of compounds within its group. CHCs usually become catabolized within the liver; they are excreted over a long period of time in comparatively high concentrations. OPAEs become catabolized and excreted much more quickly. The elimination of pesticides from the organism occurs in two ways:

- excretion of the unchanged substance;
- elimination of biotransformed substances (metabolites) through excreta; substances more poisonous than the original compound may appear.

Each compound has a specific rate of excretion: its **biological half-life**. This is the time needed to eliminate 50% of the compound (Holler 1986).

Table 7. Ways of application and concentrations of compounds used for vector control[a] (active substance) chlorinated hydrocarbons, OPAEs, carbamates, cyclic amidins and avermectins

Group of compounds	Generic name[a]	Trade name	Concentration (%) of active substance for dipping	
			First fill	Refill
Chlorinated hydrocarbons (CHC)	Toxaphen	Coopertox Alltox Phenacide Phenatox	0.375	0.5
	CHC	Lindane BHC Hexachloran Nexen Nexit	0.05	0.075
	HEDD	Dieldrin		
	Thiodan	Endosulfan		
Organic phosphoric acid esters (OPAE)	Coumaphos	Asuntol CO-Ral, Agri-Dip	0.05	0.05
	Trichlorphon	Neguvon Chlorfos, Metrifonat, Phoschlor		
	Quinthiophos	Bacdip	0.08	0.125
	Dicrotophos	Ectaphos 50 u. 100	0.05	0.05
	Chlorphen-vinphos	Esteladon 30 Supona, Birlane, Steladone	0.03	0.06
	Fenthion	Tiguvon Baytex, Lebaycid		
	Diazinon	Neocidol	0.05	0.05
	Dioxathion	Delnav, Cattle Dip	0.05	0.065
	Malathion	Malathion	0.05	0.1
	Chlorpyriphos	Dursban 26 E	0.03	0.05
	Ronnel	Nankor 4 E Fenchlorphos, Korlan, Trolene	0.2	0.25
	Crufomate	Ruelene		
	Dition	Dition M	0.05	0.075
	Tetrachlor-vinphos	Rabond Gardona		
	Dichlorvos	Vapona, Nuvan		
	Famphur	Famophos, Warbex		
	Phoxim	Valexon, Sebacil		
Carbamates	Carbaryl	Sevin, Vioxan		
	Promacyl	Promicide	0.2	0.2
Cyclic amidins (formamidins)	Amitraz	Tantik, Triatix, Triatox	0.25	0.2
	Cymiazole	Besuntol	–	
Avermectins	Ivermectin	Ivomec pour-on 5 mg/mL		

[a] Trade names possibly incomplete in terms of worldwide distribution.

Application as					
Spray	Dust	Pour/spot-on	Back-rubber	Dust-bag	Surface application within habitat (g/ha)
0.75					
0.075					0.5
					2.5
					1
0.1	5	5–10		1	0.1
0.15					0.75
0.08					
0.1					
0.05		5–20	20	5–10	
0.1	2	10		10	
0.15					
0.15					3–5
0.05/0.1					0.6–2.2 ppm
0.2			1	1	4
0.5	3	5–10	20	5–10	
	3				
				1	
				1	
		2.6 mg/kg			
0.05					
0.5					
0.2					
0.25					
0.03					
		10 ml/50–100 kg			

4.2.2.1. Acaricides and Insecticides

Acaricides are used for the control of ticks, and acarinae, which are plant pests; insecticides are used to control insects on animals and plants, the compounds used for both purposes being identical as a rule. In Table 7, most of the CHC/ OPAE group and other products are compiled under their generic and trade names. The pyrethroids are summarized in Table 8. Some of the products mentioned are no longer allowed to be applied in industrialized countries but are still used in developing countries.

For application either in the environment or on animals, compounds are formulated in such a way that a long-lasting **residual effect** is obtained. For this purpose, the products may either be applied as W.P.s or emulsions. For the pour-on method, the active substances are dissolved in oil or emulsified. A solution in water is rather active chemically and only produces a short residual effect. It is therefore of little importance for this purpose.

Synergists, when applied with an insecticide, increase the potency of the insecticide by inhibiting the action of the resistance mechanism, thereby reducing the discrimination between resistant and susceptible genotypes, much as if the insecticide alone conferred only a low level of resistance. Synergists can only be used effectively if there is a single mechanism of resistance, if resistance does not

Table 8. Ways of application and concentration of pyrethroids used for vector control

Generic name	Trade name	Concentration forDip[a]		Spray[a]
		First fill ppm	Refill	ppm
Flumethrin	Bayticol, Bayvarol, Drastic Deadline	30–40	75	40
Cyfluthrin	Bayofly, Responsar			
Deltamethrin	RU-22974, NRDC-161, OMS 168, Decamethrin, Decis, K-Orthrin	25–33	50	33
Cypermethrin	Flectron, Buffalo Fly Spray, Curatik Cattle Dip, Ectomin, Robust SBZ, Barricade S	150–200	250	200
Cyhalothrin	Grenade, Liberekto	70–100	140	100
Cypothrin	Panecto	150	225	175
Alphamethrin	Ultimate, Pavacide	70	84	75
Fenvalerate	Sumitik, Sumifly, Sumifleece	200	250	200
Permethrin	Stomoxin MO, Delice			
Flucythrinate	Guardian			
Fluvalinate	Apistan			0.0033%

[a] Formulated as EC = emulsifiable concentrate.
        b.w. = body weight.

subsequently develop to the combination of synergist and insecticide, if the synergist is stable under field conditions and has no mammalian toxicity, and if it is cost-effective. Although many synergists are known for different pesticides, few meet these criteria (Roush and Tabashnik 1990).

Taking advantage of synergism can only be a subsequent reaction to the development of resistance. No proof exists so far that the combination of different groups of compounds, e.g. OPAEs with CHCs or OPAEs with pyrethroids, can be used to prevent the development of resistance (Thullner 1995). It is the aim of the synergistic effect to amplify the spectre of possibilities for attaching the compound to the vector while at the same time reducing the toxicity for the warm-blooded animal, as probably it is the case with BARRICADE S for the control of *B. microplus*. In some cases the use of the synergistic effect can also contribute to increasing the chemical stability of the dip.

Pesticides can be applied to the animal externally as **contact poisons** for the vector, or they can destroy the vector after dermal, parenteral or oral application due to their **systemic action** on the arthropod (I/4.2.2.2). In the following discussion of the different compounds, they have to be evaluated with respect to the tolerance of the animal to the compound, the potential residues in the animal product and last but not least the effect of the compound on the arthropod.

Application as				
Dust	Pour/spot on (mg/kg b.w.)	Back-rubber (mg/kg b.w.)	Dust-bag (mg/kg b.w.)	Surface application within the habitat
	1	1		
	1	1	x	
	0.75	0.75		12–30 Residuum 0.1–0.3 knock-down
	1	1	x	
	2.5	2.5		
	2	2		
			x	

Special attention is paid to the question of whether a compound may be used for dairy animals or not.

In animal production in the tropics, the problem of the **development of resistance** within a population of vectors is of importance for the epizootiology of the disease, the economy of animal production, the environment and human health. In many areas, a resistance of ticks to numerous compounds already exists which causes considerable damage to animal production. In areas where cross-resistance between DDT and pyrethroids is present, it is rather difficult to produce dairy products without contaminating the animal product. The WHO has defined the term resistance as the ability to tolerate the toxic amount of a compound which would destroy the majority of the individuals of a normal population. The consequence is that the compound used against a particular vector within a determined area will gradually lose its effect during a certain period of time depending on the interval of generations, the size of the vector population, and the conditions of the environment, thus rendering it useless. With regard to the applied pesticides, a selection takes place within a heterogeneous population of vectors with varying sensitivity to the compound which is genetically defined. Consequently, only a few such vectors which have a high genetically fixed sensitivity survive at first, being the fairly and highly sensitive parts of the population, and, after repeated treatment the resistant parts of the population become gradually able to multiply. This occurs especially when the compound, which because of its long residual effect gradually drops under the DL_{50} for the vector, allows the multiplication of those parts of the population which have only been influenced sublethally (Fig. 33). It can be assumed that resistance is the consequence of mutation of the vector population while selective pressure is exerted through the acaricide/insecticide treatment. The emergence of the resistance in the field, however, is influenced substantially by the management of the acaricide/insecticide treatment, especially suboptimal doses and compounds with long residual effect being responsible for allowing the development of resistant strains of vectors. There may be substantial differences in the development of a site-specific resistance which can be specific to a region, a locality and even a ranch, depending on the different local approaches to, e.g., tick control (Thullner 1995). The mechanism of development of resistance against OPAEs and carbamates is caused by enhancing the activity of the acetylcholinesterase (ACE). The proportion of ACE activity within the total activity of enzymes in ticks depends on the species and strain. Consequently, strains with a high percentage of insensitive ACE are able to develop a resistance much more quickly. The ACE insensitivity is inherited and is either recessive, dominant or coupled to autosomal alleles independent of the tick genera and species. Some tick strains have a higher capacity for detoxication and slower rate of conversion which creates a further mechanism of resistance to OPAE.

The mechanism of resistance of arthropods to pyrethroids, especially permethrin, depends on the increased capacity of cis- and trans-isomers for detoxication though Nolan et al. (1989) found no indication of detoxication being the cause for resistance in *B. microplus* to pyrethroids. In the case of amidins, it is assumed that there is an alteration of sensitivity in the enzyme monoamino-oxidase, the details of the mechanisms being widely unknown.

A number of factors exists influencing the effectiveness of toxicants and opportunities for the development of resistance. Of the number of possible resistant mechanisms only those of behaviour, penetration, detoxication and insensitivity have been verified experimentally. Behaviour resistance has not yet arisen in any of the ectoparasites of interest. Penetration, as a mechanism of resistance has only be found in *B. microplus* with cypermethrin isomers. Detoxication is important in OPAE- and carbamate-resistant ticks and is supplemented through insensitivity of acetylcholinesterase to the compound. The most well-known of causative detoxication mechanisms is dehydrochlorination associated with DDT resistance. Insensitivity at first was found to be caused by a recessive gene that conveys insensitivity to DDT at the site of action, called kdr, for knock-down resistance. DDT-resistant strains of cattle ticks are cross-resistant to pyrethroids; seemingly exhibiting also a kdr-type resistance to pyrethroids. OPAE and carbamate inhibitors act against acetycholinesterase and can be coupled with reduced sensitivity. In ticks these two aspects are intimately related, being controlled by the

Table 9. Resistance of ticks to chlorinated hydrocarbons (CHC) (Georghiou and Lagunes Tejeda 1991, Holler 1986)

Species	Compound	Country
Boophilus annulatus	CHC (Lindan)	Jamaica
Boophilus decoloratus	CHC	Zambia, Kenya, Malawi, South Africa, Uganda, Zambia, Zimbabwe
	Toxaphen	Kenya, Zambia
Boophilus microplus	DDT	Australia, Jamaica, Trinidad, Guyana
	CHC	Australia, India, South Africa, Jamaica, Trinidad, Guyana, Ecuador, Columbia
	Toxaphen	Australia, South Africa
	all CHCs	Argentina, Australia, Brazil, Colombia, Caribbean, Ecuador, India, Madagascar, Malaysia, Venezuela
Rhipicephalus appendiculatus	CHC	Kenya, Tanzania, Zimbabwe, South Africa, Zambia
Rhipicephalus evertsi	Toxaphen	Kenya, Uganda, Tanzania, South Africa, Zambia, Zimbabwe
Amblyomma hebraeum	CHC + DDT	South Africa
	Toxaphen	South Africa, Swaziland
Amblyomma variegatum	CHC	Kenya, Tanzania, Uganda, Zambia
	Toxaphen	Kenya, Tanzania, Zambia

same altered active site on the acetycholinesterase (Nolan and Schnitzerling 1986).

The WARRC, FAO World Acaricide Resistance Reference Center, Berlin, Germany, has developed an acaricide resistance Test Kit. The test is based on the larval packet test which entails the exposure of larvae for 24 hours in the incubator to acaricide impregnated filter papers at different concentrations. Subsequently mortality is determined for every concentration and acaricide and a dose/mortality graph is plotted. The comparison with reference strains allows the specification of resistance and the calculation of resistance factors (Thullner et al. 1994).

Table 10. Resistance of ticks to organophosphorus compounds (Georghiou and LagunesTejeda 1991, Holler 1986)

Species	Compound	Country
Boophilus annulatus	Coumaphos	Sri Lanka
	Dioxathion	Sri Lanka
Boophilus decoloratus	Coumaphos	Zimbabwe
	Diazinon	Zimbabwe, South Africa
	Dioxathion	Zimbabwe, Zambia, South Africa
	Dicrotophos	South Africa, Zimbabwe
	Quinthiophos	South Africa
Boophilus microplus	Coumaphos	Argentina, Australia, Brazil, Mexico, Columbia, Venezuela
	Chlorphenvinphos	Argentina, Australia, Jamaica, South Africa
	Chlorpyriphos	Argentina, Australia, Brazil
	Dicrotophos	Australia, Brazil, Jamaica, South Africa
	Diazinon	Argentina, Australia, Brazil, South Africa
	Dioxathion	Argentina, Brazil, South Africa, Columbia, Venezuela
	Fenthion	Brazil
	Quinthiophos	South Africa
Amblyomma hebraeum	Chlorphenvinphos	South Africa
	Dioxathion	South Africa
	Quinthiophos	South Africa
Rhipicephalus spp.	Chlorphenvinphos	Kenya, Tanzania
	Dioxathion	Kenya, Tanzania
Hyalomma spp.	Chlorphenvinphos	Kenya, Tansania
	Diazinon	Uganda
	Dioxathion	
	Trichlorfon	

Table 11. Resistance of ticks to carbamates (Holler 1986)

Species	Compound	Country
Boophilus decoloratus	Carbaryl	Zimbabwe
Boophilus microplus	Carbaryl	Jamaica, South Africa
B.m., Biarra strain	Carbaryl	Australia, Brazil, Venezuela Columbia
B.m., Ridgelands strain	Carbaryl	
B.m., Mt.-Alford strain	Carbaryl	
B.m., Bajool strain	Carbaryl	
B.m., Bajool strain	Promacyl	Australia

Recommendations to prevent and control the development of the resistance of vectors to pesticides in connection with vector control schemes are described in I/4.2.2.2.

Table 9 shows that the resistance of ticks to CHC products has already developed worldwide. Because of the association of similar genes in the development of resistance, the resistance to DDT products and pyrethroids is often combined. DDT used to be applied indiscriminately which led to a multiplication of these genes. During the age of OPAE products it was neutralized and reduced. With the use of pyrethroids, these genes have spread again and are leading to a rapid development of resistance against pyrethroids. Permethrin and cypermethrin are especially effected, though there is no development of general resistance to pyrethroids, but only of resistance to specific products. The Malichi-tick strain which had a rather low resistance against DDT, developed a homozygous permethrin resistance under permanent permethrin application after only 9 generations (Grothe and Weck 1983).

The development of tick resistance to **OPAEs** and **carbamates** is described in Tables 10 and 11. The resistance which at first appeared in one-host ticks in Australia, Africa and South America has since spread to two- and three-host ticks in Africa. The mechanism of development of resistance against OPAEs and carbamates has already been discussed above.

In the following section, chemical characteristics, peculiarities of toxicity and pharmacology as well as pesticidal effect and indications for application of the most important and used compounds are described.

• *Chlorinated Hydrocarbons (CHCs)*

Pesticides from the CHC group are aromatic, ring-shaped compounds; their basic skeleton is derived from the benzene ring. The benzene nucleus is very stable and can only be altered if the H-atoms are changed. The isomers of some CHC compounds can have a very different effect.

CHC compounds are very stable within the environment, and their catabolism within the organism is rather slow, depending on their composition. After resorption they enter the liver where they are slowly metabolized before they are released into the circulatory system and either they are finally deposited in the fat or they are excreted. Their transformation is lengthy: for example, DDT (1,1,1-trichloro-2,2-bis(p-chlorophenyl)ethane) is transformed into DDE, and DDD into DDA which in the end is soluble in water and can be excreted with the urine. Aldrin and heptachlorine are oxidized enzymatically into products which are difficult to catabolize, while endrin (the isomer from dieldrin) is excreted quickly. It is extremely poisonous. DDT and its metabolites are passed on to the calf and to the consumer in milk.

The biological activity of CHCs destroys the nervous system of living organisms. The principle of the poison is not yet clear in detail. After oral and pulmonal intake, toxicity is at its highest. Because of the lipid solubility of the compounds, food rich in fat (milk) increases the toxicity; equally oily solutions are reabsorbed rather quickly after dermal application. Since the compounds are deposited in body fat, thin animals can easily become poisoned; fat animals fall ill if they use up their fat deposits during periods of extreme hunger and thus release the poison.

While CHCs are not used any more in western industrialized countries, and its use is forbidden in animal production, these products are still applied in the countries of the Third World, e.g. toxaphene for tick control and endosulfan for tsetse eradication. Like no other pesticide in the past, DDT has been applied on a large scale for the control of gnats, midges and tsetse flies. It has become the symbol of the danger man poses to nature. Since it is not used any more, not even for tsetse control, it is not discussed further.

HCH (1,2,3,4,5,6-hexachlorcyclohexane, BHC, HEXACHLORAN, NEXIT, JAKUTIN, LINDANE[1] was developed in 1942 and has been used since 1949, especially as γ-isomer LINDANE for mosquito and tsetse control. It has also been used as an acaricide for cattle dip (0.025–0.05%), as W.P. and as an emulsion. LINDANE is reabsorbed through the skin as well as by the internal mucosa of the organism; the speed of reabsorption depends on the formulation. The quickest effect occurs when it is bound to an oily carrier. Because of its high lipid solubility, HCH is deposited in fat, but less so in the liver, kidney and brain, and is excreted above all in milk. Like other CHCs, HCH is a nerve toxin for warm-blooded animals and a respiratory toxin for arthropods. In contrast to DDT, it is highly soluble in water and is quickly catabolized by vertebrates, turning it first into pentachlorcyclohexane, then into tetrachlor- and trichlorbenzene and finally into phenols which are excreted with urine and faeces as water-soluble sulphates or glucuronacid conjugates. HCH is eliminated completely even after long-term application.

At present, a large number of arthropods has become resistant to HCH. The development of resistance arises because the vectors are also capable of catabol-

---

[1] Trade names are written in capitals in the following text.).

izing the compound into non-toxic substances. The following arthropods are resistant to HCH: *Pulex irritans, Boophilus decoloratus, B. microplus, Amblyomma americanum, Rhipicephalus sanguineus, R. evertsi, R. appendiculatus, Dermacentor variabilis, Musca domestica, Stomoxys calcitrans, Haematobia irritans* and *Culex pipiens*. After a longer period of application, resistant strains of ticks may become sensitive again (Table 9).

Toxaphene (Octachlorcampher, COOPERTOX, ALLTOX, PHENACIDE, PHENATOX) has been used for more than 30 years as an acaricide, especially in dips, as well for controlling two- and three-host ticks (*Amblyomma, Hyalomma, Rhipicephalus*) as *Boophilus* spp. It is a complex mixture of 177 polychlorinated components with an average composition of $C_{10}H_{10}C_{18}$. The warm-blooded animal quickly catabolizes the compound into dichlorinated products; cancerogenic and mutagenic effects have been demonstrated in laboratory animals. It still appears in milk as much as 21 days after spray application; for beef cattle a withdrawal time of 6 weeks is required.

In the dip, toxaphene is applied as a 75% emulsion in a 0.25% concentration with an interval of 7 days and in a 0.375% concentration with an interval of 10 days. Higher concentrations cause poisoning. Since dips with toxaphene are often used without being changed for many months, the dip may become unstable and cause poisoning. Because of its resin-like properties, the compound remains on the coat of the animal even when it rains and thus produces a residual effect which lasts several days. Due to its stability, toxaphene is also used as a synergistic partner. In South America, Africa and Australia, *Boophilus* and *Rhipicephalus* ticks have become resistant to toxaphene (Table 9).

Dieldrin and **Aldrin** are closely related and are both excreted as dieldrin. Because of their high toxicity (higher than DDT), they are not suitable for the use on animals and are prohibited in most countries. They are, however, still used for vector control within the environment, especially for mosquitoes and tsetse flies.

Dieldrin is soluble in organic solvents, but not in water; it is used in the form of 50% W.P. and 50% emulsion. The production of dieldrin is supposed to have ceased, but old stocks may still be available. Residues in animal products are mostly due to feed which is produced from contaminated agriculture products.

Methoxychlor (1,1,1-trichloro-2,2 bis-(4-methoxyphenyl)ethanol, MARLATE, DMDT, METHOXY-DDT) is mainly used in the form of a powder or a spray to control horn fly. Like DDT, it is solvent in lipids and is therefore excreted with the milk. Storage in fat deposits is only short-lived, and half of the substance will already have been eliminated after 10 days. Therefore, a withdrawal time of 3 days in dairy cattle is considered to be enough.

In summing up, the use of CHC compounds should be restricted as much as possible:

– they should not be used on animals; if toxaphene is used, use on dairy cows should not be permitted;
– they may be used environmentally for vector control but only with great caution.

Table 12. Restrictions of application of chlorinated hydrocarbons in animal production (Holler 1986)

Group of compounds	Trade name	Withdrawal time	Applicable to dairy production	
			Possible	Impossible
CHC	Lindane	At least 60 days		+
Methoxychlor	Marlate	3 days		+
HEDD	Dieldrin	More than 7 days depending on the concentration		+
HHDN	Aldrin	More than 7 days depending on the concentration		+
Octachlor-campher	Toxaphene	At least 6 weeks		+

Table 12 shows the restrictions of the use of CHC compounds in animal production.

• *Organo Phosphoric Acid Esters (OPAEs)*

OPAE compounds are phosphates, phosphonates, phosphothionates, phosphothiolates and phosphodithiolates. They inhibit the cholinesterase which is the enzyme which splits acetylcholine, the transmitter of the subsynaptic cleft between muscle and nerve cell, into acetate and choline. These compounds have an effect on the nervous system of the arthropod as well as on the nervous system of warm-blooded animals. Arthropods are much more sensitive to OPAEs than vertebrates. The acute toxicity of OPAEs for mammals is considerably higher than, for example, DDT. Nevertheless, if the compounds are applied according to the instructions, poisoning rarely occurs. Application using the prescribed intervals and continuous application with the prescribed dose does not produce accumulation and is tolerated in the long run without any symptoms. This means that the cholinesterase level of the animal may settle down to that of a lower titre without impairing the health of the animals. Poisoning occurs if the DT_{50} for the animal is exceeded.

Depending on the formulation, most products are easily reabsorbed after dermal, parenteral and oral application. Therefore, they are suitable for use as wettable contact pesticides or as systemic-acting substances. During metabolization, part of the active substance may be deposited for a short time within the fat before it is excreted with the milk fat. This part of the substance is not involved in the systemic action.

OPAEs are metabolized quickly and, with few exceptions (fenthion), only have a short residual effect. Consequently they have a lesser tendency to develop a resistance in comparison with CHCs (Fig. 34). During the catabolism of

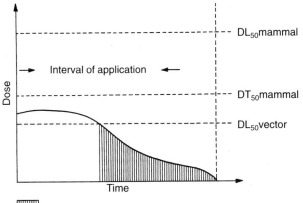

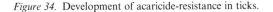

Figure 34. Development of acaricide-resistance in ticks.

OPAEs with in the organism, toxic catabolites with a high anticholinesterase activity may appear. Through degradative metabolization they are broken down into products with low toxicity and are excreted later on.

The great advantage of OPAEs over CHCs is that they are biodegraded in a short time within the environment. Last but not least, because of this, they may still be utilized and, if applied correctly, will not present any danger to the animal and or to man as a consumer of animal products.

Chlorphenvinphos   (2-chloro-1-(2,4-dichlorophenyl)vinyldiethyl-phosphate, ESTELLADON 30, SUPONA, BIRLANE, STELADONE) is used as acaricide against *Amblyomma, Boophilus, Haemaphysalis* and *Hyalomma*, and as insect-icide against horn fly, lice and mites. It should be applied where ticks have become resistant to CHC and other OPAE compounds. It is marketed as 20 and 30% emulsion and is applied to the dip in concentrations of 0.01–0.1% active substance without producing any side-effect; concentrations above 0.2% cause intoxication. If applied correctly, the compound is deposited in the fat of the target animal in low concentrations and will be excreted with the milk for a short time. 2 weeks after the application of a 0.25% spray, residues of 0.006 ppm have been found in the depot fat.

Chlorpyriphos   (0,0-dietyl-0-(3,5,6-trichloro-2-pyridyl)phosphorothionate, DURSBAN) is one of the oldest OPAE products and sold as a 26% emulsion. It is applied as acaricide, mainly against *Boophilus* (0.025–0.05% spray, 0.05% dip). The compound has only an average residual effect and should be applied every 7 days where there is a lot of challenge from ticks.

Chlorpyriphos and its metabolites are deposited in the fatty tissue and, as a consequence of considerable amounts of residue, remain in the organism and are excreted with the milk. Therefore it cannot be used in dairy production; in beef production a withdrawal time of 2 weeks, a time, during which no animal pro-duct may be consumed, has to be observed.

A product which combines chlorpyriphos with fenoxycarb is marketed as KADOX spray. The latter, fenoxycarb, is a hormonal regulator of development of insects which inhibits the development of egg, larva and pupa. The product is applied in microcapsules which slowly release the active ingredients (controlled release). It is mainly used for endo- and ectoparasite control in the faeces of pets.

Crufomate (4-tert-butyl-2-chlorophenylmethyl-methylphosphoramidate, RUELENE) is sold as a 24% oily concentrate and is applied as a 5% emulsion for pour-on application but may be used also as a spray against ticks, lice and flies. It is well tolerated and produces a residual effect of 1 week. The systemic application may be carried out in dry dairy cows and beef cattle. Lactating dairy cows cannot be treated.

Coumaphos (0,0-diethyl-0-(3-chloro-4-methyl-7-coumarinyl)-phosphorothionate, ASUNTOL, CO-RAL, AGRIDIP) has been one of the acaricides and insecticides of the OPAE group most often applied. It is sold as both a 16 and 20% emulsion and a 50% W.P. For the control of stomoxydines with a dust-bag it is applied as a 1% powder. For systemic application, a 5% emulsion is used. Coumaphos has been also applied orally (44 ppm in the feed) against haemophages.

Coumaphos has a broad range of efficiency; it is effective against ticks, stomoxydines, warble flies, fleas, lice and mites. In the past it has been especially effective against the metanymphal stages of ticks which generally are difficult to control. It is probably the acaricide which has been utilized most worldwide. In comparison with other OPAE products, coumaphos has no knock-down effect. It takes 1–4 days until the ticks die in spite of the fact that they have been damaged irreversibly immediately after application. This is a phenomenon which may mislead the non-informed observer. The formulation as a W.P. has the advantage of not evaporating and remaining stable for a long time in the dip. The emulsion is used in a lower concentration than the W.P. Like other OPAE products, W.P.s and emulsions are "stripped out" of the dip, a phenomenon which depends on the type of coat of cattle and sheep. The residual effect after spray and dip lasts 4–7 days, depending on the rainfall.

After dermal application, about 80% of the active substance is reabsorbed by cattle within 1 day and excreted in the course of a week with urine and faeces; the same is true for oral application. About 5 hours after application, sprays, dips and dust-bags produce a short lasting elevation of residues (0.2–0.25 ppm) which drop below the critical limit after 48 hours. Since the compound is not deposited in the fat, it is completely metabolized and excreted. Due to the MRL of 0.2 ppm for coumaphos and its metabolites, the compound may be applied as a spray or as a dust-bag in dairy production without any withdrawal time. There is a withdrawal time of 7 days for beef cattle.

Diazinon (0,0-diethyl-(2-iso-prophyl-4-methyl-pyrimydil-6) phosphorothionate, NEOCIDOL) is available as a 25% W.P. Since it is difficult to prepare the dip suspension, the compound has not become popular. As a spray it has mainly been used for application both in animal stables and on animals against flies. In Argentina it is used as a synergist in combination with pyrethroids to overcome

resistance. Diazinon pellets have been applied orally (1.4–2.5 mg/kg b.w.) against haematophagous flies; residues do not appear in the milk. Diazinon is well tolerated and is metabolized and excreted quickly. The MRL is set for milk at 0.02 ppm, and at 0.7 ppm for beef. After external application to dairy cattle, 3 days withdrawal time have to be observed.

Dichlorvos (0-2,2-dichlorovinyl)-0,0-dimethyl-phosphate, VAPONA, NU-VAN) is used as a 1% spray for surface application to control *Haematobia irritans, Stomoxys calcitrans* and mosquitoes. The 1% spray may also be used for short-term protection of animals. The compound may be applied in dairy production. The MRL is set at 0.02 ppm for milk, and at 0.05 ppm for beef.

Dicrotophos (0,0-dimethyl-0,0(2-dimethylcarbamyl-1-methyl-vinyl)-phosphate, EKTAPHOS) is sold as both a 50 and 100% formulation, mainly for the control of three-host ticks which are resistant to other OPAE products (*Amblyomma, Hyalomma, Rhipicephalus*), but as well it is used against *Boophilus microplus* and *B. decoloratus*. It may also be applied against lice and fleas. The efficiency of the compound is not decreased because of tick resistance to CHCs. The compound is soluble in water and therefore easy to handle. In spite of its solubility in water, the dip remains stable. The strip-out effect does not occur here. After spraying, residues of 0.02–0.04 ppm appear in the milk, but they descend below 0.005 ppm after 29 hours. Though the compound is metabolized and excreted rather rapidly it is not recommended for use in dairy production.

Dioxathion (2,3-p-dioxanodithiol-s-s-bis-(0,0-diacethyl-phosphorodithionate, DELNAV, DFF (Diluent Free Formulation) CATTLE DIP) is an acaricide which is used mainly in Africa against one-, two- and three-host ticks. It is also sold as SUPAMIX, a synergist with chlorphenvinphos, or ALTIK, which is combined with octachlorcampher. The compound is well tolerated and remains stable in the dip over a long time, but it has a considerable strip-out effect and therefore has to be supplemented regularly. The MRL for milk is set at 0.008 ppm. Because of its pattern of excretion it is not supposed to be applied to lactating cows.

Famphur (0,0-dimethyl-0-p-(dimethylsulfamyl)-phenyl-phosphorothionate, FAMOPHOS, WARBEX) is applied dermally, orally (2.6 mg/kg/b.w./d) and also parenterally (i.m.) for the control of lice and ticks of the species *Amblyomma maculatum*. The active substance is deposited in the body fat and muscle; the residues fall below 0.002 ppm only after the fourth day p.a. The compound may not be used in dairy production; the withdrawal time for beef cattle is 14 days.

Fenthion (0,0-dimethyl-0-(3-methyl-4-methylthiophenyl)-phosphorothionate, TIGUVON, BAYTEX, LEBAYCID) is mainly used systemic-dermally either as a 2, 10 or 20% oily solution against ticks, flies, midges, warble flies and lice. In Brazil it is very popular for the control of *Dermatobia hominis* which is an important fly where economic considerations are concerned, but is not relevant as vector. As spray it may be applied as a 0.05% solution, but because of its extreme systemic effect it is not suitable as dip. After pour-on application, fenthion penetrates the skin in a short time and is ingested by the parasite along with blood from the host. The oily formulation does not drip when it is poured

on the animal, and thus all of the active substance is reabsorbed. Rain does not wash the compound off either. A single dermal treatment protects against two- and three-host ticks for 20 days, against *Boophilus* for 30 days, and against flies for 10 days. It is difficult to surpass its effectiveness upon stomoxydines with any other OPAE compound. Apparently it has also a repellent effect, since cattle which are infested with *Haematobia irritans* become free of flies immediately after pour-on application. It is also very efficient against midges.

Fenthion is off limits in dairy cows since the active substance is excreted with the milk. Application of 10 mg/kg b.w. results in residues of 0.2–0.3 ppm in the milk. This is reduced to 0.05 ppm only after 6 days. If an overdose occurs, symptoms of intoxication often appear only after a few days. In extreme cases, an irreversible blockage of the cholinesterase appears as happens with folidol poisoning in humans. Treatment with atropine is ineffective in such cases.

Malathion   (S-(1,2-dicarbethoxyethyl)-0-0-dimethyl-dithiophosphate, MALATHION) is one of the oldest acaricides of the OPAE group. It is applied as an emulsion and a W.P., but because of its short residual effect, it has recently only been used for fly control in stables. The active substance is fat-soluble and is excreted partly bound to the milk fat. The compound is excreted with the milk (> 0.1 ppm) after application of 0.5%/1% emulsion/suspension, it may therefore not be used on lactating cows.

Phoxim   (0-(a-Cyanobenzylidenamino)-0',0'-diethylthiophosphate, VALEXON, SEBACIL) is marketed as an emulsion, has a low toxicity for mammals and a good acaricidal and insecticidal effect. It is used to control one-, two- and three-host ticks, flies, sheep ked, lice and mites. After spray application, it is reabsorbed by cattle through the skin and is quickly hydrolysed into non-toxic compounds which are excreted mainly with the urine. Parts of the compound are deposited up to 14 days in muscles, kidneys, liver and fat. After application of a 0.05% spray, residues of 0.13 ppm remain after 12 hours and 0.03 ppm after 24 hours. The MRL is set at 0.01 ppm. Therefore, the compound may not be applied to lactating cows.

Quinthiophos (0-ethyl-0-(8-quinolyl)-benzene-thiophosphonic acid ester, BACDIP) is available as both a 30 and 50% emulsion. In the past it has been used especially in East and South Africa for the control of three-host ticks which have become resistant against other compounds. It remains stable in the dip for 1 year, even after it has been heavily contaminated. Quinthiophos is tolerated extremely well by cattle: an overdose 80 times too high over 1 month did not produce symptoms of intoxication in calves. Therefore, a dip interval of 3–4 days is possible when there is a lot of pressure from three-host ticks without any danger of intoxication. The compound is metabolized and excreted rapidly. After 24 hours, the residues are already below the detectable limit. Manufacturing of the product has been discontinued, only remaining stock may be available in some countries.

Ronnel (0,0-dimethyl-0-(2,4,5,-trichlorphenyl)phosphorothionate, FENCH-LORPHOS, KORLAN, NANKOR 4E, TROLENE) is used mainly to control *Haematobia irritans* with spray application and systemically as an oily solution

(1%) with a back-rubber. It has also a systemic action against screw worm. As an acaricide it may be applied as a synergist together with other products. Applied externally, it has a long residual effect on *H. irritans* which lasts for 20 days. The compound is rapidly reabsorbed dermally and deposited in comparatively high concentration in the fat. Therefore, it may not be applied to lactating cows; in beef cattle, a 4-week withdrawal time has to be observed.

Tetrachlorvinphos (0-/2-chloro-1-(2,4,5-trichlorophenyl)vinyl/-0,0-dimethyl-phosphate RABOND, GARDONA) is used as an insecticide externally and for oral application (feed-through) for fly control (*H. irritans, M. domestica*). Residues are found in the milk with either of these forms of application, residues being deposited in the fat which disappear completely after only 3 weeks. The compound is hydrolysed in the liver and excreted mainly with the urine. Only small amounts appear in the faeces (6%) – this makes its application as larvicide doubtful.

Trichlorphon (2,2,2-Trichlor-1/hydroxy-ethyl)-dimethylphosphate, CHLORO-FOS, METRIFONAT, NEGUVON, PHOSCHLOR) is available as a water-soluble powder which used to be used externally as an insecticide as well as after oral application to control haemophages and endoparasites. The compound is not stable in a water-solution and is quickly broken down into the more toxic DDVP (0,0-dimethyl-0,2,2-dichlorvinylphosphate) especially within an alkaline environment. It may therefore neither be used for ground spray nor for external application to the animal. During the step-by-step degradation of trichlorphon, hydrochloric acid is produced which neutralizes the alkaline environment and blocks the further breakdown of the compound for a short time. Nevertheless, spray solutions should not be stored. The use of the compound as dip is forbidden; the same is true if the combination with coumaphos is used in the dip in order to prevent tick resistance through synergism, which has occasionally been practised in Brazil.

After external application, trichlorphon is reabsorbed through the skin and acts systemically on haemophages. For this purpose, it may also be applied in a pour-on formulation. Trichlorphon is rapidly metabolized and excreted by the organism. During the process, the more poisonous DDVP appears. It is not bound to the fat and excreted quickly with the milk. The milk is already free of residues 6 hours p.a. The MRL for trichlorphon is set at 0.05 ppm, so therefore, if it is applied correctly as a spray, a withdrawal time is not even required in dairy production. Like almost no other form of OPAE poisoning, trichlorphon poisoning can be treated excellently with atropine. Animals which are already in agony recover completely shortly after i.v. or intracardial injection.

In Table 13 the range of application and the possibilities and limits of responsible use of OPAE products, especially under the conditions of dairy production in the tropics is presented.

Table 13. Alternatives of application of organophosphorus compounds

Compound	Application				Concentration	LD_{50} mg/kg rat (oral)	Residual effect (days)
	Dip	Spray	Dust	Pour on			
Chlorphenvinphos	+	+			0.1–0.3% dip	9.6–39	7–12
Chlorpyriphos	+	+			0.05–0.25% spray 0.05% dip		4–7
Crufomate		+	+		5% emulsion	460–635	7–8
Coumaphos	+	+	+		20% emulsion 1% dust	41–230	4–6
Diazinon		+			0.03–0.08% suspension	75–110	2
Dichlorvos		+			1% spray	56–80	
Dicrotophos	+	+			0.05 + 1% suspension	10–29	7–9
Dioxathion	+	+			0.05% dip and spray		4–7
Famphur				+	45 mg/kg b.w., control of warble fly	35–60	
Fenthion		+		+	25 mg/100 kg b.w.	215–245	10–20
Malathion	+	+			0.25% emulsion + 0.50% suspension	–1370	2
Quinthiophos	+	+			0.02% dip and spray	50–150	–4
Ronnel		+			2.5–5% spray 1% backrubber	1250–2600	20 for flies
Tetrachlorvinphos		+			0.25% spray	–2300	
Trichlorfon		+		oral	0.15% for flies 2% for warble flies	560–630	< 2

Table 13. Alternatives of application of organophosphorus compounds (continued)

Compound	Ectoparasites	Application		MRL within milk (days)	Milk discard time (days)	Catabolism of residues	Applicable in dairy production			
							during lactation		before lactation	
		Animal	Animal habitat				Possible	Impossible	Possible	Impossible
Chlorphenvinphos	Ticks, flies	+	+	< 1	0	Rapid	+		+	
Chlorpyriphos	Ticks	+		> 7	14	Slow		+		withdrawal time to be observed
Crufomate	Ticks, flies	+		> 0.5	2	Medium		+	+	
Coumaphos	Ticks, mites	+		< 1	1	Rapid	+		+	
Diazinon	Flies		+	Not detectable with oral tests	0	Medium	+		+	
Dichlorvos	Flies		+	Not detectable	0	Rapid	+		+	
Dicrotophos	Ticks	+		> 1	1–2	Medium	+	+		
Dioxathion	Ticks	+		> 1	2–3	Medium		+	+	
Famphur	Flies	+		> 4	4	Slow		+	+	
Fenthion	Ticks	+		4–5	5	Slow		+	+	
Malathion	Ticks, flies	+		> 1	1–2	Medium		+	+	
Quinthiophos	Ticks	+		> 1	1–2	Medium		+	+	
Ronnel	Flies	+	+	2	2	Medium		+	+	
Tetrachlorvinphos	Flies	+	+	Not detectable	0	Rapid	+		+	
Trichlorfon	Flies	+		< 0.5	0	Rapid	+		+	

● *Carbamates*

Pesticides of the carbamate group which are also called **urethanes**, are esters from carbaminic acid chemically speaking with the group structure H_2-N-CO-O as its basic skeleton; there are three basic groups:

– dimethylcarbamates from cyclic enols;
– monomethylcarbamates from phenols;
– monomethylcarbamates from oximes.

Biologically, carbamates also inhibit cholinesterase. They bind with the active centres of the cholinesterase and block the hydrolysis of the acethylcholine. The effect of carbamates on arthropods already appears 1–30 minutes p.a. depending on the compound and the form of its application. The transmission of stimuli and impulses of the nervous system is interrupted, and after a phase of uncontrolled movements which are followed by paralysis, death supervenes. Domestic animals hardly ever reabsorb carbamates dermally.

Carbamates are metabolized through oxidation of the phenyl nucleus and of the N-ethyl group and through hydrolysis. The metabolites are exhaled as CO_2 or excreted with urine, faeces and milk; they are considerably less toxic for the warm-blooded animal as they are pesticidal for the arthropod. The oral toxicity fluctuates between < 1 and < 1000 mg/kg b.w. In the case of an overdose symptoms of OPAE-like poisoning appear. Atropine is also indicated here as an antidote. The development of arthropods' resistance to carbamates depends on their ability to hydrolyse the phenol-ring of the compounds.

Carbamates are mainly used for plant protection. When used as a ground spray, the sensitivity of these compounds to alkalis (chalked walls) has to be taken into consideration.

Carbaryl (1-naphthyl-N-methylcarbamate, SEVIN, VIOXAN) is mainly used for plant protection, but where available on the farm, it is often also applied as an acaricide and an insecticide. It is sold as a W.P. and also used as a dip. In spite of its low dermal reabsorption, residues appear in the milk after spray application (0.5%) to cattle; 0.18 ppm are still present 24 hours p.a. The MRL for milk is set at 0.1 ppm, so the compound is not to be applied to lactating cows.

Promacyl (3-methyl-5-isoprophylphenyl-N-(n-butanyl-N-methylcarbamate, PROMICIDE) has been used only in Australia as an acaricide against *Boophilus* spp. in dips and as a spray (0.2%). The compound is excreted with the milk fat and therefore not allowed to be applied to lactating cows. Though it has been rather efficient against resistant tick strains, its production has been discontinued.

Propoxur (BAYGON) is mostly used as a spray for insect control in living quarters; it has an excellent insecticidal action. It is inefficient against ticks and therefore of no importance for animal production.

● *Pyrethroids*

Pyrethroids are etheric oils and they are synthetic esters from *Chrysanthemum* acid or its derivates with both an acaricidal and insecticidal action. The chemical structure is similar to that of natural pyrethrins which are substances contained in *Chrysanthemum* spp., which have been derived from chrysanthemums for a long time. They are extracted out of flowers with a minimum content of pyrethrin of 0.3% and concentrated up to a 25% content of active substance. The most important pyrethrins are pyrethrin I + II and cinerin I + II. They come from plants in the form of trans-chrysanthemum acid and trans-pyrethrin acid in association with ketoalcohols, pyrethrols or cinerols. In contact with the arthropod, pyrethrins have a significant knock-down effect. Through contact with the active substance, the ion-exchange at the axon of the nerve of the arthropod is blocked which causes an interruption of the nerve transmission. The effect on the nervous system of the arthropod depends on the dose and is irreversible. Therefore, once paralysed, arthropods do not recover. Even with a low dose, a tick's oviposition is inhibited (Barlow and Hadaway 1975, Stendel and Hamel 1990).

Pyrethroids have the same effect on arthropods as pyrethrins, mostly with an excellent insecticidal as well as acaricidal action on ticks which have become resistant to OPAE products. There exists a partial cross-resistance to DDT, the resistance to pyrethroids and DDT being caused by the same set of genes.

Pyrethroids have a low toxicity for vertebrates: neither neurotoxic nor mutagenic effects are to be found. The fertility of treated animals is not impaired. Even after an extreme overdose, calves do not present symptoms of intoxication. Pyrethroids are, however, highly poisonous for fish and reptiles. After external application, the target animal reabsorbs hardly ever the pyrethroids, thus only a reduced residual problem exists when pyrethroids are applied. When these etheric oils are applied to the animal in high concentration and in a slow amount with the pour-on or spot-on method, they "creep" through the coat over the entire surface of the animal and reach all the ticks even on parts of the body which have not been treated. Flumethrin, as in BAYTICOL pour-on, spreads over the entire body of a treated cow as soon as 2–4 hours p.a. (Hamel and van Amelsfoort 1986). The compounds have been specifically formulated for this creeping effect in order to facilitate the fast distribution over the body surface. The animal itself also contributes to the distribution with movements of its head and tail. It is certainly not a systemic effect. Only very small amounts of the active substance are reabsorbed percutaneously, and thus a withdrawal time is generally not required after dermal application of pyrethroids. Therefore, these compounds are considered to be safe as far as residues are concerned. Nor does the pour-on application produce local skin reactions, such as inflammation etc. When applied to the environment of the animals, pyrethroids are comparatively stable.

Through combination with synergists, e.g. sesame oil, as in BARRICADE S, the toxicity of the pyrethroids for the arthropod may be enhanced considerably because the process of detoxication in the arthropod is inhibited.

The application of pyrethroids for vector control in developing countries, especially for surface application, not least of all against the tsetse fly, is impaired by the comparatively high price of this group of products.

Cyfluthrin (a-cyano-(4-fluoro-3-phenoxy)-benzyl-3-(2,2-dichloroethenyl)-2,2-dimethyl-cyclopropane-carboxylate, BAYOFLY, RESPONSAR) is closely related to flumethrin chemically. In contrast to flumethrin, its pour-on formulation (1%) is not effective as an acaricide, but is very efficient as an insecticide for the control of haematophagous and licking flies. It is a "new generation" pyrethroid and is also offered as an ear tag for the control of *Haematobia* and *Stomoxys*. As a typical pyrethroid, cyfluthrin acts immediately upon contact with the arthropod; it has a distinct knock-down effect, followed by a good residual effect which allows the extension of the intervals of application considerably (Stendel and Hamel 1990). In Germany, pour-on application has been known to control 87–98% of haematophagous flies for 30 days during the summer (Liebisch 1992).

Cyhalothrin (Cyano-3-phenoxybenzyl-3-(2-chloro-3,3,3-trifluoroprop-1-enyl)-2,2-dimethylcyclopropane-carboxylate, GRENADE, LIBEREKTO) is effective against haematophagous flies, lice, but especially against glossines. It may be used as an acaricide for dipping and spraying, whereby even resistant ticks are destroyed within 24 hours (99%). The compound is subject to the strip-out effect but its residual action is not impaired by rain.

The active substance is not deposited in the fat of the organism and will not appear in the milk if the compound is applied appropriately. No withdrawal time is required in dairy production.

Cypermethrin (alpha-cyano-3-phenoxybenzyl-2.2-dimethyl-3-(2,2-dichlor-vinyl)-cyclopropane-carboxylate, FLECTRON, BUFFALO FLY SPRAY, CURATIK CATTLE DIP, ECTOMIN, ROBUST SPZ, BARRICADE S). The products, mentioned with their trade name, are synergistic combinations of cypermethrin and chlorphenvinphos. Cypermethrin is mainly used as a spray to control haematophagous and other flies, and tsetse flies as well. As a concentrate, it may also be applied as an ear tag or as a pour-on or spot-on formulation. It is supposed to have a particularly repellent effect. For dipping, a synergistic formulation (25:128) is applied. Pure cypermethrin, even if it is applied dermally as a concentrate, does not present any danger to dairy production.

Deltamethrin (s)-alpha-cyano-3-phenoxybenzyl(1R,3R)-cis-3-(2,2-dibromo-vinyl)-2,2-dimethylcyclopropane-carboxylate, RU-22974, NRDC-161, OMS-168, DECAMETHRIN, DECIS, K-ORTHIN, COOPERS SPOT ON, DECATIX) is the pyrethroid which is used at present in Africa mostly for surface application against tsetse flies within the environment of the animals. For residual application, 12–30 g/ha are used, and 0.1–0.3 g/ha for the knock-down method. It may also be used as a spray on cattle (0.24 mg/kg/b.w.). As a pour-on (0.55 mg/kg/b.w.), it has a residual effect of 35 up to 71 days p.a. upon *G. palpalis gambiensis* (Meyer 1990). In Tanzania, regular dipping with deltamethrin on a large cattle ranch decreased the tsetse population by 90% and the mortality rate in cattle by 66% (Fox et al. 1993). After oral application deltamethrin is barely reabsorbed by mammals, but is rather excreted almost unchanged with the faeces; it appears in the milk in small amounts (Akhtar et al. 1986).

Flumethrin   (alpha-cyano-(fluoro-3-phenoxy)-benzyl-3-2-chloro-2-4-chloro-phenyl)-ethyl-2,2-dimethyl-cyclopropane-carboxylate,   BAYTICOL,   BAY-VAROL, DRASTIC DEADLINE – 1% spot-on formulation), is efficient as acaricide against one-, two- and 3-host ticks, and as an insecticide against haematophagous flies. After pour-on application, (1 mg/kg b.w.) ticks (*Amblyomma, Hyalomma, Boophilus*) and tsetse flies (*G. palpalis gambiensis, G. tachinoides, G. morsitans submorsitans*) are controlled for almost 4 weeks (Meyer 1990). It is an alternative to tick control where resistance to OPAE compounds exists. It may be applied for this purpose as a dip, a spray or a spot-on. Due to its broad therapeutic range, the spot-on formulation may be used even by people with little expertise in dairy production. The tick infestation is already reduced drastically 24–48 hours p.a.; the oviposition of the ticks is inhibited. It is enough for the pour-on to be applied only every 2 or 3 weeks along the dorsal midline of the animal in order to control *Rhipicephalus appendiculatus* and *R. zambesiensis*. The compound resists heavy rain well. After an overdose 2.5 times the normal strength, no residues were found in the milk (Duncan 1991, Hamel and Duncan 1986, Stendel and Hamel 1990).

Permethrin (3-phenoxybenzyl-[3-(2,2-dichlorvinyl)-2,2-dimethylcyclopropane-carboxylate], STOMOXIN MO) is a mixture of 25 cis-:75 trans-isomers. Cis-permethrin has a higher toxicity for vertebrates and arthropods, trans-permethrin is less toxic for vertebrates. The compound is mainly used as an insecticide to control flies (*Musca autumnalis, Hydrotaea* spp., *Moriellia* spp.), haematophagous flies (*Haematobia irritans, S. stimulans*), horse flies *(Tabanus* spp. *Hybomitra* spp.) and midges (*Culicidae, Simuliidae*). It may be applied within the environment of the animal and on the animal itself. Because of the good residual effect, surface treatment has to be repeated only after 3 weeks. In Australia, permethrin is efficient as a spray (0.0033%) against the "problem tick", *B. microplus* (98%). Permethrin may be applied in dairy production; the MRL for milk is set at 0.1 ppm.

Beside the cyclic pyrethroids described above, other products are traded worldwide which do not differ from one another in principle with regard to their range of action and the way they are applied. They are summarized in Table 14.

Meanwhile, ticks as well as haematophagous flies have developed a resistance to pyrethroids. The resistance appears not only with dermal application, but also with dip and spray. At present attempts are being made to overcome this problem through synergistic application. OPAE compounds are best suited as synergists since the resistance to pyrethroids in arthropods is genetically combined with a resistance to DDT. The following **synergistic pyrethroid-OPAE-combinations** have been used against ticks and flies in Argentina:

– flumethrin 1.2% + coumaphos 16%
– cypermethrin 5.6% + diazinon 56%
– cyhalothrin 4.0% + diazinon 62.2%
– cypermethrin 10.0% + chlorphenvinphos 40%

(Romano et al. 1982, Romano 1991), though other reports state that the combination of flumethrin + coumaphos is inefficient (Roush and Tabashnik 1990).

- *Cyclic Amidins (Formamidins)*

Cyclic amidins belong to a heterogeneous group of compounds which must be principally distinguished from the classical acaricides/insecticides as far as the way they work is concerned. These compounds, also known as detaching agents,

- interfere with the metabolism of the ticks, reduce the glycogen and glucose level and block the development of the ova;
- interfere with the respiratory enzyme system of the arthropod by blocking the NADH-fumarat-reductase;
- cause a neuromuscular blockage.

Ticks to which cyclical amidins have been applied drop off the animal a few days p.a., whereby the various species react differently to the particular compound. Apparently the ticks are still alive after treatment, but are dead in the epizootiological sense because they are unable to find a new host. Eventually the ticks are paralysed and are unable to release their hypostome from the skin of the host; they hang lifeless on the animal and become discoloured. In Australia, *B. microplus* has already developed resistance to cyclical amidins. Symptoms of overdosing in calves are: sleepness, salivation and polyuria (Gothe and Hartig 1976).

Amitraz (N-methylbis-(2,4 xylyliminomethyl)-amine, TAKTIC, TRIATIX, TRIATOX) is used for the control of one-, two- and three-host ticks, as well as of midges and lice as a 0.25% dip or spray. Being unstable in the dip over a longer period of time, the application as a spray is preferable. The dip however can be stabilized by adding lime adjusting the pH of the dip liquid to pH 12 making the compound very efficient especially against resistant *Boophilus* ticks. Amitraz has a good residual effect: with one-host ticks, the interval of application may be extended to 45 days. In contrast to most of the other cyclic amidins, the compound has a fast action: 90% of the ticks drop off the animal within 8 hours p.a. Because of its diffusion effect within the coats of the animals, which is similar to the same phenomenon of pyrethroids, the compound also becomes effective on parts of the body surface which have not been reached by the spray application. After having been used in Australia for 20 years there are fewer farms with resistance to amitraz than to pyrethroids (Thullner 1995).

Amitraz has a very low toxicity for mammals. After oral application, the DL_{50} for rats is 1900 mg/kg b.w., and 4100 mg/kg b.w. after dermal application. Horses cannot be treated with this substance. Residues are found in the milk only within 24 hours p.a., and the MRL of 0.01 ppm is not reached. The compound may be used within dairy production without a withdrawal time. The withdrawal time for beef is 24 hours.

If Amitraz is used for surface treatment, its high toxicity for fish has to be taken into account. The contamination of water must, therefore, be prevented.

Chlordimeform, formerly **chlorphenamidin**, is a monoaminooxidase-inhibitor and used to be used as a synergist with OPAE compounds as an acaricide for the control of ticks resistant to OPAEs. It is not produced any more. The same applies to **clenpyrin** (BIMARIT). After dip and spray application, it is deposited in the body fat and is excreted with the milk. This is probably the reason why it is no longer marketed (Beemann and Matsumara 1973).

Cymiazole (2-(2',4'-dimethyl-phenylimino)-3-methyl-4-thiazoline, BESUNTOL EC 25%, TIFATOL; CGA 50 439) is used as an acaricide against ticks of the genera *Boophilus, Amblyomma, Hyalomma* and *Rhipicephalus*. It may be applied as a spray (0.03%); for the control of two- and three-host ticks, an interval of 7 days between sprayings is required and in one-host ticks the producer suggests an interval of 14 days. The application as dip is not recommended possibly because of the limited stability of the compound. According to the instructions of the producer, no withdrawal time is necessary in dairy production.

- *Avermectins*

Avermectins are *Actinomyces* species which are divided into the components A and B and because of different structure, are subdivided into A 1 and A 2 and in B 1 and B 2 respectively. The action of these mycotoxins on ecto- and endoparasites was discovered in the Kitasato-Institute in Japan 1975. The compound is active against intestinal parasites, lungworms, warble flies, lice, midges and different genera of ticks. Liver flukes and tape worms are not sensitive. The active substance inhibits the transmission of stimuli between the interneurons of the ventral nerves and the motor neurons of the parasites. Furthermore, the bonding of the γ-aminobutteracid to the postsynaptic receptors is enhanced. Nematodes and arthropods lose their mobility and die.

Ivermectin (IVOMEC), a derivate of avermectin B 1, contains 80% dihydro-avermectin B 1a and a maximum of 20% dihydroavermectin B 1b. The suggested daily dose for controlling *B. microplus* is a subcutaneous injection of 0.015 mg/kg. When the treatment has been stopped, the residual effect is supposed to last for 14 days. This type of application seems to be inappropriate for tropical animal production because of the danger of transmitting vector-borne diseases. A more suitable intraruminal ivermectin slow-release device providing 90-day protection against tick infestation has been demonstrated in Kenya (Tatchell 1992). The IVOMEC pour-on application which contains 5 mg/mL ivermectin is recommended for the treatment and control of gastrointestinal nematodes, lungworms and haematophagous insects, it is not recommended as an acaricide by the producer. A withdrawal time of 42 days for beef and 28 days prior to milking is indicated. The product cannot be applied to lactating cows.

The compound is deposited for a short time in the liver, fat and only in small amounts in muscle and kidney. About 50% is excreted unchanged with the faeces. Therefore, a withdrawal time of 38 days is required for milk and beef. Because of this, it also does not make sense to utilize the compound in tropical animal production.

Table 14. Range of application, residual effect and withdrawal time of pyrethroids, cyclic amidins and avermectins[a]

Compound	Application	Indication	Residual effect (days)	Milk discarding time (days)
Permethrin	Spray	Flies	14–21	0
Flumethrin	Dip, spray, pour/spot-on	Ticks, flies	up to 30	0
Deltamethrin	Dip, spray, pour/spot-on, ear tag	Flies (ticks)	70 (*Glossina*) 120–150	0
Cyfluthrin	Dip, spray, pour/spot-on, ear tag	Flies		0
Alphamethrin	Dip, spray, pour/spot-on	Flies (ticks)		0
Cyhalothrin	Dip, spray, pour/spot-on, ear tag	Flies (ticks)	14 120–150	0
Cypermethrin	Spray, ear tag	Flies	14 120–150	0
Clenpirin	Dip, spray	Ticks	7	1
Amitraz	Dip, spray	Ticks	45	0
Cymidazole	Dip, spray	Ticks	7	0
Ivermectin	s.c. Injection	Ticks	14	38

[a] Only the action against disease vectors but not against other parasites of domestic animals has been considered.

Ways of application, residual effect and required withdrawal time for products out of the group of pyrethroids, cyclical amidins and avermectins are compiled in Table 14.

4.2.2.2. Methods of Application of Acaricides/Insecticides

Arthropods as vectors of vector-borne diseases may be controlled while feeding or during their stages of development on the host with chemical compounds through

– contact with the active ingredient after external application to the animal, either by wetting it completely or by applying a substance locally which reaches the vector by "creeping" through the coat;
– the systemic effect after dermal, oral or parenteral application.

With several compounds, a systemic effect cannot be prevented after external application (residues in the animal product). The combined contact and systemic effect may compensate for an incomplete wetting of the animal, e.g. after hand spraying.

• *External Application with Contact Effect*

Acaricides/insecticides can be applied to the animal in order to destroy the vector while feeding or during its development through contact with the compound. The vector is destroyed at the time of application if the vector is already feeding or developing on the animal, or when the vector infests the host shortly after application (residual effect). The compound may be applied externally

- with the stationary cattle or sheep dip;
- through spraying or dusting in portable or stationary installations (spray-race);
- with the pour-on/spot-on methods or by applying impregnated ear tags, neck or tail bands, provided no systemic acting compounds are used (OPAEs).

Cattle dip. For the control of vectors through complete wetting of the animal, the cattle dip, in spite of its being the oldest type of installation used, is still preferred. In principle it can be applied to control all haematophagous arthropods which transmit tropical animal diseases. Beside ticks, all other vectors only remain for a short time on the animal. Therefore, the expense of the dip in relation to its efficiency is only justified in the case of tick control. The use of dips was introduced in order to control ticks with the application of arsenic. For this it was necessary to wet the animal completely, since arsenic only has an effect if it penetrates the tracheas of the tick. The longer the immersion, the better the effect which was a further argument for dip. Arsenic was cheap, and thus the size of the dip did not matter. The volume of a dip was about 20 m^3 as a rule. Considering the cost of modern acaricides, the running costs of such a big dip would be extremely high. With a dip volume of only 7–10 m^3, dehorned cattle can easily be treated efficiently. In such a small dip the dip-liquid is, however, stressed considerably through pollution and stripping (Seifert 1971).

A principal disadvantage of a dip set-up is that it is an expensive installation which cannot be moved, and to which the animals have to be brought. During the dry season, the animals are thus put under additional stress, and during the rainy season, the pasture may be damaged. In order to reduce the distance to the dip, it has to be set up decentrally but near to the pasture which again requires considerable expenditure for maintenance and control. Maintenance and control of a dip are not as easy as often claimed, especially if development of resistance to the acaricide is to be prevented.

The site for the installation of a dip has to be chosen with water availability in mind, but on the other hand, it must be possible to dispose of the utilized dip-liquid without causing any hazard to the environment, humans or animals. Under no circumstances may the used dip-liquid be run off into a creek or a river. All acaricides used at present are poisonous to fish. The dip has to be constructed on a hill or a slope in such a way that the rainwater can be drained off and development of mud within the pens is prevented. Otherwise the animals will soil the dip very quickly. Components of a functional dip are:

- one or more pens for the collection of the animals;
- a paved pen from which the animals are forced into the dip;
- a passage to the dip;
- if possible, a hoof dip;
- a jumping ramp;
- a covered dip basin;
- a drip-pen;
- one or more pens for collection of dipped cattle.

The **fences** of the pens also have to be built in such a way that animals which may become nervous can be restrained; walls built out of set stones are the most practical because the drover can walk along them. The **pen** which leads into the dip cannot afford to have corners in which the cattle could get caught up. If the funnel which leads to the dip is connected directly to the dip on one side, and if the animals are driven along this side, they will automatically enter the dip without any resistance. If this pen is paved with rough stones the animals will lose the dirt which sticks to their hooves. The pavement has to be inclined away from the dip in order to prevent rainwater from flushing the dirt from the pen into the dip. Doors which only allow one animal at a time to pass through have to be placed at both ends of the pen in order to be able to control the entrance to the dip. The **foot-dip** is a 3–4 m long, 25 cm deep basin not only for cleaning the hooves, but also for treating hoof problems which may appear during the rainy season (copper sulphate). The **jumping ramp** is about 2 m long, paved roughly and should under no circumstances be inclined towards the dip. It should be below level of the dip-liquid and be slightly covered with it. The edge towards the dip must be rounded off in order to prevent lesions to the udder. When the animals enter the jumping ramp and step into the void in front of them they will instinctively jump and thus also immerse their heads. If in contrast the ramp is inclined towards the dip and paved smoothly, the animals will slip, try to return and squat thus hurting their udders. They would behave similarly if, as sometimes proposed, the entrance to the dip is built as a set of stairs. The **depth** of the dip has to be at least 2.10 m in order to prevent the animals from hurting themselves when they jump into the dip. The lower edges of the dip have to be rounded off in order to prevent the build-up of mud. The dip itself should be about 7 m **long** and be inclined towards the outlet for better cleaning. In order to reduce the cost of construction and maintenance during the last 20 years, shorter constructions have been designed. The "Danish" model is only 3 m long. The **width** of the dip depends on whether dehorned or longhorned cattle have to be dipped; it can be between 90–120 cm wide at surface level and should be

Figure 35. Dipping cattle (the jutting edge on the jumping ramp prevents soiling of the driving-pen).

narrowed at the floor of the dip to 70 cm wide (Fig. 35). The dip has to be covered to protect the acaricide from UV-radiation and dilution with rainwater. Coming out of the dip, the animals enter the **drip-pen** over flat steps. The dip-liquid which drips off the animals is collected and returned for sedimentation through a siphon into the dip. Whether to return liquid into the dip or not is questionable. The animals urinate and defecate frequently in the drip-off pen (OPAE-effect) after being dipped and thus soil the dip-liquid considerably. Through the strip-out effect, a certain amount of active substance of the acaricide is carried out of the dip by the animals, which means the returning liquid is understrength. If there are no means to control the concentration of the dip, it may be more practical to dispose of the drip-off liquid and make up the missing level of the dip with liquid in double concentration (Seifert 1971).

A careful **dip-management** is the prerequisite for successful vector control. Not only the preparation of the dip-liquid and replenishment but also permanent control of the dipping and protocol of the dip have to be carried out correctly.

The **preparation** of the dip-liquid has to be done according to the instructions of the producer. This is especially true for a W.P. which may first have to be prepared as a paste. In order to guarantee an even suspension of the product, the first 20 animals can be run twice through the dip thus stirring up the liquid. If the only water available has a high calcium carbonate content, the wetting capacity of the dip-liquid can be increased by adding a detergent. This will also increase the contact effect on the arthropod. If two acaricides are used as synergistic partners one must make sure that they are compatible with each other. Care should be taken when "home made" mixtures are applied which have not been composed according to the producer's instructions. Since there is usually no information available about the stability and toxicity of such combinations, refraining from such private initiatives is recommended. The biggest problem in dipping is the control of the concentration of the dip-liquid which in very few cases is possible ad hoc at the site of the dip. Arsenic could be controlled because of its specific weight which is a method which was also applied in a modified way to CHC compounds. OPAE compounds can only be controlled by means of gaschromatography or with complicated test kits. Pyrethroids cannot be controlled at all in the field. The concentration of the dip is not only reduced through the strip-out effect, but also considerably by pollution with dirt. Because of the high cost of the acaricide, the same dip-liquid is used as long as possible, over several months or even a year. The effect of W.P.s is neutralized through dirt since it produces an isolating layer between the active ingredient and the arthropod. The vectors themselves may thus also be covered by a protective layer of dirt when the animal is dipped. The high rate of evaporation during the dry season may additionally complicate the maintenance of the required concentration of active substance in the dip. The concentration will be increased up to an unknown level. Generally, it may be calculated that 1000 animals can be dipped at an interval of 3 weeks in a 10 m^3 dip for 6 months, if acaricide and water are replenished regularly. It is possible to use the double concentration of acaricide for replenishing the dip and it is thus possible to dip for a certain time without controlling the concentration. Such a procedure, however, is not without risk.

The control of the **dip-concentration** can be carried out by observation or biological test methods. The efficiency of the compound contained in the dip may be controlled

- by determining the level of contamination of the dip: after having dipped a herd of cattle, 100 mL are filled into a measure and the amount of contamination is read after 24 hours;
- by judging the percentage of damaged ticks after dipping;
- by applying the oviposition test according to Endrejat and Stubbs: fully engorged females are collected after dipping from some of the animals and stuck on to Scotch tape on their backs. If no oviposition takes place after some days, the compound has been fully effective. As a negative control, engorged females from untreated animals are observed the same way;

– by determining the residual effect of the compound: the time is taken which passes between dipping and the point where the ticks, which remain on the animal, appear damaged. Furthermore, how long it takes until new ticks appear on the animal is observed. With this procedure, the degree of infestation, the coat of the animal and the climatic conditions have to be considered;

– by applying the biological test of efficiency according to Stendel: 10 engorged ticks of the same species are wrapped into a piece of gauze which is immersed for 1 minute in the dip-liquid. Afterwards the ticks are unwrapped and set out on filter paper. The number of damaged ticks is controlled every 24 hours over 5 days. In order to bring the degree of damage up to an uniform nominal value, the amount of dead or heavily damaged ticks is multiplied by 1, the number of ticks which are damaged about 50% with 0.5 and the number of only slightly damaged ticks with 0.25. The addition of these numbers is the percentage of damaged ticks.

Example: Calculation:

After 48 hours   2 ticks are dead   2 × 1   = 2
                      4 ticks heavily damaged   4 × 1   = 4
                      2 ticks 50% damaged   2 × 0.5 = 1
                      2 ticks slightly damaged   2 × 0.25 = 0.5

                      10 ticks   = 7.5 damaged ticks from 10
                                                 = 75% damaged ticks

Using the standard curve presented in Fig. 36, the percentage of ticks damaged in the course of several days can be taken to determine the concentration of active ingredient in the dip-liquid. These curves, which are only an example, have been prepared with engorged ticks of *B. annulatus* (Seifert 1971). If they are used

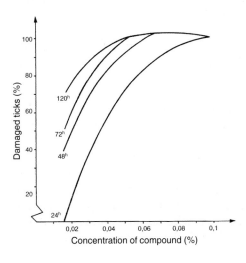

Figure 36. Standard curves for the biological dip test according to Stendel (cited from Seifert 1971).

for the example of calculation presented here with a tick damage of 75%, a concentration of active ingredient of 0.03% will be the result.

The following **general rules for dip-management** are prerequisite for an effective vector control which prevents the development of resistance and minimizes negative influences on the animals and the management of the animal production system.

The time of dipping, as well as regular vaccinations or control of endoparasites, calving time, and weaning time etc. all have to be integrated into a specific animal health scheme which has to be developed for each production system and each individual farm or ranch (III.). The intervals between dipping depend on the biology of the tick species, the characteristics of the acaricide and the climatic conditions. The best time for dipping is the early morning hours. If the animals have to be brought from far away, they have to stay overnight at the dip pen and can be dipped after having rested. It is imperative that water is provided for the animals, because thirsty cattle, especially calves, might otherwise drink the dip-liquid. If dipping is carried out while it is raining, the residual effect may be reduced considerably. Only healthy animals are allowed to be dipped. One should make sure that all the animals of a herd are dipped at the same time. The animals should preferably be sorted into categories, i.e. cows and calves get dipped separately. It is better to spray cows in an advanced state of pregnancy. The dip interval used for preventing one-host ticks may be reduced to 2 weeks during the rainy season. A dipping interval of 5 days is required against two- and three-host ticks if the residual effect of the acaricide does not allow for a longer interval. Such a short dipping interval is required since the imagines of two-host ticks may remain for only 6 days on the host, and, in the case of three-host ticks, they might only remain for 3 days on the host during each development stage. Figure 37 shows how the dip interval correlates with the development of the one-host ticks on the animal, and Fig. 38 shows the same for the two- and three-host ticks.

Spray application. When spraying in mobile or fixed installations, acaricides/ insecticides can be applied in such a way that the animal is wet thoroughly from all sides. Since the same compound does not remain in the equipment for several months as with the dips, the development of resistance can be prevented rather easily by changing the product. There is also a better chance of using compounds which may be effective against vectors other than ticks. In this way seasonal peaks of vector incidence can be taken into account.

Mobile installations are suitable for a small farm or for places where the herds would otherwise have to walk long distances in order to get to the fixed installation. The easiest way is to spray the animals with a hand sprayer. Larger numbers of animals may be sprayed with a motor driven high pressure pump which delivers the liquid to two spray guns each on one side of a chute. Thus the animals can be sprayed easily from both sides. This type of application requires a comparatively small amount of liquid. Therefore, no control of concentration is required since the liquid is prepared anew for each spray meaning that compounds with low stability may be used (trichlorfon). The drip-off-liquid is lost.

For a ranch, a **permanent spray-race** is preferred through which up to 600

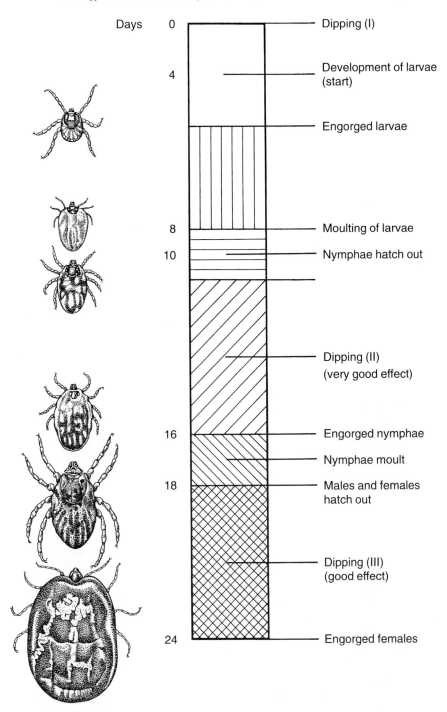

Figure 37. Correlations of dipping interval and development of one-host ticks (Wellcome 1976)

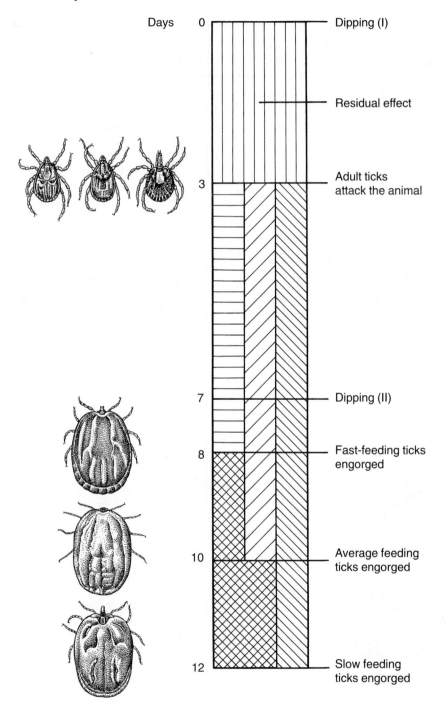

Figure 38. Correlation of dipping interval and development of two- and three-host ticks (Wellcome 1976).

animals may pass per hour; the required pump and engine can be transportable if necessary. The pump should be coupled to an engine of at least 6 h.p. and have a capacity of not less than 800 L/minute. Such an installation is a real alternative to the dip and has the advantage of a reduced water and pesticide consumption. The animals are treated with less stress, get wet from all sides and, if managed properly, the spray-race will prevent the development of resistance through below-strength acaricides. Furthermore, no danger of poisoning exists through overconcentration and the animals cannot drink the liquid.

Normally, a spray-race is integrated into a central system of cattle management, like a chute etc. This way, no special pens are required. The actual spray-race is about 4–5 m long, protected on the sides by walls, and slopes slightly towards the collection grill, through which the drip-off liquid is run off into a storage tank. This is best done through a gravel filter. The spray nozzles are placed on 3 arch-shaped tubes which are connected to each other. Instead of the sort of nozzles used for spray-guns, small pieces of sheet metal can be mounted in an inclined position (45°) in front of holes of about 3 mm in diameter. The liquid rebounds from the metal and turns into a spray if enough pressure (2.5 bar) is put to the system. This simple and cheap technique prevents blocking of the system with particles of dirt or dung. The nozzles have to be so placed that the animals get wet from all sides. It is advisable to connect the lower part of the spray tubes at a height of 25 cm above the ground so that when the animals walk through the spray-race, they will lift their feet and tails instinctively, thus allowing the spray to get to the perineum which is one of the favourite sites where ticks attach themselves (Fig. 39).

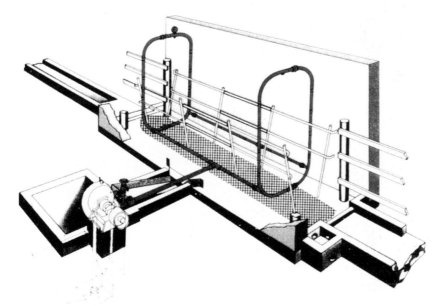

Figure 39. Scheme for a spray-race (WELLCOME 1976).

Emulsions are best suited for use in a spray-race because, in contrast to W.P.s, they do not get caught up in the filter. When preparing the spray-liquid, one can calculate that one adult animal carries 2.5 l out of the system during the dry season, and 1.5 l during the rainy season. 400 l which remain in the system have to be taken into account additionally.

Acaricides/insecticides which are used in the spray-race are the same as those used in dips. However, they usually have to be used in higher (double) concentrations, see also Tables 13 and 14. The intervals of treatment depend on the same guide-lines as for the dip.

Dusting. An alternative to the application of liquid acaricides/insecticides is dusting. All W.P.s which can be mixed with talcum in the desired ratio can be used for dusting.

Either by hand or using motor-driven devices like those used for plant protection, the powder is mixed in an airstream in order to produce the dust. Machines are available which additionally mix oil into the airstream producing an aerosol which means that the powder sticks to the coat of the animal. Either the application can be carried out in an existing chute, or the animals can be crowded into a protected corner of a pen and powdered as a group. The dense cloud of powder covers the animals completely. This method may have the same

Table 15. Ear tags applied to control haematophagous insects and biting flies (Hamel 1991)

Compound	Generic name	Concentration (%)	Weight (g)	Trade name	Duration of effect (months)
Carbamate	Diazinon	20	15	Optimizer Y-Tex	up to 5
				Terminator Fermenta	3–4
Organic phosphoric acidester (OPAE)	Methyl pirimiphos	20	9.5	Tomahawk Coopers	5
Pyrethroid + OPAEs	Cypermethrin	7	9.5	Super Max-Con Y-Tex	4
	Chlorpyrifos	5			*Haematobia* resistant
	Permectrin	10	11.5	Ear Force Ranger	6
	Chlorpyrifos	6.6		Anchor	
Pyrethroids	Permethrin	10	9.5	Atroban Coopers	5
				Wellcare Coopers	*Haematobia*
				Gard Star Plus Y-Tex	resistant
				Ear Force Anchor	
	Lambda-cyhalothrin		10	Saber Coopers	5
	Cyfluthrin	10	14.3	Cutter Gold Mobay	5
	Flucythrinat	7.5	10	Guardian Cyanamid	5
	Fenvalerat	9	10	Ectrin + Starbar + Tirade	

effect as a dip or spray-race especially during the dry season when water is scarce. It is a rather cheap method because with 30 kg of a mixture of talcum and the necessary compound, 1000 cattle may be treated. Dusting with 5% ASUN-TOL blended with talcum can protect cattle 8–10 days against *B. annulatus* and keep the animals free of flies for one week (Seifert 1971). When applying pesticides as dust, precaution must be taken to protect the personnel from aerogenic and cutaneous contact with the compound.

Pour-on, spot-on, application of impregnated ear tags, neck and tail bands. In much the same way as is described as follows for systemic application, pyrethroids, which have no systemic action, can be applied locally to the animal with an overall effect on the vectors. Surprisingly, etheric oils spread through the coat and also reach arthropods which feed at a spot which is far away from the site of application. For pour-on and spot-on, the method of application is the same as for systemic acting compounds. Pyrethroids may, however, also be used with impregnated ear tags or neck and tail bands which release the compound gradually and produce a long-lasting protection against arthropods for the whole animal. Since pyrethroids are reabsorbed very little dermally, this method can be used in dairy production. The EPA in the USA has allowed the application of cyfluthrin using an ear tag to dairy cows. Tsetse flies may also be controlled with ear tags impregnated with fenfluthrin and thus the chemoprophylaxis against trypanosomiasis is assisted (Dolan et al. 1988). Table 15 shows the currently used products together with the respective compounds and the time of protection which can be achieved (Hamel 1991).

• *Methods of Application with Systemic Action*

The principle of systemic-acting acaricides/insecticides is the gradual reabsorption by the organism and the even distribution within the system of the animal depending on the method and place of application. Only OPAE compounds are used for systemic application since they are also quickly metabolized and excreted, and become completely catabolized in the environment.

Systemic pesticides may be applied dermally, parenterally or orally. The dose has to be high enough to destroy the haematophagous insects while they are feeding. This is rather easy with **stationary vectors** since they suck blood continuously and thus get a high enough dose of poison with the blood. It is much more difficult to destroy **temporary vectors** which have to ingest enough of the poisonous compound within the given feeding time which may only be short. If they get only a sublethal dose, resistance will develop.

Systemic vector control has several advantages in comparison with dips and sprays. Under the conditions of extensive cattle production in the tropics, it is rather easy to apply the compound with the feed additive; pour-on application is equally economical and easy to carry out. Herd management becomes simpler and expensive constructions and infrastructure can be saved. If the compounds are applied precisely according to the instructions, a good residual effect will be obtained and no danger occurs to the health of the animal or the consumer of the animal product.

Application of systemic acaricides/insecticides is off limits for dairy production in order to prevent contamination of the milk. In beef cattle, the application has to be stopped 4 weeks before the animals are marketed. This will guarantee that no contaminated product is consumed. With the development of the dermal application of non-systemic pyrethroids, the use of systemic OPAE products has diminished considerably. They are, however, still discussed not only in order to explain the difference to the external application of pyrethroids, but also because they may become useful for economic reasons and because resistance to pyrethroids is emerging.

Parenteral application. The injection of systemic acaricides/pesticides may only be used if animals are brought from vector infested areas and are introduced into vector-free pastures or intensive production systems. Regular parenteral application would be dangerous because it may also transmit pathogens of vector-borne diseases.

Oral application. With oral application, a distinction must be made between the use of a single maximum dose of a compound which has to be repeated regularly, and the continuous application of the compound, i.e. with the feed. Both methods of application have to guarantee that the level of active substance within the blood of the host animal is continuously kept high enough to destroy or block the development of haematophagous arthropods. The continuous medication is more practical if the animals are treated with feed supplements for example. If this method is carried out precisely according to the instructions, the development of resistance can be prevented. With the feed-through method, the compound will inhibit the development of the larvae in the faeces.

Examples of feed additives with an acaricide/insecticide content are salt licks, feed-blocks or molasses mixtures like RUMAVITE for example. A mixture which has given excellent results for many years is composed as follows:

65% molasses
3–4% H_3PO_4 (60%)
6% urea
6% C_2H_5OH
5% coumaphos

Rest: Mineral salts, minor elements, water.

Overdosing is prevented through the acid component which lowers the pH and prevents the animals from swallowing large amounts of the feed (Seifert 1971).

The intraruminal application of an ivermectin slow-release device as reported by Tatchell (1992) has also to be grouped with the oral and systemic methods of application.

Dermal application. The application of systemic acaricides/insecticides on the **skin** has, in comparison with the oral application, the advantage of much slower reabsorption and a 5–10 times longer maintenance of the concentration of the active substance in the blood. The application is carried out with the pour-on or spot-on method which both have the same principle but differ only in concentration and technique of application. Back-rubbers and dust-bags are additional means for dermal application. With the pour-on technique, relatively low concentrations (5–10%) of an oily preparation of the compound are poured along the dorsal midline of the animal, which also may be done with a drench gun (Fig. 40). For spot-on, a gun-applicator which is set for the required dose is connected to a closed portable storage container and shoots the liquid over a limited distance on to the back of the animal. Since only limited amounts may be delivered, higher concentrations are required. Both methods can easily be applied without expensive installations. Only oily preparations which stick well to the coat are used. The dose depends not only on the body weight of the animal, but also on the breed and the production system. Cebus are 2–3 times more sensitive to OPAE products than European breeds. Products which may be applied are: NANKOR 4E, DELNAV, NEOCIDOL H, TIGUVON, RUELENE 24 R and NEGUVON. TIGUVON and RUELENE 24 R are best suited. TIGUVON has been marketed as TIGUVON spot-on 20% and is formulated in such a way that it is reabsorbed easily through the skin. It is used above all against flies. For tick control, TIGUVON may be applied as a 5–10% preparation. RUELENE 24 R is sold as a 25% oily emulsion and is used against ticks, flies and midges as a 5–10% emulsion diluted with water. The residual effect of RUELENE 24 R is shorter than that of TIGUVON. The TIGUVON treatment may be carried out in a 3–5

Figure 40. Pour-on application with drench gun.

week cycle. Cattle tolerate the compound very well; the DT_{50} for TIGUVON is 6–8 times that of the *dosis therapeutica*.

Systemic compounds can also be applied to cattle with a **back-rubber**. It is a device with which cattle can treat themselves. The back-rubber was developed in the USA and is commercially available. It can be easily reproduced since its principle is rather simple. The home-made back-rubber consists of two thick chains which are surrounded with sack cloth. A pesticide (e.g. TIGUVON 10%) is poured daily on to the cloth. The same concentrations which are recommended for the pour-on application can be applied here. The chain surrounded with sack cloth is hung between two poles in such a way that the cattle which walk through this "door" rub their backs. The back-rubber is placed in an opening through which the animals have to pass daily when they walk to the watering place, the salt lick or the feed trough. Through this continuous self-application, the pesticidal compound is reabsorbed and the required level maintained in the blood stream of the animal. The sack cloth has to be replaced monthly because they become soiled and lose their absorbing action.

The **dust-bag** is a bag filled with pesticide made from cloth of highly penetrable material. Similar to the back-rubber, it is hung between poles in a passage which cattle have to pass through. Thus they powder themselves every time they pass through this passage. ASUNTOL, NANKOR 4 E, RUELENE 24 R, NEOCIDOL H and TIGUVON can be applied for this purpose. The control of ticks, flies and mosquitoes with a dust-bag is less efficient than with a back-rubber. W.P.s act mainly as contact pesticides and only destroy those vectors which rest on the powdered spots on the animal. The dust-bag has to be replaced quite often because the underside quickly becomes clogged with a layer of dust and skin-fat (Seifert 1971).

4.3. Control of Vampires (*Desmodontidae*)

Vampires are vectors of protozoonoses, rickettsioses and especially rabies. The Latin American *Paralyssa* is only transmitted by vampires. Besides helping to prevent rabies, the control of vampires is an important tool in preventing these vector-borne diseases. Several alternatives for vector control are available.

The **destruction of the hide-outs** can be accomplished rather easily. In contrast to other bats, vampires live in small colonies, often in hollow trees or beneath rocks. Such hide-outs can be destroyed mechanically or made uninhabitable for the vampires. The resting sites of vampires can be recognized because of the characteristic excrement on the ground. With **systemically**-acting compounds, vampires have been controlled, for example in the Chaco in Bolivia. Diphenacion (2-diphenylacetylindan-1,3-dion), a rodenticide with antithrombin action has been applied intraruminally to cattle. The dose applied must be harmless for a cow, but toxic for the vampire which takes an amount of blood equivalent to several times its body weight. Cattle which have been poisoned in the afternoon are kept overnight in a pen in the area where the vampires are active. If these animals are attacked during the night by the vampires, they ingest the poisoned blood. Since a vampire often attacks an animal repeatedly over several nights, the animal can also be injected with diphenacion after it has been attacked by the vampire the first time.

Another alternative to poisoning vampires with diphenacion is to catch vampires and smear diphenacion ointment on their bellies. After they have been released, they return into their colony where they their companions clean them by licking, and thus a large part of the colony can be poisoned. Vampires can be caught by placing fish nets vertically around a pen in which cattle are held. The vampires are unable to locate the fish nets with their ultrasound sensors when they fly towards the cattle trying to attack them. They can be collected with gloved hands, smeared with the **ointment** and released to return to their colony.

With all the measures available there are for vampire control, one has to make sure that it is really vampires and not other bats which are destroyed. Furthermore, the personnel has to be vaccinated against rabies since an infection with contaminated blood through a skin lesion always is possible.

References

Aeschlimann, A., M. Brossard, T. Haug and B. Rutti (1990): A survey of tick vaccines. Anim. Res. Dev. **32**, 52–72.

Agbede, R. I. S. and D. H. Kemp (1986): Immunization of cattle against *Boophilus microplus* using extracts derived from adult female ticks: histopathology of ticks feeding on vaccinated cattle. Int. J. Parasitol. **16**, 35–41.

Akhtar, M., K. E. Hartin and H. Locksley Trenholm (1986): Fate of (^{14}C) Deltamethrin in lactating dairy cows. J. Agric. Food Chem. **34**, 753–758.

Allen, J. R. (1989): Immunology of interactions between ticks and laboratory animals. Exp. Appl. Acarol. **7**, 5–13.

Allen, J. R. and S. J. Humphreys (1979): Immunization of guinea pigs and cattle against ticks. Nature **280**, 491–493.

Amsler, S., J. Filledier and R. Millogo (1994): Attractivité pour les *Tabanidae* de différents pièges à glossines avec ou sans attractifs olfactifs. Résultats préliminaires obtenus au Burkina Faso. Rev. Élev. Méd. vét. Pays trop. **47**, 63–68.

Barlow, F. and A. B. Hadaway (1975): The insecticidal activity of some synthetic pyrethroids against mosquitoes and flies. PANS **21**, 233–238.

Beeman, R. W. and F. Matsumara (1973): Chlordimeform: A pesticide acting upon amine regulatory mechanisms. Nature **242**, 273–274.

Behrens, H., K. Damman-Tamke, W. Kirchner and A. Liebisch (1987): Studies on the strategy of fly control in piggeries. Zbl. Bakt. Hyg. A **265**, 494.

Bram, R. A. (1975): Tick-borne livestock diseases and their vectors. I. The global problem. World Anim. Rev. **16**, 1–5.

Brown, S. J. and P. W. Askenase (1986): *Amblyomma americanum*: physiochemical isolation of a protein derived from the tick salivary gland that is capable of inducing immune resistance in guinea pigs. Exp. Parasitol. **62**, 40–50.

Brown, S. J., S. Z. Shapiro and P. W. Askenase (1984): Immunization of guinea pigs with *Amblyomma americanum* derived salivary gland extracts and identification of an important salivary gland protein antigen with guinea pig anti-tick antibodies. J. Immunol. **133**, 3319–3325.

Dehoux, J. P. (1993): Lutte contre *Glossina tachinoides* au Bénin. Utilisation particulière de piège-pneus imprégnés de deltaméthrine. Rev. Élev. Méd. vét. Pays trop. **46**, 581–589.

Deportes, I., B. Geoffroy, D. Cuisance, C. J. Den Otter, D. A. Carlson and M. Ravaelle (1994): Les chimiorécepteurs des ailes chez la glossine (*Diptera:Glossinidae*). Approche structurale et électrophysiologique chez *Glossina fuscipes fuscipes*. Rev. Élev. Méd. vét. Pays trop. **47**, 81–88.

Dolan, R. B., P. D. Sayer, H. Alushula and B. R. Heath (1988): Pyrethroid impregnated ear tags in Trypanosomiasis control. Trop. Anim. Hlth. Prod. **20**, 267–268.

Duncan, I. M. (1991): Tick control on cattle with flumethrin pour-on through a Duncan applicator. J. S. Afr. Vet. Assoc. **63**, 125–127.

FAO (1990): Manual for the control of the screwworm fly *Cochliomyia hominivorax*, Coquerel. FAO, Rome.

Filledier, J. and P. Mérot (1989): Attractive power of M-Cresol 1-Octen-3-Ol in a practical device for *Glossina tachinoides* in Burkina Faso. Rev. Elev. Méd. vét. Pays trop. **42**, 541–544.

Fivaz, B. H., D. T. de Waal and K. Lander (1992): Indigenous and crossbred cattle – a comparison of resistance to ticks and implications for their strategic control in Zimbabwe. Trop. Anim. Hlth. Prod. **24**, 81–89.

Georghiou, G. P. and A. Lagunes-Tejeda (1991): The occurrence of resistance to pesticides in arthropods. FAO, AGPP/MISC/91–1.

Gill, H. S., R. Boid and C. A. Ross (1986): Isolation and characterization of salivary antigens from *Hyalomma anatolicum anatolicum*. Parasite Immunol. **8**, 11–25.

Gordon, J. R. and J. R. Allen (1987): Isolation and characterization of salivary gland antigens from the female tick *Dermacentor andersoni*. Parasite Immunol. **9**, 337–352.

Gothe, R. and M. Hartig (1976): Zur ixodiziden Wirksamkeit von Clenpyrin, Chlordimeform und Chlormethiuron gegen PE-resistente *Boopilus microplus*-Stämme. Zbl. Vet.-med. B. **23**, 243–254.

Gouteux, J. P. (1991): La lutte par piégeage contre *Glossina fuscipes fuscipes* pour la protection de l'élevage en République centrafricaine. II. Caractéristiques du piège bipyramidal. Rev. Elev. Méd. vét. Pays trop. **44**, 295–299.

Gouteux, J. P., D. Cuisance, D. Demba, F. N'Dokoue and F. Le Gall (1991): La lutte par piégeage contre *Glossina fuscipes fuscipes* pour la protection de l'élevage en République centrafricaine. I. Mise au point d'un piège adapté à un milieu d'éleveurs semi-nomades. Rev. Elev. Méd. vét. Pays trop. **44**, 287–294.

Graf, J. F. (1993): The role of insect growth regulators in arthropod control. Parasit. Today **12**, 471–474.

Grothe, R. and P. Weck (1983): Zur Problematik der Akarizidresistenz ixodider und argasider Zeckenarten. Dtsch. tierärztl. Wschr. **90**, 493–498, 534–539.

Hamel, H. D. (1991): Personal information.

Hamel, H. D. and A. van Amelsfoort (1986): Darstellung der dermalen Verteilung von Flumethrin 1% m/v mit einer Fluoreszenztechnik. Vet.-med. Nachr. **1**, 34–39.

Hamel, H. D. and I. M. Duncan (1986): Kontrolle von Rinderzecken mit Flumethrin 1% pour-on in Zimbabwe. Vet.-med. Nachr. **2**, 115–122.

Hoffmann, G. (1987): Fliegenbefall in landwirtschaftlichen Betrieben – Resistenzursachen und Bekämpfungsmethoden. Dsch. tierärztl. Wschr. **95**, 10–14.

Holler, H. -H. (1986): Die Minimierung von Pestizidrückständen in tierischen Milchprodukten tropischer Regionen. Diss. FB Agrarwiss., Göttingen.

IAEA (1990): Sterile insect technique for tsetse control and eradication. IAEA Panel Proc. Series. IAEA, Wien.

Jongejan, F. (1990): Tick/host interactions and disease transmission with special reference to *Cowdria ruminantium* (Rickettsiales). Proefschrift Utrecht.

Kaminski, R. (1983): Untersuchungen zur Biologie, Ökologie und Infektion von Tsetsefliegen (Diptera, Glossina) in einem Regenwaldgebiet Liberias. Göttinger Beitr. Land. Forstwirt. Trop. Subtrop. **2**.

Kemp, D. H., R. D. Pearson, J. M. Gough and P. Willadsen (1989): Vaccination against *Boophilus microplus*: localization of antigens on tick gut cells and their interaction with the host immune system. Exp. Appl. Acarol. **7**, 43–58.

Kemp, D. H., R. I. S. Agbede, L. A. Y. Johnston and J. M. Gough (1986): Immunization of cattle against *Boophilus microplus* using extracts derived from adult female ticks: feeding and survival of the parasite on vaccinated cattle. Int. J. Parasitol. **16**, 115–120.

Liebisch, A. (1992): Weidefliegen beim Rind und ihre Bekämpfung im Aufgießverfahren mit Bayofly Pour-On (Cyfluthrin). Prakt. Tierarzt **6**, 3–7.

Matha, V. and J. Weiser (1988): Detection of antigens common to salivary glands and other tissues of the tsetse fly, *Glossina palpalis palpalis* (Diptera: *Glossinidae*). Folia parasitol. **35**, 285–287.

McCosker, P. J. (1979): Global aspects of the management and control of ticks of veterinary importance. In: Recent advances in acarology, 2 (ed: Rodriguez, J.), Academic Press, London, 45–53.

Mehlhorn, H. and G. Piekarski (1989): Grundriß der Parasitenkunde, 3. Aufl., Fischer, Stuttgart.

Meyer, F. (1990): Simultane Zecken- und Tsetsefliegenbekämpfung bei Rindern in Burkina Faso. Diss. Tierärztl.Hochsch. Hannover.

Mitscherlich, E. und K. Wagener (1970): Tropische Tierseuchen und ihre Bekämpfung. II. Aufl., Parey, Berlin und Hamburg.

Müller, P. (1991): Pesticides in Africa. Environmental problems. Göttinger Beitr. Land. Forstwirt. Trop. Subtrop. **60**.

Needham, G. R., D. C. Jaworski, F. A. Simmen, N. Sheriff and M. T. Muller (1989): Characterization of ixodid tick salivary-gland gene products, using recombinant DNA technology. Exp. Appl. Acarol. **7**, 21–32.

Nolan, J. and H. J. Schnitzerling (1986): Drug resistance in arthropod parasites. In: Chemotherapy of parasitic diseases (eds: Campbell, W. C. and R. S. Rew), Plenum, 603–620.

Nolan, J., J. T. Wilson, P. E. Green and P. E. Bird (1989): Synthetic pyrethroid resistance in field samples in the cattle tick (*Boophilus microplus*). Austr. Vet. J. **6**, 179–182.

Opdebeeck, J. P., J. Y. M. Wong and C. Dobson (1989): Hereford cattle protected against *Boophilus microplus* with antigens purified by immunoaffinity chromatography from larval and adult ticks. Immunol. **67**, 388–393.

Opdebeeck, J. P., J. Y. M. Wong, L.A. Jackson and C. Dobson (1988): Hereford cattle immunized and protected against *Boophilus microplus* with soluble and membrane associated antigens from the midgut of ticks. Parasite Immun. **10**, 405–410.

Rand, K. N., T. Moore, A. Sriskantha, K. Spring, R. Tellam, P. Willadsen and G. S. Cobon (1989): Cloning and expression of a protective antigen from the cattle tick *Boophilus microplus*. Proc. Natl. Acad. USA **86**, 9657–9661.

Romano, A. (1991): Personal information.

Romano, A., E. Alvarez and J. Greco (1982): Evaluación bajo condiciones de campo de la acción ixodicida de un nuevo garrapaticida a base de un piretroide sintético (Flumethrin) más un compuesto organo-fosforado (Coumaphos). Gac. Vet. B. Aires, XLIV, 1078–1095.

Roush, R. T. and B. E. Tabashnik (1990): Pesticide resistance in arthropods. Chapman and Hall, New York/London.

Schaub, G. A. (1986): *Blastocrithidia triatomae* (Trypanosomatidae): a biological agent against vectors of Chagas' Disease? Zbl. Bakt. Hyg. A **265**, 489.

Seifert, G. W. (1984): Selection of beef cattle in Northern Australia for resistance to cattle tick *Boophilus microplus*: research and application. In: Impact of diseases on livestock production in the tropics (eds: Riemann, H. P. and M. J. Burridge), 553–559.

Seifert, H. S. H. (1971): Die Anaplasmose. Schaper, Hannover.

Seifert, H. S. H. (1983): Tierhygiene in den Tropen und Subtropen. Teil I: Theoretische Grundlagen. Göttinger Beitr. Land.Forstwirt. Trop. Subtrop. **3**.

Seifert, H. S. H. (1992): Tropentierhygiene. Fischer, Jena.

Shapiro, S. Z., G. Buscher and D. A. E. Dobbelaere (1987): Acquired resistance to *Rhipicephalus appendiculatus* (Acari:Ixodidae): identification of an antigen eliciting resistance in rabbits. J. Med. Entomol. **24**, 147–154.

Shapiro, S. Z., W. P. Voigt and J. A. Ellis (1989): Acquired resistance to ixodid ticks induced by tick cement. Exp. Appl. Acarol. **7**, 33–41.

Shapiro, S. Z., W. P. Voigt and K. Fujisaki (1986): Tick antigens recognized by serum from a guinea pig resistant to infestation with the tick *Rhipicephalus appendiculatus*. J. Parasitol. **72**, 454–463.

Skerman, P. J., D. G. Cameron and F. Riveros (1988): Tropical forage legumes. 2nd ed. FAO Plant

Production and Protection Series **2**, FAO, Rome, 442–443.

Stendel, W. and H. D. Hamel (1990): Flumethrin pour-on and Cyfluthrin pour-on for ectoparasite control in cattle and sheep. Medicamentum, Germed, Dresden, 10–13.

Tatchell, R. J. (1992): Ecology in relation to integrated tick management. Insect Sci. & Appl. (in press).

Tizard, I. (1992): Veterinary Immunology – An Introduction. 4th ed. Saunders Company, Philadelphia.

Trager, W. (1939): Acquired immunity to ticks. J. Parasitol. **25**, 57–81.

Thullner, F., K. Hoffmann, S. Rettig, W. P. Voigt, E. Benavides (1994): Investigation into acaricide resistance of ticks parasitising cattle with the FAO acaricide resistance test kit. J. Trop. Med. Parasitol. **45**, Suppl. II.

Thullner, F. (1995): personal communication.

Vogler, H. (1976): Versuche zur Bekämpfung der Tsetsefliege mit Fenthion und Trichlorphon "Bayer" unter Anwendung der "pour-on"-Technik. Dipl. Arb. FB Agrarwiss., Göttingen.

Weiser, A. (1995): Analyse der Tierhygiene-Situation in mobilen pastoralen Tierhaltungssystemen in der Butana/Nordost-Sudan. Göttinger Beitr. Land- Forstwirt. **101**.

Weiser, J. (1987): Pathology of diptera and their microbial control. Proc. Intnat. Conf., Ceské Budejovice, CSFR.

WELLCOME (1976): Cattle tick control. Wellcome Res. Org. Wellcome Found., London.

Willadsen, P. (1980): Immunity to ticks. Adv. Parasitol. **18**, 293–313.

Willadsen, P. and D. H. Kemp (1988): Vaccination with "concealed" antigens for tick control. Parasitol. Today **4**, 196–198.

Wong, J. Y. M. and J. P. Opdebeeck (1990): Larval membrane antigens protect Hereford cattle against infestation with *Boophilus microplus*. Parasit. Immunol. **12**, 75–83.

Part II – Animal Diseases in the Tropics

Epizootiologically, animal diseases in the tropics and subtropics can be grouped as follows:

– **Vector-borne diseases**: these are diseases bound to the tropical environment because their transmitting vectors require a tropical habitat for their development. They can only be transmitted from a live infected animal (carrier) to a susceptible one.
– **Soil-borne diseases**: this disease complex has now almost become an exclusive problem of tropical and subtropical developing countries, on the one hand because of the favourable conditions which appear either during the dry or the rainy season, and on the other hand because of geographical, biological, socio-economic and infrastructural conditions which are typical for such countries. The infection is directly connected to the presence of infected animals, carcasses or remains of infected animals as well as infected animal products. The causative pathogen is almost never transmitted from an infected to a susceptible animal, but rather from spores which originate from infected excretions, or the remains of sick or dead animals, and develop on or in the soil (soil-borne disease). These spores are usually ingested with feed. Toxicoses have a particular epizootiology.
– Numerous **contact diseases** have become almost entirely limited to the developing countries because of geographical, biological, socio-economical and infrastructural conditions. These countries represent a reservoir from where these diseases can be brought back to the industrialized countries. Therefore, they are not only a problem in terms of disease but also for the marketing of animal products. The pathogen is transmitted from a latently or acutely infected carrier through direct or indirect contact with a susceptible animal. Only few contact diseases are originally tropical diseases (African swine fever, FMD-SAT types).
– **Endoparasitoses** are favoured by the tropical climate because their intermediate hosts find excellent conditions for survival and development in the tropical and subtropical ecology. Furthermore, the socio-economic conditions of the animal holder are such that it becomes difficult to carry out

regular measures for disease control. The infection is directly connected to the presence of infected animals which expel the development stage of the parasite in their excrement and thus become the original source of infection for the susceptible animal.

-   **Plant poisoning** is gradually becoming an increasing problem for animal production in the tropics. Because of the tendency to intensify the systems of tropical animal production and the enormous variety of species of poisonous plants, animals are forced to eat plants which are poisonous. Ecological factors in the tropical environment also favour plant poisoning.
-   **Deficiency diseases** may appear because of a relative or absolute lack of vitamins, minor elements, minerals and nutrients in the diet of the animal. These deficiencies again can lead to the appearance of soil-borne diseases and plant poisoning.

In contrast, factorial diseases may only be a problem in calf-rearing and in intensive production systems like feed- and dairy-lots. They are not a typical problem in a tropical production system and may appear in the same way in a temperate climate.

The particular aspects of these disease complexes which will be discussed are aetiology, occurrence, pathogenesis, clinical and pathological features, diagnosis, treatment and control with the exception of endoparasitoses and deficiency diseases. The presentation of endoparasitoses has been omitted because their aetiology, epizootiology and measures of control differ only marginally from those in a temperate climate. In part III, however, endoparasitoses are taken into account. Those ectoparasites which are relevant as vectors of vector-borne diseases have been discussed in part I. Recognition and prevention of deficiency diseases is part of animal nutrition.

1. Vector-borne Diseases

1.1. General

The complex of vector-borne diseases is characterized by

- the vector and the mode of transmission of the pathogen;
- the particular type of host defence of the organism against those pathogens which belong to *Protozoa* and *Rickettsiales*, and which is characterized by a long-lasting and perhaps even lifelong unsterile dynamic premunity (I/2.2.3.3);
- the difference in resistance between autochthonous and exotic breeds in the tropical production system. While autochthonous animals usually only present slight clinical symptoms or none at all, exotic animals show severe disease symptoms which may eventually lead to death. Economic losses are therefore less if autochthonous animals are used within the tropical production system;
- the importance of domestic animals and game as a reservoir of infection for animals and/or humans and as an alternative feed source for the vector.

Vector-borne diseases are only those infections which require a vector for their transmission. This vector is able to take the pathogen from the system of a carrier and to implant it into the system of a susceptible animal. In the case of transovarial transmission, the infection has been taken up by the previous vector generation. Wounds (myasis) may enlarge the range of vectors to non-biting arthropods; even the patient itself may help to contract the infection by scratching (chagas disease). Humans are vectors of special importance when carrying out procedures on animals involving bleedings, and especially during vaccination if large numbers of animals are vaccinated with the same contaminated needle. A high incidence of anaplasmosis may appear following vaccination (Seifert 1971).

Vector-borne diseases can only be controlled at present with a lasting effect if one succeeds in eliminating the vector completely. This requires a considerable infrastructure and appropriate financial resources. Therefore, the reduced resistance level of exotic breeds hinders the improvement of animal production in the tropics. In contrast, the autochthonous breeds with their particular resistance to vector-borne diseases are able to produce even under the challenge of infection and obtain a considerable level of production (N'Dama/trypanosomiasis).

Causal organisms of economically important vector-borne diseases of domestic animals in the tropics are

– *Protozoa* with the families
  – *Trypanosomatidae*
  – *Piroplasmea*
– *Bacteria* with the genus
  – *Spirochaetaceae*
– Rickettsiales with the families
  – *Rickettsiaceae*
  – *Anaplasmataceae*
– viruses
  – with arbovirales
  – with the rabies virus.

With the exception of viral infections, an infection caused by one of those groups of pathogens of vector-borne diseases leads to premunity with the problems typical for this type of immunity (I/2.2.3.3).

The vectors of vector-borne diseases and their control have been described in I/4. They are compiled again in Table 16.

Table 16. Vectors of tropical diseases in humans and animals

Specific vectors	Transmitter of
Tsetse fly	Trypanosomoses
Ticks	
Boophilus	Babesioses
Amblyomma	Heartwater
Hyalomma, Rhipicephalus	Theilerioses
Mosquitoes	Malaria
Simulides	Onchocercosis
Phlebotomes	Leishmanioses
Triatomes	Chagas disease
Carnivores, vampires	Rabies
Unspecific vectors	
Ticks	*Rickettsiaceae* and viruses
Haematophagous flies	Trypanosomes, *Rickettsiaceae*, viruses
Midges and gnats	*Rickettsiaceae* and viruses
Vampires	Trypanosomes, *Rickettsiaceae*, viruses

1.2. Vector-borne Diseases Caused by *Protozoa*

Protozoa are single-celled organisms which reproduce by splitting, budding, disintegration into spores, or sexually through merging and subsequent division.

Pathogens which cause vector-borne diseases from the *Protozoa* are the family of trypanosomes and the suborder of piroplasms. The biological classification of the protozoa is presented in Table 17.

Table 17. Protozoa

Tribe	Order	Suborder	Family	Genera	Species
Flagellata	*Kineto plastidae*	*Trypano somatina*	*Trypano somatidae*	*Trypano soma*	
					T. b. brucei
					T. congolense
					T. dimorphon
					T. evansi
					T. equinum
					T. uniforme
					T. vivax
					T. b. gambiense
					T. b. rhodesiense
					T. simiae
					T. cruzi
Sporozoa	*Haemo- sporidia*	*Haemo- sporina*	*Piroplasmea*	*Babesia*	
					B. bigemina
					B. divergens
					B. bovis
					s. *major*
					s. *argentina*
					s. *berbera*
					B. equi
					B. ovis
					B. motasi
					B. trautmanni
					B. perroncitoi
				Theileria	*T. p. parva*
					T. p. lawrencei
					T. annulata
					T. mutans
					T. hirci
					T. ovis
					T. taurotragi
					T. velifera

1.2.1. Trypanosomoses[1]

Aetiology

Trypanosomatidae are blood and tissue parasites on humans and animals which are transmitted mainly by haematophagous insects from host to host.

The intracellular stages of development (Fig. 41) do not carry a flagellum (*amastigote* form). The flagellate form (*epimastigote* form) appears in the gut of invertebrates; in the blood of the vertebrates, the flagellum is connected to the cell of the parasite with an undulating membrane.

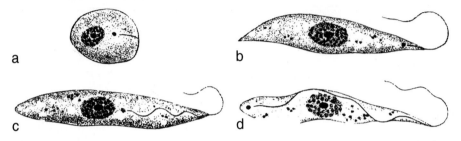

a   b

c   d

Figure 41. Forms of *Trypanosomatidae.* a. amastigote (micro-kryptomastigote,) b. promastigote, c. epimastigote, d. trypomastigote (Mitscherlich and Wagener 1970).

A special feature of the *Trypanosomatidae* is the change of shape during the cycle of development which mostly continues over a period spent on two hosts. During this change of shape, differently structured stages with differently inserted flagellae follow each other (polymorphism), or equally flagellated stages appear which are either slender or stout (pleomorphism).

In Table 18, the difference between *Trypanosoma* and *Leishmania* arising from their different development within the circulatory system, the tissue of the vertebrate as well as in the gut of the invertebrate vector are described.

The *Trypanosoma* form is of special veterinary importance since the genus *Trypanosoma* causes the most important economic damage of all trypanosomoses.

The *Trypanosoma* is covered by a single cell membrane; the stages which circulate in the blood have an additional 10–15 nm thick surface coat which apparently protects the parasite against the resistance mechanisms of the host. Microtubuli which are situated in a spiral-like form underneath the membrane are the supporting skeleton of the pathogen. The African *Trypanosoma* species are able to change their surface coat continuously (variable surface glycoprotein/VSG). The spontaneous variation of the surface coat blocks the efficiency of the specific host antibodies. Because of this characteristic of the

---

[1] Because of the complex character of this disease, its presentation is different from those of the following infectious diseases

Table 18. Characteristic forms of development of *Leishmania* and trypanosomes in the vertebrate and non-vertebrate vector (modified according to Mitscherlich and Wagener 1970 and Mehlhorn and Piekarski 1989)

Pathogen	Flagellate, trypo-mastigote form inside vascular system of the vertebrate	Flagellate, epi-/promastigote form in the gut of the non-vertebrate vector	Non-flagellate, amastigote form in the tissue of the vertebrate

Leishmania
  donovani
  tropica
  brasiliensis
		+
		(promastigote)
			+

Trypanosoma *cruzi*	+	+	+

Trypanosoma
  theileri
  lewisi
	+
	When dividing
	epimastigote
		+	

Trypanosoma
  vivax
  uniforme
  congolense
  dimorphon
  simiae
  suis
  b. brucei
  b. rhodesiense
  b. gambiense
	+	+	

Trypanosoma
  evansi
  equinum
	+	No development
		in the vector	

Trypanosoma *equiperdum*	+	No vector, venereal infection	

pathogen, the prospects of developing a protective vaccination against trypanosomes with VSG (*T. brucei* group, *T. congolense, T. evansi, T. equiperdum*) seems to be hopeless.

The division of trypanosomes between the less pathogenic section of *Stercoraria* (exception *T. cruzi*) and the pathogenic *Salivaria* is shown in Table 19.

All stages of *Trypanosoma* carry a flagellum which does not protrude over the edge of the cell (micromastigote) in the *amastigote* form. The flagellum which is inserted into a flagellar pocket is anchored with the basal apparatus into the plasma; it carries an axial rod. The flagellum serves as an instrument for mobilization and for sticking to the invertebrate host. *Trypo-* and partly also *amastigote* forms carry an undulating membrane which is only partly connected

Table 19. Characteristics of the sections *Stercoraria* and *Salivaria* of mammalian trypanosomes (modified according to Mitscherlich and Wagener 1979, Mehlhorn and Piekarski 1989)

Characteristic	Stercoraria	Salivaria
Free flagellum	Always present	Present or missing
Kinetoplast	Big, non terminal	Terminal or subterminal
Multiplication in the mammal	Periodical, typically as an epi- or amastigote form (except *T. cruzi*)	Always in the trypomastigote form
Pathogenicity	Low or none	Pathogenous
Development in the gut of the vector	Hindgut (except *T. rangeli*)	Frontal part (except *T. evansi, T. equinum, T. equiperdum*)
Transmission	Contamination (except *T. rangeli*)	Bite (except *T. equiperdum*)
Belonging subgenera	*Megatrypanum* *Herpetosoma*	*Duttonella* *Nanomonas* *Pyknomonas* *Trypanozoon*

to the cell wall. The kinetoplast of the flagellum may lose its DNA through the action of trypanocides and thus its ability to change its shape within the vector.

Trypanosomes take their nutrients from the substratum in which they parasitize. This is done through phago- or pinocytosis or through a cytostome on the membrane of the flagellar pocket. Blood forms have a glycolytic metabolism, and those which live in the vector have an oxidative metabolism.

Trypanosomes always multiply by dividing themselves lengthwise whereby they double their basal apparatus and kinetoplast. The trypanosomes of the *T. brucei* group are able to multiply sexually. Most species live exclusively extra-cellularly, but only the *amastigote* stages live intracellularly (Mehlhorn and Piekarski 1989).

So far, the grouping of animal diseases caused by trypanosomes is rather confusing since it is done partly according to aetiological criteria, partly according to epizootiological ones, and also partly according to those criteria which are dependent upon the vector (inter alia Brown et al. 1990). It is taxonomically note-worthy that *T. b. gambiense* and *T. b. rhodesiense* which are important human pathogens are now grouped as subspecies to the species *T. brucei*. The species pathogenic for animals has been denoted *T. b. brucei*. This is also a hint about the importance of animal reservoirs for *T. b. rhodesiense* and *T. b. gambiense*.

In Table 20, the pathogenic trypanosomes are compiled together with data about particular morphological features and their epizootiological and epidemiological characteristics.

Transmission

In animals, trypanosomes are either transmitted acyclically by haematophagous flies of the genus *Tabanus* (Surra, Mal de Caderas, South American trypanosomoses) or cyclically by the tsetse fly. Acyclically, the pathogen can only be carried over a short distance since it will survive only for a short time on and in the proboscis of the tabanide. In contrast, the transmission by tsetse flies is a complex mechanism in which the tsetse fly remains a lifelong carrier. The transmission and the interchange of hosts of African trypanosomiasis which is transmitted by the tsetse fly can be summarized as follows:

– The tsetse fly gets infected with the trypomastigote blood form which loses its surface coat in the goitre of the fly and, while remaining there for at least one hour, restructures its mitochondrium.
– The trypanosomes enter the midgut where they transform through lengthwise division into the epimastigote form in the cardia.
– The trypanosomes penetrate the haemocoel via the peritrophe membrane and the midgut epithelium and move from there to the saliva gland of the tsetse fly where they develop into the metacyclic infectious trypomastigote form which has now got its surface coat; the trypanosomes are haploid. Because of the complicated development of the trypanosomes within the tsetse fly, only about 0.1–0.4% of the flies are infected and thus are potential vectors of trypanosomiasis.
– After the vertebrate host has been infected by the tsetse fly, syngamy takes place; the trypanosomes become diploid and multiply through lengthwise division.

Prenatal infection with trypanosomes is also possible, but this is not considered important. Lions, hyaenas, dogs and cats can be infected with *T. brucei* by feeding on the carcasses of infected animals (Brown et al. 1990).

Pathogenesis

The pathogenesis of the trypanosomoses is rather complex and depends on the species of the transmitting vector as well as on the resistance of the host. The real cause which leads to the death of the animal is as ever not fully understood. On the one hand it is believed that the parasite releases toxic substances when it is destroyed within the circulatory system which damage the lining of the blood vessels. In some cases the sudden release of large amounts of such toxins triggers a chain reaction which produces a shock-like syndrome. Therefore, the damage to the host does not depend on nutrients being taken away by parasites but rather on the production of toxic substances (Mehlhorn and Piekarski 1989). With this theory, the typical symptoms of trypanosomiasis, such as cachexia, oedemas, anaemia and nervous symptoms can be explained. On the other hand, the pathogenic effect of the pathogen on the organism may be understood as a syndrome, the components of which are as follows:

Table 20. Biology of *Salivaria* and *Stercoraria* of mammalian trypanosomes and their range of diseases (modified according to Mehlhorn and Piekarski 1989)

Species	Length (μm)	Vertebrate host	Disease
Salivaria			
T. b. brucei	25–42	Equines, ruminants, pigs, rodents	Nagana
T. b. gambiensis	16–31	**Humans**, monkeys, dogs, pigs, antelopes	Sleeping sickness
T. b. rhodesiense	20–30	**Humans**, (rats artificially)	Sleeping sickness
T. congolense	9–18	Ruminants, predators	Nagana
T. simiae	12–24	Sheep, goats, monkeys	Deadly, rarely acute
T. vivax	20–27	Ruminants, equines	Souma
T. evansi	18–34	Ruminants, dogs	Surra
T. equinum	20–30	Equines, cattle, water hogs	Mal de caderas
T. equiperdum	18–28	Equines	Dourine
Stercoraria			
T. cruzi	16–20	**Humans**, domestic, wild animals	Chagas disease
T. rangeli	25–32	Rats, **humans**	Apathogenous
T. theileri	25–120	Cattle	Apathogenous
T. melophagium	25–70	Sheep	Apathogenous
T. lewisi	25–35	Rats	Apathogenous

- Pancytopenia as the result of the direct influence of the parasite on the cells or the phagocytic defence reaction of the organism which may be on an autoimmunological basis. The resulting anaemia which appears with the progressing parasitaemia is the classical symptom of the disease.
- Metabolic effects of the *Trypanosoma* which withdraw essential nutrients and produce toxic metabolites. The consumption of glucose, production of pyruvate and deaminization of the amino acids tyrosine and tryptophan seem to

Symptoms	Occurrence	Vector	Transmission mode
Fever, meningo encephalitis, paresis	Tropical Africa	*Glossina spp.*	Cyclical
Swelling of neck lymph nodes, oedemas. meningo-encephalitis	West Africa	*G. palpalis,* *G. tachinoides*	Cyclical
Fever, sleepiness and as above	East Africa	*G. morsitans*	Cyclical
Fever, excitation, anaemia	Tropical Africa	*G.* spp.	Cyclical
As above	East Africa	*G.* spp.	Cyclical
Becomes chronic, self recovery	Tropical Africa	*G.* spp.	Cyclical
Fever, oedemas anaemia	India, Africa, Sibiria, Australia, Latin America	*Tabanus* spp. *Stomoxys* spp. vampires	Mechanical-acyclical
Fever, anaemia	Latin America	*Tabanus* spp.	As above
Swelling of genitals, paresis	Mediterranean region, India, Java, America	–	During mating
Oedema, myocarditis, CNS-symptoms	South America	*Triatoma* spp. *Rhodnius* spp	Faeces
–	South America	*Rhodnius* spp.	Cyclical, faeces
–	Cosmopolitan	*Tabanides*	Faeces
–	Cosmopolitan	*M. ovinus*	Faeces
–	Cosmopolitan	Rat fleas	Faeces

be especially important. Intermediate metabolic disorders are the result.
- The actions of the secretions, such as acid phosphatase which activate the complement system, and pharmacokinetic active substances like serotonine and kinine also have a direct pathogenic effect.
- A trypanosome-induced hypothyroid status which may play a role in the impairment of mitochondrial ATPase activity, a key enzyme in energy metabolism (Lomo et al. 1995).

–   The action of biologically active lipids which are released when the pathogens are autolysed, and which together with free fatty acids, lead to cell damage and immunosuppressive effects.

–   Mechanical cell and tissue damage caused through the active mechanical invasion of the extraordinarily strong and mobile pathogens. Additional cell damaging lysosomal enzymes and the remains of autophagosomes are excreted in the flagellar pocket. Pinocytosis from the flagellar pocket blocks the plasma proteins of the host.

–   The consumption and/or depression of resistance factors during the course of the disease, especially of haemolytic complement and C_3. The exhaustion of the complement system adversely affects the direct defence mechanisms of the resistance system but also indirectly affects the immunological host defence.

–   Immunological mechanisms, characterized by the ability of the pathogen to change its surface-coat-antigen continuously, thus already exhausting the antibody production of the host during the IgM phase (only with *Salivaria*). The phenomenon known as antigenic variation seems to be the major reason why the pathogenic trypanosomes evade the host defence mechanisms. Variation occurs spontaneously and does not require an antibody as a stimulus. The number of antigens which a trypanosome may express is unknown but, theoretically, could run into many hundreds. In addition to the multiple variable antigen types expressed during a single infection, there is also evidence for the existence of different strains or serodemes, each capable of expressing a different repertoire of variable antigens. There is also a number of metacyclic variable antigen types, but these are more limited in number than those found in the blood stream trypanosomes (Brown et al. 1990, Hörchner 1983).

The immunosuppressive action of the trypanosoma exceeds the actual course of the trypanosomiasis. Secondary infections from other haematozoarias and other pathogens are activated; it may remain unclear which pathogen has killed the patient. On the other hand, the challenge to the organism by the infection with trypanosomes hinders any effective measure of immunoprophylaxis against other infections. Thus it has been shown that cattle infected with *T. congolense* do produce a worse immune response after the application of *B. anthracis* spore suspension than non-infected animals (Mwangi et al. 1990).

Importance, Characteristics and Occurrence of Trypanosomoses
Nagana, Surra and South American trypanosomoses are discussed below.

• *Nagana*

Nagana is widespread in Africa south of the Sahara, with exception of South Africa and Namibia.
    Causal organisms are mainly three *Trypanosoma* species: *T. vivax, T. congolense* and *T. b. brucei. T. vivax* and *T. congolense* which develop mainly in the

blood plasma, and *T. b. brucei* which develops additionally in the tissue liquids.

Nagana is transmitted **cyclical-alimentarily** by haematophagous flies of the genus *Glossina* (tsetse fly). Infected glossines are vectors all their lives. The pathogenicity of these three *Trypanosoma* species for cattle depends on the strain, the number of transmitted trypanosomes, on ecological factors, the transmitting tsetse species, other simultaneously appearing infections, and the genetically determined susceptibility of the particular cattle breed. The age and the condition of the animals are also important.

Beside the tsetse fly, the appearance of game is of great importance for the persistence of the disease within an infested area. Highly resistant wild ruminants harbour the trypanosomes and being alternative hosts for the tsetse fly become a reservoir of infection for the tsetse fly and contribute to its survival.

The infection with *T. vivax* is widespread: if transmitted by *G. palpalis*, it shows a light and chronic course, but if as in East Africa, *G. pallipides* becomes the vector or if it is *G. morsitans* or *G. tachinoides* as in West Africa, the disease occurs with high fever, oedemas in the subcutis and causes death after 3–4 weeks.

The *T. congolense* infection shows the most severe course. High fever is a symptom for the heavy multiplication of these, the smallest pathogenic *Trypanosoma* species within the blood up to 3 weeks after natural infection. Further symptoms are coarse hair, rapid emaciation and progressive anaemia which lead to death as well after the acute as after the chronic course.

The *T. b. brucei* infection usually shows a mild and almost symptomless course. West African autochthonous taurine breeds have a natural resistance to this disease (N'Dama cattle in the tsetse belt) (I/2.2.3.3).

• *Surra*

Surra appears in the tropical regions of Asia, in North Africa, Middle and South America. The causal organism is *T. evansi* but *T. vivax* in Latin America as well. Within the tsetse belt, superinfection with the causal organisms of Nagana can occur.

The disease is transmitted **acyclical-alimentarily** by haematophagous insects (*Tabanidae*, *Culicidae* and sometimes *Muscidae*). The trypanosomes ingested by these vectors from an infected host survive only 15 minutes inside the hypostome. Therefore, the infection will not be carried over long distances and has a higher incidence in stables. In South America *T. evansi* and *T. vivax* are also transmitted by vampires (*Desmodus* spp.).

Surra is, economically speaking, an important disease in camels in the Sahel which may lead to severe symptoms. In contrast, Surra regularly only shows light symptoms in cattle; exotic animals which are highly productive will be affected severely.

• *South American Trypanosomoses*

The South American trypanosomoses are the Mal de Caderas and the Mal de Chagas.

Mal de Caderas has been known in Latin America for more than 100 years and appears in equines especially during the rainy season in the hot and humid areas from Nicaragua down to Argentina.

The causal organism of Mal de Caderas is *T. equinum* which now is considered identical to *T. evansi*. Vectors of this organism are stomoxydines, tabanides and vampires. The mode of transmission is acyclical-alimentary. Mal de Caderas affects mainly equines. In ruminants, especially in cattle, the disease only appears with slight symptoms or none at all. Though the disease is of little economical importance for the cattle industry, cattle can be a reservoir of infection for equines since they can harbour the pathogens for up to 110 days p.i. in their blood (Mitscherlich and Wagener 1970).

Mal de Chagas (Chagas disease) appears in Mexico and in Middle and South America, but most of all in Brazil. The causal organism is *T. cruzi* which multiplies intracellularly in the reticulo-endothelial system of the muscles especially in the myocardium. The vectors are haematophagous hemiptera of the family *Reduviidae*. The transmission is cyclical-excretoric. Adult cattle are not infected by this disease; only calves are susceptible. Dogs and cats as well as armadillos are also reservoirs of infection. The Chagas disease is an important human disease.

Clinical Features of Trypanosomiasis

Trypanosomiasis develops in ruminants depending upon the species and strain of *Trypanosoma*, the vector and the resistance of the affected breed either as an acute or chronic infection. Animals exposed to infection by tsetse flies develop patent infections after incubation periods of variable length not only depending on the strain and species but also on the number of trypanosomes introduced by the tsetse flies. Trypanosomes appear in the blood of most animals exposed to infection with *T. vivax* after 8–10 days and with *T. congolense* after 12–16 days. The prepatent period with *T. brucei* is less well defined but the parasites can sometimes be detected in the blood as early as 3–4 days after exposure to the infection.

The invasion of the blood by actively dividing trypanosomes is associated with increased body temperature, and the initial parasitaemia and fever usually persist for several days before a trypanolytic crisis occurs: the parasites become well reduced in the blood and the temperature returns to normal. The first trypanolytic crisis is usually followed by further intermittent periods of parasitaemia, associated febrile attacks and remissions of infection. The subsequent course and outcome of the disease varies considerably and is influenced by the breed, background and management of the stock concerned, the nature and severity of the trypanosome challenge, the pathogenicity of the infecting trypanosome and the period of exposure to infection.

Trypanosomiasis is more commonly seen as a chronic disease with intermittent fever, and increasing degree of anaemia and progressive loss of condition. Infected animals are listless, their coats lack lustre, they lose weight, become easily exhausted, and lag behind the herd. Surface lymph nodes are enlarged and prominent. Cattle infected with *T. vivax* often show signs of photophobia and excessive lacrimation. Severe trypanosomiasis resulting from repeated exposure to infection leads to increasing weakness, debility and emaciation, and results in recumbency and death in many cases after periods of 1–6 months.

Trypanosomiasis is not invariably a fatal disease. Some animals, particularly those of trypano-resistant breeds, given good feed and management may recover after transient infections which last a few weeks, particularly after limited periods of exposure to infection with low numbers of trypanosomes and strains of low virulence. Occasionally, animals recover more slowly over periods of weeks or months from more severe trypanosome infections which progress through a state of premunity to complete self-cure (Brown et al. 1990).

Reproductive disorders are a common occurrence in human and animal trypanosomiasis and include irregular oestrus, abortion, neonatal death and infertility. The duration of the disorders varies considerably depending on the animal species and breed and the species and strain of trypanosomes. Thus, N'Dama cows infected with *T. vivax*, *T. congolense* and *T. brucei* resume an oestrus cycle 3.5, 9 and 16 months respectively after diminazene aceturate treatment. Following *T. congolense* infection, goats and Boran cattle have anoestrus due to persistent *corporal lutea* for 53–97 days and 21–73 days respectively. Sheep artificially infected with *T. vivax* show anoestrus for 40–96 days p.i. Compared to control animals, infected ewes have prolonged low levels of plasma progesterone until they either recover or die (Elhassan et al. 1994).

The course of the disease in **equines** is characterized after a prepatent period of only 1 week by high remittent fever, excessive lacrimation with anaemic and icteric conjunctives, photophobia, sometimes corneal opacity, accumulation of pus in the anterior chamber of the eye, frequent respiration and pulse as well as pounding of the heart, oedema of the lymph nodes and ventral surface of the abdomen, scrotum or vulva, staggering and prostration.

In **dogs**, the infection with *T. brucei* will lead to death after 1–3 weeks. Infections with *T. simiae* in **pigs** are also usually particularly severe and short, and infected animals die 12–36 hours after the onset of signs, which include high fever, respiratory distress and prostration.

In **camels**, acyclical Surra transmitted by haematophagous insects shows either a patent or chronic course. After a prepatent period of 2–3 weeks, the animal's temperature rises up to 41 °C and the febrile phase becomes intermittent with lapses of 8–14 days; prostration, a lustreless coat, debility, refusal to eat, emaciation and progressive anaemia are additional symptoms. The patent course can last for 3–4 months; death is the consequence of secondary infections, pneumonia and enteritis. Chronically, the disease can last for years with fever appearing at monthly or even longer intervals. Anaemia, oedema of the head, mandibular space, neck and dewlap, as well as the ventral surface of the abdomen, limbs, scrotum and vulva, emaciation and loss of hair are typical features of the infection in the camel.

Game, especially elephants and buffaloes can succumb to acyclical Surra with deadly consequences.

Pathology

The pathogenesis and pathology of tsetse-transmitted trypanosomoses are complex and not fully understood, and differ according to the species causing the infection. *T. vivax, T. congolense, T. simiae* and *T. suis* are essentially parasites in the blood plasma, but *T. brucei* is more widely distributed in the host, infecting blood plasma, the intracellular fluids of the connective tissue of various organs, and the extracellular fluids of the body cavities.

Tissue damage due to trypanosomosis is probably multifactorial in aetiology, but the underlying feature is the progressive anaemia throughout the course of the disease, even in the later stage when the parasites are present in either very low numbers or apparently absent altogether. In the live animal infected with trypanosomes, the level of anaemia estimated by measurement of erythrocyte numbers, packed cell volume percentage and blood haemoglobin content can be related to the severity of the clinical disease.

In the early stage of the disease, which can last up to 12 weeks in cattle, the main features are a fluctuating parasitaemia together with anaemia. The cause of the anaemia is complex but is believed to be haemolytic, caused primarily by erythrophagocytosis due to stimulation and expansion of the mononuclear phagocytic system throughout the reticulo-endothelial system. As a result, splenomegaly occurs. The severity of the anaemia is directly related to the level of parasitaemia, and trypano-tolerant breeds of cattle such as the West African N'Dama are able to suppress the parasitaemia following infection, thereby reducing the resulting anaemia and attendant clinical signs.

If animals survive the early parasitaemic phase, the disease becomes chronic during which the parasitaemia is low and the parasites are difficult to detect, or are even apparently altogether absent. Despite this, the anaemia persists.

Post mortem examination of animals after acute trypanosomosis may show extensive small haemorrhages involving mucous and serous surfaces, areas of emphysema in the lungs and mild gastroenteritis. After more chronic infections, the carcass may be anaemic and emaciated, with an enlarged spleen and lymph nodes. Subcutaneous oedema and accumulations of pericardial and thoracic fluid containing trypanosomes are found particularly in horses and dogs infected with *T. brucei*.

Histologically, aggregates of *T. congolense* occur in the capillaries of the heart, skeletal muscle and brain causing impairment of the microcirculation and frequently the development of focal polioencephalomalacia in the brain of cattle. Death caused by *T. vivax* may be associated with extravascular coagulation of the blood elements and aggregates of trypanosomes in the vascular system. *T. brucei* infections result in the extravascular accumulations of trypanosomes associated with cellular infiltration and necrosis in many tissues and organs, notably heart, skeletal muscles, brain, pericardial and serosal surfaces. Examinations of lymph nodes and spleens from infected animals reveal a progression of changes after different periods of infection. Increased numbers of lymphoblasts and marked plasma cell hyperplasia occur in the early stages of infection, but these cells are depleted as infections progress and are replaced by macrophages and reticular cells. Such cellular changes have been linked in experimental studies with the production of abnormal amounts of immuno-globulin and the occurrence of immunosuppression in trypanosome-infected animals (Brown et al. 1990).

Diagnosis
The definite diagnosis of trypanosomiasis depends on the detection of the parasite. It can be accomplished through

– direct and/or
– indirect demonstration

of the parasite. In spite of certain differences between the different species, **direct demonstration** of the parasite can generally be accomplished with

– a blood smear in the form of a wet film with or without concentration, e.g. by centrifugation in a haematocrit capillary (HCT), the silicone centrifugation technique (SCT) (Nessiem 1994), the dark ground buffy coat technique (DG) or by separating trypanosomes from blood by anion-exchange chromato-graphy (e.g. diethylaminoethyl cellulose);
– stained (Romanowski or Giemsa stain) blood smears as either thick or thin films; a lymph node biopsy may be useful (*T. vivax*);
– transmission of blood to splenectomized calves or small laboratory animals.

Since, depending on the species of trypanosomes, the pathogens are either found in the peripheral blood (*T. congolense*) or in the blood of the large vessels, col-lecting the blood has to be done accordingly. It also has to be taken into consideration that the activity of the different species depends on the time of day. The methods of direct parasite detection have undoubtedly limits in relation to the numbers of different species of trypanosomes in a blood sample. In order of decreasing sensitivity, the results in an evaluation trial were as follows: DG>HCT>thick film>thin film>wet film (Luckins 1992 (table 21)).

The differentiation of the species is only partly possible within the wet or stained blood smear because of the morphological characteristics of the parasite (Fig. 42 on page 1 annex).

A large number of serological tests has been used to indicate infections with trypanosomes **indirectly**. However, few of them have found practical application. The most commonly used techniques are

– immunofluorescence agglutination test (IFAT),
– enzyme-linked immunosorbent assay (ELISA),
– card agglutination test (CAT) (*T. evansi*).

With the exception of the CAT, these methods also detect non-agglutinating antibodies which only appear as long as the pathogen is present in the organism.

The persistence of such serologically detectable antibodies will be terminated through a *sterilisatio magna* after a certain time. With the mentioned serological techniques, the responsible *Trypanosoma* species may be identified within certain limitations. Interpretation of the results are made difficult because antigens from different trypanosome species show considerable cross-reactivity, and antibodies persist for several months after trypanocidal drug treatment (*sterilisatio magna*). The most promising test apparently is the ELISA. Species-specific monoclonal antibodies are currently being developed which should allow preparation of defined antigens for use in assays for antibody detection. In addition, mono-clonal antibodies can be used in a sandwich-ELISA to detect trypanosomal antigens and thus the presence of active infections (Brown et al. 1990). The monoclonal sandwich-ELISA has proven to be especially efficient in detecting *T. evansi* in camels (Diall et al. 1992). A promising antigen detection test or antigen trapping ELISA which acts as a sandwich-ELISA with monoclonal antibodies circulating antigens (*Trypanosoma*) is now available for *T. evansi*, *T. congolense* and *T. vivax*. Specific circulating antigens can be detected in cattle from 8–14 days after infection, but within 14 days of treatment they are not longer detectable. Therefore, this test seems to be an important tool for controlling the efficiency of trypanocidal treatment, or whether or not a treatment has elimin-ated premunity (Nantulya et al. 1989, Nantulya 1990).

Pathogenic trypanosomes must be distinguished from *T. theileri* which occurs in a high proportion of cattle in some areas. It is recognized by its large size (60–70 μm), pointed posterior, subterminal kinetoplast, very creased undulating membrane, and free flagellum.

Treatment

Table 22 shows trade names and generic names of currently used drugs for the treatment of trypanosomal infections in domestic animals. No effective treatment is available for *T. simiae*. Almost no new drugs for treatment or prophylaxis have been developed for nearly 30 years and some have either been withdrawn or are contraindicated because of resistance to them in recent years. Consequently, those remaining in use require careful management in order to minimize resistance problems (Brown et al. 1990). CYMELARSAN is a new trypanocide effective against *T. evansi* in camels, cattle and horses and *T. b. brucei* in humans and domestic animals. It is related to the drug ARSOBAL which is the only treatment available for the late stage of sleeping sickness in humans. The compound belongs to a chemical group which has not previously been used against animal trypanosomiasis; it is chemically: bis(amino-ethylthio)-4 melaminophenylarsine dihydrochloride. It is presented as a sterile freeze-dried powder for injectable solutions (Raynaud et al. 1989).

Control

Prophylactic measures for the prevention of trypanosomiasis are

– vector control,

Table 21. Sensitivity of different parasitological diagnostic techniques for the detection of *Trypanosoma* spp. (Brown et al. 1990)

Numbers of trypanosomes (ml^{-1})	DG			HCT			WF			TF			I
	Tb	Tv	Tc	Tb	Tv	Tc	Tb	Tv	Tc	Tb	Tv	Tc	Tb
5×10^4	+	+	+	+	+	+	+	+	+	+	+	+	+
1×10^4	+	+	+	+	+	+	+	−	+	+	+	+	+
5×10^3	+	+	+	−	+	−	−	−	−	+	−	−	+
1×10^3	−	+	+	+	+	−	−	−	−	−	−	−	+
5×10^2	−	+	+	+	−	−	−	−	−	−	−	−	+
1×10^2	−	−	−	−	−	−	−	−	−	−	−	−	−

DG = dark ground phase contrast microscopy
HCT = haematocrit centrifuge technique
WF = wet film
TF = thick film
I   = rodent sub-inoculation

– chemoprophylaxis and
– keeping trypano-resistant breeds.

Immunological host defence mechanisms based on innate resistance and naturally acquired immunity only protect those autochthonous animals against economically important losses which have been born in the area where the respective *Trypanosoma* species and strain are prevalent (I/2.2.3.3).

Immunity against trypanosomoses contains all the problems typical for premunity in vector-borne diseases caused by protozoa and is additionally complicated through the variability of the causal organism as explained above. Though it is known that artificially infected, relatively resistant breeds do not present severe clinical symptoms after repeated challenges, no applicable method for artificial premunization is available at present. Young animals doubtless have a higher relative resistance and/or passive maternal immunity which allows them to get a premunity through field infection. Depending on their level of resistance, this unsterile immunity may even be sustained under stress conditions (I/2.2.3.3).

Vector Control
The control of the vector is the most efficient method for controlling trypanosomiasis. The methods for vector control within the habitat and on the animal have been described in chapter I/4.2.

Chemoprophylaxis
Chemoprophylaxis of trypanosomiasis has been taken as a real alternative in disease prevention. The pharmacodynamics of the applied trypanocides, however, are mostly unknown. There is very little information about the catabolism

of the drugs in the organism of the animal and the effect of those catabolites on animal products. Therefore, chemoprophylaxis should only be a temporary measure within the conception of an ethical animal health scheme (I/3.2). Table 22 presents the specificity of common trypanocides. Figure 43 schematically describes the course of the concentration of the drug and, in contrast, the titre of the pathogen during repeated prophylactic treatment. In Table 22, two groups of trypanocides can be recognized: on the one hand, those drugs mostly with a therapeutic effect (antrycide-dimethylsulphate, diminazene aceturate, ethidium, novidium and suramin), on the other hand, drugs with a prophylactic effect (antrycide-prosalt (R.F.)), isometamidium, prothidium suramin and the suraminates of ethidium and metamidium). Though diminazene is considered a therapeutic drug because of its way of acting, it is also being applied more and more as a prophylactic.

After prolonged application, trypanocides develop resistance in the causal organism. The only alternative for circumventing this problem is the alternating application of products which should not have a chemical relationship and thus are not subject to a cross-resistance. Known prophylactic products do not fulfil this prerequisite, and therefore diminazene aceturate offers itself as a secondary "sanative" trypanocide. It has no prophylactic action but rather eliminates those *Trypanosoma* strains which have become resistant to the primary prophylactic. After eliminating the resistant strain, the primary drug may be applied again. A combination of diminazene aceturate (7 mg/kg b.w.) with DL-alpha-difluoro-

Table 22. Drugs used in the control of tsetse-transmited trypanosomosis in domestic animals (Brown et al. 1990)

Drug	Proprietary preparations	Host	Indications (trypanosomes)
Curative drugs			
Diminazene aceturate	Berenil Trypazen	Cattle	*T. vivax, T. congolense, T. brucei*
Homidium chloride	Ethidium "C" Novidium	Cattle, horses	*T. vivax, T. congolense, T. brucei*
Homidium bromide	Ethidium bromide		
Isometamidium chloride	Samorin Trypamidium	Cattle, horses,	*T. vivax, T. congolense, T. congolense, T. brucei*
Suramin	Naganol	Horses, camels, dogs	*T. brucei*
Prophylactic drugs			
Isometamidium chloride	Samorin Trypamidium	Cattle	*T. vivax, T. congolense, T. brucei*
Quinapyramine-suramin complex		Pigs	*T. simiae*

Quinapyramine dimethyl sulphate 10 g	Inject subcutaneously behind the ear
Suramin anhydrate 8,9 g	at 20–40 mg/kg b.w.
Distilled water q.s. 200 ml	for quinapyramine sulphate

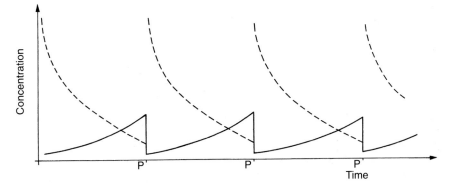

Figure 43. Course of the concentration of active substance and titre of parasites during repeated prophylactic treatment (Rüchel 1974).

methylornithine (2% mixed with drinking water) has been tried in laboratory animals to overcome resistance against classical trypanocides (Onyeyili et al. 1991). The intravenous application of isometamidium chloride has not only a very good therapeutic but also a considerable prophylactic effect of not less than 4 weeks in Boran cattle in Kenya (Münstermann et al. 1992). Appropriate combinations of drugs are shown in Table 23.

Depending upon the threat of infection, the treatment has to be repeated in irregular intervals. With a high incidence of infection, prothidium may be applied in an interval of 2 months, and diminazene aceturate may be applied twice a year in order to eliminate resistant strains. Diminazene aceturate is injected directly before or after the application of prothidium. With high incidence of infection, antrycide-prosalt is applied every 2 months, or prothidium is applied every 15 weeks, and diminazene aceturate is applied once a year in order to control resistance. For medium incidence of infection, antrycide-prosalt is applied every 3 months, or prothidium is applied every 5 months; resistance control with diminazene aceturate is done every 2 years. In the meantime, resistance against diminazene aceturate has also emerged; it can be overcome with an increased dose of 7.0–10.5 mg/kg b.w.

In chapter I/2.2.3.3, the consequences of the *sterilisatio magna* on the phenomenon of premunity have been discussed. This problem has to be taken into

Table 23. Combination of chemoprophylactics for preventing the *T. vivax* and *T. congolense* infection

Chemoprophylacticum	Alternative trypanocide
Antrycide-prosalt (RF.)	Isometamidium, diminazene
Isometamidium	Diminazene
Prothidium	Diminazene, isometamidium
Metamidium-suraminate	Diminazene

account especially if high doses of diminazene aceturate are applied.

Table 24 shows how clinical Nagana can be suppressed through the combination of appropriate trypanocides according to the respective challenge of infection while at the same time trying to reduce the danger of developing resistance.

During the last 30 years, attempts have been made to improve the group of the diamidines regarding their efficiency as trypanocides. There have been trials to introduce SHAM, a derivate of salicylic acid and glycerol and DFMO, difluorethylornithine as trypanocides. There is no information available about whether CYMELARSAN (bis(amino-ethylthio)-4-melaminophenylarsindihydrochlorite) can be used as a chemoprophylactic or not.

In principal, the application of trypanocides as chemoprophylactics should be restricted to a temporary intervention. Only animals which are exposed to the infection with trypanosomes during certain periods (transportation, fattening) should be protected. Under such conditions, *sterilisatio magna* may be acceptable if complete vector control is carried out within the production system at the same time.

Use of trypano-resistant breeds. Based on actual experience in the field, the introduction and keeping of trypano-resistant West African taurine cattle breeds seem to be an alternative biological method to preventing clinical trypanosomiasis and thus economic losses for the animal holder. N'Dama cattle and animals of other West African nondescript local breeds possess an increased titre of resistance factors (lysozyme, haemolytic complement C_9 and the third complement component C_3) and are better able to stabilize the balance of the host-parasite relationship known as premunity. Consequently they are in a better position to replace the protective mechanisms which are continuously consumed during the course of infection (I/2.1.3; Lampe-Schneider 1984, Seifert 1983). It must however be taken into account that only those autochthonous taurine

Table 24. Chemoprophylaxis of *T. vivax-* and *T. congolense* infection with combinations of trypanocides under varying exposure to infection

Exposure to infection							
Very high	P	P	P+B	P	P	P+B	
High	A	A	A	A	A	A+D	
		P		P		P+D	
Medium	A		A		A		A
	Diminazene every 24 months		P		P		
			6			12	
						months	

D = Diminazene
P = Prothidium
A = Antrycid-prosalt

breeds which have been born within the infected area and have a chance of actively acquiring their premunity under the protection of the maternal passive immunity tolerate the infection. As well, the cellular defence system of the West African taurine breeds seems better equipped to resist the infection with trypanosomiasis (Hörchner 1983). Whenever cross-breeding schemes with West African taurines are started, one should be aware that the special resistance potential of these breeds is lost rather quickly and is rarely maintained within combination crosses (I/2.1.3).

1.2.2. Piroplasmoses

Piroplasmidae are single-celled organisms which only survive for a short time outside an animal cell. The reproduction of piroplasms takes place either in the host (mammal) or cyclically inside the vector (tick). Therefore, piroplasms are bound to both, and cause an infection which also adversely affects the vector. Two important families of pathogens of the protozoa which are pathogenic for domestic animals are the *Babesidae* and *Theileridae*. *Babesia* and *Theileria* have a different cycle of development in the mammal and the tick.

The *Babesia* parasites multiply inside the erythrocytes of the mammal and are first found there a few days after infection. In the tick, they divide inside the gut-epithelium and from there enter the ovaries and the ova and thus infect the next tick generation. The relationship between pathogen and vector is cyclical-transovarial. While the tick is feeding, the *Babesia* which have entered the saliva gland reproduce prior to their delivery to the next host. However, there always remains a reservoir of the organisms in the delivering tick.

The *Theileria* reproduce in the mammal after the infection, at first inside the lymphocytes, and only enter the erythrocytes during a second phase. The reproduction inside the lymphocytes is a schizogenia (development of the Koch's spheres), whereas they divide when they are in the erythrocytes. Inside the tick, the organisms reach the gut where most of the *Theileria* perish and only a few are able to penetrate the wall of the gut and reach the saliva gland. The next time the tick feeds, they multiply there within the secreting cells. When the tick next moults, these cells are shed and thus no reservoir of infection remains in the tick. It is prerequisite that the tick does not get a new infection with "alien" *Theileria* prior to feeding. Nevertheless, the pathogen-vector relationship in *Theileria* remains cyclical.

1.2.2.1. Babesioses

Aetiology and Occurrence
Babesioses are tick-borne diseases in domestic, wild and laboratory animals as well as in humans, caused by the genus *Babesia*. Of more than 70 known species, 18 cause diseases in domestic animals, notably in cattle, sheep, goats, horses, pigs, dogs and cats.

In Table 25, the most important causal organisms of babesioses in domestic animals together with their principle hosts and reservoirs of infection have been compiled.

The genus *Babesia* belongs to the phylum *Apicomplexa*, class *Sporozoasida*, order *Eucoccidiorida*.

Because of their size, the *Babesia* can be divided into the so-called

- large *Babesia* (about 2 × 4–5 μm): *B. bigemina, B. caballi, B. canis, B. major, B. motasi, B. trautmanni*, and
- small *Babesia* (about 1 × 2 μm: *B. bovis, B. divergens, B. equi, B. ovis*.

Single *B. ovis* organisms are round, oval or irregular in shape while paired forms are piriform or club-shaped. The angle between the paired organisms is often, but not invariably, obtuse. Single forms of *B. bigemina* are elongated or amoeboid in shape and contain fine cytoplasmic filaments. Paired forms are typically piriform with an acute angle between the merozoites (De Vos and Potgieter 1994).

The diseases caused by *Babesia* occur worldwide, their distribution being dependent on a wide variety of ticks through which they are transmitted, and the presence of the appropriate habitat for these vectors. As presented in Table 25, all domestic animals are susceptible to the different species; game is an important reservoir of infection.

Epizootiology, Pathogenesis and Clinical Features
The three most important infections in farm animals are presented here.

• *Texas Fever*
(Mal de brou, Tristeza, Tocazón, Texasfieber)

Texas fever which is caused by *B. bigemina* occurs worldwide within tropic-subtropic grazing grounds (Table 25). Vectors of the disease are mainly one-host ticks of the genus *Boophilus* (*B. annulatus, B. decoloratus, B. microplus*). However, two-host ticks (*Rhipicephalus bursa, R. evertsi*) and three-host ticks (*R. appendiculatus, Haemaphysalis punctata*) may also transmit the disease.

After the tick has caused an infection, the *Babesia* actively penetrate the erythrocyte where they divide into the typical pear-shaped bodies in pairs which converge at the thin end to a pointed angle. They reproduce asexually so quickly that the parasitaemia and the destruction of the erythrocytes which result as a consequence progress at such a speed that the haemoglobin cannot be catabolized; this results in haemoglobinuria. The *Babesia* which do not participate in the asexual multiplication develop into egg-shaped or ball-like gamonts which do not develop further until they are ingested with the erythrocytes by ticks through feeding. The tick gets infected only during the last 24 hours of feeding. At this time, the environment in the tick-gut is appropriate for the survival of the *Babesia* because the proteases are reduced up to 90%

Table 25. The common babesioses of domestic animals

Causal organism	Vertebrate host	Disease	Principal vectors	Occurrence
Babesia bigemina	Cattle, buffalo, cervides, antelopes, gazelles	Texas fever	*Boophilus microplus, B. decoloratus, B. geigyi*	Africa, Australia, Asia, Central and South America
B. bovis s .*argentina* s. *berbera*	Cattle	Redwater	*B. microplus, B. annulata, B. geigyi, Ixodes ricinus*	As for *B. bigemina*
B. major	Cattle	Babesiosis	*Haemaphysalis punctata*	Europe, North Africa
B. divergens	Cattle	Redwater	*I. ricinus*	Northern Europe
B. jakimovi	Cattle	Babesiosis	*I. ricinus*	Northern Russia
B. ovata	Cattle	Babesiosis	*H. longicornis*	Japan
B. ocultans	Cattle	Babesiosis	*Hyalomma marginatum rufipes*	Southern Africa
B. caballi	Equines	Babesiosis	*Dermacentor (anocentor) nitens, Hyalomma* spp., *Rhipicephalus* spp.	Worldwide
B. equi	Equines	Babesiosis	As for *B. caballi*	As for *B. caballi*
B. motasi	Small ruminants	Babesiosis	*Dermacentor* spp., *I. ricinus* *Haemaphysalis* spp.	Europe, southern Russia, Mediterranean countries, Near East, Africa
B. ovis	Small ruminants	Babesiosis	*Rhipicephalus* spp.	Southern Europe, North Africa, Near East, northern South America
B. trautmanni *B. perroncitoi*	Pigs	Babesiosis	*Rhipicephalus* spp. *Haemaphysalis* spp. *Dermacentor* spp.	Africa, CIS, southern Europe
B. canis	Dogs	Babesiosis	*Rhipicephalus* spp. *Haemaphysalis* spp. *Dermacentor* spp.	Tropics and subtropics
B. gibsoni	Dogs	Babesiosis	*R. sanguineus, Haemaphysalis bispinosa*	South and East Asia, Southern Europe, North America
B. felis	Cats	Babesiosis	Not known	Africa, southern Asia
B. herpailuri	Cats	Babesiosis	Not known	Africa

amongst other things. After lysis of the erythrocyte membrane within the tick-gut, the gamonts develop out of the gametes. They are phagocytized by the gut-cells where two gametes at a time become one immobile zygote. A mobile kinet develops out of the zygote. The kinets enter the other organs like muscles, epidermis and Malpighi vessels via the haemolymph where they multiply asexually. Since the *Babesia* invade also the ovary they are transmitted transovarially to the next tick generation. Once the infected tick is attached to a host, the kinets multiply asexually (sporogony) within the saliva glands and become infectious sporozoites which can be transmitted to a new host by the thousands. In contrast to *B. bovis*, *B. bigemina* is not transmitted by larvae but only by nymphs and imagines (Mehlhorn and Schein 1984).

The incubation period of Texas fever is about 1–2 weeks. The **clinical symptoms** are more severe in older animals than in the young. The acute course of the disease shows pyrexia with temperatures from 40–42 °C which may be sudden in onset, causing severe anaemia and swelling of spleen and liver, jaundice and haemoglobinuria. The destruction of erythrocytes causes anaemia and tissue anoxia, and the release of excessive amounts of haemoglobin results in icterus (jaundice) and haemoglobinuria (redwater). The latter is regarded as pathognomonic and thus has given rise to the descriptive term of the disease, redwater.

Animals become listless, anorexic, anaemic and jaundiced. Other signs are haemoglobinuria, dehydration, general weakness and intestinal paralysis; constipation and diarrhoea may appear. Severe cases of the acute disease may end terminally after 3–6 days with CNS signs (ataxia, flaying limbs, coma) associated with sludging of parasitized red blood cells in brain capillaries. The mortality rate can reach up to 50% (Brown et al. 1990, Seifert 1992).

The basis of the nature of the disease is the destruction of the erythrocytes by the multiplying pathogens which leads to a severe anaemia (haematocrit < 10). Through the catabolism of the remains of the red blood cells by the cellular resistance system of the organism, immunocompetent cells are apparently stimulated which consequently leads to an autoimmune erythropenia. This mechanism of autoimmunity can explain the lasting anaemia which may persist even after the causal organisms have been therapeutically eliminated. With its toxic catabolites the haemoglobin which is released during the catabolism of the erythrocytes causes the distinct icterus, haemoglobinuria, petechiae on the serosae, parenchymal inflammation of the kidneys, spleen tumours, and CNS signs. The direct cause of death are anaemia and the collapse of the circulation system.

The course of the disease may be apoplectic, acute or chronic depending on the level of resistance of the animal and species and strain of the causal organism. Animals from highly productive breeds, especially Holstein-Friesian, often die apoplectically if they fall ill at an age of 8–24 months. Calves of these breeds may also show distinct symptoms of the disease, and the outcome may even eventually be fatal. During the chronic course of the disease the autoimmune anaemia can appear, i.e. the destruction of the erythrocytes continues after the parasitaemia has disappeared (Friedhoff 1987). Chronically ill animals may either recover only after a long time or not at all. In contrast, young animals of autochthonous and adapted breeds may experience only slight pyrexia with no or only slight icterus and sporadic haemoglobinuria after a mild infection during the dry season. Haemoglobinuria is often the only apparent symptom of the infection, the animals not showing reduced feed consumption or other general symptoms.

After recovery, the animals become premune, and thus animals which have recovered remain reservoirs of infection and may relapse under stress. Exceptions to this are such cases which have been treated until *sterilisatio magna* has been obtained.

In the course of the infection, a humoral immunity (IgG) develops which is passed on passively with the colostrum to the calf (IgG1) which may be protected for up to 6 months. With defined antigenic fractions of the pathogen, it has been possible to produce a considerable humoral antibody production which however has been protective up to only a mild degree (decoy mechanism). These are antibodies which can be recognized with the usual serological methods but are not an indicator about the protective immune situation of the animal. After surviving an infection, the animals are protected for 2–3 years against a challenge from *B. bigemina* whereby it is supposed that a cross-immunity to *B. bovis* exists. Likewise, *B. bovis* antigen protects against a *B. ovis* infection in sheep. There is no information whether the duration of the immunity is maintained through the persistence of the causal organism (Wright 1990).

Young stock in enzootic areas is protected better than with all other vector-borne diseases through natural premunization which develops under the protection of maternal passive antibodies and distinct youth resistance. Autochthonous cattle breeds of the tropics, especially Cebus, have a considerably higher level of resistance than European breeds. However, where new species are brought into an area which has so far not been infected, adapted animals may also suffer tremendous losses (*B. argentina* in Criollo cattle in the Bolivian Chaco). These empirical experiences contradict reports about a cross-immunity between *B. bovis* (s. *argentina*) and *B. bigemina*.

• *Redwater, Blackwater*
(Haemoglobinuria enzootica, Babesiosis bovum, Mal de brou, Hematinuria, Weiderot, Maisseuche)

Redwater appears where humid pastures are surrounded by brush or forests which are the ideal habitat for *Ixodes*, *Rhipicephalus*, *Haemaphysalis* and *Boophilus*, the vectors of this disease. The infection occurs in a belt from Russia through southern Europe as far as France. It is prevalent in Asia, Africa, Australia and Middle and South America. The causal organisms are *B. divergens* and *B. bovis*, s. *argentina*, s. *berbera*, s. *major*. Other *Babesia* species like *B. caucasica*, *B. occidentalis* and *B. colchica* are considered individual causal organisms or synonyms of the above mentioned main pathogens.

The symptoms of redwater are similar to Texas fever but less severe. As with redwater, icterus and haemoglobinuria are pathognomonic symptoms. Additional CNS signs may appear which can be confused with those of rabies.

Recently it has been reported that also humans may be infected by *Ixodes ricinus* with *B. bovis*, *B. divergens* and *B. microti*.

• **Equine babesiosis** is caused by *B. caballi* and *B. equi*. *B. caballi* resembles *B. bigemina*; *B. equi* is only 2 μm long and has a rounded amoeboid or pear-shaped form. The latter often appear in the "Maltese cross" form which means they are joined by their pointed ends. In contrast to most *Babesia* spp. *B. equi*, like *B. microti*, is transmitted transstadially and not transovarially, and there is, as with *Theileria*, an exoerythrocytic cycle in the lymphoid mononuclear cells. Such small *Babesia* may thus prove indeed to be members of another genus, perhaps somewhere between *Babesia* and *Theileria*.

The infection is transmitted by ticks of the genera *Dermacentor*, *Hyalomma* and *Rhipicephalus*. All equines are susceptible including zebras (*Equus burchelli*) which can become carriers of the infection. The same is true for horses which recover from the disease and enter the status of premunity. Prenatal infection of horses has been reported more often than it has in the case of canine or bovine babesiosis, but it is exceptional.

Equine babesiosis occurs worldwide where the tick vectors of *B. caballi* and *B. equi* exist; it has been reported in North, Central and South Africa, the USA, Central America and South America, southern and eastern Europe including the CIS, Asia Minor, the Middle East and India. *B. equi* apparently is more widespread than *B. caballi*.

The clinical and pathological features of the disease are similar to those of bovine babesiosis (Brown et al. 1990).

Origin, transmission and occurrence of **ovine**, **caprine** and **canine** babesiosis are summarized in Table 25. The diseases correspond in their clinical and pathological features with most characteristics as described for Texas fever in cattle. The same is true for the epizootiology and diagnosis of these diseases while special precaution has to be taken with treatment (see below).

Pathology
Haemolytic anaemia is the outstanding feature of clinical pathology, the red blood cell destruction occurring more rapidly with a *B. bigemina* infection than with *B. bovis* infection and being accompanied by a precipitous fall in PCV. Osmotic fragility of the red blood cells increases during the acute phase of the infection, and serum haemoglobin levels are high in acute cases. Severe anaemia is particularly evident in protracted cases, while very acute cases may die with little evidence of anaemia. In cattle which survive the infection, recovery is slow and it may take several weeks before normal red blood cell counts are restored.

Lesions seen at necropsy resemble those in animals which have died after an anaemic crisis. There is evidence of severe haemolysis, such as a pale carcass, watery blood and haemoglobinuria. Haemorrhages in internal organs and splenomegaly are more marked with *B. bovis* than with *B. bigemina* infection, but pulmonary oedema is regularly found. Icterus of all serosae and body linings is always pronounced with all *Babesia* infections. The liver is swollen and may be yellowish-brown, with the gall bladder containing copious amounts of thick granular bile, especially so with *B. bovis* infection. The kidneys and lymph nodes are also enlarged. Pulmonary oedema may be present but is uncommon. The grey matter of the cerebrum and cerebellum has a characteristic cherry-pink colour in cattle affected by the cerebral infection of *B. bovis*.

Microscopically, sludging of parasitized red blood cells in the peripheral circulation and evidence of vascular stasis are striking in acute *B. bovis* infections but are not features of the *B. bigemina* infection. Other lesions which are present in *B. bovis* and *B. bigemina* infection are

- degeneration and necrosis of the epithelium of the convoluted tubules in the kidneys and accumulation of hyaline or granular casts in the tubular lumens;
- centrilobular hydropic or fatty degeneration to extensive centrilobular and midzonal hepatic necrosis and bile stasis;
- marked congestion of the sinusoids of the spleen and a reduced ratio of white to red pulp with the germinal centres containing few cells; extensive necrosis of the red pulp with large thrombi in *B. bigemina* infection;
- oedematous and congested sinuses in the lymph nodes and depletion of lymphocytes in the germinal centres;
- marked distension of the capillaries of the brain by parasitized red blood cells in *B. bovis* infection;
- haemorrhages in the myocardium and hyaline degeneration of some myocytes;
- degeneration of skeletal muscle fibres in the hind limbs (De Vos and Potgieter 1994).

Diagnosis

Diagnosis of the babesioses is often possible based upon the clinical and pathological features. A fever, associated with haemoglobinuria, anaemia, jaundice and constipation with the demonstration of *Babesia* spp. in the red blood cells is considered diagnostic. In acute cases, a thin blood smear from peripheral blood (ear vein) stained with a Giemsa or Romanowski stain will confirm the diagnosis. With remission, the number of parasites diminishes and can be so sparse as to be difficult to detect. This is particularly so with *B. bovis*, which shows a marked tendency to accumulate in capillaries, especially those of the brain. During a *post mortem*, brain crush preparations will often reveal capillaries packed with red cells and parasites. For chronic infections, acridine orange staining and microscopic scanning with UV light will be helpful to find the parasites.

Shortly after appearance of the clinical symptoms, antibodies against *Babesia* spp. can be demonstrated serologically. The following procedures have been and are still applied:

– complement fixation test (CFT)
– indirect fluorescent antibody test (IFAT)
– indirect haemagglutination test (IHA)
– other agglutination tests (latex, slide, card, capillary)
– radio-immunoassay (RIA)
– enzyme-linked immunosorbent assay (ELISA).

CFT, capillary agglutination and IFAT are methods which also help to recognize chronically infected animals (carrier). In the USA, IFAT antigen slides of *B. canis* are commercially available for the detection. A combination of radial diffusion and CFT where the red blood cells are poured into the agarose holes after the serum sample allows the examination of large amounts of samples (Rodriguez et al. 1988). With the rapid agglutination in latex (RAL), a highly specific simple and practical test for field work for detecting *Babesia* spp. is available (Sanchez et al. 1991). The CFT is still a very reliable method for recognizing carrier animals. Most recently, the ELISA seems to have the most potential. With present antigens it is, however, less specific than CFT or IFAT, giving limited discrimination between the different pathogenic species of *Babesia*. However, the prospects of obtaining defined antigens is immediate when they are prepared by fractionation of the whole organism or by genetic engineering with cloned recombinant material. Such antigens should provide a durable test of high sensitivity and specificity (Brown et al. 1990).

Serodiagnosis is an invaluable tool in epizootiological studies of bovine babesiosis. Cattle which have been infected may have antibodies which are detectable for many years after infection and for some time (months, years) after the parasite has been eliminated. Thus the presence of antibodies will show that an animal has been infected and probably that it is immune, though not necessarily that it is carrying the infection. Seroenzootical surveys which give a prevalence rate in cattle across age cohorts can help define the epizootiological balance in the population and determine strategies for controlling the disease (Brown et al. 1990).

Furthermore, anticipated progress in diagnostic technology includes the development of an antigen detection ELISA, using monoclonal antibodies to detect circulating antigen in infected cattle, and the development of DNA probes to detect minuscule amounts of parasite DNA in an infected animal. Sensitivity and specificity studies have shown that 12–100 pg purified *B. bovis* DNA may be detected using probes containing repetitive sequences of *B. bovis* DNA. This amount would correspond to that found in 10–50 µL of blood infected with 0.01% of parasitized erythrocytes (Aboytes et al. 1991).

For differential diagnosis haemoglobinuria caused by *C. haemolyticum*, anthrax and pasteurellosis have to be excluded through bacteriological examination.

Treatment

Treatment of acute babesiosis if carried out on time has a favourable prognosis if appropriate drugs are administered promptly. Because anaemia will be severe in acute infections, supportive therapy and management are important. Fluid replacement and even blood transfusion may be indicated in extreme cases. If so, a single early administration of a large volume of blood will avoid the problems of blood matching (Brown et al. 1990). Several effective drugs are available which may also be applied for chemoprophylaxis (Table 26). However, when choosing the type and dose of a drug, it must also be considered whether a *sterilisatio magna* is to be obtained, or if the premunity is to be kept intact as it would be the case with adapted or autochthonous animals. If the treatment leads to a *sterilisatio magna*, the animal will become fully susceptible after the protective antibodies which have been induced by the infection have disappeared. This may even lead to fatal consequences in autochthonous cattle which lose the premunity they have acquired under the protection of maternal antibodies.

Large species of *Babesia* are more sensitive and hence more responsive to chemotherapy than small species. Thus *B. bigemina* and *B. major* require lower

Table 26. Drugs for the treatment and chemoprophylaxis of babesioses; dosage and range

Proprietory name	Chemical substance	Dose/100 kg b.w. and range of specificity		
		B. bigemina B. bovis syn.argentina B. divergens B. caucasica	B. equi	B. ovis B. motasi
Acaprin, Babesin, Babesan, Pirevan, Akron, Baburon, Pyroplasmin	Methyl-chinolyl-methylsulphate-urea 5%	1.5–2 ml	0.75–1,2 ml	2 ml
Gonacrine	Acridin derivates			
Berenil, Ganaseg	Diamidine-diazo-aminobenzol	2.5–3.5–7 mg	Toxic	3 mg
Diampron	Amicarbalit-di-isothionate	0.5 g		
Imidocarb	Imidazolin-carb-anilide	0.2–0.36 mg		0.2 mg
Imizol	– dipropionate or – dihydrochloride			
Lomadine	Diamidine derivate			
Trypan blue				
Haemosporidin				
Novoplasmin				

doses for both clinical cure and elimination (*sterilisatio magna*) of the parasite than *B. bovis* and *B. divergens* do. Selection of drug and dose rate are important, as most babesicides are quite toxic (Brown et al. 1990). Special care has to be taken with horses and dogs which are very sensitive to diminazene aceturate (see Table 26).

Control
Babesiosis can be prevented through vector control, chemo- and immuno-prophylaxis.

Vector Control
All babesioses depend on the cyclical transmission by ticks. One-host ticks transmit cyclic-transovarially, two- and three-host ticks cyclically and eventually transovarially. Therefore, vector control is the most efficient method for preventing babesiosis (I/4.2). Calves especially from autochthonous breeds possess an innate resistance enhanced by maternal antibodies which develops into a relatively stable premunity under field challenge. Adapted breeds can thus survive and produce within tick-infested areas. Methods for vector control have to be adapted to the conditions of the respective production system, especially the level of resistance and immunity of the animals kept. This will be discussed further in part III. The essentials are, however, already summarized here:

–   In extensive production systems with autochthonous or adapted animals, the mechanism of natural premunization of the animals ought to be taken advantage of and supported through management measures. By placing the calving season at the end of the rainy season, calves may be premunized during the dry season under the protection of maternal passive immunity while getting infected only during reduced tick activity and thus being able to build up their own immunity. Tick control may only be used to reduce tick infestation during the rainy season, but should not lead to a complete elimination of the infectious challenge transmitted by the ticks.
–   With intensification of the production system and increase of productivity through breeding, especially the introduction of high producing breeds, the intensification of tick control has to be carried out as a collateral measure! Where pure-bred exotic breeds can be kept, it should be possible to guarantee an efficient vector control. Since it is difficult to apply acaricides to the animal in dairy production because of considerations of human health, the principle of zero-grazing should be considered to keep the ticks away from dairy animals (I/4.3.2.1).

Chemoprophylaxis
Chemoprophylaxis may be a measure to prevent babesiosis, especially if animals are kept in an intensive production system and have to be freed from a latent infection. The reduction and/or elimination of the immunity can be accepted under such conditions.

In Table 26, the most important therapeutic and chemoprophylactic drugs are compiled. When these compounds are applied, it has to be clear if through this intervention

- a *sterilisatio magna* should or may be obtained,
- the relationship between pathogen and host should be influenced in order to stabilize the existing premunity,
- the infection should be prevented or the clinical symptoms be made less severe and the acquisition of premunity be supported.

Both diminazene and imidocarb are prophylactic if used in high doses, diminazene giving some protection for 2–3 weeks and imidocarb for 6–8 weeks. With both drugs, one has to know whether the existing infection should be eliminated (diminazene 5–7 mg/kg b.w., imidocarb 0.5 mg/kg b.w.) or only suppressed (diminazene 2.0–3.5 mg/kg b.w., imidocarb 1 mg/kg b.w.). Imidocarb seems more useful than diminazene for chemoprophylaxis; cattle have been protected under field conditions with one dose of 2 mg/kg b.w. for 15 weeks. It has been possible to show that the imidocarb application not only prevents the development of the infection in the animal, but also damages the *Babesia* which have been ingested by the tick. Therefore, continuous chemoprophylaxis with imidocarb may gradually reduce the level of tick infection and thus the challenge of infection (Kuttler 1975).

The alternatives to the chemoprophylaxis of babesiosis presented here have to be taken into account when interventions of treatment are carried out. When applying ACAPRIN as indicated in Table 26, the premunity will not be lost even if the treatment is repeated several times. Therefore, in spite of its poor tolerability the drug may still be indicated under certain conditions. In contrast, treatment with a therapeutic dose of diminazene (5–7 mg) will lead to *sterilisatio magna* and destruction of premunity.

Immunoprophylaxis
The immunoprophylaxis of babesiosis is determined by the problems of premunity described in chapter I/2.2.3.3. In extensive production systems with adapted cattle of a high level of resistance, this dynamic immunity almost guarantees a mild course of the infection with few losses in areas where the disease is enzootic. The permanent presence of the pathogen within the organism enhances the activity of the resistance factors, especially complement, and the activity of the cellular systems. Thus the balance between pathogen and host is favoured towards the host. As already described, the persistence of the antibodies is bound to the persistence and activity of the *Babesia*. A sterile immunity will only be temporary and, depending on the type of antigen and the added adjuvants, will last only for a certain time. A sterile immunity is not able to prevent field infection, but may only prevent or reduce clinical symptoms of the disease. Under a field challenge, an unsterile premunity will always be overlaid with a superinfection of field strains of babesia, no matter how the

sterile immunity has been induced. The artificially produced sterile immunity will thus prevent the appearance of clinical symptoms similar to maternal passive antibodies in the calf. Since the causal organism is implanted by the vector into the system of the susceptible animal by force, its organism is unable to prevent the infection through a temporary sterile immunity, no matter whether or not this infection will lead to an acute disease or to premunity.

All methods of artificial immunization against *Trypanosoma, Piroplasmidae* and most organisms of the order *Rickettsiales* depend on this principle. Accordingly, only the alternatives of active immunization as summarized in Table 27 are available for babesiosis. Here a broad spectrum of techniques is described which includes simple but risky interventions as well as a development towards modern methods which apply soluble antigens obtained in tissue culture. In the mean-

---

Table 27. Methods of artificial premunization/immunization against babesioses

---

Premunization with live antigen

– under maternal protection through passive colostral antibodies

  • as a desired natural field infection of calves after birth, at best during the dry season with low tick activity; applicable to autochthonous and adapted breeds;

– under chemoprophylactic protection: application of imidocarb or diminazene aceturate at the moment of infection or when the first clinical symptoms appear:

  • as desired field infection

  • as artificial infection with antigen produced from

    – homogenized ticks suspended in saline solution;
    – blood from naturally or artifically infected calves (the youth resistance of the calf is supposed to reduce the pathogenicity of the pathogen);
    – blood from donor cattle which are kept in stables protected against arthropods and are controlled against being infected with other blood parasites. The stabilates may be conserved by deep-freezing;

– of a minimal infective dose;
– of a relatively apathogenous *Babesia* species as, e.g., *B. ovis* against *B. bigemina*;
– which has been attenuated through passages in splenectomized calves;
– which has been attenuated through radiation.

Immunization with dead antigens which consists of killed parasites and/or soluble antigenic fractions obtained from

– erythrocytes from living infected splenectomized animals and inactivated through treatment with

  • beta-propiolactone 0.5% and addition of Al-hydroxide as adjuvant (Hinaidy 1981),
  • formaline 0.3% and addition of Al-hydroxide as adjuvant (Hinaidy 1981),
  • freeze-drying (Hinaidy 1981),
  • ultrasound and subsequent concentration of the parasite and extraction of two antigenic fractions by precipitation with protamine-sulphate from the supernatant and sediment (Wright 1983);

– infected erythrocytes maintained in tissue culture in which the *Babesia* have developed into merozoites and have delivered a soluble glycoprotein fraction of their surface antigen to the nutrient solution which is supposed to have antigenic properties (Levy et al. 1980).

---

time, antigens have been synthesized through recombinant DNA techniques which may eventually provide protection against several *Babesia* species. With a very low dose of the synthetic antigen, a high multivalent protective immune response is supposed to be obtained (Wright 1990).

As is explained in detail in part III, the immunoprophylaxis of the babesioses has to be seen in connection with the breed, the production system and the level of education and the socio-economic situation of the animal holder. At present, "the" vaccine to prevent this disease complex does not exist.

Principally with the alternatives for immunoprophylaxis of babesiosis one has to differentiate between

- methods of herd management which may lead to natural premunization of young animals while protected by the maternal colostrum;
- the iatrogenic application of a live and perhaps virulent antigen with collateral chemoprophylactic protection of the animal;
- the application of *Babesia* strains which have been attenuated through radiation and/or passage in splenectomized calves;
- a true vaccination with a dead antigen to obtain a temporary sterile immunity which will later on turn into an unsterile premunity under a field challenge without any or with only a few complications. It will, however, have all the consequences of the premunity, a fact which has to be taken into account if exotic highly productive breeds have to be vaccinated.

As far as dead antigens are concerned, most of the authors cited in Table 27 have worked with *B. bovis*, s. *argentina* and *B. divergens* but not with *B. bigemina* which is an important limitation as far as the applicability of their results is concerned.

The application of dead antigens may be an alternative in intensive production systems of the tropics where the intention is to eliminate the disease complex in the long run.

In South America, imported exotic cattle are quite often premunized with un-controlled stabilates from ticks or infected animals. Whether intentionally or unintentionally, mixed infections with *Babesia* and *Anaplasma*, perhaps also with *Eperythrozoon* and even some viruses may appear. Such interventions have, of course, to be treated with great reservation. These methods of artificial premunization mostly lead to severe diseases in the "vaccinated" animal. If they survive they may remain with a chronic anaemia because of the mechanism of autoimmunity. Often up to 20% of these animals die if premunized under field conditions. The question is whether or not it is economical to acquire an expensive animal and lose it through such a "measure of prevention" or at least have to deal with a considerable loss of productivity. A cheaper crossbred which does not require such an intervention would often be a more economical investment. On the other hand, the use of attenuated or dead vaccines will not prevent superinfection through field strains which in the end will also lead to a loss of productivity.

Table 28. Theilerioses of domestic animals – causal organisms, occurence and vectors

Causal organism	Susceptible species	Disease	Vector	Occurrence	Commentary
T. p. parva	Cattle, buffalo (*S. caffer*)	ECF	*Rhipicephalus appendiculatus*	East-, Central Africa	Subspecies cannot be separated serologically (Dolan 1989)
T. p. bovis	As above	ECF	*R. zambesiensis*		
T. p. lawrencei	As above	ECF/Corridor disease			
T. taurotragi	Cattle and other Bovidae	Sometimes pathogenous	*R.* spp.	Africa	Parasite of antelopes
T. velifera	Cattle, buffalo (*S. caffer*)	Apathogenous	*Amblyomma* spp.	Africa, south of the Sahara, Caribbean	
T. annulata	Cattle, buffalo (*B. bubalis*)	Tropical theileriosis	*Hyalomma* spp.	North Africa, southern Europe, Near and Middle East, southern CIS	
T. orientalis	Cattle	Oriental theileriosis	*Haemaphysalis Amblyomma*	Worldwide	*T. mutans* and *T. sergenti* in Britain and Australia
T. mutans	Cattle, buffalo (*S. caffer*)	Benign bovine theileriosis	*Amblyomma* spp.	Africa, south of the Sahara probably Caribbean	
T. camelensis	Camels	Unknown	Unknown	Africa, parts of CIS	
T. hirci s. lestoquardi	Small ruminants	Malignant ovine theileriosis	*Hyalomma* spp.	North Africa, South-East Europe, Near and Middle East	
T. ovis	Small ruminants	Benign theileriosis	*Dermacentor, Haemaphysalis, Ornithodorus, Rhipicephalus spp.*	Worldwide	Relevant because of differential diagnosis to *T. hirci*
T. recondita	As above				
T. separata	As above				

1.2.2.2. Theilerioses

Theileria are causal organisms of East Coast Fever (ECF, tropical theileriosis, benign bovine theileriosis, oriental theileriosis and malignant ovine theileriosis) (Table 28). Babesioses differ from theilerioses because of the development of the pathogen within both host and vector as already described (II/1.2.2), and because of their longer prepatent period, different pathogenesis and clinical symptoms.

ECF, caused by *T. p. parva*, is the most severe cattle disease caused by piroplasms. This can be demonstrated by the high mortality rate in affected cattle from Ethiopia to Swaziland and from Nigeria to Zaire. The yearly losses are calculated to be about 1 million animals. It is supposed that in Kenya 10% of heifers produced through A.I. die every year. The increasing number of losses is considered to be caused by the "tick control down" (Dolan 1983, 1989).

Aetiology and Occurrence

Table 28 shows the occurrence of the economically important theilerioses in domestic animals and their vectors. The characteristics of the *Theileria* have been compiled in Table 29. Because of their heterogeneous shape, the *Theileria* are difficult to describe morphologically. The piroplasma forms which appear in the erythrocytes after staining with May-Grünwald or Giemsa look like

- long, straight rods, $1.5–2.0 \times 0.5$ μm,
- long, comma-like rods, $1.0–2.0 \times 0.5$ μm,
- big, oval microbes, $1.0–2.0 \times 0.7$ μm,
- big, round microorganisms, $1.0–2.0$ μm, and
- big, anaplasma-like forms, $0.5–1.0$ μm.

Table 29. Characteristis of *Theileria* spp.

Species	T. p. parva	T. p. lawrencei	T. annulata	T. mutans	T. lestoquardi	T. ovis
Shape of piroplasms	80% oblong	55% round or oval	80% round or oval	55% round of oval	80% round or oval	80% round or oval
Infestation of erythrocytes	Heavy 50–80%	Slight up to 5%	Heavy up to 95%	Slight about 10%		
Diameter of schizonts	8 μm	5 μm	8 μm	8 μm	8 μm	8 μm
Infestation of lymphocytes with schizonts	Heavy more than 60%	Slight up to 5%	Heavy	Slight up to 5%	Heavy	Slight
Anaemia	Present	Mostly absent	Present	Slight	Present	Slight
Mortality	90–100%	80%	10–90%	Below 1%	45–100%	0

The protoplasma in *Theileria* stains clear blue. The nucleus can be recognized as a small, crimson red-stained granula at the broad end of the rod of the oval forms, and at the edge of the round microorganisms. Almost no cytoplasm can be seen in the anaplasma-like forms.

The pathogens are difficult to differentiate morphologically: they can be characterized using serological methods, and also by their pathogenicity and specificity to certain species of domestic animals. The classification of the *Theileria* species is still disputed last but not least because of modern methods of diagnosis (monoclonal antibodies).

The occurrence of theilerioses and the nomenclature of the *Theileria* is presented in Table 28 (Dolan 1989). This is different from previous presentations. *T. mutans, T. orientalis* and *T. sergenti* form the most confusing grouping. It has now been shown that *T. mutans* previously found in Great Britain is identical with *T. mutans* found in Australia and *T. sergenti* described in Japan, but different from *T. mutans* diagnosed in Africa. Therefore, *T. mutans* is now supposed to be the causal organism of the benign bovine theileriosis in cattle and buffaloes (*S. caffer*) in Africa and in the Caribbean. In comparison, the cosmopolitan *Theileria* which occurs in cattle is called *T. orientalis* and includes those which used to be called *T. mutans* in Britain and Australia as well as *T. buffeli* and the East Asian *T. sergenti*.

T. p. parva is the classical causal organism of East Coast Fever (ECF) and is transmitted between cattle by *Rhipicephalus appendiculatus*. *T. p. lawrenci* used to be considered the causal organism of corridor disease: it is transmitted by the same tick, and also by *R. zambesiensis* from buffaloes (*S. caffer*) to cattle. In principal there is no difference between ECF and corridor disease. The subspecies of *T. parva* cannot be distinguished from one another either morphologically or serologically. In contrast to *T. p. parva*, *T. p. lawrenci* causes an infection which is equally pathogenic but presents a comparatively lower number of schizonts and piroplasms. After it has passed through cattle, *T. p. lawrenci* behaves like *T. p. parva*; *T. p. bovis* causes a less severe disease than *T. p. parva* (January disease). It is also transmitted by *R. appendiculatus* and occurs in Zimbabwe and possibly elsewhere in the high veldt of Central and East Africa without buffaloes being present. Cattle are carriers of this organism, and while the mortality rate is lower than from infection with *T. p. parva*, it may be as high as 30% when epizootics occur. The carrier status of cattle and buffaloes infected with these two parasites of the *T. parva* complex, and their potential to transform into a form indistinguishable from *T. p. parva* puts control measures based on a sterile immunity in question (Brown et al. 1990, Uilenberg et al. 1982).

T. annulata is the causal organism of the Mediterranean Coast Fever/tropical theileriosis in cattle and buffaloes (*B. bubalis*). The disease occurs in Southern Europe, North Africa and Egypt as far down as the Sudan as well as in the Near and Middle East including India and Central Asia. Recently it has also been identified farther east in Malaysia, Vietnam and South China. The vectors are ticks of the genus *Hyalomma*.

T. taurotragi is infectious for cattle, sheep and goats as well as game *Bovidae* in Africa with low pathogenicity for cattle. Apparently the original host is the eland antelope (*Taurotragus oryx*) in which it may cause an infection with fatal outcome. The pathogen is an example of a group of parasites of wild ungulates which may complicate the epizootiology of ECF. *T. taurotragi* can be differentiated from *T. mutans* serologically; in the past both species have been confused with each other. *R. appendiculatus, R. pulchellus* and other *Rhipicephalus* species are vectors of this pathogen.

T. mutans is transmitted by ticks of the genus *Amblyomma* and has until recently been considered benign; it has now been shown to be significantly pathogenic for cattle. Both, cattle and buffaloes, are known to be carriers and the organism and the anaemia it induces may complicate clinical theileriosis caused by more pathogenic species (Brown et al. 1990).

T. velifera is an apathogenic parasite in cattle and buffaloes (*S. caffer*) in Africa. Like *T. mutans*, it has been found in the Caribbean. The pathogens can be separated because of morphological distinctions (a veil on the piroplasma form).

T. hirci is highly pathogenic for sheep and goats in South East Europe, North Africa, the Near and Middle East as well as in the Caucasian area. Ticks of the genus *Hyalomma* are *T. hirci* vectors. Nowadays, *T. hirci* is called *T. lestoquardi*.

T. ovis, T. recondita and *T. separata* are non-pathogenic *Theileria* of small ruminants. There may even be more than these three.

T. camelensis has been found in the CIS, North Africa and in the Near East. The vector is supposed to be *H. dromedarii*. Some authors doubt whether this is a species on its own.

The genus *Cytauxzoon*, with exception of some parasites of *Felidae* in North America, is synonymous with *Theileria*. The classification of some *Babesia* species, which like *B. equi* produce schizonts inside the lymphocyte, is under discussion. Possibly they will be included into the *Theileria* in the future (II/1.2.2.1).

The occurrence of theilerioses and their epizootiological situation is not static, especially in Africa. Figure 44 demonstrates the possible distribution of *T. parva* at present. In 1976, ECF, for example, was a serious problem in the eastern and northern provinces of Zambia, while in 1977–78 severe outbreaks of the disease appeared in the southern provinces, and in the central province in 1980. Since then, the disease has spread all over the country within the autochthonous cattle population. Meanwhile, ECF has appeared in West Zaire, an area which had previously been free of the disease; the same is true for Togo (Dolan 1989).

Epizootiology and Pathogenesis
As characteristic of *Piroplasmidae*, the development of the *Theileria* is divided into

– syngamy inside the vertebrate host and
– gamogony and sporogony inside the transmitting tick.

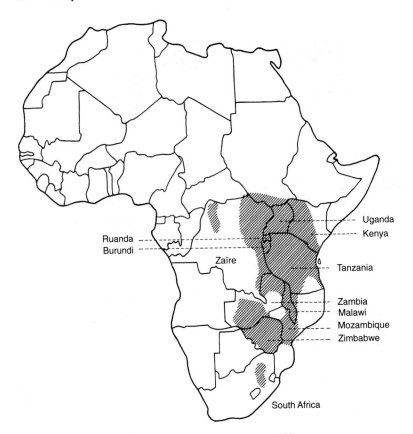

Figure 44. Incidence of *Theileria parva* in Africa (Dolan 1989).

After being ingested by a tick the organisms enter the gut where most of the *Theileria* perish. Only a few are able to penetrate the gut wall and reach the saliva gland. During the tick's next feeding period, they multiply inside the secreting cells of the saliva gland. These cells are shed during the next moulting period and thus no reservoir of infection remains in the new tick stage. The relationship of pathogen:vector remains cyclical however. The development inside the tick is as follows (Fig. 45): inside the tick gut, the few surviving *Theileria* differentiate into anisogametes which develop into the zygote. The zygote penetrates the epithelial cells of the gut; there the mobile kinets emerge which invade the saliva gland via the haemolymph and invade the acinary cells of the E-type. The kinets develop into multi-nuclear sporoblasts which again produce a complicated syncytium, the processes of which almost completely fill out the enlarged saliva gland. Finally, the complex sporozoites become separated. One sporoblast may produce up to 40000 sporozoites. The *Theileria* are only transmitted transstadially; the kinets only enter the saliva gland when the tick has completed its moulting. There in the saliva gland, the development of the sporozoites only takes place once the tick

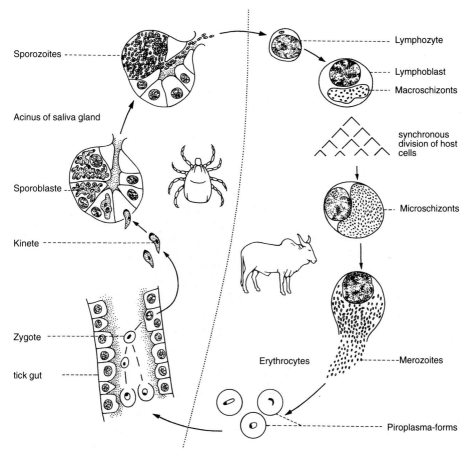

Figure 45. Development cycle of *Theileria parva* (Dolan 1989).

starts feeding, or is stimulated in some other way. Only those sporozoites transmitted by the tick during feeding are able to survive. If the tick does not get a chance to feed, the organisms can survive inside the saliva gland until they perish. Only about 1–2% of *R. appendiculatus* are infected with *T. p. parva*; about 10–30% of *Hyalomma* spp. carry an infection with *T. annulata* in enzootic areas. A high rate of infection also adversely affects the survival of the tick (Walker 1990).

As already mentioned above, the kinets of *Theileria* in contrast to those of *Babesia* are unable to penetrate the ovaries. In spite of the cyclical pathogen:vector relationship, the *Theileria* cannot be transmitted transovarially. Therefore, *Boophilus* ticks are unable to transmit *Theileria*. Furthermore, the tick stage which follows in two- and three-host ticks after moulting is unable to transmit the infection if it did not get a chance to become infected again during its last blood meal.

In contrast to babesioses, another special feature of theilerioses is the fact that the sporozoites of the *Theileria* enter the lymphocytes which are immediately concentrated at the site of the tick bite the pathogens thus being protected against the humoral antibodies of the host. On the other hand, the chance of being infected by *Theileria* is hindered because macrophages which are stimulated by the proteins of the tick saliva are ready to remove the *Theileria* at the site of infection. In addition, enzymes which are contained in the saliva of the tick damage the sporozoites.

It has also been demonstrated that *T. annulata* mainly invades the cells of the MHC class II and not the T-cells. After invading these lymphocytes inside the organs of the entire RES (reticulo-endothelial system), the sporozoite develops into the trophozoite which becomes the macroschizont (Koch's sphere) after division of the nucleus, and transforms the host cell into a large lymphoblastoid cell which divides at the same time as the macroschizont. Because of the lymphoproliferation induced by the organism, an enormous, uncontrolled multiplication of the parasitized cells takes place. *T. p. parva* joins with the mitotic spindle which makes sure that the schizonts are delivered to the next generation of daughter cells. With that, the clone of originally ingested parasites can expand. Merogony only takes place within a part of the parasitized cells whereby the merozoites develop from the microschizonts (Fig. 46 on page 1 annex) which are set free by the lymphocytes and at the same time, the host cell perishes (McKeever and Morrison 1990, Tait and Hall 1990). The merozoites invade the erythrocytes and complete their cycle inside the host by developing into the characteristic piroplasms. After penetrating the erythrocytes, the organisms are at first covered by the membrane of the erythrocyte, but after the decay of the host cell, they emerge freely into the plasma. Some of the very small (0.5–1.0 μm long) merozoites change into ovoid forms. Only these ovoid stages are able to develop within the gut of a feeding tick (Mehlhorn and Piekarski 1989). Though the lymphoproliferation has pathogenic consequences for the infected animal, it is possibly the anaemia caused by the invasion and destruction of the erythrocytes which determines the clinical picture. Differences exist here between *T. parva* and *T. annulata* as well as *T. mutans* and *T. hirci*. While *T. parva* is more pathogenic to the RES, anaemic symptoms are prominent where the latter is concerned.

The immune response of the organism to the *Theileria* infection is complicated. The mechanisms of host defence against the infection are possibly not consistent amongst the different *Theileria* species. Furthermore, the genetically determined resistance of the autochthonous breeds in the enzootic areas has a considerable influence on the development of the immune response. The specific basis of this resistance, however, is unknown. The immune response of the organism acts on the one hand against the extracellular stages (sporozoites or merozoites) and on the other hand against the antigen of the macroschizonts and piroplasmatic stages on the surface of the invaded cells. Empirical observations have shown that such an immune reaction towards homologous and partly heterologous strains of *Theileria* takes place (Tait and Hall 1990). Though humoral antibodies against all stages of development of the pathogen can be demonstrated serologically, apparently they only have an unimportant protective function. Hyperimmune serum of infected animals does not protect cattle against an experimental challenge. Monoclonal antibodies and complement, however, are able to depress and lyse schizonts in infected cells; monoclonal antibodies can also neutralize sporozoites. Apparently the cellular immune response to the infection of the lymphocytes with the merozoites is the real protective mechanism and is thus the basis for the known post-infectious immunoprotection. Lymphocytes infected with schizonts stimulate the proliferation of autologous peripheral blood mononuclear cells (PBM) (Emery and Morrison 1990). From these, only the cytotoxic T-cells are active against the specific homologous *Theileria* clone which has caused the infection. This explains why only artificial premunization against the homologous strain has had a protective effect so far (McKeever and Morrison 1989). Apparently the antigen receptor of the cytotoxic T-cells recognizes the specific antigenic peptides of the pathogen. Similar host defence mechanisms have been found for *T. annulata*. It has been shown under experimental conditions that two peaks of the cytotoxic cell population appear during infection on the one hand, and during convalescence on the other. These T-cells are stimulated during a challenge. This immune mechanism, however, could not be demonstrated when stimulated cells were transferred to immunologically naive animals (Trait and Hall 1990).

Survival of theileriosis leads to premunity with all consequences typical for this mechanism of immunity (Dolan 1983). Animals which have recovered after treatment or have been premunized artificially may be carriers for years and deliver the causal organism to "clean" ticks (Dolan 1989). The premunity only protects against the homologous *T. p. parva* strain.

The susceptibility of different cattle breeds to different *Theileria* species does not show the distinct differences between autochthonous and exotic breeds as it is the case with babesiosis (Fivaz et al. 1989).

It is unclear whether maternal colostral passive antibodies allow calves to build up a premunity. As has already been discussed, protective humoral anti-bodies have not been found so far. It is well known that heavy losses through ECF occur in calves from autochthonous breeds (Ankolé in Burundi and Boran as well as Maasai in Kenya). In Zambia, up to 30% of calves of local cattle have also been lost through infection by *T. p. lawrenci*. The severity of the disease is quantum-dependent, but one tick will transmit sufficient sporozoites to kill a susceptible cow. Both *B. taurus typicus* and *Bos t. indicus* from areas where the disease is not enzootic are susceptible, though *B. t. indicus* is comparatively more resistant at end-point dilutions of sporozoite suspensions. This, coupled with their relative resistance to ticks, makes *B. t. indicus* significantly more resistant than *B. t. typicus* to ECF. Despite this mortality rate, figures of 97% have been recorded in Boran cattle introduced to ECF enzootic areas (Brown et al. 1990). The morbidity may be up to 100%. In enzootic areas the losses are less. Never-theless up to 50% of autochthonous calves may be lost. Whether incomplete passive immunoprotection through the mother and/or bad management of the calves (withdrawal of milk) are responsible remains unclear. Last but not least,

ECF is economically so important for Africa because of the high calf mortality which appears in systems of traditional and intensive cattle production.

Clinical and Pathological Features
The five most important *Theileria* infections which occur in farm animals are presented here.

● *East Coast Fever (ECF)*
(Fièvre cotière, Theileriosis bovum, Corridor disease, Rhodesian Theileriosis, Ostküstenfieber).

Clinical Features
The clinical picture of the *Theileria* infection is characterized by the development of the pathogen in the mammal in two phases. The infection with *T. parva* leads to an acute lymphoproliferative disorder which may have an acute, subacute or mild course. While *T. p. parva* and *T. p. lawrenci* mainly are responsible for a severe course of the disease, *T. p. bovis* causes only slight symptoms. This may lead to confusion with *T. orientalis*.

8–16 days after the attachment of infected ticks to a host, schizonts appear in the lymph node, draining the site where they are attached. In the case of *R. appendiculatus*, this is most commonly the parotid lymph node: at this time it becomes both hyperplastic and enlarged. 1–3 days later, the animal becomes febrile and other clinical signs appear, thus giving an apparent incubation period of 9–18 days, most commonly 14–16. The fever rises up to 40–42 °C and continues on this level until shortly before death. Other superficial lymph nodes become enlarged, the body temperature rises to 41–42 °C and there is a fall in milk yield. Early in the disease there is little decline in condition or inappetence, but these develop late in the disease, when there is a rapid deterioration in condition. It is at this point that the disease is most frequently recognized for the first time, by which time prognosis is poor. By day 16, the first piroplasms appear in the blood. From this day on, the animal is infectious for ticks. Signs other than hyperthermia and lymphadenopathy are irregular. Respiratory distress, with evident pulmonary oedema and pneumonia is the most common sign. At the beginning of the febrile phase, constipation is usually evident. 6–8 days later, atony of the rumen, violent and even blood-stained diarrhoea with tar-like faeces, posterior paresis and wasting of muscles with occasional involvement of the CNS ("turning sickness") appear. Mucosal petechiation of the lower gum, under the tongue and of the vulva may be evident. Photophobia, ophthalmia and a milky infiltration of the eye is not uncommon. Nasal secretion and lacrimation and occasionally oedema of the eye lids appear. The animal collapses with signs of lung oedema which leads to dyspnoea and foamy nasal secretion and dies, most commonly between 18–26 days after infectious ticks have first attached themselves to a host (Brown et al. 1990, Mitscherlich and Wagener 1970).

8 days after infection, cells infected with schizonts can already be found in the regional lymph nodes of the site where the animal has been infected by the tick.

2–3 days later, they appear generalized in the lymph tissues of the organism. The consequence is swelling of the head lymph nodes (*Ln. parotidei* and *Ln. retropharyngei*) which are closest to the preferred sites of infection by the ticks. With continuing infection, the parasitosis of the cells increases and 5–7 days later, 30% of the cells of the lymph nodes may already be parasitized. The multiplication of the pathogens inside the RES may at first be apparent and the febrile phase eventually begins with a delay of up to 2 weeks. 2–4 days after the beginning of the febrile phase the merozoites have already been released from the microschizonts and have invaded the erythrocytes. The parasitaemia may increase in such a way that almost each red blood cell may contain 1–2 or more of the pathogens. Severe leukopenia and thrombopenia usually appear.

Subacute ECF may appear in enzootic areas in calves which have been born to immunized dams. It has a less severe course than the acute ECF. The febrile stage is continuous or irregularly intermittent and continues for 5–10 days. The animals usually survive the subacute disease. If the course is mild, the fever only lasts for 3–7 days; the animals are frail and the swelling of the lymph nodes is only temporary (Mitscherlich and Wagener 1970).

Pathology
With Giemsa stain, the parasites become evident inside the cells of the entire lymphoid system as well as in other tissues, such as the lung and intestinal tract. The multiplication of the organisms in the internal organs causes alterations of the RES, swelling of the lymph node marrow, degeneration of the liver and infiltration with lymphoma-like foci, pale discolouration of the kidneys and penetration with nodules and white or red spots (lymphoproliferative foci in the renal cortex). The following are pathognomonic signs: swollen lymph nodes, oedema of the lung, the subcutis and the intestinal mediastinum as well as increased amber-coloured liquid in the body cavities, epicardial blood splashing, serosal petechiae and ecchymoses on the peritoneum over the rumen, spleen and gall bladder and frequently, haemorrhages in the muscles, subcutaneous tissue and myelin sheaths of nerves. Punched erosions of the abomasum are pathognomonic as well.

The involvement and state of the lymphoid organ may vary between infections caused by different parasite stocks and is dependent on the length of disease reaction. There may be generalized hyperplasia of lymph nodes, spleen, and lymphoid tissue in liver, kidneys and gut. Alternatively, despite aberrant proliferation, the lymph nodes and spleen may be "exhausted"; the spleen is dry and aplastic, and the lymph nodes are oedematous (Brown et al. 1990, Mitscherlich and Wagener 1970).

● *Tropical Theileriosis*
(Mediterranean coast fever, Egyptian fever, Mittelmeerküstenfieber)

Tropical theileriosis is caused by *T. annulata* (syn. *T. dispar, T. sergenti, T. turkestanica*), the schizonts of which are found in the lymphoid tissue and are morphologically similar to those of *T. parva*. However, with a *T. annulata* infection, the erythrocytic stages are predominantly round (annular) or oval in form. The piroplasms appear in the erythrocytes shortly after the schizonts are first detected and in fatal cases may be present in very large numbers. In chronic or subacute infections they persist in the blood for years. In such established infections, the few piroplasms which can be seen in the blood films are also infectious for ticks (Brown et al. 1990). The incubation period is 9–25 days. The disease may have a hyperacute, acute or chronic course. Infections in calves in areas where the disease is enzootic take the form of a mild fever, but a mortality rate as high as 25% has been reported (Brown et al. 1990). The hyperacute course has a fatal outcome after 3–4 days, but most often the disease takes an acute course. The temperature during 5–20 days intermittently rises up to 42 °C. 2 days before the onset of the fever, macroschizonts can already be found in punctures of the lymph nodes; 3–5 days later, the first piroplasms appear in the red blood cells. In contrast to ECF, a distinct anaemia develops: the number of red blood cells may drop lower than 3 million/mL^3. Other than with ECF, the pathogenesis is determined by the invasion of the erythrocytes with the piroplasms. The conjunctivae are pale, icteric and show petechiae; bilirubinaemia and -uria are always present, but haemoglobinuria only occasionally. The cattle show general malaise, inappetence, atony of the rumen, nasal and eye secretion; the eye lids and lymph nodes are swollen. Blood-stained diarrhoea appears, the faeces are mixed with mucus. If the number of red blood cells decreases further, the disease will end fatally after 1–2 weeks. The mortality rate depends on the relative resistance of the breed and may be between 20–75%. The chronic course is protracted: beside intermittent fever, icterus and anaemia appear. After the disease has lasted for 4 weeks the animals may recover. There is evidence that recovered animals develop a schizont carrier state (premunity) which persists for years, recrudescence of schizonts being detectable after splenectomy.

The *pathological features* depend on the course of the disease. Petechiae are found on the mucous and serous membranes especially of the pharynx, larynx, frontal and maxillary sinus, trachea, bronchi as well as being found subpleurally; the liver is enlarged, pale brown or yellow, and friable; the spleen is enlarged and soft and the malpighian corpuscles are prominent; the lymph nodes are enlarged, oedematous, frequently hyperplastic and hyperaemic; the kidneys are pale and sometimes show lymphoproliferative foci manifested as pseudo infarcts; petechiae appear in the mucous membrane of the bladder; haemorrhagic ulcera in the abomasum and the intestine with necrotic centres and haemorrhagic marginal zones are common, and the lungs are oedematous and congested. The heart has petechiae and ecchymoses of the epi- and endocardium and the adrenal cortex may be severely infiltrated and haemorrhagic. The intima of the vessels and the serosae are icteric (Brown et al. 1990, Mitscherlich and Wagener 1970).

• *Benign Bovine Theileriosis*
(Tzaneen disease, Mild gallsickness, Gondériose, Pseudoküstenfieber).

Benign theileriosis caused by *Theileria mutans* is a disease in cattle which is almost always mild in enzootic areas. Acute disease outbreaks only appear if cattle from non-infected areas are introduced into enzootic regions and are exposed to a massive tick infestation. Characteristics of the causal organism, the distribution of the disease, the vectors and susceptible species are presented in Tables 28 and 29. The incubation period following natural infection lasts 14–22 days, after which a slight pyrexia may appear lasting for several weeks, showing malaise, variable but sometimes severe anaemia and slight swelling of the lymph nodes. Macroschizonts are found in the superficial lymph nodes which are swollen. The schizonts, which rarely are present in the erythrocytes, are characteristically different from those of *T. parva*, appearing 12–18 days after the sporozoites are inoculated by the vector. Piroplasms are more easily found from about the same time after infection, and are the pathogenic form of the parasite.

Other bovids may become infected and be carriers, and therefore wild ungulates act as a potential reservoir of the disease. African buffalo are proven carriers and *T. mutans* gets transmitted from buffalo to cattle by *A. cohaerens* (Brown et al. 1990).

Sick animals show diarrhoea and loss of condition; the mortality rate in cattle is below 1%. A special form of benign bovine theileriosis is "turning sickness": there is no fever, and death follows within the following 2–21 days. This is caused by the invasion of *T. mutans* into the brain capillaries. Surviving the disease will lead to a long-lasting premunity, and the animals remain as reservoirs of infection.

Pathological features of benign bovine theileriosis are swelling of the spleen and lymph nodes as well as ulceration of the abomasum and haemorrhagic spots on the mucosa of the intestine. Hyperplastic lymphoid nodules are rarely found in the kidneys. With turning sickness, thrombosis of the cerebral vessels with haemorrhages is evident along with meningoencephalitis and demyelination. It is diagnostically important that the mortality rate is low as well as the fact that from the piroplasms found in the erythrocytes, about 45% are rod-shaped and 55% oval or round. *T. mutans* can be transmitted to small ruminants but at the same time, no piroplasms appear in the erythrocytes (Mitscherlich and Wagener 1970).

• *Oriental theileriosis* is caused by *T. orientalis* (*T. sergenti*, *T. buffeli*, *T. mutans*) and is transmitted by *Haemaphysalis* spp. and *Amblyomma* spp. The disease occurs worldwide, but is only recognized as being of significance in imported cattle in the Far East (Japan, Korea). In many instances, the causal organism has been called *T. mutans*, but recent serological studies have shown that the Japanese, Korean, Australian and British parasites are antigenically indistinguishable from each other, and yet are distinct from African *T. mutans* (see above aetiology) (Brown et al. 1990).

• *Malignant Ovine Theileriosis*
(Malignant ovine and caprine theileriosis, Kwaadaardige Theileriose van skape en bokke, Theilériosis du mouton et de la chèvre, bösartige Theileriose der kleinen Wiederkäuer)

Malignant ovine theileriosis is caused by *T. hirci* and transmitted cyclically by *Hyalomma* spp. Nowadays, *T. hirci* is called *T. lestoquardi*. Other non-pathogenic *Theileria* which appear in small ruminants are *T. ovis, T. recondita* and *T. separata*; diagnosis of malignant ovine theileriosis may be confused by the widespread presence of the benign species (Table 28).

The disease is very important in Iraq, Syria, Jordan and Iran and is probably enzootic from North Africa to India. It may be considered to be equivalent to the disease which is caused by *T. annulata* in cattle. The disease may be highly pathogenic in susceptible sheep. In Iraq for example, losses of up to 40% have been attributed to theileriosis in autochthonous Awassi sheep moved from one area to another; losses may be higher in imported sheep. The parasite is also apparently widely distributed. Losses of the magnitude of those described in Iraq have not been reported elsewhere (Brown et al. 1990).

Depending on the susceptibility of the animal, the disease may take either an acute or subacute course. With the acute form, pyrexia appears about 14 days p.i. and may last 4–5 days. Both schizonts and piroplasms are incriminated in the pathogenesis, though the disease, in fatal cases at least, is almost invariably associated with anaemia. The animals show malaise, nasal secretion, dyspnoea, atony of the rumen, oedema of the mandibular space and in advanced cases icterus. Haemoglobinuria often appears. The superficial lymph nodes are swollen, and the animals stagger when they walk. The mortality rate is high. The subacute form is mild and the animals usually recover, but they become anaemic and need weeks to regain their former condition. Lambs from adapted breeds which recover acquire a relatively stable premunity.

Pathological features of malignant ovine theileriosis are petechiae spread over the pleura, the peritoneum, epi- and endocardium as well as on the mucosa of the gall bladder, the abomasum and caecum; oedema of the lung, hydrothorax, hydropericardium and ascites may appear. The spleen and lymph nodes are swollen; lymphomas are often found in the kidneys as is the case in cattle.

Diagnosis of Theilerioses
The direct demonstration of the causal organism is achieved using Giemsa-stained thin blood smears for piroplasms, and a lymph node biopsy or organ smears for schizonts. A blood smear is an essential early step if there is any evidence of lymph node enlargement, and then a lymph node biopsy is necessary. Between these two procedures it is necessary to find schizonts to make a definite diagnosis. In clinical disease these should be common for *T. p. parva* but less so for *T. p. bovis* and perhaps rare for *T. p. lawrencei*. Piroplasms are rare or absent in *T. p. lawrencei* infections, but will be numerous in *T. p. parva* and *T. p. bovis* at a late stage of the disease (Brown et al. 1990).

There is a number of serodiagnostic techniques using antigens prepared from the two major stages of the vertebrate cycle, namely schizonts from infected lymphoblastoid cell lines grown *in vitro*, and piroplasms from infected blood. The tests most widely used are the indirect fluorescent antibody (IFA) test, applying cultured schizonts or intra-erythrocytic piroplasms as antigen, and the indirect haemagglutination (IHA) test, using a piroplasm antigen (Brown et al. 1990). With these tests it is possible to identify the different species of *Theileria*. The demonstration of antibodies, however, does not provide information about whether the causal organism is present in the animal. This would be possible by means of a sandwich ELISA with monoclonal antibodies which is targeted either against the schizonts or the piroplasms. Attempts have also been made to develop highly sensitive test methods by cloning the genes of the parasite through DNA hybridization in order to identify the organisms within the blood or the tissue of the infected animal (Dolan 1989, Tait and Hall 1990).

Other tests which have been developed but which are not widely used are the complement fixation test (CFT), the capillary agglutination (CA) test and the enzyme-linked immunosorbent assay (ELISA). This last test, with increased specificity, holds most promise for future seroenzootical studies.

A strong antibody response is almost invariably recorded with both IFA and IHA tests 21–28 days following infections, but these antibodies do not persist for the duration of the immune state of the animal, declining to insignificant levels 6 weeks to 9 months after a single infection. Antibodies are usually absent during a clinical reaction, almost invariably so in fatal cases. A repeated challenge to immune cattle with a homologous strain of *T. p. parva* does not induce an anamnestic response unless the schizonts become established. High antibody levels may be induced in susceptible cattle by inoculation with dead antigens but these cattle are not immune to ECF.

Serological tests are of restricted value in disease surveys because of the above mentioned limitations. Moreover, immunologically distinct strains of *Theileria* cannot be distinguished with any of these tests. Indeed, *T. p. parva* cannot be differentiated serologically from *T. p. lawrencei* or *T. p. bovis*.

There are indications that cell-mediated immune techniques, monoclonal antibodies, isoenzyme electrophoresis and DNA probe technology may help in the identification of *T. parva* strains or subspecies. However, this still depends on very expensive cross-immunity trials with immunization and challenges to cattle (Brown et al. 1990).

Treatment of Theilerioses
It used to be very difficult to treat theilerioses, especially the *T. p. parva* infection. To a certain degree, the synergistic combination of tetracycline with diminazene aceturate was effective. Only since the introduction of quinazolinone halofuginone-lactate and halofuginone-hydrobromide for the treatment of the *T. annulata* infection (Schein and Voigt 1979), which were known as coccidiostatica so far, has an efficient therapy of theilerioses, including the *T. p. parva* infection, been made possible. Primaquine-phosphate (2mg/kg b.w.) and other 8-

aminochinolines are also effective against the piroplasma forms of *T. p. parva*. A further group of chemotherapeutica which are effective against *Theileria* spp. are the hydroxynaphthochinones menoctone, parvaquone and buparvaquone. With parvaquone (20 mg/kg b.w.) 90% and with halofuginone-lactate (1 mg/kg b.w.) 80% of naturally infected animals have been treated effectively (Dolan 1989). While previously only those animals which still were in the lymphocytal phase of infection could be efficiently treated with tetracycline combinations, it is now also possible to treat the invasion of the erythrocytes by the parasites with these modern drugs. Brown et al. (1990) give a summary of the actual regimens used for the treatment of patent theilerioses:

1. Parvaquone 1 × 20 mg/kg or 2 × 20 mg/kg i.m. at 48-hour intervals.
2. Buparvaquone 1 × 5 mg/kg or 2 × 2.5 mg/kg i.m. at 48-hour intervals.
3. Halofuginone 1 × 1 mg/kg orally.

As ever, tetracyclines and diminazene aceturate are used for the treatment of theilerioses. The reaction of *T. p. parva* to the application of the mentioned drugs is different. The use of monensin and salimomycin in therapeutic doses has in contrast to other recommendations (Boch and Supperer 1983) caused toxic symptoms (Dolan 1989).

Control of Theilerioses
The prevention of theilerioses can be achieved through vector control, with chemo- and immunoprophylaxis.

Vector Control
As with other piroplasmoses, theilerioses can also be prevented through rigourous vector control. Since the theilerioses are transmitted only by two- and three-host ticks, the intervals of treatment have to be shortened accordingly which makes tick control more expensive and complicated (I/4.3).

In principal, the methods of tick control for theilerioses also have to be adapted to the specificities of the respective production system. Within traditional production systems one should try to assist the natural premunization of the calves through reduction but not complete elimination of the vector tick. Sporadic or temporary interventions of tick control should by no means prevent a limited field infection completely as long as the calves are protected through their special youth resistance and/or maternal antibodies.

In contrast, because of the low resistance of exotic highly productive breeds, these animals have to be protected through a persistent and regular application of acaricides to the animal (I/4.3), or be kept in a production system which does not allow for a tick invasion (III). The present problem of keeping "grade cattle" in East Africa lies in the policy that the programme of "up grading" adapted breeds continues, but few means or infrastructure are available to carry out a regular acaricide application in intervals of 4–6 days. This problem is further complicated by the danger which a weekly acaricide application to lactating cows presents to human health because of the residues of the applied compounds which are excreted with the milk. It is also known that a high level of contamination with CHC and OPAE compounds has been reached through the high frequency of acaricide application in the case of beef cattle. This is especially true because the required withdrawal time before slaughtering is not observed. The application of pyrethroids may open new perspectives for the control of theilerioses as long as they remain effective against two- and three-host ticks (I/4.3.1.4).

Chemoprophylaxis

The use of chemoprophylaxis for preventing theileriosis by temporarily maintaining a therapeutic effective blood level of a drug within the organism is not looked upon as an independent measure. In contrast, chemoprophylactic intervention is applied especially in young animals to assist the organism to establish a premunity against the disease. It would, however, be feasible to use chemoprophylaxis in order to obtain *sterilisatio magna* whenever animals are brought into an intensive production system in order to free them from the burden of chronic infection and prevent outbreaks of patent cases, similar to the practice recommended for babesiosis. The following drugs can be applied for chemoprophylaxis of the theilerioses:

- oxy- and chlortetracycline, perhaps in a synergistic combination with diminazene aceturate;
- diminazene aceturate on its own to a certain extent;
- halofuginone-lactate (Chema et al. 1987, Schein and Voigt 1979, Voigt and Heydorn 1981);
- monensin;
- parvaquone;
- primaquine-phosphate.

The combination diminazene aceturate-oxytetracycline (10:20 mg/kg/b.w.) is supposed to eliminate the carrier state. With 5:10 mg/kg/b.w., the acquisition of naturally or artificially acquired premunity can be assisted, a result which can also be obtained with 1.2 mg/kg b.w. halofuginone (Boch and Supperer 1983). With primaquine-phosphate the gametocytes can be eliminated and thus the reservoir of infection inside the ticks can be controlled (Zhang Zhong Hang 1987).

Immunoprophylaxis

Theilerioses are of great economic and socio-economic importance especially for small-holder animal production in East Africa. Because of the difficulties in carrying out efficient tick control schemes in traditional production systems, efforts have been made for a long time to premunize animals which are exposed to theileriosis. By 1911 almost 300000 animals had already been infected in South Africa with homogenates from lymph nodes and spleen tissue of infected animals. Later on, infectious blood was used for artificial premunization. Large numbers of the animals premunized with these crude methods always died. By culturing *T. annulata* in tissue cultures as it has been established successfully in several countries of the Near and Middle East, it is now possible to produce attenuated stabilates which can be used for the premunization of animals exposed without provoking the disease. Through cultural passages of cell lines infected with schizonts from *T. annulata*, strains have been selected which are unable to produce the erythrocytic stage in calves but do produce an infection of the lymphocytes resulting in a protective immune response. Whenever an animal is infected with the cells of the cell line which itself are infected with macroschizonts, the schizonts are transferred in a way which is so far unknown into the monocytes of the hosts which, together with the pathogen, produce an associated antigen and start to proliferate because of the infection. In a later stage of infection, the schizonts differentiate inside the lymphocytes and produce merozoites the clones of which are unable to invade the erythrocytes. The immune response built up by the host organism is targeted as much against the original infected cells which were injected as it is against the infected cells of the recipient host organism (Tait and Hall 1990).

The tissue culture vaccine of *T. annulata* is able to prevent clinical cases of tropical theilerioses but not the transmission of infection through ticks. From the epizootiological point of view, it has to be taken into account that such a measure of prevention will contribute to increasing the already existing threat of infection by enhancing the infection rate of the tick population.

The multiplication of *T. p. parva* on tissue culture is much more difficult to establish. So far it has not been possible to produce a stabilate which could induce a challengable immunity either by attenuating the antigen in tissue culture or by deactivating the pathogen. Whether only the broad heterogenicity of the field strains and their unspecificity against the applied vaccine strain or peculiarities of the biology of the pathogen have been responsible for this is not yet known. Meanwhile, an attempt has been made to produce a recombined vaccine from fractions of the antigen which are able to stimulate a cytotoxic T-cell immune response (McKeever and Morrison 1990).

Thus far it has not been possible to produce an antigen *in vitro* for protecting small ruminants against infection with *T. hirci*.

In contrast to the efforts to induce a challengable immunity against ECF with *in vitro* produced antigens, it was demonstrated as far back as 1953 that calves survive a challenge after natural premunization with ticks infected with *T. p. parva* of a homologous strain along with a simultaneous application of chlortetracycline. Meanwhile this experience has been developed into a method of artificial premunization which resists field challenge and takes into account the antigenic heterogenicity of the field strains. For this purpose, deep frozen stabilates are prepared from local ticks which contain sporozoites of a defined species; *T. p. parva* sometimes is combined with *T. p. lawrencei*. Through infection with the stabilate and simultaneous application of depot oxytetracycline (LA), a premunity is induced which resists field challenge. It has been

Table 30. Methods of artificial premunization of cattle to prevent theilerioses

Inoculum	Organism	Origin	Chemoprophylaxis and measures to enhance host defence
Schizonts in lymphocytes	*T. parva*	Suspension of spleen and lymphnodes from dying infected calves	
Natural infection	*T. parva*	Infected ticks	– Oxytetracycline L.A. 20 mg/kg i.m. on day 0 – Oxytetracycline short acting 5–10 mg/kg i.m. on days 0 and 4 – Buparvaquone 2.5–5 mg/kg i.m. on day 0 – Parvaquone 10–20 mg/kg i.m. on day 8
Sporozoites from saliva glands of ticks – in infective dose – in reduced number (diluted) – in reduced number (radiation)	*T. parva*	Tissue culture of saliva gland cells from ticks	– Application of *Corynebacterium parvum (T. annulata)* – Chlortetracycline – Oxytetracycline L.A. – Buparvaquone – Parvaquone
Schizonts	*T. parva*	Cultured lymphocytes	Tetracycline L.A., Halofuginone-lactate
Schizonts	*T. annulata*	Cultured lymphocytes	No chemoprophylaxis required

Day 0 being the day on which sporozoites are inoculated

possible to reduce the mortality rate of calves in Zambia from 30 to 5%. Buparvaquone and halofuginone can also be combined as a chemoprophylactic with a stabilate (Dolan 1989).

In Table 30, the methods of artificial premunization of cattle against theilerioses, especially ECF, are summarized.

The problems involved when trying to premunize animals artificially in order to prevent theilerioses are characteristic for the complex mechanism of a host defence which is mobilized during infection.

Since quite a few trials of artificial premunization have produced contradictory results, and because of empirical experience with the disease in the field, it should not be forgotten that ecological factors typical for the respective production system as well as the resistance of the breed play an important role in the development of a naturally acquired or artificially induced premunity. Therefore, any measure of artificial premunization should not be oriented only towards the antigen application and perhaps chemoprophylaxis, both of which lie within the responsibility of the veterinarian. One should also put emphasis on counselling and supporting the animal holder especially in order to change the management of the mother cows and calves in such a way that they may get a natural premunization under better maintenance conditions. Even if there is no specific passive immunity against theileriosis, there will be a passive immunoprotection against other opportunistic infections. This is especially true for those traditional production systems where the calves are separated from the mother in order to get more milk from a cow which in many cases is not fed appropriately. The introduction of foster cows and better feeding of the dam and its calf could possibly favour the mechanism of natural premunization, especially if, for example, traditional methods of tick control (manual removal of ticks) are applied simultaneously.

Management Methods to Prevent Theileriosis

Because of the cycle of development of *Theileria* inside the tick, the tick stage which emerges after moulting is free of *Theileria* if the host on which the tick has previously fed does not harbour the infection. As shown in Table 28 *Theileria* are rather host-specific. This means there is the possibility of ridding those two- and three-host ticks which are prevalent on the infected pasture of their *Theileria* infection. By presenting an alternative host, the tick population can in this way be rid of the particular *Theileria* species which is pathogenic for cattle for example.

Under practical conditions, cattle have to be removed from the infected pastures for 18 months. Meanwhile, the ticks on the pasture can be starved out. If it were possible that this policy might fail because of the presence of game, alternative hosts to cattle can be offered, e.g. horses. By feeding on the alternative host, two- and three-host ticks will be rid of their original infection. Small ruminants are also now considered to be susceptible to *T. p. parva*. Therefore, they cannot be used for this prevention scheme.

The cattle which have been removed from the infected pasture have to be rid of their latent infection (carrier state) through application of chemotherapeutics in order to prevent a renewed infection of the cattle when returning to the original pasture. Similar measures are required if new animals are brought in. Such a prevention scheme is, of course, only possible in a intensive production system. Regular vector control and serological testing have to supplement such a policy.

1.3. Vector-borne Diseases Caused by *Rickettsiales*

In section 9, Bergey (1984) classifies the *Rickettsiales* and *Chlamydiales* in the following order

– I. *Rickettsiales*
– II. *Chlamydiales.*

The following, as important causal organisms of vector-borne diseases of domestic animals in the tropics, belong to the *Rickettsiales*

- from tribe II *Ehrlichieae*, the genus V *Cowdria*,
- from family III *Anaplasmataceae*, the genus I *Anaplasma*, the genus III *Haemobartonella* and the genus IV *Eperythrozoon*.

Bergey (1984) classifies *Rickettsiales* as mainly rod-shaped, coccoid and often pleomorphic Gram-negative microorganisms with typical bacterial cell walls and no flagella which multiply only inside host cells. They may be cultivated in living tissues such as embryonated chicken eggs or metazoan cell cultures. Except for binary fission, which is common to all members of this order, there are notable exceptions to any one of the characteristics listed above. For example, micro-organisms are included which appear ring-shaped in stained preparations, or have a flagellum, or are Gram-positive, or multiply on bacteriological media of moderate complexity. All are regarded as parasitic or mutualistic. The parasitic forms are associated with the reticulo-endothelial and vascular endothelial cells or erythrocytes of vertebrates and often with various organs of arthropods which may act as vectors or primary hosts. They may cause diseases in humans or in other vertebrate and invertebrate hosts. The mutualistic forms in insects are regarded as essential for the development and reproduction of the host.

Like viruses rickettsiales are living organisms which require another living cell for their survival and multiplication, a feature which makes them obligate parasites. It is exceptional for them to be found extracellularly in the vector. It is typical for this group of pathogens that they consist, as it is known for viruses, out of smaller subunits, initial bodies which as a cluster form the inclusion body. The initial body is the actual infectious agent which, after leaving the cell in which the inclusion body developed, infects another cell. Many members of the *Rickettsiales* need an intermediate host and vector as an epidemiological/epizootiological link (vector-borne diseases).

Though no metabolism has been proven for *Rickettsiales*, they are sensitive to antibiotics and ingest antibiotics selectively from the parasitized cell. In contrast, viruses are not sensitive to antibiotics, and being $< 0,1$ μm are filterable, which *Rickettsiales* are not usually.

Most *Rickettsiales* are typical causal organisms of vector-borne diseases, i.e. they may only infect a vertebrate by mediation of a vector. An exception is *Coxiella burnetii* from the *Rickettsiae* which only requires a vector in order to be transmitted between animals but may be transmitted to humans by animal products or dust.

Most *Rickettsiaceae* which are causal organisms of vector-borne diseases can be transmitted acyclically and cyclically as well as transovarially. Only a few vectors of only one genus of arthropods can transmit only a particular pathogen (*Cowdria*) while others may be transmitted by a large number of haematophagous as well as by licking arthropods (*Anaplasma*). It is important that in most cases the *Rickettsiales* may also be transmitted by humans during interventions conducted upon the animal which may lead to the transfer of tiny particles of tissue (vaccination, application of ear tags, castration, dehorning etc.) The vampire *Desmodus* spp. can also become an important vector in tropical America.

Table 31. Susceptibility of *Arthropodes*, *Aves* and *Mammalia* to *Rickettsiales*

Causal organism/ disease	Arthropod host pathogenous to	Vertebrate host	
		pathogenous to	apathogenous to
Rickettsiaceae			
R. prowazekii Epidemic typhus (spotted fever)	Fleas *(Xenopsylla)* *Amblyomma,* *Hyalomma* spp apathogenous for lice: *Pediculus vestimenti* *Pediculus capitis* *Phthirius pubis*	Humans, monkeys, rabbits, rats, guinea-pigs	Mice, cattle, small ruminants
R. typhi Murine typhus	Fleas (*Xenopsylla cheopsis, Pulex irritans*) ticks, mites, bugs, lice[a]	Humans, monkeys, guinea-pigs, rabbits, rats, mice	Rabbits, wild rodents, dogs, cats, donkeys
R. tsutsugamushi Scrub typhus	Larvae of mites (*Trombicula akamushi*)	Humans, monkeys, guinea-pigs, rats, white mice	Rats, field vole (birds)
R. rickettsii Rocky mountains spotted fever	Ticks, bugs, lice[a]	Humans, monkeys, goats, sheep, dogs, guinea-pigs, rats, mice	Wild boar, horses
R. conorii Tick-bite fever	Ticks, lice[a]	Humans, monkeys, guinea-pigs, rats, mice	Dogs, donkeys, mules, small rodents
R. australis Queensland tick typhus	Ixodes	Humans	O
R. acari Rickettsialpox	Ticks, lice	Humans, white mice	Rats, mice, guinea-pigs
Coxiella burnetii Coxiellosis (Q fever)	Ticks, lice, fleas, bugs	Humans, guinea-pigs, white mice	Humans, monkeys, dogs, cattle, sheep, goats, birds, rodents
Ehrlichiae			
Ehrlichioses			
E. canis Canine Ehrlichiosis	Unknown	Dogs, hyenas, monkeys	Jackals
Cytoecetes s. E. ondiri Bovine petechial fever	Unknown	Cattle	Wild ruminants
E. (C.) phagocytophilica Tick-borne fever	Ticks (*Ixodes ricinus*)	Cattle, sheep, goats	Wild ruminants
Cowdria ruminantium Heartwater	Ticks (*Amblyomma* spp.)	Cattle, sheep, goats, camels	Calves, lambs, autochthonous cattle breeds

Table 31. (continued)

Causal organism/ disease	Arthropod host pathogenous to	Vertebrate host	
		pathogenous to	apathogenous to
Anaplasmataceae			
Anaplasmoses *A. marginale*	Ticks, haematophagous flies, midges, gnats	Cattle (*B. taurus t., B. t. indicus,*) buffaloes, bison	Sheep, goats, wild ruminants
A. centrale	Ticks[a]	O	Cattle
A. ovis	Ticks[a]	Sheep, goats	Wild ruminants
Eperythrozoon Eperythrozoonoses			
Ep. ovis	Haematophagous arthropods	Sheep	O
Ep. suis	Haematophagous arthropods	Pigs	O
Ep. wenyonii	Haematophagous arthropods	Cattle	Cattle, wild ruminants
Haemobartonella Haemobartonelloses			
H. bovis	Ticks	Cattle	O
H. sturmannii	O	Buffaloes	O
H. canis	Fleas[a]	Dogs	Dogs, cats
H. felis	O	Cats	O

O = No information available
[a] = Experimentally

It is at least true for *Bartonellaceae* and *Anaplasmataceae* that recovery from infection and disease leads to a typical premunity with all the consequences characteristic for this type of immunity. It used to be thought that the premunity would only continue for a limited time following *sterilisatio magna*; nowadays it is believed that a sterile immunity may persist for several months (Wright 1990). For *C. ruminantium* it is known that only a premunity caused by living *Rickettsia* gives a certain immunoprotection, which means that the principle of the particular pattern of premunity known from *Protozoa* also remains true in this case.

For the control of this disease complex it is an important common property of *Rickettsiales* that they are sensitive to antibiotics especially tetracyclines and imidocarb and that already with a relatively low dose the *sterilisatio magna* may be obtained. Through the elimination of the pathogen, however, the stimulant for the immune system is removed and thus as well diagnostic as protective antibodies will disappear after certain time.

In Table 31, the complexity of the diseases caused by *Rickettsiales* together with the causal organisms important for man and animals are summarized.

Important causal organisms of vector-borne diseases of domestic animals in the tropics are:

- from the *Rickettsiaceae*, the genera
- *Coxiella*: Q fever
- *Cowdria*: Heartwater
- *Ehrlichia*: Ehrlichioses
- from the *Bartonellaceae/Anaplasmataceae*, the genera
- *Haemobartonella*: Haemobartonelloses
- *Eperythrozoon*: Eperythrozoonoses
- *Anaplasma*: Anaplasmoses.

1.3.1. Coxiellosis (Q Fever)

Aetiology and Occurrence

The causal organism of Q fever, *Coxiella burnetii*, is a non-motile obligate intracellular organism which completes its life cycle within the phagosomes of infected cells. It is similar to the *Ehrlichieae* in that it remains within the phagosomes after internalization and that all stages of its development are accomplished within the phagosomes.

Though the organisms do not always stain well with the Gram method they are on the basis of their ultrastructure and chemical composition regarded as Gram-negative. They appear as purple pleomorphic coccobacilli, the most common form of the organism being a rod-shaped, small bacterium about 250 nm wide and 500–1250 nm long.

At least three morphological variants of *C. burnetii* have been described: the terminal large cell variants (LCVs), the small cell variants (SCVs) and the dense intracellular structures within the LCVs regarded as endospores or initial bodies which, as an infectious particle, is filterable through a pore size of 100 nm. *C. burnetii* has two antigenic surface phases (phase I and phase II); they differ in their pathogenic and immunogenic properties and undergo an antigenic phase variation when serially passaged in embryonated eggs.

C. burnetii does not grow in artificial media. The organism can be cultivated in embryonated eggs, primary cells, several cell lines, including Vero cells and a number of macrophage-like cell lines.

In contrast to the other *Rickettsiales*, *C. burnetii* has a high tenacity, it survives in blood and other organs for weeks, dried in dust and wool as well as in tick faeces up to 2 years; in milk and dairy products it will be killed at 70–80 °C.

Although the clinical disease is limited largely to humans, a wide range of domestic and free living ungulates are naturally infected. The disease is described here because Q fever is a zoonosis which occurs worldwide but especially in the tropics, affecting a very wide range of hosts in cattle, buffaloes, camels, sheep, goats, horses, pigs, dogs, cats and wild and domestic birds which can be infected without any clinical signs. Cattle, sheep and goats are the most important reservoirs for human infections. In blood samples from intensively and extensively kept cattle, camels and sheep from Senegal, Togo, and also Bangladesh, an average of 30% infected animals could be found (Seifert 1989). *C. burnetii* has been reported from over 50 countries in different parts of the world (Woldehiwet and Aitken 1993).

Epizootiology and Pathogenesis
C. burnetii is unique among the members of the family *Rickettsiaceae* in its non-dependence on arthropod transmission and its remarkable resistance to adverse environmental conditions. That resistance and the presence of small dense cell variants (inclusion bodies) has led some investigators to suggest that it has an endospore form. The organism is readily excreted in milk, urine, faeces and uterine discharge of affected cattle, sheep, goats and other ungulates, although the period of shedding is shorter in sheep than in cows. *C. burnetii* is present in very high numbers in the amniotic fluid, the placenta and foetal membranes of parturient ewes, goats and cattle (Woldehiwet and Aitken 1993).

While people become infected by inhaling aerosols generated from body fluids or contaminated dust after desiccation of the primary source, ticks are the important vectors of *C. burnetii* in between animals, especially in the tropics. Because of its unusual wide range of hosts, to which belong wild living mammals – especially rodents, birds and arthropods – a cycle apparently exists within the population of wild animals. Such natural foci are the reservoir of infection for humans and domestic animals. In the tropics, ticks from the genera *Amblyomma, Dermacentor, Haemaphysalis, Hyalomma, Ixodes, Ornithodorus, Rhipicephalus* and *Argas* are important vectors. The infection through the tick is alimentary-excretorical and may also occur aerogenically when the dried faeces of the ticks in the coat of the animal becomes the source of infection. Humans get infected from domestic animals mainly orally with milk and dairy products and/or aerogenically through infected particles of dust from the coat of the animals. Therefore, Q fever cannot be considered solely a vector-borne disease.

The infection of domestic animals in the tropics will mostly occur unnoticed, but is important since it is a hazard to human health because of infected animal products.

Interaction of *C. burnetii* with the host's immune system is complex and still poorly understood. Of central significance is the organism's ability to grow and multiply within phagolysosomes and its propensity to establish persistent infection. Chronic infections are believed to be a result of immunological reactions and/or defects. Infection with or vaccination against *C. burnetii* induces both humoral and cell-mediated immunity. The latter probably effects elimination of the pathogen, whereas specific antibodies serve to accelerate the process (Woldehiwet and Aitken 1993).

Clinical and Pathological Features
The animal most susceptible to the infection with *C. burnetii* is the guinea-pig. In this case, the incubation period, which varies from a day up to several weeks, is

reversed in proportion to the infective dose. The infection results in a long-lasting septic fever phase after only a day or perhaps weeks during which the agent develops inside the endothelial cells of the vessels and serosae of the body cavities as well as in the blood, kidney, spleen and brain.

Afterwards, a stage of generalization develops analogously in ruminants which is followed by a localized infestation of lung, udder and testes. In pregnant animals, the pathogen enters the uterus which leads to abortion. Aborted fetuses appear normal but the heavily infected placenta exhibits intercotyledonary, leathery thickening and a discoloured exudate of variable consistency.

The commonest expression of disease is late abortion in ruminal animals, particular sheep and goats, either sporadically or in sudden outbreaks. The birth of weak, nonviable offspring also occurs in infected flocks. However, abortion due to *C. burnetii* is influenced by climate and systems of animal husbandry, occurring more frequently in warm dry areas as well as when animals are closely herded. Abortion is more common in goats and sheep than in cows but endemic infection in dairy cattle is considered to reduce fertility (Woldehiwet and Aitken 1993).

Cattle and small ruminants which have recovered from the infection will maintain a premunity for several years during which abortion and sterility may occur.

Diagnosis
The organism can be demonstrated directly in impression smears or histological sections prepared from the placenta and from uterine discharges from aborting animals. With a Machiavello or modified Ziehl Neelson stain, large clusters of red-coloured coccobacilli can be found. The direct and indirect immunofluorescence technique and immunoperoxidase technique are also suitable for demonstrating the pathogens directly.

The organisms can also be demonstrated directly by inoculating chick embryos or susceptible cell lines with discharges from infected animals.

The complement fixation and capillary agglutination tests are the most common tools for demonstrating antibodies against *C. burnetii*. The antigen is prepared from infected yolks. In recent years immunofluorescence and ELISA have replaced these tests.

Modern DNA probes and PCR will probably overtake these tests in the future. Restriction endonuclease digestion of various strains of *C. burnetii* has been used to differentiate new isolates. The presence or lack of plasmids is an additional typing aid (Woldehiwet and Aitkin 1993).

Treatment
As with all infections caused by *Rickettsiales*, Q fever can also be treated with tetracyclines and imidocarb. *C. burnetii* is also sensible to chloramphenicol and lincomycin.

Control
Hygienic measures such as destruction of placenta, contaminated bedding and

dung (composting) and disinfection of utensils are important to control the disease within a farm. Separation of infected animals is also important.

Vector-control will reduce Q fever especially in dairy production in the tropics.

Parenteral chemoprophylaxis as well as continuous oral application of tetracycline can be used to eliminate carriers from infected herds. In Germany, the aim is still to cull all positive animals (Schließer and Krauss 1982).

Trials to prevent the disease with inactivated and live vaccines are still inconclusive since it has not been proven whether a premunity induced by vaccination is able to prevent the excretion of *C. burnetii* with the milk. Vaccine production has also been hampered by the phase variation of *C. burnetii* when grown in embryonated eggs or tissue culture. In general, vaccines prepared from phase II are less protective than those prepared from phase I but the latter is more virulent. Refined vaccines containing specific protective epitopes may be expected in the future (Woldehiwet and Aitkin 1993).

1.3.2. Heartwater
(*Rickettsiosis ruminantium*, Cowdriosis, Herzwasser)

Aetiology and Occurrence

Cowdriosis is caused by a rickettsial organism found in the ruminant host in groups or clusters in the cytoplasm of endothelial cells of blood vessels as well as in epithelial cells of the gut and the gut lumen of the *Amblyomma* vector. Originally named *Rickettsia ruminantium* it has been subsequently renamed *Cowdria ruminantium*.

The diameter of individual organisms is rather uniform, but differs from cluster to cluster, ranging from 0,5–3.0 μm, which indicates a developmental cycle. *C. ruminantium* divides using binary fission in intracytoplasmic vacuoles, resulting in large groups of inclusion bodies. After 3–4 days these develop into smaller intermediate bodies with an electron-dense core which then condenses into electron-dense elementary bodies which after the rupture of the cell become the infective units.

The organism grows on primary endothelial cells of ruminants which now also allows the development of specific diagnostic probes.

C. ruminantium does not survive long outside the host: at room temperature, blood loses its infectiousness in less than 12 hours, and on ice after 8 hours; it can be kept frozen at –78 °C and lower using DMSO as a cryoprotectant (Uilenberg and Camus 1993).

Because of the difficult diagnosis the geographic distribution of the disease has only gradually been recognized. Detected at first at the end of the last century in South Africa it is now found in most parts of Sub-Saharan Africa including the neighbouring islands of Madagascar, La Reunion, Mauritius, Zanzibar, the Comores, and Sao Tomé. Since 1980, its presence in the Antilles has been confirmed, where it had already apparently been brought to from Senegal as early as 1830 (Uilenberg et al. 1983).

Epizootiology

Cattle, sheep, goats and domestic African buffalo are ruminants susceptible to *C. ruminantium*. Whether or not camels become infected is not yet clear. African antelopes also contract fatal heartwater. In the Americas, the white tail deer (*Odocoilius virginianus*) may be a reservoir of infection. Other species of *Bovidae* and *Cervidae* as well as several wild rodents, laboratory mice and rats can be infected experimentally (Uilenberg et al. 1983, Uilenberg and Camus 1993).

The only known natural vectors of *C. ruminantium* are *Amblyomma* ticks which transmit the agent transstadially either from the larval to the nymphal stage or from the nymphal to the adult stage. Infection acquired in the larval stage may persist even if the nymph is fed on a non-susceptible animal and may be transmitted by the adult. At present there are 10 African *Amblyomma* species, and 3 American species which can also transmit the infection experimentally. The African species are: *A. hebraeum, A. variegatum, A. pomposum, A. gemma, A. lepidum, A. tholloni, A. sparsum, A. astrion, A. cohaerens* and *A. marmoreum*; the American species are: *A. maculatum, A. cajennense* and *A. dissimile*. Not all *Amblyomma* species are equally important as vectors; *A. hebraeum* is probably more effective in transmitting the agent than *A. variegatum*. It now appears likely that *C. ruminantium* is injected into the host with the tick saliva and develops in epithelial gut cells in a cycle similar to that in mammalian endothelial cells (Bezuidenhout 1988, cited from Uilenberg and Camus 1993).

African wild ruminants were considered to be the wild reservoir and natural hosts of *C. ruminantium*. It has also been proven that *C. ruminantium* persists in domestic ruminants in the absence of wild reservoir hosts. *Amblyomma marmoreum* (tortoise tick) can acquire the infection from experimentally infected tortoises, guinea fowls and scrub hares and transmit the infection to ruminants. The possible role of free-living rodents in the epizootiology of heartwater is unclear; *C. ruminantium* can survive for weeks in the back-striped mouse (*Rhabdomys pumilio*) and the multimammate mouse (*Mastomys coucha*) (Uilenberg and Camus 1993).

The tick activity dependent on the seasonal influences determines the occurrence of the disease. In regions with a pronounced dry season, infections mostly occur during the rainy season.

It is feared that birds, which apparently spread the infection by carrying infected ticks from Guadeloupe through the Caribbean Islands, may bring the infection to the mainland (Uilenberg et al. 1983).

Pathogenesis

The pathogenesis of the disease is not yet clear. It has been suggested that after infection, the organisms develop in the macrophages of the lymph nodes; they occur in typical clusters in endothelial cells of blood vessels where they apparently undergo a development cycle. The pathogens also occur in the blood plasma, presumably as a result of lysis of the host cell, and they have also been shown to circulate in granulocytes.

Hydropericardium might lead to cardiac insufficiency, while hydrothorax and pulmonary oedema might cause respiratory difficulties. Lung oedema is often so pronounced in peracute heartwater that it may be responsible for sudden death by asphyxia.

The number of rickettsiae found in brain capillaries is not correlated with the intensity of the central nervous symptoms. Toxins have been suggested as an explanation but have never been demonstrated convincingly. The demonstration of lipopolysaccharide peaks in acute heartwater have led to the suggestion that the endotoxic activity of LPS may play a role in the pathogenesis of cowdriosis (Uilenberg and Camus 1993, Van Amstel et al. 1987).

Clinical Symptoms

After an incubation period of between 1 week and more than a month, depending on the transmitting tick stage, either a peracute, acute or chronic course of disease develops depending on the resistance and immunity of the animal. Young animals of all breeds are highly resistant up to an age of 3 weeks; the same is true for Iranian sheep (Uilenberg et al. 1983). The clinical symptoms may vary according to the strain of *C. ruminantium* and to the genetically determined level of resistance. Imported breeds of cattle, goats and less commonly sheep are especially susceptible while subclinical and mild cases are common in local autochthonous ruminants. However, even in enzootic areas a very high mortality rate in local breeds of goats may occur. Sheep are less susceptible than goats, but in heartwater-free areas even African breeds may suffer severe losses.

During the peracute course the disease starts with a sudden rise in body temperature up to 42 °C and leads to death with convulsions already after 36 hours. During the acute course distinct disease symptoms develop: high temperature, loss of appetite, listlessness and gradual development of nervous symptoms, dyspnoea, accelerated pulse and respiration, wet bronchial rales, a moist cough, red conjunctives, foul smelling diarrhoea and finally nervous symptoms (staggering, circling movements, "goose-stepping", other manifestations such as kneeling or leaning or pushing against a wall and occasional aggressive behaviour as well as chewing spasms). The animal has often an anxious and haggard facial expression, is hypersensitive, and shows exaggerated response to touch and sound. Blinking the eyes is reported to be one of the most common symptoms. Pregnant animals often abort. Death appears with the animal in lateral recumbency, exhibiting pedalling movements, opisthotonus, nystagmus, delated pupils and chewing movements.

Apart from the typical acute case, subacute or chronic cowdriosis also occurs with less pronounced symptoms and a higher recovery rate (Mitscherlich and Wagener 1970, Uilenberg and Camus 1993).

Pathology

Haematological changes are not spectacular. There is usually a degree of anaemia and eosinopenia. Splenomegaly and the characteristic amber-coloured, seldom blood-stained transudates in the pericardium and the peritoneal cavities are typical *post mortem* lesions. Oedema of the lung often appears. Froth may be found in the respiratory passages, often spilling from the nostrils. Lung oedema is usually very marked in peracute heartwater and may be the only visible lesion in such cases. Further typical features are oedema of the mediastinum, liver congestion, distension of the gall bladder, degeneration of the heartmuscles and kidneys and swelling of the lymph nodes. Often petechiae and haemorrhages are present in various organs. The mucosa of the abomasum can show hyperaemic patches; cattle often show haemorrhagic enteritis. In the brain macroscopic changes are mostly limited to congestion of the meningeal blood vessels, oedema of the meninges and petechiae.

The development of lesions depends on the strain of the pathogen and the resistance of the breed. In goats *post mortem* lesions may not be present at all.

Histopathological lesions are usually limited to the presence of typical intra-cytoplasmic clusters of rickettsiae in endothelial cells of the brain (Mitscherlich and Wagener 1970, Uilenberg and Camus 1993).

Diagnosis
The diagnosis of heartwater is extremely difficult. Though symptoms are often typical and even when nervous signs are present, many other diseases have to be considered for differential diagnosis (rabies, babesiosis, theileriosis, tetanus, listeriosis, plant poisoning). It will only occasionally be possible to make a precise diagnosis because of *post mortem* inspection of the carcass.

The direct demonstration of the organism from blood smears is not possible. The organisms occur in the granulocytes of affected animals and may be demonstrated within 1–2 days by maintaining separated granulocytes at 37 °C *in vitro*. It is possible to demonstrate the organisms in smears of the brain cortex. Cerebral cortex can be obtained with the syringe and needle inserted through a hole in the scull. Histological sections of the brain and kidney may also be used to demonstrate the organisms. The best staining method to use is Giemsa.

With the polymerase chain reaction (PCR) it may become possible to detect reliably carrier animals. The signal obtained by DNA hybridization in plasma of infected sheep has also been improved with the PCR.

Through xenodiagnosis with *Amblyomma* ticks on non-infected small ruminants it can be shown whether or not heartwater is present in a particular area. By inoculating randomly collected blood into susceptible small ruminants, it can be proven whether or not the infection is present in the local animal population (Mitscherlich and Wagener 1970, Uilenberg and Camus 1993).

Treatment see Chemoprophylaxis.

Control
Vector control as well as chemo- and with limitations immunoprophylaxis are suitable tools for the control of heartwater.

Vector Control
In keeping with the development of the production system, systematic tick control

can prevent heartwater. However, as with no other vector-borne disease, the special mechanisms of premunity acquired by the calf have to be taken into account. Only two alternatives exist to prevent heartwater through tick control:

- uncompromising systematic tick control and maintenance of infection-free herds;
- reduction or tolerance of tick infestation and maintenance of the natural infection of young animals while protected through youth resistance and maternal colostrum.

When applying methods of vector control, the special biology of the three-host tick *Amblyomma* has to be taken into account and attention has to be paid to the possibilities of reinfestation of an area through small rodents as well as birds.

Chemoprophylaxis
Sulphonamides are no longer used to prevent or treat heartwater. Instead, tetracyclines (chlortetracycline, oxytetracycline, rolitetracycline and doxycycline) are the only drugs which are effective for treatment and prevention of the disease. Long-acting preparations are especially advantageous for chemoprophylaxis. If the disease appears, protecting the entire herd with 5 mg/kg b.w. long-acting oxytetracycline for instance is recommended. Thus multiple losses are prevented and an existing premunity is kept in tact at the same time; such a preventive measure may also eventually be applied through feeding. For treatment of sick animals, oxytetracycline has to be applied as early as possible in a dose of 20 mg/kg b.w.

Immunoprophylaxis
The "infection and treatment" method of premunization first used half a century ago so far remained the only practical and effective means of premunization against heartwater. It is determined by the following problems:

- that no attenuated antigen was available;
- that a premunity could only be obtained with a live pathogenic antigen;
- that with the artificially produced premunity, only the onset of clinical symptoms through field infection, and not the superinfection with the field strain can be prevented;
- that through artificial premunization, additional reservoirs of infection and sources of infection for the ticks are produced.

The method is essentially crude, unsuitable for application to large numbers of animals and is reasonably safe only under close supervision. At present in Africa and also in the Caribbean, artificial premunization is carried out with the BALL-3-strain from the Veterinary Research Laboratory, Onderstepoort, S.A. The vaccine consists of blood from controlled donor animals which is preserved deep-frozen. Material prepared from homogenized, pre-fed infected ticks is also used and may offer certain advantages (Bezuidenhout 1981). Because it may cause serious shock in kids and lambs, it has been withdrawn from use. Trials are under-way to use infected endothelial cell cultures in order to replace infected whole blood (Brett and Bezuidenhout 1989).

5 mL infected whole blood which can only be kept for a short time at cool temperatures after defreezing usually has to be applied i.v. to the animal; the best place for application in Cebus is the ear vein. As soon as a rise of temperature appears after an expected incubation time of 12–22 days, 10 mL/kg b.w. oxytetracycline has to be applied daily, beginning on the second day of the febrile phase until the temperature disappears. It is unclear if the premunity is retained or only a temporary sterile immunity is obtained with such a high dose of tetracycline in all cases. Reports about losses through such methods of artificial premunization are contradictory (Brown et al. 1989, Uilenberg 1983). In South Africa, a slow release preparation of doxycycline (Doximplant) subcutaneously implanted at the time of intravenous injection of infective material is reported to make monitoring of the immunized animals unnecessary (Olivier et al. 1988).

Recently, a Senegalese stock of *C. ruminantium* was attenuated by passage in bovine endothelial cell cultures; it was not virulent in sheep and goats but induced a partial immunity to homologous challenge (Jongejan 1991, Gueye et al. 1994). In the meantime, the Senegalese stock has been similarly attenuated in Guadeloupe.

As field experience has shown, especially in the case of heartwater, the natural premunization of young animals under the protection of maternal colostral antibodies against the pathogen prevalent in the respective area is the best method to prevent the disease. Difficulties appear where animals are brought to other areas and/or have been imported to the infected region. Losses are also frequent where Government farms produce breeding stock under vector control and later on distribute these animals to the peasant farmer.

If a genetic marker of resistance or susceptibility can be identified, it should be feasible to select animals for resistance to cowdriosis (Uilenberg and Camus 1993).

1.3.3. Ehrlichioses of Ruminants

Domestic animals in the tropics and subtropics can be affected through 3 different infections caused by *Rickettsiaceae* of the tribe *Ehrlichiae*:

–   Bovine petechial fever (BPF, transmissible petechial fever, Ondiri disease, Ondiritis, Petechialfieber des Rindes), caused by *Cytoecetes ondiri*;
–   the *Ehrlichia bovis* infection of cattle;
–   the *Ehrlichia ovina* infection of sheep.

The *Cytoecetes* (*Ehrlichia*) *phagocytophilica* infection of ruminants (tick-borne fever) occurs in the temperate climatic zones of the world.

Aetiology
Ehrlichiae are polymorphic typical rickettsiae forming from polymorphic initial bodies (0.3–2.0 μm) larger, mostly lengthy colony-like inclusion bodies (3.0–4.0 μm). The initial bodies are the infectious units and can also occur singularly. In blood smears and preparations from spleen kidney and liver, the organisms are found in the monocytes, granulocytes and lymphocytes. Some members of the *Ehrlichiae* multiply in the granulocytes, but others prefer mononuclear white blood cells. In ruminants, *Cytoecetes phagocytophilica* and *C. ondiri* occur in granulocytes while *E. bovis* and *E. ovina* invade mononuclear cells.

As typical vector-borne diseases, ehrlichioses are partly transmitted through ticks of the genera *Hyalomma*, *Rhipicephalus* and *Amblyomma*. In BPF the transmission is still under discussion: biting flies and mites have been investigated and it is unclear whether several genera of arthropods are involved.

Occurrence, Epizootiology, Pathogenesis, Clinical and Pathological Features
The three most important *Ehrlichiae* infections occurring in the tropics are described below:

● *Bovine Petechial Fever (Ondiri Disease) (BPF)*

BPF is caused by *Cytoecetes ondiri* and occurs in the Highlands of Kenya at altitudes over 1500 m in imported cattle and their crosses with autochthonous breeds. It is assumed that the disease also is enzootic in similar altitudes of the neighbouring countries. The disease only occurs in cattle which are grazing in shady bushbuck habitats. The infection may appear sporadically during the whole year in particular in exotic breeds; Sahiwal are supposed to be especially susceptible. The vector has not been identified so far. The bushbuck (*Tragelaphus scriptus*) has been recognized as a reservoir of infection. Since disease outbreaks also occur after pastures have been kept unused over longer periods of time, it has to be found out whether domestic animals or game are the source of infection. Transmission is readily achieved experimentally when exotic cattle are injected with infected blood (Scott and Woldehiwet 1993).

The rickettsiae appear after the onset of the clinically apparent disease in the circulating granulocytes and monocytes; they multiply in the spleen and are carried with the blood into other organs.

After an incubation period of 7–21 days, the disease can show either an inapparent, subacute, acute or peracute course, the latter leading to a deadly outcome.

Acute reactions begin with high prodromal fever (41.6 °C) and a few other signs, as agalactia in lactating cows. The animals become dull, lag behind the herd and lie down instead of grazing. The next day, fine petechiae emerge in the hyperaemic visible mucous membranes of the naso-oral and urogenital tracts, and the conjunctivae. They are particularly common and prominent on the under-surface of the tip of the tongue. Within hours the fine petechiae enlarge and suffuse into ecchymoses. Although individual petechiae fade in about 48 hours, they are continually replaced and observed for up to 10 days. The fever lasts for 4 days and is accompanied by muscular tremors, depression, anorexia, staring coat and profuse serous lacrimo-nasal discharges. The blood is slow to clot. In some severe fatal cases the so-called "poached-egg eye" develops in which the conjunctival sacs are grossly swollen and everted around a tense, protuberant eyeball. Free blood appears in the tar-like faeces which smell fetid. The palpable lymph nodes can be slightly swollen. Affected animals cease to ruminate, grind their teeth, lose condition and stand motionless. The fatality rate can be up to 50%. If respiratory distress develops as a result of pulmonary oedema, the animals collapse and die in 2–3 days. The usual course is 4–11 days and recovery to full health is slow. Pregnant animals will abort (Scott and Woldehiwet 1993).

The gross pathology of BPF is characteristic: multiple petechial haemorrhages everywhere, profuse haemorrhages with gelatinous oedemas in submucosal and subserosal surfaces, subcutaneous and intramuscular tissues, in the serosa of the alimentary tract and also the cortex of the kidney and the respiratory tract; especially the wall of the gall bladder is severely affected. Cardiac splash haemorrhages appear sub- and endocardially; a mottled yellow appearance of the liver is typical in chronic cases. In the CNS, slight meningeal congestion and haemorrhages occasionally appear in the dura mater (Scott and Woldehiwet 1993).

- *Ehrlichia bovis Infection of Cattle*

Ehrlichia bovis is transmitted by ticks of the genera *Hyalomma*, *Rhipicephalus* and *Amblyomma*. The organisms occur as round or irregular shaped intracytoplasmic structures, ranging from 2 to over 10 µm in diameter, and consisting of numerous small granulations. The development cycle for *E. bovis* comprises inclusion bodies of 2.0–8.0 µm consisting of initial bodies of 0.2–0.8 µm in diameter.

E. bovis is found most numerously in the lungs, in the capillaries of the meninges, in the kidneys, the liver, the spleen, the circulating blood, the lymph nodes and other organs; it occurs in monocytes or macrophages but not in the endothelial cells of blood vessels (Uilenberg 1993).

The disease is reported from the Mediterranean, Africa and South America. In most countries, *E. bovis* is not considered as a cause of serious problems, while serious disease outbreaks in Cebu cattle have been reported from Mali, the Central African Republic, the Ivory Coast and Senegal. The Fulani call the disease "Nofel", meaning ear; it occurs mainly during the rainy season or the beginning of the dry season (Uilenberg 1993).

After an incubation period of about 14 days a longer lasting febrile phase appears which leads to dullness and loss of condition. The peracute course, however, can lead to a fatal outcome especially in calves. In the most characteristic acute form, one ear, less often both, is held down over the parotid region and the head is tilted towards that side. The animal shakes its head frequently, does not eat and is indifferent to its surroundings. Nervous symptoms similar to heartwater develop, but there are no terminal convulsions. Few animals recover from this form. Death occurs 12–48 hours after the first symptoms. Other clinical signs include swelling of the lymph nodes, mucosal congestion and lacrimation. Losses of 10 to over 20% have been reported in animals whose resistance has been lowered due to unfavourable climatic conditions or other causes.

The main macroscopical lesions are swelling of superficial lymph nodes, particularly the parotid gland and an increase in the cerebrospinal fluid. In fatal cases, the lesions can resemble those of heartwater (Mitscherlich and Wagener 1970, Uilenberg 1993).

- *Ehrlichia ovina Infection of Sheep*

Ehrlichia ovina is transmitted by the ticks *Rhipicephalus bursa* and *R. evertsi*; it multiplies like *E. bovis* in the monocytes or macrophages. The clinical symptoms and pathological features resemble those of the *E. bovis* infection but nervous symptoms and oedemas are more prominent. The disease has been reported from Algeria, Anatolia, Mali, Sri Lanka, Senegal, Sudan, South Africa and Namibia (Mitscherlich and Wagener 1970, Uilenberg 1993).

Diagnosis

The causal organism of BPF can be found with Giemsa-stained blood films and impression smears from a cut surface from the spleen taken within an hour or two after death. Since parasitaemia is on the wane when severe signs of illness have developed, blood or a suspension from the lung or spleen should be inoculated intravenously into susceptible sheep or cattle from which blood films have to be examined daily for 10 days (Scott and Woldehivet 1993).

E. ovis and *E. bovis* can be found by staining blood or organ smears with Giemsa as soon as the affected animal's temperature rises, but more readily so from the third day of fever (Uilenberg 1993).

Treatment

The ehrlichioses can be treated with oxytetracyclines, the treatment being most effective if it is carried out at the onset of the parasitaemia. In BPF rickettsial drugs are of little value when clinical signs already have appeared. Dithiozemi-carbazone has also been applied for the treatment of BPF but been withdrawn from the market because of adverse side-effects (Scott and Woldehivet 1993, Merck 1991).

Control

Systematic vector control can prevent the appearance of *Ehrlichia* infections. Apparently the best way to prevent the disease is to guarantee that the animals have been premunized naturally under the protection of the passive antibodies provided by their dams. The diseases, therefore, can be best controlled by an appropriate management of cattle within the production system. If the disease appears in a herd, multiple losses can be avoided by applying long-acting oxytetracycline to all exposed animals. No effective way of vaccination or artificial premunization is available at present other than artificial premunization as has been proposed for heartwater.

1.3.4. Infections Caused by *Anaplasmataceae*

Bergey (1984) groups the genera *Haemobartonella* and *Eperythrozoon* as well as *Anaplasma* to the *Anaplasmataceae*. In order to keep the present description clear those infections relevant in the tropics caused by the genera *Haemobartonella* and *Eperythrozoon* are discussed together in a separate chapter which is followed by the real anaplasmoses.

1.3.4.1. Infections Caused by Haemobartonella spp. and Eperythrozoon spp.

Three species of *Haemobartonella* are currently recognized, *H. felis* in cats, *H. canis* in dogs and *H. muris* in rodents; the classification of *H. bovis* remains in doubt.

From the genus *Eperythrozoon* 5 species are recognized: *E. suis, E. parvum, E. wenyonii, E. ovis* and *E. coccoides*. Furthermore *E. teganodes* and *E. tuomii* have been described (Kreier and Ristic 1984). Eperythrozoonoses with importance in the tropics are caused by *E. wenyonii* in cattle, *E. ovis* in sheep and *E. suis* in pigs. *E. parvum* causes a benign disease in pigs.

The organisms which are rather specific to each species of domestic animals can easily be demonstrated with the usual staining methods for blood smears. They appear as basophil, polymorphic (haemobartonellae are elongated, and eperythrozoon are round or ring-shaped) organisms which appear on the surface or at the edge of the red blood cells. They may surround the host cell completely but can also appear free in the plasma and phagocytized inside the macrophages. As typical rickettsiae, the organisms are surrounded by a simple membrane and composed of initial bodies which explains their polymorphism. According to the arrangement of the subunits organisms which are either elongated, rod-shaped, ovoid, round, or ring-shaped appear. The initial bodies of *E. parva* are supposed to be filterable. Inside their cytoplasm *Eperythrozoon* contain microtubuli, ribosomes and filaments but, like all rickettsiales, no nucleus.

Haemobartonellae and eperythrozoon are transmitted by arthropods, the precise range of vectors still being unknown.

Generally, haemobartonelloses and eperythrozoonoses are mild infections which present only a slight febrile reaction but lead to a more or less pronounced anaemia. Under challenge, a febrile icterohaemorrhagy with anaemia and haemoglobinuria can appear. If this will lead to an autoimmune anaemia or not is unclear. Only the malign eperythrozoonosis of pigs causes serious economic losses.

Table 32. Bartonelloses and eperythrozoonoses in domestic animals

Disease	Causal organism	Morphology	Occurrence	Clinical symptoms
Haemobarto-nellosis in cattle	*Haemobartonella* (*Bartonella*) *bovis* s.B. *sergenti*	Polymorph on the surface of erythrocytes	Algeria, Ruanda, Israel, Australia, USA, Mexico	Fever, anaemia, dyspnoea, constipation
Haemobarto-nellosis in buffaloes	*H. sturmannii*	Polymorph on the surface of erythrocytes	Israel	Febrile anaemia
Eperythrozoono-sis in cattle	*E. wenyonii*	Ring-shaped, coccoid, filamen-tous on erythrocytes	Africa, Near East, Japan, Australia, USA, Northern Europe	Slight anaemia
Eperythrozoono-sis in sheep	*E. ovis*	Round to cob-shaped, comma-shaped on erythrocytes	Worldwide	Remittant fever, anaemia
Malign epery-throzoonosis in pigs	*E. suis*	Ring-shaped, coccoid-ovoid on erythrocytes	Europe, USA, Africa	Febrile anaemia
Benign epery-throzoonosis in pigs	*E. parvum*	Ring-shaped, stretched on erythrocytes	Europe, USA, Africa	Symptomless or slight anaemia under stress

Haemobartonelloses and eperythrozoonoses are, however, important in the tropics because a superinfection with another vector-borne disease can be complicated by a haemobartonellosis and/or eperythrozoonosis. When animals are premunized artificially with blood obtained from uncontrolled donor animals superinfection with haemobartonella and/or eperythrozoon can occur.

The disease complex has to be taken into account especially in view of differential diagnosis. In Table 32, those haemobartonelloses and eperythrozoonoses which are of importance in domestic animals have been compiled.

Diagnosis
Haemobartonellae and eperythrozoon can be demonstrated in the red blood cells with the usual staining techniques, the May-Grünwald-Giemsa method being the most efficient. Acridine orange staining with ultraviolet microscopic inspection is also an effective means of demonstrating the organisms. Attempts at serodiag-nosis with the Coombs test and ELISA have been unsuccessful (Nash and Bobade 1993).

Treatment
Oxytetracycline is effective against *Haemobartonella* and *Eperythrozoon*. It can be used for the treatment and chemoprophylaxis of the infection and also for obtaining *sterilisatio magna*, if required. The same goes for imidocarb.

Control
Though the range of vectors is unknown, methods of vector control may assist in controlling the infection where required.

1.3.4.2. Anaplasmoses
(Anaplasmosis, Gallsickness, Anaplasmose)

The following species belong to the genus *Anaplasma* of *Rickettsiales* of the family *Anaplasmataceae*:

– *A. marginale*, which causes malign bovine anaplasmosis;
– *A. centrale*, which causes benign bovine anaplasmosis;
– *A. ovis*, which causes anaplasmosis of small ruminants.

Aetiology and Occurrence
As a typical member of the *Rickettsiales*, the *Anaplasma* consists of initial bodies 0.3–1.0 μm in diameter which in a number of 1–8 compose the inclusion body. The initial bodies pass through filter membranes of 0.65 μm pore size and are the actual infectious units. As inclusion bodies they are found in the red blood cells but can also appear free in the plasma and the bone marrow. Only the inclusion bodies can be demonstrated with the light microscope. The initial bodies are surrounded by a membrane thus forming the inclusion body. It is unclear whether it is produced by the organism or if it is a sequestrum of the red blood cell. The matrix which surrounds the organism can also be, at least partly, material from the erythrocyte. The initial body itself consists of DNA and RNA; the DNA is double stranded. An enveloping cover and an inner double plasma membrane have been demonstrated electron microscopically. The initial bodies are supposed to exist in two variants, with or without a central mass.

Malignant bovine anaplasmosis occurs worldwide in the tropics and subtropics. Because of the broad range of vectors and the difficulties of efficient vector control, the disease is an important economical factor even in countries with a developed veterinary service. Especially where animal production is supposed to be increased by importing exotic breeds, anaplasmosis may become the limiting factor of such a measure. About 300 million cattle worldwide are exposed to anaplasmosis. The losses attributed to bovine anaplasmosis in the United States have been estimated at US$ 100 million (Wanduragala and Ristic 1993).

A. centrale is of no importance as the causal organism of a serious disease. The worst thing which this organism, which is located centrally in the red blood cells, will cause is a mild anaemia. *A. centrale* has become important because it is supposed to produce a cross-immunity to *A. marginale* and therefore has widely been used for artificial premunization against *A. marginale*.

Anaplasmosis of small ruminants is a serious infection, especially in exotic sheep and goats in the Mediterranean, Near East and several African countries. Autochthonous breeds rarely show clinical symptoms.

Epizootiology
As with no other vector-borne diseases with anaplasmosis, the resistance of the breeds is inversely proportional to their level of productivity. Exotic dairy breeds especially show a high rate of mortality, in particular in Latin America. Though no antigenic variation has been demonstrated so far in precise scientific experiments (Callow et al. 1976), we have found that antigen made of *A. marginale* strains from the U.S.A. does not react with immune sera from animals which have recovered from anaplasmosis in Togo. It is unclear whether the much milder course of disease of animals in Africa in comparison with the mostly severe anaplasmosis in Latin America is the consequence of the different resistance of breeds or the different pathogenicity of the organism. It has been reported that taurine breeds are more susceptible than Cebus, but it has also been observed that no differences exist (Seifert 1971, 1995, Wilson and Ronhardjo 1983).

For the resistance of the animals to the organisms, it is important whether or not the calf has been allowed to become premunized under the protection of the passive antibodies of its dam or has been able to be premunized gradually through a field infection, and if its level of production allows it to keep its premunity in balance (I/3.1.2). There are many observations from Latin America which show that Cebu crosses (Pitangeiras in Brazil) also fall ill and die as early as 2–3 months of age because they have not got enough maternal antibodies after birth thanks to management failures (Seifert 1995).

For the potential of resistance of the respective animal population it depends whether the enzootic stability of the disease is kept in balance or is disturbed through the introduction of exotic animals or changes within the production system (III).

Vectors of *Anaplasma* are ticks, insects and vampires (*Desmodus* spp.) (Tables 33 and 34). Furthermore, anaplasmosis can be transmitted easily iatrogenically and through bleeding manipulations performed on the animal (castration, application of ear tags).

Transmission through ticks is supposed to take place

- transovarially and cyclically;
- interstadially, which seems to play an unimportant role since the interstages rarely feed on cattle;
- rarely through males.

Under experimental conditions, about 20 species of ticks have been shown to be capable of transmitting anaplasmosis (Wanduragala and Ristic 1993).

Table 33. Summary of insects which are supposed or known to transmit anaplasmosis grouped into the zoological system (Seifert 1971)

Suborder	Family	Tribe	Species
Nematocera	*Culicidae*	*Anopheles*	*A. earlei*
			A. crucians W.
			A. quadrimaculatus
			A. aegypti
			A. campestris
			A. cataphylla
			A. cinereus
			A. communis
			A. dorsalis
			A. fitchii
			A. flavescens
			A. hexodontus
			A. increpitus
			A. infirmatus
			A. melanimon
			A. niphadopsis
			A. punctor
			A. schizopinax
			A. sollicitans
			A. spencerii idahoensi
			A. sticticus
			A. taeniorhynchus
			A. vexans
		Culex	*C. erraticus*
			C. salinarius
		Culiseta	*C. inornata*
		Mansonia	*M. perturbans*
		Psorophora	*P. ciliata*
			P. columbiae
			P. confinnis
			P. cyanescens
			P. discolor
			P. ferox
			P. howardii
	Ceratopogonidae	*Culicoides*	*Culicoides* spp.
	Simuliidae	*Simulium*	*S. vittatum*
Brachycera	*Tabanidae*	*Tabanus*	*T. abactor*
			T. abdominalis F.
			T. americanus For.
			T. atratus F.
			T. equalis Hine
			T. erythraeus
			T. fumipennis
			T. fusciostatus
			T. lineola F.
			T. nigrovitatus
			T. oklahomensus
			T. proximus
			T. stygius

Table 33. (continued)

Suborder	Family	Tribe	Species
			T. sulcifrons
			T. subsmilis
			T. trimaculatus
			T. venustus
			T. wilsoni
		Chlorotabanus	*T. crepuscularis*
		Leucotabanus	*L. annulatus*
		Chrysops	*C. lasiophthalma*
			C. carbonarius
			C. discalis
			C. flavidus
			C. fulvaster
			C. mitis
			C. sequax
		Atylotus	*A. incisuralis*
		Hybomitra	*H. frontalis*
			H. fulvilateralis
			H. opaca
			H. repestris
			H. sonomensis
			H. tetrica
		Haematopota	*H. americana*
		Symphoromyia	*S. hirta*
			S. johnsoni
Cyclorrhapha	*Muscidae*	*Musca*	*M. autumnalis*
			M. domestica
		Stomoxys	*S. calcitrans*
		Haematobia	*H. irritans*
	Hypodermatidae	*Hypoderma*	*H. bovis*
			H. lineatum
Siphonaptera		*Tunga*	*T. irritans*

Anaplasma marginale bodies have been demonstrated in the gut contents and malpighian tubules of engorged *D. andersoni* nymphs, and morphologically distinct stages of the parasite have been demonstrated in the midgut of these ticks, suggestive of a developmental cycle in the vector. Infected adult ticks after 8 days of feeding had small numbers of *A. marginale* in the acinar cells of the salivary gland which coincided with a decrease in numbers in the gut, indicating that transmission occurred via the salivary gland (Wanduragala and Ristic 1993).

The knowledge of these mechanisms is important when carrying out prophylactic measures. Areas infected with ticks are free of anaplasmosis after 6 years at the earliest when rotating grazing is applied and all natural tick hosts are eliminated. This is quite different from the system of controlling babesiosis (Argentina). If two- and three-host ticks are prevalent, one has to take into

Table 34. Transmission of *A. marginale* by ticks and the occurrence of ticks as vectors of anaplasmosis (Seifert 1971)

Vector	Number of hosts	Occurrence
Amblyomma americanum	3	Southern USA, Central America, Northern South America
A. cayennense	3	Southern USA, Central America, Northern South America
A. maculatum	3	Southern USA, Central America, Northern South America
A. ovale	3	Southern USA, Central America, Northern South America
Argas persicus	Many	CIS, Australia, China, India, Iran, most parts of Africa and USA
Boophilus annulata	1	USA, Mexico, West Coast of South America
B. australis	1	Australia, Brazil
B. calcaratus	1	CIS
B. decoloratus	1	Most parts of Africa
B. microplus (s. *fallax*)	1	East and South Africa, Mauritius, Madagascar, Australia, Eastern South America, Middle America, Florida
Dermacentor albipictus	3	USA
D. occidentalis	3	USA
D. variabilis	3	USA
D. venustus (s. *andersoni*)	3	USA
Haemaphysalis cinnabarina punctata	3	Bulgaria, Yugoslavia, Portugal, Turkey, CIS, Northern Africa
Hyalomma lusitanicum (s. *anatolicum*) (s. *excavatum*)	3	Northern Africa, Portugal, Turkey, CIS, Near East
H. aegypticum	3	Northern Africa
H. dromedarii	3	CIS, Afghanistan, India, Indonesia, Near East, Northern-, East Africa
H. scupense	1	CIS
Ixodes	3	Bulgaria, Yugoslavia
I. ricinus	3	Turkey, CIS, China, Algeria, USA
I. scapularis	3	USA
Ornithodorus lahorensis		USA
Rhipicephalus bursa	2	South-East Europe, Portugal, CIS, Near East, Northern Africa
R. sanguineus	3	South-East Europe, Portugal, CIS, Asia and Near East, most parts of Africa, USA
R. simus	3	Most parts of Africa
R. turanicus	3	CIS
R. evertsi	2	Arabia, Central-, East- and South Africa, Madagascar, Mauritius

account that these vectors can be brought back into cleaned areas out of infected regions, through birds for instance.

Insects as vectors of anaplasmosis are much more important than it has been

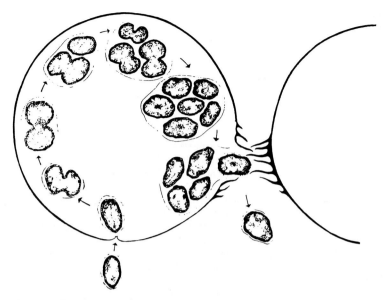

Figure 47. Development of *Anaplasma marginale.*

1 Initial bodies (IB); 2/3 Binary division of IB; 4/5 Non-mature IB; 6 Mature IB; 7 Distribution of IB through microfibrilles and blood plasma, transfer to another erythrocyte (Ristic 1960, cited from Seifert 1971).

thought previously. This is especially important since methods of controlling ticks do not usually affect insects. Haematophagous flies in particular are important vectors of anaplasmosis in intensive cattle production systems (Seifert 1971).

Most domesticated and wild ruminants are apparently a reservoir of infection for *Rickettsiales* of the genus *Anaplasma*. In America, deer are considered to be carriers of *Anaplasma*. Under the conditions of cattle production in South America, the importance of deer as a reservoir of infection is considered to be of little importance since deer and cattle rarely graze together on the same pasture, which makes the acyclic-mechanical transmission through insects almost impossible (Seifert 1971).

Reports about cases of anaplasmosis in young calves which have been kept in cowsheds and have not had an opportunity to be naturally infected by their dams suggest the possibility of intrauterine infection (Salabarria and Pino 1988).

Pathogenesis
The initial body as an infectious unit is able to penetrate the membrane of the erythrocyte from inside and outside in order to invade new erythrocytes. The initial bodies multiply by dividing and can remain for certain time in the shape of a dumbbell. Fully developed inclusion bodies only appear at the end of the incubation period (19–21 days) which then can be demonstrated with the light microscope. Prior to that, mixtures of naked initial bodies and complete inclusion bodies may appear in the erythrocytes (Fig. 47).

The metabolic rate of the blood of an artificially infected animal at the peak of infection with *A. marginale* becomes doubled; the intake of glycine by the erythrocytes can be increased 100-times. The erythrocytes are destroyed by the breakdown of the transport of cations as well as by disorders of the membrane permeability because they are unable to maintain their shape and volume. This leads to a pronounced anaemia.

Beside the direct destruction of the red blood cells caused by the *Anaplasma*, an autoimmune anaemia develops during the course of infection. Thus an anaemic phase of the disease may develop during convalescence which is not accompanied by parasitaemia. Through the erythrophagocytosis triggered initially by the parasites the immune system has apparently been stimulated to produce haemagglutinins which are able to destroy non-infected erythrocytes through opsonization. This phenomenon has to be taken into account when methods of artificial premunization are applied which use virulent or live attenuated *A. marginale* as an antigen. In these cases, a reaction of the immune system of the infected animal will also appear and lead to autoimmune anaemia and thus to a reduction of production level. It is, however, supposed that the autoimmune immunity in anaplasmosis is a temporary process which will disappear after complete recovery of the animal.

The immune response to the *Anaplasma* infection is both humoral and cell mediated. Though antibodies alone apparently cannot protect animals against anaplasmosis, they are able to neutralize the infectiousness of the organism *in vitro*. Whether a similar situation can occur *in vivo* is unclear. Removal of parasites from the circulation takes place by phagocytosis of the entire infected erythrocyte. If the initial bodies are not exposed to the antibody, and parasite antigens are not present on the infected erythrocytes, neutralization, lysis or antibody-mediated phagocytosis might not occur. Autohaemagglutinins and opsonins directed against the erythrocyte membrane may be responsible for the erythrophagocytosis observed in the bone marrow of both infected and noninfected erythrocytes.

Calves immunized with a killed vaccine showed leukocyte migration inhibition (LMI) which appears to correlate with protection. Animals with an early LMI response show less severe clinical signs. The LMI response to an inactivated antigen is less than to a live antigen (Wanduragala and Ristic 1993).

Repeated treatment with tetracyclines can lead rather quickly to a *sterilisatio magna* which has the consequence that a sterile immunity will persist only for a limited time and the animals become susceptible to a new infection afterwards. In contrast, Wright (1990) suggests that animals which have recovered from an infection with *A. marginale*, and have been treated in such a way that the organism have been eliminated, have become immune to a homologous field infection and will only fall ill with slight symptoms after a heterologous infection. He cites the work of Magonigle and Newby (1984) who have demonstrated an immunity lasting 8 months after an infection with *A. marginale*. The responsible immune mechanisms are described as follows:

- blockage by the antibodies of the site where *A. marginale* inclusion body binds to the surface of the red blood cell;
- lysis of the initial bodies through antibodies and/or complement;
- phagocytosis and intraerythrocytal destruction of the initial bodies and/or infected erythrocytes. There are signs that macrophages and/or T-helper-cells assist in the immune defence (Seifert 1971, Wanduragala and Ristic 1993).

Clinical Features
Anaplasmosis can have three typical courses:

- **peracute** in animals of an age of 12–36 months or also older imported animals which have not had a chance to become premunized as calves under maternal passive antibody protection and have been exposed to sudden massive infection. The course of the disease is patent, often with CNS symptoms which resemble rabies; the animals are usually found when already in agony. In some cases, the original high body temperature has already decreased to subnormal levels. Extreme anaemia is always present;
- **acute** in animals of any age group with the typical clinical picture of anaplasmosis. After an incubation period of about 3 weeks, high fever and gradually increasing anaemia develop. The clinical signs are those which are attributable to severe anaemia. They include depression, weakness, high body temperature, laboured breathing, inappetence, dehydration, constipation and abortion in pregnant cows; in bulls abnormal sperm morphology and transistory testicular degeneration will appear. During the febrile stage there is suppression of rumination, dryness of the muscles, loss of appetite, dullness and depression. There is frequent urination, but the urine is not coloured. After about one week after the onset of the clinical symptoms – high fever, the animal showing a decreasing level of red blood cell count, the temperature returning to normal and becoming subnormal – the disease will lead to a fatal outcome if no treatment is applied;
- **chronic**, after recovering from the acute course, or in relatively resistant young or autochthonous animals. The chronic course, which often can take months, may lead to erroneous diagnosis since the symptoms are not characteristic and the causal organism will only seldom be found microscopically. Chronic anaplasmosis should always be considered if a pronounced anaemia and slight fever appear and no other blood parasites or massive endoparasitic infection are present. The "brisket disease" described in the Andes of South America could partly be recognized as chronic anaplasmosis. The chronic course of disease can also be confused with plant poisoning which can affect erythropoiesis (Seifert 1971).

After recovery from an infection with *A. marginale* without treatment, a long-lasting premunity will appear which may persist during the remaining life of the animal. Such animals will rarely recover their previous level of production (Wilson and Ronhardjo 1983). Feed-lot animals especially should be slaughtered as soon as possible. The persisting autoimmune anaemia with the low red blood cell count is certainly responsible for this problem. In contrast, animals which soon recover their previous level of red blood cell count will have a favourable prognosis.

Pathology

The macroscopic lesions found *post mortem* are those associated with acute anaemia. The blood is watery. The mucous membranes show a pronounced pallor and are rarely icteric. The spleen is often greatly enlarged and its cut surface shows dark pulp and enlarged splenic follicles. The liver can be enlarged with rounded borders and the cut surface appears brownish-yellow to sulphur-yellow. The gall bladder is enlarged and abstracted with dark, thick bile (gallsickness). Petechiae can be observed on the epicardium, pericardium, pleura and diaphragm. The lymph glands are enlarged.

Microscopic hyperplasia of the bone marrow, decrease of lymphoblasts in the spleen and increased vacuolation and degeneration of reticular cells, reduction of the white pulp and hepatic changes which consist of leukocytic infiltration of Glissons capsule, dilation of portal veins and swelling of endothelial cells are all evident.

The haematological picture is characterized by the extreme anaemia, the erythrocyte count can drop below 2 million and the haemoglobin content below 3 g/100 mL of blood. Erythrocytes show increased osmotic and mechanical fragility along with anisocytosis, poikilocytosis and polygromasia. Leukocytosis and bilirubinaemia appear during the acute stage of disease (Wanduragala and Ristic 1993).

Diagnosis

The case history is important for the diagnosis of anaplasmosis. Information about breed, origin of the animals, season and weather during the outbreak of disease, history of the disease within the production system, changes in the pasture management, production and reproduction status of the animals as well as previous sanitary measures (vaccinations) can be important for diagnosing the disease.

The causal organism can be demonstrated directly, either

– microscopically,
– or by infecting susceptible (splenectomized) animals.

The diagnosis by microscopic examination of thin blood smears stained with Giemsa, May-Grünwald or Wright, though convenient, becomes less reliable where dye deposits may be mistaken for *A. marginale*. The slides have to be prepared carefully and the dye-solutions buffered appropriately. With these conventional staining methods the *Anaplasma* inclusion bodies appear as a dense, round, blue mass of varying size (0.8–0.9 μm diameter) at the margin of the erythrocyte whereby the cell wall characteristically bulges out. In this way the organisms are only to be found during the acute stage of disease from day 10–20 (Fig. 48 on page 1 annex).

Use of acridine orange stain which reacts with the DNA of the *Anaplasma* or the nucleic acid of immature erythrocytes and observation with ultra violet light can improve detection. With this method, the *Anaplasma* inclusion bodies appear as bright orange coloured points at the margin of the slightly green erythrocyte. Even animals in the chronic course of disease can eventually be detected with this method. After haemolyzing the erythrocytes, the *Anaplasma* bodies can be recognized with the phase contrast technique (Seifert 1971). Direct immuno-fluorescence is another method for demonstrating *Anaplasma* bodies provided conjugated immune serum is available.

If it is important enough to demonstrate that anaplasmosis is prevalent in a herd, the infection of splenectomized calves with suspension either of erythrocytes, spleen or liver tissue or samples obtained by puncture of the bone marrow may become necessary. The organisms then can be found microscopically after only 10 days p.i. in the febrile phase of infection (Seifert 1971).

Indirectly, anaplasmosis can be detected through the presence of antibodies. The most widely used serological test is the indirect fluorescence antibody test (IFAT) which has, however, the drawback of non-specific reactivity. Three other tests, complement fixation (CFT), capillary tube agglutination (CTA) and card agglutination (CAT) which use relatively crude antigens have been used successfully worldwide. The CAT has special advantages for testing large amounts of animals under field conditions (Seifert 1971). Now ELISA, Dot-ELISA and radio-immunoassay are also being applied successfully. Nowadays, DNA probes are used in hybridization assays to diagnose *A. centrale* and *A. marginale* infections in cattle and in tick midgut tissue. A RNA probe developed from a fragment of the gene coding for a surface protein was able to detect infections in carrier cattle with low parasitaemia (Wanduragala and Ristic 1993).

Treatment

As a typical representative of rickettsiales, the *Anaplasma* is sensitive to tetracyclines. Since introducing tetracyclines during the 50's into the treatment of anaplasmosis, these drugs together with imidocarb have replaced the previously used arsenicals and antimalaria drugs. For the treatment of acute anaplasmosis, chlortetracycline is recommended in a dose of 5.0–10.0 mg/kg b.w., oxytetra-cycline in a dose of 4.0–6.0 mg/kg b.w. and imidocarb at a dose of 5.0 mg/kg b.w. Severe cases have been treated successfully by applying the tetracycline pre-paration REVERIN intravenously. Chronic anaplasmosis can be treated and/or the carrier state be eliminated by applying chlortetracycline 3.0–10.0 mg/kg b.w., or 22 mg/kg b.w. on 5 consecutive days, or the administration of 2 doses of long-acting oxytetracycline 7 days apart, and also by feeding chlortetracycline at 1.0 mg/kg b.w. for 120 days or 2.0 mg/kg b.w. oxytetracycline over 2 weeks (Seifert 1971, Wanduragala and Ristic 1993).

Control

As with no other vector-borne disease, the structure of the production system and its management methods are relevant for deciding which scheme of control can be applied to prevent anaplasmosis (III). Within an intensive production system, anaplasmosis can be eradicated by detecting the carrier animals with an appropriate serological method and eliminating the carriers through treatment as described above. At the same time vector control has to be carried out systematically. Prerequisite is of course the existence of an efficient animal health control infrastructure. As soon as the infection in the stock has been eliminated, animals which are bought are controlled accordingly and if vector control is carried out appropriately, the ranch in question will remain free of infection. It is not necessary to carry out regional campaigns; it is possible to maintain control even in single ranches (Seifert 1971) (I/4.3). Wherever expensive highly productive animals represent a large amount of capital, this alternative should be preferred instead of trying to protect the animals through immunoprophylactic measures which by all means will lead to a reduction of the production level, perhaps through autoimmune anaemia. Efficient and reliable methods for the diagnosis and treatment of carriers are available (see above).

Under the conditions of an extensive production system with autochthonous or adapted breeds, a state of enzootic stability of anaplasmosis should be maintained. This can be best obtained if calves are allowed to contract a mild field infection between birth and weaning during the dry season when vector activity is reduced. Thus a challengable premunity will develop under the protection of the passive maternal immunity (III/3).

Wherever both of these alternatives can or should not be applied, two other schemes of control of anaplasmosis may be considered:

- chemoprophylaxis by periodically applying chemoprophylactica (LA tetracycline). Because of the contamination of the animal product with the chemoprophylactic, this method can only be applied when it is guaranteed that the compound will be withdrawn in time before the animal product is brought to market (feed-lot);
- immunoprophylaxis.

Chemoprophylaxis
Chemoprophylactic methods for the control of anaplasmosis can be applied in order to

- eliminate carrier animals within a scheme of eradicating anaplasmosis from the production system;
- remove carriers from animals which are brought into the production system (feed-lot);
- protect animals temporarily, e.g. during transportation while staying in the feed-lot or while on exhibitions, against infection.

Chemoprophylaxis can be carried out either through a single application of a high dose of long-acting oxytetracycline or imidocarb (both 10 mg/kg b.w.) or by feeding. With oral application carriers can be eliminated with 2 mg/kg b.w. oxytetracycline/day in 16 days and 10 mg/kg b.w. chlortetracycline/day in 20 days.

Immunoprophylaxis

It is well known that animals which recover and become premune are immune to a challenge with a homologous strain. This knowledge has led to the practice of premunization of cattle which is carried out in quite a number of both rather crude and more sophisticated ways, and is nowadays gradually being replaced by methods of vaccination with non-living antigens. According to the present state of knowledge, the following methods for the immunoprophylaxis of anaplasmosis are applied (McHardy 1983, Seifert 1992, Wanduragala and Ristic 1993, Wright 1990):

- Premunization with a minimal infective dose: calves can be infected with a dose of 1×10^4–1×10^6 kryopreserved *A. marginale* organisms without presenting clinical symptoms. The infective dose depends on the *Anaplasma* strain, the breed and age of the animal and the local condition, therefore, it is difficult to standardize. This method is not widely used under practical conditions.
- Infection with *A. centrale*: the artificial premunization with *A. centrale* has been used in order to protect animals against clinical symptoms of the field infection with *A. marginale* especially in Australia and South Africa, less so in Israel, and also lately in South America. *A. marginale* and *A. centrale* are supposed to have a common antigenic fraction. Since *A. centrale* produces an infection with only mild clinical symptoms in cattle, the effort is made in this way to produce a persisting premunity which will protect the animal against a field infection with a pathogenic field strain of *A. marginale*. As a vaccine, cooled or deep-frozen blood of those donor animals which are guaranteed to be free of other blood infections is used. Not only is the *A. centrale* premunity supposed to protect against clinical symptoms of the *A. marginale* field infection, it is also supposed to reduce the anaemia which is the consequence of the *A. marginale* infection (Abdala et al. 1990, Anziani et al. 1987, Wright 1990).
- Infection with attenuated *A. marginale* strains: through the application of an *A. marginale* strain which has been attenuated through passages in deer and radiation, a premunity can be produced which in a considerable number of animals can prevent clinical symptoms after a super-infection with the field strain. The stabilates have to be kept deep-frozen and can only be thawed shortly before application. It is difficult to apply such a system under field conditions in developing countries. An alternative is to infect sheep or goats with the vaccine strain and to vaccinate the animals with the blood of these donor animals. It has, however, to be guaranteed that these donors are free of other infections (Ristic and Sibinovic 1968, cited from Ristic 1968).
- Infection with virulent *Anaplasma* strains in combination with chemo-prophylaxis: since it is well known that animals which have recovered from anaplasmosis become premune, it has become a worldwide practice, which is especially applied in Latin America, to provoke an artificial infection and control it through treatment. This technique is applied in a number of variations and often combined with artificial premunization against babesiosis.

Quite often this crude method of premunization leads to losses and at the very least to permanent impairment of the health of the "vaccinated" animals. Either the stabilates used for premunization are prepared carelessly when a tick tissue suspension taken from a tick which is thought to be infected, or blood of animals from the field which are also presumed to be infected is used. Accompanying infections with other organisms of vector-borne diseases or even viruses are often provoked. Ordinarily, the infected animals are treated after the first clinical symptoms appear, usually on the 19th or 29th day p.i. with a low dose (2.5 mg/kg b.w.) of imidocarb or tetracycline, and the treatment is repeated until the febrile reaction disappears. It remains unclear here whether a *sterilisatio magna* and only a short lived premunity is produced. One always has to take into account that a pronounced auto-immune anaemia will be provoked which will reduce the productivity of an exotic animal considerably.

– Immunization with inactivated antigen: stabilates containing inactivated *Anaplasma* organisms either bound to red blood cells or free of cells have been used for years as antigens to immunize against anaplasmosis. If components of the red blood cells are contained in the vaccine, isoantibodies can be produced by the vaccinated animal which can lead to allergic reactions. Nowadays, *Anaplasma* organisms are cultivated on tissue cultures and *Anaplasma-albumin*-complexes are produced which induce antibodies to act against *A. marginale* in the vaccinated animal. An oil adjuvant vaccine made of lyophilized whole cell *A. marginale* organisms, called ANAPLAZ, has been marketed for years in the U.S.A. Two doses of the vaccine are administered subcutaneously at 4–19 week intervals. It produces a limited protection against the heterologous field infection. Since it also contains parts of the erythrocytes as contaminates, it produces anti-erythrocyte-isoantibodies which provoke erythrolysis in vaccinated calves. The vaccine has hardly been used outside the U.S.A. Another vaccine recently made available commercially (AM-VAX), contains purified initial bodies of *A. marginale* and is devoid of erythrocyte components that contain blood group antigens. Furthermore, the cattle utilized for vaccine production are screened for blood groups in order to select animals with less common blood groups. Several other killed vaccines consisting of isolated *A. marginale* initial bodies have been used.

In the meantime, synthetic antigens have been produced which possess epitopes of *A. marginale* and *A. centrale*, but also of different pathogenic field strains of *A. marginale*. In the U.S.A. the antigens AM 105 and 106 have been produced and similar activities are also in progress in Australia (CSIRO) (Wright 1990).

As described in detail under I/3.1.2, the sterile immunity initiated by inactivated vaccines lasts in diseases caused by organisms which produce a premunity for a rather short time. An antibody level which protects against the clinical symptoms of the field infection can only be obtained with regular boosting. Since such a scheme would be too difficult to manage and too expensive, the expectation is to obtain a natural premunity through field infection while the animal is protected by the sterile antibodies produced by the inactivated antigen. However, one has to take into account that such a premunity which finally appears is also a challenge for the organism and has the respective consequences.

All methods of immunoprophylaxis of anaplasmosis contain the hazards and problems already described under I/2.2.3.3 and 3.1.2, because

- no method is able to protect the animal against the super-infection with the field strain, and thus the animal remains exposed to the autoimmune anaemia which eventually appears, as well as to an acute infection which can develop under stress;
- the artificial distribution of *A. marginale* increases the challenge of infection by enhancing the amount of organisms in the vector and reservoir of infection (carrier animal), thus changing the epizootiological balance in favour of the pathogen;
- through wrong or lacking diagnosis, the infection may be introduced into areas which have been previously free of anaplasmosis.

1.4. Vector-borne Diseases Caused by Bacteriae[2]

• *Avian spirochaetosis*
(Avian borreliosis)

Aetiology and Occurrence
Avian spirochetosis is caused by *Borrelia anserina* (s. *Spirochaeta anserina*, s. *gallinarum*, s. *marchouxis*, s. *anata*) and is only transmitted by ticks of the genus *Argasidae*.

The causal organism, a blood inhabiting, actively mobile spirochete, *Borrelia anserina*, is 0.02–0.3 µm wide and 8.0–20.0 µm long and consists of 5–8 loosely arranged coils which in older cultures can also appear coccoid in shape. No reliable data are available concerning *in vitro* cultivation. It can be propagated in embryonating duck or chick embryos, or young ducks or chicks. Diverse immunological and serological types of *B. anserina* have been found in many areas.

The disease occurs worldwide, but generally in temperate or tropical regions, wherever the vectors are found, especially in the Mediterranean, Africa, South America and U.S.A. as well as in Asia.

---

[2] Consequent logical systematically vector-borne diseases caused by bacteria should have been grouped prior to those caused by *Rickettsiales*, but because of didactic reasons they come afterwards, since as in vector-borne diseases caused by viruses, a sterile immunity appears in animals which recover from the disease.

Epizootiology, Pathogenesis, Clinical and Pathological Features
Vectors of the disease are *Argas persicus*, *A. reflexus A. miniatus* and *A. sanchezi*. *Ornithodorus moubata*, principally a tick of mammals, is also supposed to be a vector. The relationship between pathogen and vector is cyclical-transovarial, the mode of transmission alimentary. The borreliae ingested by the tick migrate through the gut about 14 days p.i. into the salivary gland where they can persist for 6–7 months. The ticks do not get rid of the infection after a blood meal, so this allows the transovarial transmission. Other vectors than ticks, lice, mosquitoes and inanimate objects can transmit the spirochete mechanically. The infection is supposed to be possible through ingestion of infected feed or infected ticks but this mode of transmission seems to be only of minor importance.

The incubation period is 4–9 days during which the organisms multiply in the liver, spleen and bone marrow and afterwards in large amounts in the blood. The disease is acute, and the mortality rate can be 100% in a few days. In chickens, signs are usually less dramatic, and mortality is about 10% during the clinical course of the disease. Some affected quail or chickens may die without obvious signs of disease or weight loss. Typical clinical features are characteristic for septicaemia: high fever, anaemia, weakness, paralysis, lethargy and death after 4–6 days. Infected birds discharge characteristic droppings that are streaked with urates surrounded by a watery ring. Chronically affected birds are listless and anorectic; they appear hunched-up with their necks retracted and their eyes partially closed.

In contrast to the vector-borne diseases described previously, animals which recover from the disease obtain a challengable sterile immunity, but only against the homologous type of borrelia.

The primary lesions are found in the lower third of the small intestine, caeca, and liver. Lesions in the intestine and caeca vary from punctate haemorrhages to ulceration. The well-defined ulcers vary in size and may be 5 mm in diameter. The larger ulcers may show yellow diphtheric membranes with a depressed centre and raised edges. Perforating ulcers are frequent and cause local or diffuse peritonitis. Liver lesions appear as yellow isolated foci or irregularly shaped yellow areas in the parenchyma. The only other organ that may show lesion is the spleen; it may be enlarged and either haemorrhagic or necrotic.

Diagnosis
The organisms can easily be demonstrated microscopically with the usual staining techniques in the blood of sick or dead animals. Borreliae may also be isolated in embryonating duck or chick embryos and the infection can be demonstrated serologically.

Treatment and Control
Several chemotherapeutic agents are effective. The most widely used are penicillin derivatives, but streptomycins and tetracyclines are also effective. The antibiotics can be completely efficacious if the regimen is instituted when the number of spirochetes per oil-immersion field is low or moderate; however, if large numbers

of spirochetes are present in the blood stream when chemotherapy is begun, the sudden liberation of large quantities of spirochetal degradation products can result in more deaths than no treatment at all would.

Control must first be directed against the biological vector. *Argas* ticks are notable for their long life span, ability to survive for extended periods without a blood meal, efficiency in transmitting the spirochetes, and an ability to remain securely hidden in cracks and crevices often beyond the effective reach of pesticides. Accordingly, control is difficult. A combination of tick eradication and immunization offers the most effective means of control.

Immunization can be highly successful and, next to eradication of the biological vector, is the preferred method of control. Bacterins prepared from infective blood have been used with success. The most widely used bacterins are egg-propagated products composed of yolk material containing the spirochetes, but also whole-egg propagated bacterins have been used successfully; usually 1–2 i.m. injections suffice. Formalin (0.2%) is mostly applied to inactivate the spirochetes. Extreme care must be employed in using the appropriate serotype(s) of the spirochete in any given locality. Little if any cross-protection is afforded to different serotypes (Merck 1991, Mitscherlich and Wagener 1970).

1.5. Vector-borne Diseases Caused by Viruses

The virus infections presented in the following are epizootiologically true vector-borne diseases, i.e. with few exceptions the transmission is only feasible by means of a vector which is able to take up the organism from the system of a carrier and to implant it into the system of a susceptible organism (I/4). Thus the diseases presented in this chapter may not only be controlled by eliminating the specific vector but also in contrast to vector-borne diseases caused by *Protozoa* and *Rickettsiales* much better through vaccination with live or inactivated vaccines since the respective viruses do produce an active sterile immunity. In Table 35, those vector-borne diseases in domestic animals caused by viruses which are of economic importance in the tropics are compiled.

1.5.1. Arboviroses

Characteristically, *Arboviridae* can only be transmitted through an arthropod ("arthropod-borne"). The *Arboviridae* belong partly to the *Togaviridae*; these are the smallest enveloped RNA viruses. They are divided into *Alphaviridae* (previously arbo-A-viruses) and *Flaviviridae* (previously arbo-B-viruses). *Alphaviridae* have a diameter of 45–75 nm and a density of 1.25 g/mL; they can be transmitted by mosquitoes (*Nematocera*). All viruses grouped in this family are related antigenically and do not have any serological relationship to other *Togaviridae*. Alphaviruses mainly cause encephalitides in humans and animals; important pathogens of tropical diseases like those of the American equine

Table 35. Vector-borne virus infections of domestic animals[a]

Disease	Affected species	Systematic grouping of virus	Vector	Occurrence
Arboviridae				
Horse sickness	Equines	*Reoviridae* G. *Orbivirus*	*Culicoides* spp. *Aedes aegypti Culex pipiens*	Africa
Bluetongue	Ruminants	*Reoviridae*	*Culicoides* spp. *Aedes* spp. *Melophagus ovinus*	Worldwide
Rift Valley fever	Ruminants	not classified	*Aedes* spp *Culex* spp.	Eastern/ southern Africa
Nairobi sheep disease	Small ruminants	*Bunyavirus*	*R. appendiculatus/bursa, A. variegatum*	East Africa, Zaire
American horse encephalitides	Equines	*Togaviridae* G. *Alphavirus* type WEE type EEE type VEE	*Culex tarsalis* *Culiseta melanura* *Culex, Anopheles, Mansonia, Aedes spp.*	America
Crimean-Congo haemorrhagic fever	Ruminants	*Flavivirus*	*Hyalomma* spp.	Africa, East Europe, Asia
Akabane disease	Ruminants	G. *Alphavirus* *Simbu virus*	*Culicoides* spp.	Israel, Australia, Japan
Bovine ephemeral fever	Cattle	*Rhabdo* virus	*Culicoides* spp.	Africa, Australia, Asia
Wesselbron disease	Sheep	*Flavivirus*	*Aedes caballus* *A. circumluteolus*	South Africa, southern Africa
Rabies virus				
Rabies	All vertebrates	*Rhabdovirus*	Wild and domesticated carnivores, vampires	Worldwide

[a]Though African swine fever has the characteristics of a vector-borne disease when transmitted by *Ornithodorus moubata porcinus* from wild to domestic pigs it is not classified to the complex of vector-borne diseases because it is also transmitted like a contact disease, thus the particular characteristics of the vector-borne disease complex are irrelevant for control measures of African swine fever

encephalomyelitis belong to this group. Flaviviruses have a diameter of 25–70 nm and a density of 1.25 g/mL. They are transmitted partly by mosquitoes, partly by ticks. All species grouped in this family are related antigenically and can be serologically differentiated from other *Togaviridae*. Important pathogens of tropical diseases of humans and animals belong to *Flaviviridae*, i.e. yellow-fever, dengue, louping ill, Wesselbron-virus infection. Because of morphological properties, *Rubivirus* and *Pestivirus* are also grouped into this family; they are, however, transmitted through contact.

A further family of *Arboviridae* are the *Bunyaviridae*. They represent the previously called *Bunyavera* subgroup of arboviruses. Approximately 130 species of viruses divided into 30 serodemes belong to this largest group of animal viruses which occur especially in the tropics in domestic and game animals.

Rhabdoviridae can also be transmitted by arthropods and other vectors. They are joined systematically because of their rod-like quiver shape (Gk. *rhabdos* = rod). They are single stranded RNA viruses with a helical nucleocapsid which has a diameter of 15 nm; it is screw-like and forms a hollow cylinder. On their surface, *Rhabdoviridae* carry a seam of projections, their density being 1.2 g/mL. The multiplication of virus takes place inside the cytoplasm, it has a cytopathogenic effect and produces cytoplasmic inclusion bodies. The causal organism of ephemeral fever and rabies belong to this group (Rolle and Mayr 1993).

1.5.1.1. *African Horsesickness (AHS)*

(Pestis equorum, Peste africaine du cheval, La peste equine, Peste africana del caballo, Peste equina, Pferdeziekte, Paardenziekte, Perdesiekte, Pferdesterbe)

Aetiology and Occurrence
AHS is caused by a double-stranded RNA virus which belongs to the genus *Orbivirus* from the family *Reoviridae*; it shares many properties with other orbiviruses such as bluetongue and the equine encephalosis virus. The virions are 70–80 nm in diameter and have an icosahedral symmetry. They contain 10 double-stranded RNA genome segments encapsulated within a double-layered capsid made up of 32 capsomeres, each comprising 7 structural proteins.

9 serotypes of AHS virus are known which have no significant intratypic variation; cross-neutralization between serotypes 1 and 2, 3 and 7, 5 and 8 and 6 and 9 occurs.

The virus is relatively impervious to heat: it does not become inactivated after exposure to temperatures of 55–75° for 10 minutes and persists for 3 months at 4 °C, but at –25 °C a marked reduction in virus titre occurs if the virus is not diluted in a stabilizer. Minimal loss of titre occurs in lyophilized vaccine kept at 4 °C. Putrid blood remains infective for more than 2 years. The optimal pH for virus survival is 7.0–8.5.

The disease has been around for centuries in Africa. It was mentioned in an Arabic scripture as early as 1327. During the past centuries, epizootics have occurred regularly. The main enzootic area has always been on the African continent from South Africa, where it appears in the northern parts of Transvaal and Natal every summer, up to as far as 5° North. From there the disease has spread repeatedly to the Near East and the Mediterranean. Outbreaks of AHS have also appeared in the CIS and on the Indian continent. In 1966 AHS reached Spain from Morocco being quickly terminated, but in 1987 and 1990, AHS occurred again in Spain. The source of infection was suspected to be zebras imported from Namibia; in 1989 it spread to Portugal.

Epizootiology

AHS primarily affects equines, horses being the most susceptible (mortality rate 70–95%), mules less so (50–70% mortality rate), while donkeys and zebras are very resistant, most infections being subclinical. There are, however, autochthonous horses which have been in North and West Africa since at least 2000 BC which have apparently acquired natural resistance to AHS. In addition to equines, dogs can contract a highly fatal form of AHS. It is doubtful whether they are reservoirs of infection since *Culicoides* do not feed on them.

Foals born to immune mares acquire passive immunity by the ingestion of colostrum soon after birth.

The virus is transmitted mechanically by *Culicoides* spp., *C. imicola* being the most significant vector in Africa. Whether the mode of transmission is cyclical or not is not yet clear. *Culicoides* midges transported by wind are supposed to be responsible for the spread of the disease over the sea, i.e. from Turkey to Cyprus in 1960 and from Morocco to Spain in 1966. The disease can also be carried over long distances if infected vectors are transported in vehicles, even by aircraft. The role of other insects like *Anopheles stephensi*, *Culex pipiens*, *Aedes aegypti*, biting flies (*Stomoxys calcitrans*) and tabanides as vectors of AHS remains doubtful; they may play a minor role. The same is true for ticks. The incidence of disease increases at the onset of the rainy season when the conditions for the multiplication of the vectors improve.

There is no confirmed information about a reservoir of infection; animals recovered from the disease do not remain carriers of the virus. There may, however, be a continuous transmission cycle of AHS virus in tropical areas of Africa between *Culicoides* midges and wild or domestic equine species or other wild reservoir hosts (Coetzer and Erasmus 1994, Seifert 1992).

Pathogenesis

Depending on the virulence of the virus and susceptibility of the animal, the incubation period varies between 3–10 days. The course of the disease can be peracute, acute or pulmonary ("Dunkop"), subacute, cardiac or oedematous ("Dikkop"), mixed (pulmonary and cardiac) and atypical chronic (horsesickness fever). The factors determining the course and severity of infection are not fully understood.

After infection, initial multiplication of virus occurs in the regional lymph nodes and is followed by a primary viraemia with subsequent infection of target organs, namely, the lungs and lymphoid tissues throughout the body. Virus multiplication at these sites give rise to a secondary viraemia of variable duration; in horses it is generally not higher than 10^5 $TCID_{50}$/mL and may last 4–8 days, but will not exceed 21 days; the virus is closely associated with the erythrocytes in the blood (Coetzer and Erasmus 1994, Seifert 1992).

Clinical Features

The peracute and pulmonary forms are characteristic for the course of the disease during severe epizootics when fully susceptible horses are affected. Fever may be the only sign for a day or two, reaching a maximum of 41 °C or sometimes higher. At first, the appetite of the animal remains good in spite of high fever and respiratory distress. Characteristic signs of these forms of AHS are severe dyspnoea, paroxysms of coughing, and a discharge of large quantities of frothy, serofibrinous fluid from the nostrils which also may appear only after death. Dyspnoea and hyperpnoea are rapidly progressive; the respiratory rate may reach 75 b.p.m. In the terminal stages, the nostrils are flared, the mouth open, the tongue protrudes, the head and neck are extended and the animal sweats profusely. Death can occur very swiftly: sometimes an apparently healthy horse suddenly becomes severely dyspnoeic and dies. The prognosis for horses suffering from the pulmonary form is extremely grave: less than 5% recover. In recovering horses, the fever gradually subsides but the breathing remains laboured for some days.

The subacute or cardiac form ("Dikkop") is characterized by subcutaneous oedemas chiefly around the head and neck. The same is true for the more protracted course of the disease. The febrile reaction reaches its maximum later, and may remain high for 3–6 days before declining. Some animals develop only a mild fever. The oedematous swelling of the head and neck usually appears late in the course of disease; as the swellings increase, dyspnoea and cyanosis may supervene. Varying degrees of swelling of the supraorbital fossae and other parts of the head are evident in affected horses, in severe cases the eyelids, lips, cheeks, tongue, intermandibular space, and sometimes also the neck, chest and shoulders are involved (Fig. 49). The mortality rate is about 50%, death usually occurring within 4–8 days of the onset of the febrile reaction.

The mixed form, although it is the most common form of AHS, is rarely diagnosed clinically. Horses affected show signs either of respiratory distress followed by oedematous swellings, or of the cardiac form before suddenly developing respiratory distress from which they may die. The mortality rate is approximately 70%. Death usually follows 3–6 days after the onset of the febrile reaction.

The atypical chronic form or horsesickness fever usually occurs in horses immune to one or more serotypes of the AHS virus or in animals with a high level of relative resistance (donkeys, zebras). The rather mild course of the disease is not usually diagnosed clinically. Usually the animals recover completely after having fever for about a week; loss of appetite, congestion of the conjunctivae, slightly laboured breathing and increased heart rate may be transient signs (Coetzer and Erasmus 1994, Seifert 1992).

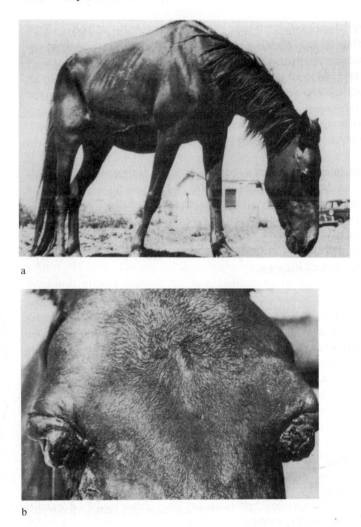

a

b

Figure 49. African horse sickness. a Oedemas on the head, thorax and belly, subacute case.
b Oedemas of the lids and prolapse of conjunctivae (photos: S. Gürtürk, from Mitscherlich and Wagener 1970).

Pathology

The pathological features are most prominent in the pulmonary and cardiac form. The most striking features of the pulmonary form are severe oedema of the lungs and hydrothorax. Several litres of pale yellow fluid which may coagulate on exposure to air are found in the thoracic cavity. Epi- and endocardial haemorrhages of the endocardium may be evident. Congestion of the mucosa of the stomach and patchy congestion and petechiation of the serosa and sometimes of the mucosa of the intestine appear. The liver may be slightly enlarged and congested; there is usually some degree of ascites.

The most characteristic signs of the cardiac form are the distinctly yellowish gelatinous oedema of the subcutaneous and intramuscular connective tissues of the head and neck which in severe cases extends to the back, shoulders and chest. The eyelids, supraorbital fossae, lips, cheeks, tongue, and intermandibular space are commonly involved. The tongue may be severely swollen, cyanotic and have mucosal petechiae on its ventral surface. Lesions in the heart are more severe than described for the pulmonary form and accompanied by severe hydropericardium. Similarly, the lesions in the gastrointestinal tract are usually more severe than in the pulmonary form.

In the mixed form lesions typical for the pulmonary and cardiac form of AHS appear.

Histopathologically execudative pneumonia, congestion of alveolar capillaries and arterioles and venules as well as perivasculitis are evident. Degeneration and necrosis of myocytes, oedema of the myocardium with infiltration of mononuclear cells, plasma cells, siderocytes and polymorphnuclear leukocytes as well as lysis of necrotic myocytes appear.

Diagnosis

Wherever AHS is enzootic the epizootiology, clinical signs and macroscopic lesions are sufficiently specific to allow its diagnosis. The chronic form is more difficult to diagnose. The virus can be isolated from blood during the febrile stage or from specimens of the lungs, spleen and lymph nodes on tissue culture (BHK21, Vero, MS), by intracerebral inoculation of suckling mice (0.03 cm^3) which fall ill 4–22 days p.i., or inoculation in embryonated hens' eggs.

Serologically, the isolates of the virus can be identified through complement fixation (CF), agar gel immunodiffusion (AGID), and both direct and indirect immunofluorescence (IFA) or ELISA. Serotyping of the AHS virus is performed using virus neutralization (VN) on tissue culture or in mice. The disease can be diagnosed indirectly by demonstrating antibodies against AHS virus in recovered animals by means of CF, AGID, IFA, VN and ELISA tests. A new generation of immune assays, genomic probes and PCR-based assays are being developed (Coetzer and Erasmus 1994, Seifert 1992).

Treatment

An aetiological therapy of AHS is not feasible. Sick animals have to be kept carefully and treated symptomatically.

Control

Because of the broad range of vectors which transmit AHS, it is difficult to prevent the disease through vector control. Combating vectors may, however, become relevant if AHS has been spread to areas which have previously not been infected.

Infection of susceptible horses can be prevented to a large degree by stabling them some hours before sunset as *Culicoides* midges are nocturnal and are not inclined to enter buildings.

Since the demonstration in the early 1930s that the AHS virus can be attenuated by serial intracerebral passage in mice the immunization of horses against the disease has become feasible. In enzootic areas in Africa annual vaccination of horses is a very practical means of control, although it cannot be relied upon fully, and only horses which have received three or more courses of immunization are usually well protected against the disease. Multivalent vaccines containing attenuated strains are applied. In South Africa, two vaccines with strains 1, 3, 4 and 5 and 2, 6, 7 and 8 are available. A slight temperature response may appear between 5 and 13 days after inoculation as a result of low level virus replication in immunized animals. The ensuing viraemia will be followed about a month p.i. by a sterile immunity. There are also inactivated vaccines on the market containing in aluminium hydroxide absorbed virus. It is reported that a disease outbreak occurred as a consequence of the application of an attenuated vaccine (Hassanain et al. 1990).

In most instances, the level of antibodies acquired from colostrum correlates well with the level of antibodies in the sera of the mares. Because of the passive immunity acquired by foals born to immune mares, foals should not be immunized before they are 6 months of age.

Through application of recombinant DNA technology, subunit vaccines are expected to become available for the AHS virus within the foreseeable future (Coetzer and Erasmus 1994, Seifert 1992).

1.5.1.2. Bluetongue (BT)

(Febris catarrhalis ovium, Fièvre catarrhale du mouton, Ovine catarrhal fever, Bloutong, Lengua azul, Blauzunge)

Aetiology and Occurrence
BT is an arthropod-borne viral disease in domestic and wild ruminants, particularly sheep, caused by the BT virus which belongs to the genus *Orbivirus* in the family *Reoviridae*. From the BT virus 24 serotypes are known which have, however, a common antigenic component and are all interrelated in a complex network of cross-relationships. No scientific evidence for an antigenic drift has been postulated. The double-stranded RNS virus has a diameter of 100 nm, its outer layer envelopes a capsid which consists of 92 capsomeres and has a diameter of 60 nm. At a pH of 6.4, the BT virus is inactivated rather quickly while it remains relatively stable up to a pH of 8.0. It is readily inactivated by disinfectants containing acid, alkali, sodium hypochlorite and iodophores.

BT was first described in South Africa where it has probably been endemic in wild ruminants since antiquity. While initially thought to be confined to sheep and game on the African continent, the first outbreak outside Africa was confirmed in Cyprus in 1943 and since then in the last 50 years BT has been reported in many European, Asian and American countries as well as in Australia. BT is mostly prevalent where climatological factors which favour the development of the vector appear, i.e. in the tropics and subtropics. More and more BT has also developed into an enzootic of cattle which causes considerable economic losses.

Epizootiology
BT is enzootic in many areas especially in Africa and shows an inapparent course in autochthonous breeds of sheep in particular. If exotic breeds are imported, severe outbreaks of the disease can occur. Autochthonous goats are also highly resistant to BT. The same is true with African wild ruminants which play a special part as carriers of BT. While African antelopes do not develop the disease clinically, the white-tailed deer (*Odocoileus virginianus*), pronghorn (*Antilocapra americana*) and desert bighorn sheep (*Ovis canadensis*) of the North American continent may develop it severely. At first it was supposed that cattle are rarely infected and fall ill in old enzootic areas. In the meantime, it has become obvious that BT also causes serious lesions and economic losses in cattle which show symptoms different from sheep.

BT is transmitted by the midge *Culicoides* spp., the most important being *C. imicola* in Africa and in the Middle East, *C. variipennis* and *C. insignis* in North America and *C. fulvus* in Australia. The virus multiplies in the salivary glands and thus the transmission is cyclical but apparently not transovarial. Infected midges remain infective for the rest of their lives. Acyclic-mechanical transmission probably occurs with the sheep ked *Melophagus ovinus*, the tick *Ornithodorus coriaceus*, and blood-sucking flies such as *Stomoxys* and *Tabanus* spp. Factors such as flight periodicities, biting activities, host preferences, preferred habitats for oviposition and climate have a direct influence on the incidence and spread of BT. The disease may be carried to other areas by winds, blowing infected midges over vast distances. BT is not contagious, and infected animals secrete and excrete only very little virus; transplacental infection may, however, occur.

Outbreaks of the disease in sheep usually occur simultaneously in widely separate localities in late summer or autumn, suggesting that populations of BT virus-infected midges build up in the primary cycle involving cattle or wild animals during spring and early summer, and that sheep become infected in a secondary cycle as a result of "spill-over". In endemic areas, serotypes are distributed randomly and the dominant serotypes are largely determined by herd immunity (Seifert 1992, Verwoerd and Erasmus 1994).

Pathogenesis
The average incubation period is about a week; the course of the disease depends on the virulence of the virus type as well as on environmental influences (weather), and stress factors (walking) may influence the outbreak of the disease. BT virus probably multiplies at first in the regional lymph nodes before spreading to the rest of the body. Viral replication occurs primarily in endothelial cells and pericytes of capillaries and small blood vessels which leads to degenerative and necrotic lesions and finally to vascular occlusion, stasis and exudation, which eventually gives rise to hypoxia, oedema and haemorrhage, as well as to secondary lesions in the overlying epithelium. The severity of the secondary lesions is influenced by mechanical stress and abrasion, and severe lesions develop mainly in tissues exposed to the environment, such as the oral mucosa and skin of the coronary border of the hooves. Exposure of the infected

animals to sunlight exacerbates the severity of the disease. The mortality rate can escalate dramatically when infected sheep are exposed to cold wet conditions.

Following initial replication in lymphoid tissues and endothelial cells, the BT virus appears in the circulation 3–6 days after infection, the viraemia reaching its peak 7–8 days p.i. and persisting for 6–8 days, though rarely longer than 14 days. In cattle, however, the viraemia may last up to 49 days and possibly longer. In the blood the virus is primarily associated with erythrocytes and, to a lesser extent, with the buffy-coat fraction.

The pathogenesis of the infection in cattle differs from that in sheep. Rapid accumulation of IgE antibodies may result in an immediate hypersensitivity reaction involving the release of histamine, prostaglandins and thromboxane A2 (Seifert 1992, Verwoerd and Erasmus 1994).

Clinical Features

The extreme variability in the clinical manifestation of BT, not only between different ruminant species but also between different breeds of sheep is the feature of the disease. The course of the disease in sheep can vary from peracute to chronic with a mortality rate of between 2 and 30%. Peracute cases usually die within 7–9 days of infection, mainly as a result of lung oedema and eventually asphyxia. They may show very few signs of illness prior to death. In chronic cases, death can result from secondary bacterial pneumonia and exhaustion, or recovery can be very protracted. Mild cases usually recover rapidly and completely. In the acute case the temperature will rise up to 42.5 °C and the mucosa of the mouth become reddened which leads to catarrhal inflammation of the bucal and nasal mucosae, and in some cases also of the conjunctivae. The animals become noticeable on the pasture because they froth at the mouth while chewing; they are unable to take feed. Watery clear liquid appears in the nostrils which in the further course of disease becomes slimy and mixed with blood. Oedema of the tongue, lips, face, eyelids and ears develop about 48 hours after the onset of fever. Submandibular oedema may be marked and occasionally extend down the neck to the axillae. Lips, cheeks, the palate, tongue and larynx are swollen, the mucosa of the mouth turns blue-red (bluetongue). The epithelium comes away from the lips and secondary infections appear. The mucosa of the mouth mostly comes off in scraps from the tongue and the palate, especially behind the incisors and at the upper pad, and thus bleeding and sore spots appear. The necrotizing mouth lesions result in foetid breath. Because of inflammation of the nasal mucosa, the nostrils are blocked with dried secretions. In the further course of the disease, the characteristic inflammation of the feet appears especially in younger animals, usually as soon as the lesions in the mouth have past their peak and the febrile reaction has reached its end. Hyperaemia of the coronary bands and petechiae under the periople, which later becomes streaky in appearance as a result of haemorrhage into the fine medulary canals of the growing horn, give rise to a red zone or band in the horn of the hoof. The lesion is most pronounced on the pads of the feet. The feet are warm and painful, and affected animals are reluctant to move; they walk with a stiff gait and show varying degrees of lameness.

Sometimes severely affected animals try to walk on their knees. In animals which recover, the band of discolouration on the hooves grows out and a split in the hoof may develop, with the old horn eventually sloughing after 3–4 months. Inability to move and recumbency may also be exacerbated by emaciation and muscle lesions. Depending on the virus type the lesions may be more severe in the mouth or on the feet. Because of degenerative alterations of the neck muscles, some animals may develop torticollis. Females abort and become anoestric; rams become also temporarily infertile.

In cattle, the disease also shows a febrile course, possibly with distinct symptoms of pododermatitis. The economic important losses, however, are abortion in pregnant cattle and infertility; the calves may be born deformed (arthrogryposis, hydranencephalus), or they are weak. Characteristically, calves from infected dams are immunotolerant to the homologous virus. The virus is excreted with the sperm of infected bulls (Seifert 1992, Verwoerd and Erasmus 1994).

Pathology
The pathological features of BT result from the clinical picture described above. Oral lesions consist of hyperaemia, oedema, cyanosis and haemorrhage of the mucous membranes. Destruction of epithelial cells gives rise to excoriations and ulcerations on the inside of the lips, dental pad, cheeks and tongue being turned into diphtheric necrosis because of secondary infection.

Hyperaemia, petechiation, erosions and ulcerations of the mucosa of the forestomachs, particularly of the papillae, ruminal pillars, reticular folds and oesophageal groove are common. Lesions in the small intestines vary from mild localized areas of hyperaemia to severe catarrhal or haemorrhagic lesions extending into the larger intestine.

Hyperaemia, oedema and petechia occur in the mucosae of the nasal cavity, pharynx and trachea as well as in the lungs. Severe hyperaemia and oedema of the lungs accompanied by copious amounts of froth in the trachea and hydrothorax occur, especially in peracute and acute fatal cases.

The pathological changes of the skeletal musculature are associated with severe loss of condition, weakness, and a slow, protracted recovery period. These are petechiae, ecchymoses and gelatinous oedema of the intramuscular connective tissues, particularly in the neck and the dorsal thoracic region. Small localized areas of degeneration and necrosis appear as greyish-white foci. A typical feature is the streaky degeneration of the heart muscle (tiger heart) as is also common in FMD (Fig. 50 on page 1 annex).

Diagnosis
Where BT is enzootic and outbreaks occur at the beginning or during the rainy season, it is rather easy to make the diagnosis from the clinical signs and lesions in affected sheep. In other ruminant species serological identification is required since the disease is mostly subclinical. Through injection of blood obtained from infected animals into susceptible sheep, it can be demonstrated that the disease is transmitted by parenteral infection while secretions from the mouth will not

trigger the disease when transferred to the oral mucosa of susceptible animals.

The virus can be isolated from infected animals during the febrile reaction by inoculation of embryonated hens' eggs. After one passage on embryonated hens' eggs it also can be grown on tissue culture (BHK21, mouse L– and Vero cells). The identification of the virus usually is done with the CFT and direct fluorescence though both tests do not distinguish between different serotypes of BT virus. Neutralization tests are used for serotyping the virus. Without prior isolation the BT virus can now be demonstrated directly with an indirect peroxidase-antiperoxidase (PAP) test. Genomic probes also have been developed for this purpose which also can identify the BT virus type (De Mattos et al. 1989). A capture ELISA has also been used to detect virus in cell cultures and insect tissues. For rapid diagnostic purposes, indirect or competitive ELISA is applied and is able to detect group-specific antibodies (Seifert 1992, Verwoerd and Erasmus 1994).

Treatment

It is not possible to achieve a causal treatment of BT. Symptomatically the lesions in the mouth can be treated with mild astringents and chemotherapeutic agents. Spraying the mouth with ENTOZON by means of a spray-gun has been very effective. This prevents the animal from becoming unable to feed, becoming emaciated, and dying of recumbency (Seifert 1962).

Control

Schemes for the control of BT have to take into account the possible virus reservoir, the vector and the target animal.

Since a wide variety of game can be a reservoir of infection in enzootic areas, the only alternative to control the disease is vaccination. The virus used for vaccines used to be attenuated in passages through embryonated hens' eggs. It is now grown in BHK21 cells. Because of different serotypes, polyvalent vaccines have been applied. In South Africa, 3 pentavalent vaccines have been developed and are administered to sheep at 3 week intervals and repeated annually. Pregnant ewes should not be immunized during the first half of pregnancy, as brain defects may result in the foetus. There are also indications of temporary infertility in both ewes and rams vaccinated for the first time. Another disadvantage using live attenuated vaccine is the possibility of reassortment and recombination between attenuated and virulent strains as well as the possibility of spreading the vaccine virus through insects. Therefore, the effort is made to produce efficient recombinant vaccines which may protect against all serotypes.

If BT appears in an area which has previously been free from the disease, the infection can be controlled by combating the vector efficiently. It has been shown that when the entire stock of sheep on a farm is powdered with 5% coumaphos contained in talcum, the spread of the disease can be controlled (I/4.3.2.1) (Seifert 1962).

BT becomes more and more important in the international trade of cattle. Stringent serological controls are carried out to prevent carrier cattle from being traded.

1.5.1.3. Rift Valley Fever (RVF)

(Enzootic hepatitis, Fiévre de la Valée du Rift, Fiebre del Valle de Rift, Slenk-dalkoors, Rifttalfieber)

Aetiology and Occurrence

RVF virus belongs to the family *Bunyaviridae* and the genus *Phlebovirus*; these are spherical virions with a diameter of 80–120 nm and a host-cell-derived, bilipid-layer envelope through which virus coded glycoprotein spikes project. They have a three-segmented, single stranded RNA genome with a total molecular weight of 4.0–7.5 × 10^6 Da and contain three mature structural proteins, two envelope glycoproteins G1 and G2, a nucleocapsid protein N and minor quantities of viral transcriptase, or L (large) protein. No significant antigenic differences have been detected between RVF isolates. RVF virus is very stable in serum and at temperatures lower than –60 °C or after freeze-drying, and in aerosols at 23 °C and 50–85% relative humidity. It is inactivated by lipid solvents, formalin and pH below 6.8. The virus grows readily in all continuous cell lines and embryonated chicken eggs as well as in a variety of laboratory animals.

RVF occurs in southern and western Africa from Zambia to Namibia and from South Africa up to Mozambique, Kenya and also Egypt. Outbreaks have also been reported from the Sudan and Mauritania, and serological evidence of the occurrence of RVF has been recorded in several other countries on the mainland of Africa and in Madagascar (Seifert 1992, Swanepoel and Coetzer 1994).

Epizootiology

Enzootics have always tended to occur in eastern and southern Africa, and have usually been associated with above average rainfall at irregular intervals of 5–15 years or longer. RVF is transmitted by midges of the genus *Aedes*, *Culex*, *Anopheles*, *Eretmapodites* and *Mansonia*. It is not yet clear whether rodents are vertebrate hosts. It has, however, be confirmed that the virus is enzootic in livestock areas and is maintained by transovarial transmission in aedines, thus probably being maintained during interenzootic periods with a low level of transmission to livestock. Infection rates in vector populations may be quite low, usually below 0.1%, but enormous numbers of aedines emerge from flooded areas and vertebrates are subjected to a high mosquito-biting frequency. The virus may be spread to previously non-infested areas by flying insects carried at high altitudes by strong winds over long distances (450–500 km). Trading infected sheep and cattle can also spread the disease.

Ruminants, small rodents, monkeys, cats as well as humans are susceptible to RVF, and thus RVF is a zoonosis. The susceptibility to RVF depends on the resistance of the species and breed which is also determined by the age of the animal and the virulence of the virus strain. Lambs and kids as well as small rodents are highly susceptible while older ruminants are more resistant. Equines develop only low grade viraemia without clinical signs; antibodies have also been detected in camels which have aborted as well as in water buffaloes.

A high degree of herd immunity arises in locations where infection is most intense and it can be surmised that this must be one of the factors which contributes to the abatement of enzootics. Because of the massive immunity produced in recovered animals which is also transferred passively with the colostrum from the dam to the calf and from the ewe to the lamb, enzootics appear only in intervals of 4–7 years.

Humans become infected from contact with infected tissues or from mosquito bites. The contact infection is the result from the contact with razed skin, wounds or mucous membranes. Though low concentrations of virus have been found in milk and body fluids it appears that there may have been a connection between human infection and the consumption of raw milk (Seifert 1992, Swanepoel and Coetzer 1994).

Pathogenesis
After infection the virus spreads from the initial site of replication to critical organs such as the spleen, liver and brain which are either damaged by the pathogenic effects of the virus or immunopathological mechanisms, or else there is recovery mediated by non-specific and specific host responses. The virus is conveyed from the inoculation site by lymphatic drainage to regular lymph nodes where there is replication and spill-over into the circulation which leads to viraemia and systemic infection, already detectable 16 hours p.i. in artificially infected lambs, and in older animals after 2 days. Lesions in target organs in the acute disease are produced by the direct lytic effect of the virus on infected cells. Haemostatic derangement occurs in the fatal hepatic syndrome as a viral haemorrhagic fever with a tendency to bleed. Factors contributing to fatal outcome in the hepatic form of the disease are anaemia, shock and hepatorenal failure. Ocular lesions may occur as a complication of RVF in humans as well as encephalitis. Abortion is the usual outcome of infection in pregnant sheep, cattle and goats following on from the death of the foetus with lesions in the foetus (Seifert 1992, Swanepoel and Coetzer 1994).

Clinical Features
Signs of the disease in domestic ruminants tend to be non-specific. During epizootics, however, the simultaneous occurrence of numerous cases of abortion and disease in ruminants, together with disease of humans tends to be characteristic of RVF. After an incubation period which may be as short as 12 hours in lambs but up to 96 hours in adult cattle and sheep, fever develops which may exceed 41 °C, being often biphasic with a remission after 12–18 hours, following the initial rise of temperature and then subsiding a few hours prior to death. Affected animals are listless, disinclined to move or feed and show evidence of abdominal pain. Progressive loss of condition, anorexia, staggering, mucopurulent secretion from the nostrils, diarrhoea and abortion may occur, the latter being often the only symptom in adult animals. The course of the disease in lambs is usually peracute and they survive rarely more than 24–36 hours, the mortality rate being 90%, while in adult sheep and cattle which often have only

an inapparent infection, rarely more than 10–30%.

In humans, the infection is mostly inapparent but may be associated with a moderate to severe, non-fatal influenza-like illness. In some cases haemorrhagic fever may develop with a fatal outcome; beside ocular lesions, encephalitis and haemorrhagic manifestations may occur (Seifert 1992, Swanepoel and Coetzer 1994).

Pathology

The most prominent pathological features are the necrotic lesions on the liver which not only appear in all animals but also in humans. The organ becomes discoloured and is covered with numerous small greyish-white necrotic foci, oedema and haemorrhages are found in the wall of the gall bladder. Hepatic lymph nodes may appear, and icterus is evident in adult animals. The necrotic foci in the liver are the consequence of a coagulative necrosis of the hepatocytes with the production of intracytoplasmic hyaline inclusion bodies. In lambs, these foci may also coalesce into larger white-yellowish necrotic areas. These lesions are also found in the aborted foetuses of cattle and sheep. Spleen tumours are apparent especially in sheep presenting small petechiae at the edges, and which also appear in the capsule of the kidney. Petechiae are also present subepicardially in the coronary region and subendocardially in the left ventricle. Petechiae may also appear in the meninges as well as perivasculary in the lung, along with lung oedema and emphysema. Petechiae and ecchymoses are also present in the mucosa of the abomasum which is filled with a dark chocolate-brown content. In all animals, the peripheral and visceral lymph nodes are enlarged, are oedematous and may have petechiae. The carcass decays rather quickly which is supposed to be a consequence of the severe liver damage (Seifert 1992, Swanepoel and Coetzer 1994).

Diagnosis

The epizootiology, clinical and pathological features of RVF are pathognomonic. High losses in lambs and calves with low mortality in adult animals, accumulated abortions in sheep and cattle and influenza-like symptoms in humans, perhaps in combination with widespread death in wild rodents are important diagnostic hints.

The virus can be directly demonstrated in the blood of infected animals by inoculating suckling mice through the intracerebral route; they die 36–72 hours p.i. showing the pathognomonic disseminated necrotizing hepatitis.

The virus can also be isolated in a variety of cell cultures including Vero, CER, BHK21, mosquito line cells and several primary cells, and can be identified by IF test or serum neutralization already after 24 hours.

Antibodies to RVF virus can be demonstrated by CF, ID, ELISA, indirect IF, haemagglutination-inhibition (HAI), reversed passive haemagglutination-inhibition, radio-immunoassay, neutralization of the cytopathic effect, plaque reduction and neutralization tests in mice.

The viral antigen can also be detected rapidly in impression smears of infected tissues with IF, CF and ID, and the antibodies in serum by reversed passive haemagglutination or ELISA (Seifert 1992, Swanepoel and Coetzer 1994).

Treatment
There is no treatment for sick animals.

Control
An efficient control of mosquitoes, e.g. by applying phosphoric acid esters in powder form to sheep during the rainy season (I/4.3.2.1) can prevent the appearance of the disease. Furthermore, the movement of stock from low-lying areas to well-drained and wind-swept pastures at higher altitudes, or the confinement of animals to mosquito-proof sheds may be measures of management to reduce the incidence of RVF. The breeding management might also take into account the seasonal activity of vectors and prevent the lambing and/or calving season during the rainy season.

The most effective way to protect livestock against RVF remains vaccination. The mouse neuroadapted Smithburn strain of RVF virus is used for vaccine production in South Africa and Kenya, while wild strains of RVF virus are used for preparation of formalin-inactivated cell culture vaccines in South Africa and Egypt. The Smithburn vaccine induces durable immunity in sheep but may cause abortions or teratology of the foetus. It is suspected, but not proven, that the virus could revert to full virulence if passaged through vectors which become infected when feeding in the viraemic stage following vaccination. Therefore, in countries where the disease is not proven to be enzootic, inactivated vaccines should be used. The Smithburn RVF vaccine induces poor antibody response in cattle which therefore should be immunized with formalin-inactivated vaccines and be boosted 3–6 months after initial vaccination which has to be repeated by annual boosters before the rains are due (Seifert 1992, Swanepoel and Coetzer 1994).

1.5.1.4. Nairobi Sheep Disease (NSD)

Aetiology and Occurrence
The causal organism of NSD is a virus which belongs to the family *Bunyaviridae*, genus *Nairovirus*. It is related to the Ganjam virus which causes a tick-borne infection in sheep and goats in India and the Dugbe virus, a tick-borne infection in cattle in West Africa. The NSD virus is sensitive to lipid solvents and detergents and is rapidly inactivated at high and low pH values.

The virus replicates in primary or secondary cell cultures derived from sheep and goats testes and kidney, and in continuous cell lines of BHK and the vector tick, *Rhipicephalus appendiculatus*. In the infected cells a distinct cytopathic effect develops, and with haematoxylin-eosine staining, typical eosinophilic inclusion bodies which contain large clusters of polyribosomes can be found. Infant mice are highly susceptible to NSD virus.

NSD is enzootic in sheep and goats in Kenya, Uganda and Zaire. It also occurs in other parts of East and Central Africa and is thought to extend as far north as Ethiopia and Somalia and as far south as Botswana and Mozambique; it has not been diagnosed in South Africa.

Epizootiology
Sheep and goats are the natural vertebrate hosts of NSD virus; cattle, buffalo, horses and pigs are refractory to infection. Strangely enough, Maasai sheep are more susceptible to infection than imported Merino sheep. No evidence exists so far that antelopes or wild rodents are involved in the maintenance cycle of the NSD virus. Humans are susceptible to NSD and may show fever, joint aches and general malaise, NSD thus being a zoonosis.

The main vector of NSD virus is *Rhipicephalus appendiculatus*, but *R. pulchellus*, *R. simus* and *Amblyomma variegatum* can also transmit the disease and are invertebrate hosts of the virus. *R. appendiculatus*, epizootiologically the most important vector, can get infected during all stages of development and transmit the virus after moulting. Transovarial transmission has only been demonstrated in *R. appendiculatus*. Though sick animals excrete the virus with faeces and urine, the virus can only be transmitted through a vector. The prevalence of NSD virus antibodies in sheep and goats corresponds roughly to the geographic distribution of *R. appendiculatus*.

Outbreaks of NSD may arise either as a result of the movement of susceptible animals into areas which have been previously free from infection or from the incursion of infected ticks into NSD-free flocks or areas. The mortality rate in the field may be as high as 70–90% for indigenous breeds of sheep and 30% for exotic and crossbreeds. Passive colostral immunity protects the lambs from the lethal effects of infection and allows them to develop active immunity, thus being able to survive in enzootic areas. Goats are regarded as being less susceptible than sheep but the mortality rate of indigenous breeds may be as high as 90% (Seifert 1992, Terpstra 1994).

Pathogenesis and Clinical Features
The virus reaches the target organs via the blood stream probably after replication in the regional lymph nodes. Posterior virus replication takes place in liver, lungs, spleen and other organs of the RES, having predilection for the vascular endothelium.

After an incubation period of about 4–6 days, clinical signs begin with a steep rise in temperature up to 42.2 °C which persists for 1–7 days and are accompanied by leukopenia and viraemia, the latter disappearing within 24 hours after the temperature has returned to normal. The animals are dull and inappetent, standing with drooping heads while straining continuously, and sometimes groaning with each expiration, showing tachycardia. Diarrhoea with faeces which are watery, dark-green and foetid, and contain mucous and blood is a prominent symptom. Mucopurulent, often blood-tinged, nasal discharge may occur and conjunctivitis may also be present. The outer genitalia become oedematous and pregnant ewes may abort. After recumbency, the disease may have a fatal outcome about 2 weeks after the onset of the clinical symptoms, depending on whether the course of disease is peracute (2 days) or acute (11–14 days). Goats show similar symptoms which however are less severe.

Pathology
In spite of the short course of disease, the carcass can be dehydrated, the hind quarters soiled with blood and faeces and presenting dried crusts around the eyes and the nostrils. The necropsy picture is typical of a haemorrhagic diathesis with catarrhal or haemorrhagic inflammation of the gastrointestinal and female genital tracts. Multiple haemorrhages of different size appear in the mucosae of the entire intestinal tract also as streaky bleedings in the longitudinal faults of the mucosa. These are the outstanding features observed at necropsy. Subserosally, haemorrhages may occur in the caecum, the gall bladder, below the capsule of the kidneys, in the mucosa of the female genital tract, and in the lower respiratory tract. The mucosa of the nasal cavity is often congested and shows catarrhalic inflammation. The heart, containing unclotted blood is pale and flaccid with petechiae in the epicardium and ecchymoses in the endocardium. The spleen and lymph nodes may be swollen, the mesenteric nodes being grossly oedematous and enlarged (Seifert 1992, Terpstra 1994).

Diagnosis
Taking into account the epizootiology of NSD as well as the clinical and pathological features which are pathognomonic, diagnosis is easy where the disease is known to be enzootic. The virus can be demonstrated either by inoculation of susceptible target animals, inoculation in mice (brain) or isolation in tissue culture where the virus can be identified by IF or CF tests. The viral antigen can even be detected in the spleen, lung tissue and mesenteric lymph nodes of sheep that have suffered an acute fatal infection, by immunodiffusion tests, the method being able to be applied in field laboratories.

Antibodies for NSD virus can be detected by CF, virus neutralization, agar-gel immunodiffusion, indirect immunofluorescence and indirect haemagglutination tests. An ELISA has been developed but its application in the field still has to be assessed (Terpstra 1994).

Treatment
There is no treatment for NSD.

Control
NSD can be prevented through efficient vector control. Since the disease is enzootic in traditional nomadic and small-holder animal production systems, there is rather little chance of establishing such control schemes. The same is true for proposals to starve out the ticks from infested pastures by keeping the animals away from them. Because the disease can be transmitted to humans and because it has serious consequences for the economy of traditional animal production systems, prophylactic measures are urgently required.

Two types of vaccine – a modified live virus (MLV), attenuated by serial mouse brain passage, and an inactivated oil-adjuvant vaccine – have been used experimentally. Although a single dose of MLV vaccine rapidly induces immunity, the strain has lost some immunogenicity and annual revaccination of sheep is necessary to maintain full protection. The MLV strain causes viraemia, but reversion to virulence is impossible as the strain cannot be transmitted by *R. appendiculatus*. On the other hand, two doses of the inactivated vaccine are required for optimal stimulation of antibody production, and the duration of immunity appears to be limited (Seifert 1992, Terpstra 1994).

1.5.1.5. American Arboviral Encephalomyelitides of Equidae

(Equine encephalomyelitis, Equine encephalitis, Eastern equine encephalomyelitis (EEE), Western equine encephalomyelitis (WEE), Venezuelan equine encephalomyelitis, Encephalomielitis equina, Amerikanische Pferdeenzephalomyelitiden)

Aetiology and Occurrence
Equine encephalomyelitis (EE) is an enzootic mosquito-borne virus infection caused by an alphavirus which belongs to the family *Togaviridae*. Three virus types are important pathogens for equines:

- the Western equine encephalomyelitis virus (WEE) which causes a comparatively mild course of the disease;
- the Eastern equine encephalomyelitis virus (EEE) which is the cause of the most severe form of the disease;
- the Venezuelan equine encephalomyelitis virus (VEE) which causes generalized symptoms of infection.

The causal organisms are enveloped viruses containing RNA which can be differentiated immunologically, but contain a common group-specific antigen for the alphaviruses. A close relationship exists between EEE and VEE which is the cause of cross-immunity. All virus types have haemagglutinating properties and have little resistance against environmental influences, and thus are not stable outside their mammalian hosts and vectors.

The American equine encephalomyelitis occurs only on the American continent where the WEE virus has the highest incidence. It is not only important in North America but also in the northern parts of South America and the Caribbean. Up to 4000 cases of equine encephalomyelitis occur annually in the USA, the majority being due to WEE. The virus has caused severe enzootics in the western USA, close to the Mexican border, in Central America especially in the swampy areas of Columbia and Panama, in Ecuador at the Peruvian border, as well as in the irrigated areas of intensive agriculture in Peru and in Argentina. The EEE virus occurs in the east of the USA and the Caribbean; the VEE virus is enzootic in Central America, Florida, as well as in South America, Brazil, Columbia, Ecuador and Peru.

Another virus which causes equine encephalitis is the Japanese encephalitis virus which is a flavivirus from the family *Flaviviridae*; it is recognized throughout the Far East. Mortality in horses is low (< 5%), and it causes abortion in pigs without other clinical signs. In addition, other flaviviruses isolated from encephalitic *Equidae* are those of louping ill, Murray Valley and West Nile encephalitides. The Maindrain virus, a *Culicoides*-transmitted *Bunyavirus* has been isolated from encephalitic horses in the western USA.

Epizootiology

Under natural conditions, equines and last but not least humans, are susceptible to all the encephalomyelitis viruses of equines; the virus can be transmitted artificially to pigs, squirrels, hares and deer. Infections from the EEE virus have also led to severe losses in pheasants. Clinical symptoms of the disease develop also in young animals of other poultry. Wherever American equine encephalitis is enzootic, it is an important and feared zoonosis. Outbreaks of the disease in humans are mostly closely related to enzootics in equines.

The EEE, WEE and sylvatic VEE viruses are transmitted to mammalian hosts by biting insects, principally mosquitoes; WEE is mainly transmitted by *Culex tarsalis*, EEE by *Culiseta melanura* and *Aedes* spp., VEE by the genera *Culex*, *Anopheles*, *Mansonia*, *Psorophora* and *Aedes*. The sylvatic VEE virus has also been isolated from swallowbugs (*Oeciacus vicarius*). The transmission by mosquitoes is cyclical and the virus persists in the salivary glands. Transmission by arthropods other than mosquitoes is probably unimportant. Wild birds serve as a principal reservoir of the EEE and WEE viruses. Forest rodents and wild birds are the probable reservoir of the sylvatic VEE virus; the reservoir of the enzootic VEE virus or its origin are unknown. The viruses' cycle is natural and silent between wild birds and mosquitoes. In North America, the viruses lie dormant in the winter in hibernating reptiles and amphibians. Minor natural hosts include wild rodents, skunks and foxes as well as domestic cats and dogs. Reservoir hosts tend to develop viraemia with blood titres adequate to infect mosquitoes, and actively contribute to the cycle of virus survival. Horses are considered dead-end hosts for the WEE and sylvatic VEE viruses; horses with an EEE infection may develop viraemia adequate to infect vectors, but probably do not contribute significantly to viral transmission or persistence. Horses are the most important amplifiers of epizootic/epidemic VEE virus and produce a viraemia adequate to infect mosquitoes. Unlike the EEE and WEE viruses, the epizootic/epidemic VEE virus may, on occasion, also spread between horses and to humans by contact or aerosol. These diseases occur more frequently in pastured than in stabled horses and are concentrated in areas having the appropriate combination of susceptible reservoir hosts and mosquitoes. Epizootics/epidemics of EEE and WEE tend to occur in mid to late summer and/or during the rainy season in tropical America (Merck 1991, Seifert 1992).

Pathogenesis

After infection, a biphasic course of disease develops. The virus multiplies at first in the regional lymph nodes from where it reaches the visceral target organs where the second cycle of replication takes place. When this occurs, an infection of the CNS may appear with the consequence of unreversible cell damages which lead to a fatal outcome of the disease. Often, however, the infection can show an inapparent course (WEE). The EEE virus infection can cause a mortality rate between 75 and 90%, the WEE 20–50% and VEE 50–75%; mules and donkeys are much more resistant. Recovered animals possess a long-lasting, sterile immunity. Because of this, enzootics develop in enzootic areas only in periods of 4–6 years whenever the immunity in the population of horses has declined.

Clinical Features

The sequential clinical and serological events following infection with EEE, WEE, or VEE viruses are similar. After an incubation period of 1–3 weeks the infection develops a febrile course with varying clinical pictures. The peracute form may be asymptomatic and already lead to death after a few hours. During the acute or subacute form central nervous disturbances appear after the second febrile phase: restlessness, excitability, dullness, impairment of vision, reduced reflexes, grinding of teeth, yawning, a pendulous lower lip, inability to swallow, photophobia, head-pressing, anorexia, irregular gait, wandering, circling, incoordination and a characteristic unnatural positioning of the limbs. There is severe depression and affected horses adopt a characteristic straddling stance with the head lowered and ears drooping. Many horses are unable to swallow water and when they raise their heads, the water pours out of the mouth. Severe pruritus may lead to self-mutilation. As with Borna disease, horses tolerate having their front legs crossed. The animals may even sit down like a dog and remain sleepily that way for some time. The disease will seldom last longer than a week; with EEE it is even shorter. After becoming unable to get up again the animal will die under occasional convulsions. Mildly affected animals may slowly recover in a few weeks but may have residual brain damage (dullness, dementia) and have been referred to as "dummies". VEE mostly shows only general symptoms, perhaps diarrhoea and nervous symptoms, which even sometimes may not appear (Merck 1991, Seifert 1992).

Pathology

No characteristic gross lesions are observed. Microscopically, haemorrhages and degeneration of neurons occur in the cerebral cortex, thalamus, hypothalamus, and other parts of the CNS. Gliosis, perivascular cuffing with polymorphnuclear (especially in EEE and VEE) and mononuclear (especially in WEE) cells, microglial proliferation, and meningitis may be present. Inclusion bodies are only present in Borna disease (Merck 1991).

Diagnosis

A presumptive diagnosis can be based on clinical signs, history, and seasonal occurrence, and is aided by knowledge of enzootic areas or known epizootic/epidemic activity of a virus type. Demonstration of typical histological lesions of a viral encephalitis strengthens the diagnosis. Diagnostic specificity results from virus neutralization, haemagglutination inhibition, or complement fixation tests during acute and convalescent phases. Because of the high mortality rate and rapid death of affected horses, it is difficult to obtain paired sera. Neutralizing antibodies are detectable about the time signs of CNS dysfunction occur, and viraemia terminates 4–5 days after infection. A 4-fold rise in the titre between acute and convalescent sera, or a very high titre (particularly of IgM) in an unvaccinated animal is considered positive. Diagnosis is confirmed by isolation and identification of virus from brain or blood. With VEE and EEE viruses, isolation is difficult after CNS signs are seen; blood should be collected for virus isolation from clinically normal, febrile horses in the same or an adjacent pasture (Merck 1991).

Treatment
Since no specific antiviral drugs are available, supportive treatment and intensive nursing care aid in the recovery of mild cases. Though tetracyclines or sulphonamides have no antiviral action they may have a favourable effect on the course of disease.

Control
Wherever an enzootic occurs, its spread can be controlled by combating mosquitoes quickly and efficiently, e.g., by applying pyrethroids to the exposed animals (I/4.3.2.1). Drainage and insecticide treatment of mosquito-breeding areas, and application of repellents to horses are also effective control measures. Valuable horses should be removed from pastures and stabled.

For immunoprophylaxis, inactivated chick embryo or cell-culture origin vaccines are now used almost exclusively and generally are considered effective. Monovalent, bivalent (EEE and WEE), and trivalent (EEE, WEE, VEE) vaccines are available. The monovalent VEE vaccine is an attenuated virus of cell-culture origin and should not be used in pregnant mares or young foals. The attenuated VEE vaccine has been largely replaced by a formolized product incorporated into a trivalent vaccine with EEE and WEE antigens. The vaccine should be given approximately one month before the mosquito season and, where the mosquito season is long, should be repeated within the year. In areas where mosquito activity is year round, foals should be vaccinated when 3, 4, and 6 months old, and annually thereafter.

1.5.1.6. Crimean-Congo Haemorrhagic Fever (CCHF)

Aetiology and Occurrence
CCHF is caused by a virus which is classified in the genus *Nairovirus*, family *Bunyaviridae*. 28 serotypes in 6 serogroups are currently recognized. The nairoviruses are aspherical, 90–120 nm in diameter and have a host-cell-derived bilipid-layer envelope. The genome consists of single-stranded RNA and has three segments with a total molecular weight of $6.2–7.5 \times 10^6$. The virus is sensitive to lipid solvents and low concentrations of formalin and beta-propiolactone; it is

labile in infected tissues after death. CCHF virus replicates in a wide variety of primary- and line-cell cultures; it can also be isolated and titrated by intracerebral inoculation of suckling mice.

A disease called Crimean haemorrhagic fever was first described in humans on the Crimean Peninsula in 1944. In 1969 it was shown that the agent of Crimean haemorrhagic fever was identical to the Congo virus isolated in 1956 in a child in the Congo, now Zaire. Since then, the CCHF virus or its antibodies have been found in many countries of eastern Europe, Asia and Africa.

Epizootiology/Epidemiology

The CCHF virus is transmitted by ticks of the genus *Hyalomma*, wild vertebrates, especially large herbivores, but small mammals and ground-frequenting birds are also an important reservoir of infection. The coincidence in distribution of CCHF virus and *Hyalomma* ticks strongly suggests that members of this genus are the most important vectors of the virus.

CCHF virus causes only inapparent infection or mild fever in livestock. Humans become infected either from tick bites or also when they come into contact with viraemic blood of young animals in the course of performing procedures, such as castrations, vaccinations, inserting ear tags or slaughtering, the infection being acquired through contact of viraemic blood with broken skin (Swanepoel 1994). Thus CCHF has to be considered as a zoonosis.

Pathogenesis

The CCHF virus probably undergoes replication at the site of inoculation being spread haematogenously and borne by the lymph to the visceral organs, the major site of replication. Capillary fragility is a feature of the disease and there is evidence of the formation of circulating immune complexes with complement activation which would contribute to damaging the capillaries and lead to renal and pulmonary failure. Endothelial damage accounts for the occurrence of a rash and contributes to haemostatic failure through stimulating platelet aggregation and degranulation, with consequent activation of the intrinsic coagulation cascade. Damage of the liver impairs synthesis of coagulation factors (Swanepoel 1994).

Clinical Signs in Humans

Since CCHF mostly occurs inapparently in domestic animals but is an important zoonosis, the clinical signs in humans are described. After an incubation period of 1–3 days, but occasionally 7 days and more, severe headaches accompanied by dizziness, neck pain and stiffness, sore eyes, photophobia, fever, rigor and chills set in. Further symptoms can be general myalgia and malaise, nausea, a sore throat, vomiting, abdominal pain followed by lassitude, depression and somnolence. A petechial rash appears on the trunk and limbs by the 3rd to 6th day of illness which may be followed by large bruises and ecchymosis. Sometimes a haemorrhagic tendency is evident. The mortality rate can be 30%, and death generally occurs between the 5th and 14th days of illness (Swanepoel 1994).

Pathology
This being a veterinary text book, the pathological features of the disease in humans are omitted but can be found by Swanepoel (1994).

Diagnosis
One ought to suspect CCHF in humans if a severe influenza-like illness with a sudden onset and a short incubation period, usually less than a week, occurs in persons exposed to tick bites or fresh blood and other livestock tissues. Laboratory diagnosis of the infection consists of isolation of the virus from serum or plasma or the demonstration of antibodies, the techniques being similar to those described for Rift Valley fever.

Control
CCHF can be controlled through the application of acaricides to livestock which of course are difficult under extensive farming conditions. Personnel involved with livestock should be aware of the disease and should take precautions to limit or avoid exposure to ticks and fresh blood or other tissues of animals.

In eastern Europe, inactivated vaccines to protect humans have been prepared from infected mouse brain (Swanepoel 1994).

1.5.1.7. Arboviroses of Local Occurrence

• Kisenyi Disease in Sheep

In Zaire, a viral disease in sheep has been recognized which causes high mortality and shows similar clinical and pathological features as with NSD, but cannot be transmitted to goats. As with NSD, *R. appendiculatus* is supposed to be the vector. The name of the disease has been taken from the district where it was described for the first time (Mitscherlich and Wagener 1970).

• Akabane Disease

Aetiology and Occurrence
The causal organism of Akabane disease is only one member of the Simbu group which belongs to the family *Bunyaviridae*, all of which may produce congenital defects. They are single-stranded RNA viruses with spherical enveloped virions, 90–100 nm in diameter.

Most of Africa, Asia (excluding Russia) and Australia may be regarded as enzootic for Akabane virus. The American continent and the island countries of the Pacific are free of infection.

Epizootiology, Pathogenesis, Clinical and Pathological Features
The virus has been isolated from mosquitoes in Kenya and midges (*Culicoides*), at first in Zimbabwe and later in other African countries and as well as in Australia and Japan, but so far it has not been transmitted experimentally to any vertebrate

by any species of insect, though is has been shown to multiply in experimentally infected *C. brevitarsis*.

In some cases, the virus causes abortions in the third trimester of pregnancy but mainly it produces congenital defects, principally arthrogryposis and hydranencephaly in both large and small ruminants. So far clinical cases have been found in Australia, Israel and Japan in calves, lambs and kids. If the animals have been infected in a late stage of pregnancy, calves may be born alive but blind and with signs of incoordination.

The pathological lesions are encephalomyelitis, hydranencephalus and cavitation of the cerebellum. Cows which have been infected earlier give birth to calves with arthrogryposis, torticollis, kyphosis, scoliosis and neurogenic atrophy of the muscles because of destruction of the motor neurons. Difficulties arise during parturition because of the cerebral lesions in the calf. In Australia, microencephaly occurs in sheep. Abortion is a characteristic symptom.

Imported animals which have been brought into enzootic areas are especially vulnerable. If young animals become infected, a massive immunity develops which protects the animals from giving birth to malformed calves and/or lambs when they become reproductive.

Diagnosis and Control

The disease can be diagnosed because of its epizootiology and because of the gross and histological pathological picture. Virus neutralizing antibodies can also be determined serologically.

To prevent the disease through vector control is extremely difficult since the range of potential vectors is poorly defined. In Japan, an inactivated vaccine has been produced from virus grown in hamster lung cell cultures. An inactivated virus vaccine is also under development in Australia (Merck 1991, St.George and Standfast 1994).

• Bovine Ephemeral Fever (BEF)
(Three-day-sickness, Lazy man's disease, Drie-Daeziekte, Fièvre éphémère, Fibre efimera, Dreitagekrankheit)

Aetiology and Occurrence

BEF is caused by a virus which belongs to the family *Rhabdoviridae*. It is a single-stranded negative-sense RNA virus with a bullet-shaped morphology. Being susceptible to both high and low pH and to atmospheric temperatures, it does not survive long outside its vertebrate or invertebrate host. There is no evidence of immunogenic diversity. The virus does not replicate in tissue cultures of bovine origin and loses immunogenicity in cultures of cell lines.

BEF is widespread in Africa including Madagascar but occurs also in the Near and Far East as well as in Australia.

Epizootiology, Pathogenesis, Clinical and Pathological Features

BEF is an arthropod-borne viral disease of cattle and water buffalo (*Bos bubalis*);

antibodies have also been found in African buffalo (*Syncerus caffer*) and a number of antelopes. The virus is also pathogenic for mice.

The BEF virus is transmitted by midges from the species *Culicoides* and *Anopheles*. Therefore, the highest incidence of the disease appears during the rainy season when outbreaks of the disease with a morbidity of up to 90% may suddenly occur. As little as 0.005 mL of infected blood are enough to be infectious.

BEF is mostly a self-limiting, disabling disease of a few days duration with rapid convalescence even in animals which are severely affected. After an incubation period of 2–3 days, high fever suddenly sets in (41.5 °C) followed by dyspnoea, secretion of the nostrils and eyes, swelling of the eyelids as well as malaise and a reduction in milk production. The animals groan and grind their teeth. Atony of the rumen and tympany as well as constipation and diarrhoea are typical symptoms. After the first day of the febrile reaction myalgias, swelling of the joints, lameness, and as a consequence the typical stiff gait appear. Affections of the muscles of the pharynx, sometimes through oedemas cause difficulties in swallowing and may in some cases cause pneumonia because of choking.

Sick animals keep to themselves on the pasture just lying around, and it is difficult to force them to get up; they walk in pain and with a bent back, although the head and neck are kept straight.

The disease does not last usually longer than 3–4 days. After recovery, milk production may be recuperated completely if no complications occur. Bulls remain infertile for weeks after having recovered. Mortality is restricted to adult animals of exotic highly productive breeds and rarely is more than 2%.

Lesions which can be observed at necropsy are tumours of the spleen and lymph nodes, increased and reddish-stained pericardial liquid as well as catarrhalic gastrointestinal inflammation.

Diagnosis
The disease is characterized by its epizootiology and clinical features. The BEF virus can be isolated in tissue cultures or by intracerebral inoculation of baby mice which die 3 days p.i. showing convulsions and paralysis in the caudal area of the body. Susceptible cattle may also be infected with blood from a suspected BEF carrier. Serological confirmation of BEF may be obtained by the detection of a rise in the neutralizing antibody titre within the sera collected from individual animals during acute illness and again 2–3 weeks later.

Treatment
BEF is one of the rare viral diseases where treatment is possible. Rest, parenteral rehydration, anti-inflammatory treatment (steroids) and, if signs of hypocalcaemia are observed, Ca-borogluconate treatment are effective.

Control
Since the vectors involved in transmitting BEF have not been identified completely there is no way of interrupting the cycle of the virus within the vector stage of its life cycle through vector control.

In enzootic areas, dairy- and feed-lot herds as well as valuable breeding stock can be immunized with attenuated tissue culture vaccines containing an adjuvant. In Japan, vaccination with attenuated virus is boosted by inoculation with a formaldehyde-inactivated virus (Merck 1991, Seifert 1992, St. George 1994).

- *Wesselbron Disease (WSL)*
(Wesselbron virus infection)

Aetiology and Occurrence
WSL is caused by a flavivirus with a diameter of about 30 nm which has a particular tropism to the cells of the liver parenchyma. The virus can be grown in tissue culture, embryonated hens' eggs, guinea-pigs and rabbits. Though it has properties similar to the RVF virus, adult mice in contrast are not susceptible.

The history of WSL is interwoven with that of RVF; it was described for the first time in the Wesselbron district of the Oranje Free State of South Africa. Since then, serological surveys have provided evidence that WSL also occurs in many African countries as well as in Madagascar where it has also been isolated from vertebrates and arthropods.

Epizootiology, Pathogenesis, Clinical and Pathological Features
WSL is an acute arthropod-borne viral infection in sheep, cattle and goats, being the most pathogenic for new-born lambs, kids and pregnant ewes; it causes non-clinical infections in non-pregnant adult sheep, goats, cattle, horses and pigs. In humans, the infection causes a short influenza-like febrile disease. The virus is transmitted by aedine mosquitoes. Ixodic ticks may also be vectors, and thus it has to be considered a zoonosis.

The virus possesses hepatotropic and neurotropic properties and shares many clinical and pathological features with RVF while it is suggested that the pathogenesis of the two diseases is similar.

In new-born lambs and kids, the incubation period of 1–3 days is followed by non-specific signs of illness, biphasic fever, anorexia, listlessness, weakness and increased abdominal respiration. Death may supervene within 72 hours of infection. In pregnant animals, abortion or retention and mummification of the foetus can occur. In humans, typical signs of influenza set in after an incubation period from 2–7 days. Most patients recover from the acute illness in 1–3 days.

During necropsy, lambs, kids and calves show severe icterus, slight to moderate hepatomegaly, the liver being yellowish to orange-brown and sometimes mottled with congested patches. Petechiae and ecchymoses are found in the mucosa of the abomasum, the contents of it being chocolate-brown because of partially digested blood. The lymph nodes are swollen. Histologically, lesions are found which are similar to RVF.

Diagnosis
The epizootiology of the disease and its occurrence during the rainy season in new-born lambs or kids are indications for the possible presence of WSL.

The virus can be isolated from blood and visceral organs by intracerebral inoculation of new-born mice and on tissue cultures (BHK21 cells).

Serodiagnosis of WSL is based on haemagglutination inhibition (HAI), complement fixation (CF) and virus neutralization (NT) tests.

Control

Management of lambing sheep and goats combined with vector control (dusting) may contribute to the prevention of WSL.

The most efficient method of control is vaccination with an attenuated WSL virus vaccine prepared from a strain attenuated by 145 passages in new-born mice. The vaccine virus is grown in BHK21 cells (Seifert 1992, Swanepoel and Coetzer 1994).

1.5.2. Rabies
(Lyssa, Rage, Rabia, Tollwut)

Following the structure of the content of this book into epizootiological complexes, rabies has been grouped with the vector-borne diseases since it is mainly transmitted by a real vector. A special feature of this mode of transmission is that the vectors become ill themselves and because of the psychological changes induced by this disease, are additionally stimulated to transmit the infection by biting.

Aetiology

Rabies is caused by a virus of the family *Rhabdoviridae* and the genus *Lyssavirus*, now also called *Lyssavirus* 1. The antigenic structures of rabies virus biotypes are stable in nature and not easily affected by passage in laboratory hosts or cell cultures. The rabies virus is bullet-shaped, measures 180 × 75 nm and consists of a nucleocapsid, size 160 × 50 nm, which is composed out of three proteins (N, NS and L) surrounded by a bilayer lipid envelope derived from host cell membranes (M), through which flattened spikes or peplomeres project; it contains a RNA genome. The outer virus envelope consists of a glycolipid (G-antigen) which can be isolated from the lipid cover of the virus. It represents the haemagglutinating as well as immunizing component of the antigen.

Though being sensitive to sunlight and ultraviolet radiation, the rabies virus is rather resistant within the environment: autolysis and putrefaction destroy the infectivity of the virus rather slowly, and carcasses may contain the live virus for up to 90 days. The virus is inactivated in an acid range, and therefore acid disinfectants are best suited for decontamination. Infectivity is destroyed also quickly by 0.2% quaternary ammonium compounds, 1% soap solution, 5–7% iodine solution, 45–70% alcohol, 50% ether, heating at 56 °C for 30 minutes, and after a few hours, by low concentrations of formalin or betapropiolactone.

Under natural conditions the rabies virus appears as "street virus" as well as "forest" and "vampire" virus (*Virusparalyssa*). The rabies virus when modified through intracerebral passages in rabbits is called *Virus fixe*; the *Flury virus* as well as the strains LEP, HEP and Kelev have been attenuated through passages in the embryos of hens' eggs. Strains which have been attenuated through passages in tissue culture are for example ERA, SAD, PV Paris and Wirab. The attenuated strains have lost their virulence partly or completely for humans or animals. These virus types are immunologically uniform.

Rabies virus can be grown rather easily on BHK21, CER and Nil2 (all of hamster origin), Vero cells and neuroplastoma cell lines of murine and human origin (Seifert 1992, Swanepoel 1994).

Occurrence
Countries free of rabies are mainly islands and peninsulas, like Great Britain, Ireland, Iceland, Sweden, Norway, Denmark, Portugal, Spain, Gibraltar, Malta, Albania, Cyprus, Bahrain, Qatar, Hong Kong, Malaysia, Singapore, certain Indonesian and Philippines islands, the Republic of Korea, Japan, Australia, New Zealand, Fiji, Hawaii and certain other western Pacific and Caribbean islands, as well as Libya, Cape Verde, Sao Tomé, Comoros, Mauritius and Antarctica, while rabies is present in the rest of the world. In Oman and the U.A.E. which are supposed to be free of rabies (Swanepoel 1994), the infection has been observed in dromedaries (Wernery and Kaaden 1995). In Europe, there were already descriptions of outbreaks of rabies involving dogs and foxes as well as wolves in the 11th and 13th centuries. The rabies transmitted by vampires (*Paralyssa*) occurs in Latin America and is characterized by its particular mode of transmission (Seifert 1992, Swanepoel 1994).

Epizootiology
In principle, all warm-blooded mammals as well as birds are susceptible to infection with the rabies virus; cold-blooded animals do not show symptoms of rabies but are able to excrete the virus. Most wild living carnivores are extremely susceptible, especially also those in the tropics (foxes, coyotes and jackals); skunks, raccoons, cats, cattle, mongooses and most rodents are highly susceptible; dogs, sheep, goats, horses and primates, including humans are moderately susceptible, while the opossum has a low susceptibility. The dose for infecting a fox is $10^{0.5}$ MICLD$_{50}$, for cattle $10^{3.5}$ and for dogs 10^6. Rabies has also been described in dromedaries showing symptoms which are similar to those described below for cattle (Wernery and Kaaden 1995).

Infected animals excrete the virus with the saliva which may happen already before the animals show clinical symptoms (in dogs 5 days before). Urine, milk, faeces, mucus of the trachea and blood may also contain virus but are of no epizootiological importance (vector-borne disease). The transmission from animal to animal and possibly on to humans takes place through biting thus implanting the virus with the infected saliva into the system of the susceptible organism. Reports of the non-biting transmission of rabies have only dealt with sporadic incidences. Wounds which are possibly open may be infected through saliva or aerosols which contain the virus. The severity, location and multiplicity of bites inflicted on the victim also influence the outcome of potential exposure to infection, and bites on the head and neck are generally associated with the shortest incubation periods and the highest mortality rates. The severity and location of bites are in turn influenced by the relative sizes of the vector. Herbivores and humans are final hosts and are of no importance as vectors. The cycle of infection is only maintained by carnivores or vampires. Depending on the animal species which is present locally as the main vector and reservoir of infection, three epizootiological/epidemiological complexes of rabies have to be distinguished:

- urban rabies, sustained by dogs and cats, whereby in 90% dogs are the main vectors. The final links of the chain of infection are other domestic animals, but especially cattle and humans;
- sylvatic rabies, which is sustained by wild living carnivores: the fox in Europe, furthermore skunks and raccoon in America, foxes and wolves in Asia, and jackals and predatory cats in Africa. In the Caribbean mongoose are vectors of rabies. Sylvatic rabies can become urban rabies thanks to dogs and wild carnivores which enter urban areas;
- vampire rabies (*Paralyssa*) or *Rabia paresiante* in Latin America which is characterized by its particular mode of transmission and the typical course of disease especially in cattle. *Rabia paresiante* is transmitted by blood-licking bats of the order *Chiroptera* mainly *Desmodus rotundus* (vampire). It was suspected as early as 1911 that blood-licking bats are vectors of rabies in cattle in Brazil. It was not until 1934 that it could be proven. In the pasture ranges east of the Andes as well as in Central America and in tropical South Mexico, virgin forests and savannahs are more and more being occupied by animal production. Cattle have probably been infected by dogs which leads to the infection of vampires which feed on infected cattle. The infected vampires come down with symptoms of "raging fury" but may also only fall ill latently. Like carnivores, rabid vampires change their behaviour, lose their shyness and also seek out cattle on pastures in developed areas for their blood meal which they normally would avoid because of their natural shyness. It has also been reported that infected vampires enter the huts of peasants and attack sleeping people. *Paralyssa* appears only regionally on the range where vampires are infected. Apparently the population of vampires needs a certain level of infection until an outbreak of rabies in cattle can occur. Suddenly an enzootic with a high mortality rate flares up within the susceptible cattle population which later on gradually subsides and is followed by an interenzootic phase without cases of rabies. Only whenever a susceptible population of cattle has been built up again which is not immunized through a natural infection, a new disease outbreak will be due. Since the animal holders are aware of the particularities of the course of *Paralyssa*, they vaccinate only as long as the infection is present. During the interenzootic phase, vaccination is mostly neglected which is why a new outbreak of rabies may appear as a consequence (Delpietro and Nader 1988).

Prior to the introduction of vaccination of cattle against rabies, 20–50% of the cattle which were kept extensively on pastures infested with infected vampires east of the Andes, Central America and Mexico as well as Trinidad, died. Nowadays, rabies has also been transmitted from vampires to fruit- and insect-eating bats. Therefore, since the 1950s bat rabies has also occurred in the USA where vampires do not exist. Because vampires live in colonies in caves, niches of rocks or hollow trees, the rabies virus is easily spread within the colony, thus increasing the pressure of the infection on the population of domestic animals.

Pathogenesis

The most superficial portal for entry of rabies virus into the nervous system is the sensory nerve ending of the epithelial and subepithelial tissues of the skin and mucous membranes, when transmission results from superficial bites, the licking of mucous membranes or shallow skin wounds and abrasions, and ingestion or inhalation of infected material. The incubation time depends on the amount of virus which has been inoculated through the bite. It can fluctuate between a week and almost a year. Since vampires feed for a rather long time on an animal, they are able to implant large amounts of the virus into the wound. Therefore, the incubation time can be very short in *Rabia paresiante*, perhaps only 24 hours in small ruminants (I/4.1.6).

After implantation with the infected saliva, the virus enters the nerve endings immediately and leaves the site of inoculation rapidly, accumulating at motor nerve endings within one hour of inoculation. From there the virus progresses centripetally to the CNS, the transport being passive of genome-containing particles, through the retrograde axoplasmic flow of the central nervous system. During this process the first replication of virus takes place. The spread of virus in the spinal cord proceeds via axons and dendrites, and the process is thought to involve either prior maturation of virions through budding on intracytoplasmic membranes, or through the direct transfer of the genome-containing moiety through membrane fusion at synaptic junctions. Infection may occasionally be limited to the spinal cord. Spread of infection is rapid within the brain. Through the dendrites of the ganglionic cells, the virus progresses from cell to cell and is also disseminated by the liquor through the brain. After further intensive virus replication, the virus gets back to the periphery and invades the saliva glands, actively through replication in the nerve cells, and passively, centrifugally along the nerve tracts. The virus replication in non-neural tissues takes place in the cornea ephitelium, the saliva glands and cells of the brown fat tissue. After the centrifugal phase, the virus appears in all organs of the body. Apparently the haematogenous spread of the virus is of no importance to the pathogenesis of the disease. So far it is not fully understood why by the oral or intranasal infection, e.g. during oral vaccination of foxes a challengable immunity is produced. The mechanism of generalization of such an infection is probably different from the natural infection. It has, however, been shown that rodents, other laboratory animals, foxes, skunks and kudus are susceptible to oral infection, but that infection occurs with greater efficiency when there are mouth lesions, and that virus replication occurs first in epithelial cells or tonsils and sensory organs, such as taste buds, and later in the nerve bundles of the submucosa of the oral cavity and tongue (Seifert 1992, Swanepoel 1994).

Clinical Features
The classical course of rabies is characterized through three stages:

- the prodromal stage, at the beginning, characterized by a change of behaviour of the animal which often may be only slight and thus may be overlooked. Shyness, nervousness, irritation, difficulties in swallowing and sometimes salivation appear. These symptoms last only for a few days;
- the excitation stage characterized by restlessness, excitement, aggressiveness, a mania for biting and snapping – during this stage the saliva is highly infective. If the syndrome of the disease is determined by excitement, this stage is also called "raging fury";
- the paralysis stage which, with classical rabies, only appears shortly before death. At this stage, paralysis of the muscles of the face, trunk and the limbs appears. If the symptoms of paralysis are prominent, this stage is called "silent fury".

The rabies transmitted by vampires, the paralytical rabies or *Rabia paresiante*, shows a special course in cattle and small ruminants with pre-dominant symptoms of paralysis. A few days after being bitten by the vampire – and this can be recognized by the distinct parallel tear-shaped lesions on the brisket, stifle fold or laterally at the croup caused by the incisors of the vampire – only a few typical symptoms like digestive disorders, reduced feed consumption and milk production, tympany and obstipation appear. These are followed by a curious behaviour of the animals: curiosity, sniffing, enhanced excitability, over-sensitive hearing and enhanced sensitivity of the skin and horns. The animals become restless, show flehmen, yawning, they throw their head back, kick their limbs and bellow loudly. They stand with a bent back, strain permanently whereby air is sucked into the rectum and again pressed out. After perhaps only short excitation, paresis and paralysis of the hind legs develop. The animals try to get up on their fore legs but break down on their swaying hind quarters. Finally, the distinct position of sitting like a seal is assumed whereby the hind legs are extended both caudally and laterally in an unphysiological position (Fig. 51 on page 1 annex). At the same time, the patients may salivate profusely. The course of disease only lasts for a short period (3–4 days) and the animals die lying sideways with rowing movements of their legs. The animals are often found dead in almost a pit which they have burrowed into the soil of the paddock during agony. Liquid secretions appear around the mouth and anus.

Pathology
Post mortem the distinct situation of the carcass may be obvious. With paralyssa the characteristic lesions caused by the teeth of the vampire are easily to be

identified. Furthermore, lesions are found which have been caused because the animal has broken down repeatedly and has damaged its skin especially around the coxae. The animals may also be emaciated and there may be a self-inflicted injury at the site of infection in carnivores.

There are no consistent macroscopic lesions to be found at necropsy; the only visible abnormality may be congestion of the blood vessels of the leptomeninges.

Significant microscopic lesions are found in the CNS and cranial and spinal ganglia. They consist of perivascular cuffing, focal and diffuse gliosis, neuronal degeneration, and intracytoplasmic inclusions, the "Negri" bodies, in neurons. These are sharply defined, round, acidophilic inclusions, which measure 2.0–8.0 µm in diameter, but they assume an elongated shape in axons and dendrites. They may contain a basophilic internal structure, or one or more vacuoles, and are sometimes surrounded by a clear halo (Seifert 1992, Swanepoel 1994).

Diagnosis

Rabies should always be suspected when animals behave strangely and become aggressive. In the living animal, a presumptive diagnosis can be established by demonstrating the antigen on the cornea of the eye by means of the Schneider cornea test (cited from Rolle and Mayr 1993). In this test, the virus is demonstrated with direct immunofluorescence in an impression smear of the eye.

Post mortem, the isolation of virus by intracerebral inoculation of brain suspension into weaned mice was the standard diagnostic method prior to the adoption of the immunofluorescence test. To perform this test, a 10% suspension in saline solution from hyppocampus, medulla or salivary glands adding 10% rabbit serum is injected intracerebrally or i.m. in amounts of 0.03 mL into white mice; 5–20 days p.i. paralysis of the hind legs develops.

The "Negri" bodies can also be easily found histologically with acid fuchsin-methylene blue or similar stains, while now the standard method of diagnosis to identify the rabies virus antigen is in impression smears of fresh brain (cerebrum, cerebellum, hyppocampus, medulla, thalamus and brain stem) by immuno-fluorescence. The test which takes only 1–3 hours to perform is comparable in sensitivity to mouse inoculation. Rabies virus can also be found in impression smears stained by the Sellers method.

The virus may be isolated in a variety of line cell cultures. Since the rabies virus has almost no cytopathic effect, the virus isolation has to be confirmed with ELISA or immunofluorescence. For epidemiological purposes viral cDNA prepared from infected tissue by reversed transcription and PCR is analysed using the endonuclease restriction mapping method or by nucleotide sequencing in order to identify the biotype of the virus (Sacramento et al. 1991).

Serological methods to demonstrate the virus indirectly are ELISA, indirect immunofluorescence for which now monoclonal antibodies are available, complement fixation, haemagglutination-inhibition, radio-immunoassay, enzyme-linked immunoassay and neutralization. Test kits for performing the ELISA are available commercially (Seifert 1992, Swanepoel 1994).

Treatment
There is no treatment for animals which come down with rabies. Sick animals have to be put down.

Control
As a means of vector control vaccination of pet dogs and cats is mandatory in many countries. In Germany, Switzerland, Italy, Austria, Luxembourg, Belgium and France, an oral vaccination for foxes with the attenuated SAD-B19 rabies virus strain, distributed as a bait in enzootic areas, has been very successful and has virtually eliminated fox rabies in several areas of these countries. In Latin America, the elimination of vampire colonies has been an effective tool in controlling *Rabia paresiante* (I/4.3). Campaigns carried out regularly have considerably reduced the incidence of rabies, in the Bolivian Chaco for instance.

For active vaccination of exposed domestic animals, a range of highly effective, safe and thermostabile inactivated vaccines, prepared from virus grown in a variety of primary and line cell cultures is available. Virus strains which have been attenuated through passage in embryonated hens' eggs (Flury, LEP, HEP, Kelev) are now being replaced by those attenuated through passages on tissue cultures (ERA, Wirab, PV Paris). Vaccines which have been prepared from virus, multiplied in the brain of mice, goats and sheep which are still manufactured in some developing countries should not be used any more since they contain neural tissue which may produce serious allergic reactions after vaccination. Because of a higher safety margin, inactivated vaccines are still recommended in developing countries. For the vaccination of cattle in areas where *Rabia paresiante* is enzootic, only inactive vaccines should be applied in order to prevent the vaccine virus from getting back into the vampire and perhaps reverting back into a virulent strain. Recombinant vaccines made from harmless viruses incorporating nucleotide sequences coding for the G-protein of the rabies virus are not yet commercially available.

Active immunization of animals can only be a prophylactic measure which has to be carried out before an infection takes place or is supposed to have occurred. All domestic animals can be protected by vaccination. *Rabia paresiante* is best prevented when weaned calves are vaccinated at an age of 6–8 months with a booster vaccination after 6 weeks. An inactivated tissue culture vaccine should be used. The vaccination has to be repeated in intervals of 1–2 years. Cows calving for the first time have to be given a booster in order to obtain an efficient passive colostral immunization for their calves. Where an efficient vaccine is used, it can be enough to give the calves and the heifers a booster shot before calving in pasture ranges infested with vampires. Later on, this initial immunoprotection is boosted naturally through sporadic infections by vampires. In Bolivia and Brazil, cattle are vaccinated at present with an attenuated live vaccine against *Paralyssa*. It is claimed that this produces a protection which lasts for 3 years. Since *Paralyssa* appears in cyclical enzootics, it is necessary to continue vaccination even if the rabies apparently has disappeared during an interenzootic phase.

References

Abdala, A. A., E. Pipano, D. H. Aguirre, A. B. Gaido, M. A. Zurbriggen, A. J. Mangold and A. A. Guglielmone (1990): Frozen and fresh *Anaplasma centrale* vaccines in the protection of cattle against *Anaplasma marginale* infection. Rev. Elev. Méd. vét. Pays trop. **2**, 155–158.

Aboytes, R., G. M. Buening, J. V. Figueroa and C. A. Vega (1991): El uso de sondas de ADN para el diagnóstico de hemoparásitos. Rvta. Cub. Cienc. Vet. **22**, 173–181.

Anziani, O. S., H. D. Tarabla, C. A. Ford and C. Galleto (1987): Vaccination with *Anaplasma centrale*: Response after an experimental challenge with *Anaplasma marginale*. Trop. Anim. Hlth. Prod. **19**, 83–87.

Bergey, H. D., N. R. Krieg and J. G. Holt (1984): Bergey's Manual of Systematic Bacteriology, 1, 2nd ed., Williams & Wilkins, Baltimore/London.

Bezuidenhout, J. D. (1981): The development of a new heartwater vaccine using *Amblyomma hebraeum* nymphae infected with *Cowdria ruminantium*. In: Tick Biology Control (eds: Whitehead and Gibbson), Rodes University, Grahamstown, S.A., 41–45.

Boch, J. and R. Supperer (1983): Veterinärmedizinische Parasitologie. 3. Aufl. Parey, Berlin/Hamburg.

Brett, S. and J. D. Bezuidenhout (1989): Mass production of *Cowdria ruminantium* in tissue culture with the aim to replace the current heartwater blood vaccine with a tissue culture vaccine. Newslett. Onderstepoort vet. Res. Inst. **7(3)**: 9.

Brown, C. C., L. L. Logan, C. A. Mebus and K. Nagorski (1989): Protection of goats against caribbean and african heartwater isolates by the Ball 3 heartwater vaccine. Trop. Anim. Hlth. Prod. **21**, 100–106.

Brown, C. G. D., A. G. Hunter and A. G. Luckins (1990): Diseases caused by protozoa. In: Handbook on animal diseases in the tropics (eds: Sewell and Brocklesby), 4th ed., Baillière Tindall, London.

Callow, L., Q. C. Quiroga and P. J. McCoisker (1976): Serological comparison of Australian and South American strains of *B. bigemina* and *A. marginale*. Int. J. Parasit. **6**, 3O7–31O.

Chema, S., R. S. Chumo, T. T. Dolan, J. M. Gathuma, A. D. Invin, A. D. James and A. S. Young (1987): Clinical Trial of Halufuginon Lactate for the Treatment of East Coast Fever in Kenya. Vet. Rec. **129**, 575–577.

Coetzer, J. A. W. and B. J. Erasmus (1994): African horsesickness. In: Infectious diseases of livestock (eds: Coetzer, Thomson and Tustin), 1, 460–472, Oxford University Press, Cape Town/R.S.A.

De Mattos, C. C., C. A. De Mattos, B. I. Osborne, C. A. Dangler, R. Y. Chuang and R. H. Doi (1989): Recombinant DNA-probe for serotype-specific identification of bluetongue virus. Amer. J. vet. Res. **50**, 536–541.

Delpietro, H. A. and A. J. Nader (1988): La rabia de los herbívoros transmitida por vampiros en el noreste argentino. Rev. sci. tech. Off. int. Epiz. **1**, 177–187.

De Vos, A. J. and F. T. Potgieter (1994): Bovine babesiosis. In: Infectious diseases of livestock (eds: Coetzer, Thomson and Tustin), 1, 278–294, Oxford University Press, Cape Town/R.S.A.

Diall, O., V. M. Nantulya, A. G. Luckins, B. Diarra and B. Kouyate (1992): Evaluation of mono- and polyclonal antibody-based antigen detection immunoassays for diagnosis of *Trypanosoma evansi* infection in the dromedary camel. Rev. Elev. Méd. vét. Pays trop. **45**, 149–153.

Dolan, T. T. (1983): Discussion Theileriosis. Proc. IV Inter. Konf. Inst. Trop. Vet. Med. Orlando, Florida, USA.

Dolan, T. T. (1989): Theileriasis, a comprehensive review. Rev. sci. Tech. Off. int. Epiz. **8**, 11–36.

Elhassan, E., B. O. Ikede and O. Adeyemo (1994): Trypanosomosis and reproduction: I. Effect of Trypanosoma vivax on the oestrous cycle and fertility in the ewe. Trop. Anim. Hlth. Prod. **26**, 213–218.

Emery, D. L. and W. I. Morrison (1980): Generation of autologous mixed leucocyte reaction during the course of infection with *Theileria parva* (East Coast fever) in cattle. Immunology **40**, 229–237.

FAO/OIE (1972): Ad hoc consultation on control of protozoal tick-borne diseases of cattle, EAVRO, Muguga.

Fivaz, B. H., R. A. I. Norval and J. A. Lawrence (1989): Transmission of *Theileria parva bovis* (Boleni strain) to cattle resistant to the brown ear tick *Rhipicephalus appendiculatus* (Neumann). Trop. Anim. Hlth. Prod. **21**, 129–134.

Friedhoff, K. F. (1987): Transmission of *Babesia*. In: Babesiosis of Domestic Animals and Man (ed: Ristic, M.), Boca Raton, Florida, CRC Press.

Hassanain, M. M., A. I. Al-Afaleq, I. M. A. Soliman and S. K. Abdullah (1990): Detection of African horsesickness (AHS) in recently vaccinated horses with inactivated vaccine in Qatar. Rev. Elev. Méd. vét. Pays trop. **1**, 33–35.

Hinaidy, H. K. (1981): Die Babesiosen in Österreich. IV. Versuche mit Totimpfstoffen. Berl. Münch. tierärztl. Wschr. **94**, 121–125.

Hörchner, F. (1983): Neue Erkenntnisse über Pathogenese und Bekämpfung der Trypanosomiasis der Haustiere. Die Blauen Hefte, 67.

Jongejan, F. (1991): Protective immunity to heartwater (*Cowdria ruminantium* infection) is acquired after vaccination with *in vitro*-attenuated rickettsiae. Infect. Immun. **59**, 729–731.

Gueye, A., F. Jongeljan, M. Mbengue, A. Diouf and G. Uilenberg (1994): Field trial of an attenuated vaccine against heartwater. Rev. Élev. Méd. vét. Pays trop. **47**, 401–404.

Kreier, J. P. and M. Ristic (1984): Genus IV. Eperythrozoon Schilling 1928, in Bergey's Manual of Systematic Bacteriology, 1, 726–729. Williams & Wilkins Co., Baltimore/London.

Kuttler, K. L. (1975): The effect of Imidocarb treatment on Babesia in the bovine and the tick. Res. Vet. Sc. **18**, 198–2OO.

Lampe-Schneider, G. (1984): Untersuchungen zur Bedeutung der dritten Komplement-Komponente (C_3) als Parameter der Resistenz bei autochthonen und exotischen Rindern in Westafrika. Diss. FB Agrarwiss., Göttingen.

Levy, M. G. and M. Ristic (198O): Babesia bovis: Continuous cultivation in a microaerophilous continuous culture. Science, **2O7**, 1218–122O.

Lomo, P. O., D. W. Makawiti and V. N. Konji (1995): Thyroid status and adenosine triphosphatase activity in experimental *Trypanosoma congolense* infection in rabbits. Brit. Vet. Journ. in press.

Luckin, A. G. (1992): Methods for diagnosis of trypanosomiasis in livestock. WAR/RMZ **70/71**, 15–20.

Magonigle, R. A. and R. J. Newby (1984): Response of cattle upon reexposure to *Anaplasma marginale* after elimination of chronic carrier infections. Am. J. vet. Res. **45**, 695.

Manickam, R., S. Dhgar and R. P. Singh (1983): Protection of Cattle against Theileria annulata infection using Corynebacterium parvum. Trop. Anim. Hlth. Prod. **15**, 2O9–213.

McHardy, N. (1983): Immunisation against Anaplasmosis – a review. Prev. Vet. Med. 2.

Mckeever, D. J. and W. I. Morrison (1990): *Theileria parva*: The nature of the immune response and its significance for immunoprophylaxis. Rev. sci. tech. Off. int. Epiz. **9**, 2, 405–421.

Mehlhorn, H. and E. Schein (1984): The piroplasms: Life cycle and sexual stages. Adv. Parasitol. **23**, 37–103.

Mehlhorn, H. and G. Piekarski (1989): Grundriß der Parasitenkunde, 3. Aufl. Fischer, Stuttgart.

Merck (1991): The Veterinary Manual, 7th ed., Merck & Co.Inc., Rahway, USA.

Mitscherlich, E. and K. Wagener (197O): Tropische Tierseuchen und ihre Bekämpfung. 2. Aufl. Parey, Berlin/Hamburg.

Münstermann, S., R. J. Mbura, S. H. Maloo and K.-F. Löhr (1992): Trypanosomiasis control in Boran cattle in Kenya: A comparison between chemoprophylaxis and a parasite detection and intravenous treatment method using isometamidium chloride. Trop. Anim. Hlth Prod. **24**, 17–27.

Mwangi, D. M., W. K. Munyua and P. N. Nyaga (1990): Immunosuppression in caprine Trypanosomiasis: Effects of acute *Trypanosoma congolense* infection on antibody response to Anthrax spore vaccine. Trop. Anim. Hlth. Prod. **22**, 95–100.

Nantulya, V. M. and K. J. Lindqvist (1989): Antigen detection enzyme immunoassays for the diagnosis of *Trypanosoma vivax, T.congolense* and *T.brucei* infections in cattle. Trop. Med. Parasitol. **40**, 267–272.

Nantulya, W. M. (1990): Trypanosomiasis in domestic animals: The problems of diagnosis. Rev. sci. tech. Off. int. Epiz. **9**, 357–367.

Nash, A. S. and P. A. Bobade (1993): Haemobartonellosis. In: Rickettsial and chlamydial diseases of domestic animals (eds: Z. Woldehiwet and M. Ristic), Pergamon Press, New York, 89–129.

Nessiem, M. G. (1994): Evaluation of the silicone centrifugation technique in the detection of *Trypanosoma evansi* infection in camels and experimental animals. Trop. Anim. Hlth Prod. **26**, 227–229.

OAU/STRC (1978): Information leaflet, 26, 1O. Rev. sci. tech. Off. int. Epiz. **8**, 11–36.

Olivier, J. A., A. P. Lötter, W. G. E. Schwulst and O. Matthee (1988): Chemoprofilakse van hartwater met behulp van 'n onderhuidse pil of pasta. Newslett. Onderstepoort vet. Res. Ins. **6(1)**, 7–9.

Onyeyili, P. A., G. O. Egwu, L. T. Zaria and B. A. Orjiude (1991): DL-alpha-difluoromethylornithine (DFMO)-Berenil combination: therapeutic and prophylactic activity against *Trypanosoma brucei brucei* infection in mice. Rev. Elev. Méd. vét. Pays trop. **44**, 443–445.

Reynaud, J. P., P. L. Toutain, T. Baltz and K. R. Sones (1989): Plasma kinetics, toxicity and tolerance of Cymelarsan in horses, cattle and camels. Twentieth ISCTRC Meeting, Mombasa, Kenya, April 10–14.

Ristic, M. (1968): Anaplasmosis. In: Infectious Blood Diseases of Man and Animals (eds: Weinman, D. and M. Ristic), Academic Press, New York/London, 473–542.

Rodriguez, O. N., P. Rodriguez, A. Rivas and L. Espaine (1988): La prueba de inmunodifusión radial simple – fijación del complemento. Consideraciones sobre su aplicación en el diagnóstico de la anaplasmosis y babesiosis bovina. Revta. Cub. Cienc. Vet. **19**, 3, 171–178.

Rolle, M. and A. Mayr (1993): Mikrobiologie, Infektions- und Seuchenlehre. 6. Aufl. Enke, Stuttgart.

Rüchel, W. M. (1974): Chemoprophylaxe der bovinen Trypanosomiasis. Diss. FB Agrarwiss, Göttingen.

Sacramento, D., H. Bourhy and N. Tordo (1991): PCR technique as an alternative method for diagnosis and molecular epidemiology of rabies virus. Molec. Cell. Prob. **5**, 229–240.

Salabarria, F. F. and R. Pino (1988): Vertical transmission of *Anaplasma marginale* in cattle infected during late stages of pregnancy. Rev. Cub. Cienc. Vet. **3**, 179–182.

Sanchez, P. A., M. Alvarez and C. O. Cordoves (1991): Aglutinación rápida en látex: Nueva técnica para el diagnóstico de la babesiosis en Cuba. Rvta. Cub. Cienc. Vet. **22**, 31–34

Schein, E. and W. P. Voigt (1979): Chemotherapy of bovine theileriosis with halofuginone. Acta trop. **36**, 391–394.

Schliesser, Th. and H. Krauss (1982): Bekämpfung des Q-Fiebers. Tierärztl. Prax. **1O**, 11–22.

Scott, G. R. and Z. Woldehiwet (1993): Bovine petechial fever. In: Rickettsial and chlamydial diseases of domestic animals (eds: Z. Woldehiwet and M. Ristic), Pergamon Press, New York, 255–268.

Seifert, H. (1962): Beobachtungen über die Epidemiologie der Piroplasmose an der Küste und in den Tälern der Cordillere Nord-Perus unter den Bedingungen des Wechsels zwischen Trocken- und Regenzeit. Zbl. Vet. Med. **1O**, 989–998.

Seifert, H. S. H. (1962): "Blue tongue" in einer Corriedale Schafzucht der Kordillere Nord-Perus. Vet. Med. Nachr. **3**, 161–166.

Seifert, H. S. H. (1971): Die Anaplasmose. Schaper, Hannover.

Seifert, H. S. H. (1983): Serumlysozym, hämolytisches Komplement und C_3 als Parameter der relativen Resistenz von autochthonen Rinderrassen. Fortschr. Vet. Med. **15**, 174–185.

Seifert, H. S. H. (1989): Wasser und Vektorenseuchen. Göttinger Beitr. Land. Forstwirt. **41**, 131–142.

Seifert, H. S. H. (1992): Tropentierhygiene. Fischer, Jena.

Seifert, H. S. H. (1995): Unpublished empirical findings.

Soltys, M. A. (1983): A review of studies on immunization against protozoan diseases of animals. Proc. IV. Int. Konf. Inst. Trop. Vet. Med. Orlando, Florida, USA.

St. George, T. D. (1994). Bovine ephemeral fever. In: Infectious diseases of livestock (eds: Coetzer, Thomson and Tustin), 1, 553–562, Oxford University Press, Cape Town/R.S.A.

St. George, T. D. and H. A. Standfast (1994): Diseases caused by Akabane and related Simbu-group viruses. In: Infectious diseases of livestock (eds: Coetzer, Thomson and Tustin), 1, 681–687, Oxford University Press, Cape Town/R.S.A.

Swanepoel, R. (1994): Crimean – Congo haemorrhagic fever. In: Infectious diseases of livestock (eds: Coetzer, Thomson and Tustin), 1, 721–729, Oxford University Press, Cape Town/R.S.A.

Swanepoel, R. (1994): Rabies. In: Infectious diseases of livestock (eds: Coetzer, Thomson and Tustin), 1, 491–566, Oxford University Press, Cape Town/R.S.A.

Swanepoel, R. and J. A. W. Coetzer (1994): Rift Valley fever. In: Infectious diseases of livestock (eds: Coetzer, Thomson and Tustin), 1, 688–717, Oxford University Press, Cape Town/R.S.A.

Swanepoel, R. and J. A. W. Coetzer (1994): Wesselbron disease. In: Infectious diseases of livestock (eds: Coetzer, Thomson and Tustin), 1, 663–670, Oxford University Press, Cape Town/R.S.A.

Tait, A. and F. R. Hall (1990): *Theileria annulata*: Control measures, diagnosis and the potential use of subunit vaccines. Rev. sci. tech. Off. int. Epiz. **2**, 387–403.

Terpstra, C. (1994): Nairobi sheep disease. In: Infectious diseases of livestock (eds: Coetzer, Thomson and Tustin), 1, 718–722, Oxford University Press, Cape Town/R.S.A.

Uilenberg G. and E. Camus (1993): Heartwater. In: Rickettsial and chlamydial diseases of domestic animals (eds: Z. Woldehiwet and M. Ristic), Pergamon Press, New York, 293–332.

Uilenberg, G. (1993): Other Ehrlichioses of ruminants. In: Rickettsial and chlamydial diseases of domestic animals (eds: Z. Woldehiwet and M. Ristic), Pergamon Press, New York, 269–279.

Uilenberg, G., N. Barre, E. Camus, M. J. Burridge and G. J. Garris (1983): Heartwater in the Carribean. Prev. Vet. Med. **2**, 255–267.

Uilenberg, G., N. M. Perié, J. A. Lawrence, A. J. de Vos, R. W. Paling and A. A. M. Spanjer (1982): Causal agents of bovine Theileriosis in Southern Africa. Trop. Anim. Hlth. Prod. **14**, 127–140.

Van Amstel, S. R., Guthrie, A. J., Reyers, F., Bertschinger, H., Oberem, P. T., Killeen, V. M. and O. Matthee (1987): The clinical pathology and pathophysiology of heartwater: a review. Onderstepoort, J. vet. Res. **54**, 287–290.

Verwoerd, D. W. and B. J. Erasmus (1994): Bluetongue. In: Infectious diseases of livestock (eds: Coetzer, Thomson and Tustin), 1, 443–459, Oxford University Press, Cape Town/R.S.A.

Voigt, W. P. and A. O. Heydorn (1981): Chemotherapy of Sarcosporidiosis and theileriosis in domestic animals. Zbl. Bact. Orig. **A 250**, 256–259.

Walker, A. R. (1990): Parasitic adaptations in the transmission of Theileria by Ticks – a review. Trop. Anim. Hlth. Prod. **22**, 23–33.

Wanduragala, L. and M. Ristic (1993): Anaplasmosis. In: Rickettsial and chlamydial diseases of domestic animals (eds: Z. Woldehiwet and M. Ristic), Pergamon Press, New York, 65–87.

Wilson, A. J. and P. Ronhardjo (1983): Some factors affecting the control of bovine anaplasmosis with special reference to Australia and Indonesia. Prev. Vet. Med. **2**, 121–134.

Woldehiwet, Z. and I. D. Aitken (1993): Coxiellosis (Q fever). In: Rickettsial and chlamydial diseases of domestic animals (eds: Z. Woldehiwet and M. Ristic), Pergamon Press, New York, 131–151.

Wright, I. G. (1990): Immunodiagnosis of and immunoprophylaxis against the haemoparasites *Babesia sp.* and *Anaplasma sp.* in domestic animals. Rev. sci. tech. Off. int. Epiz. **2**, 345–356.

Wright, I. G., B. V. Goodger and D. F. Mahonex (1983): Protection of cattle against Babesia bovis infection. Proc. IV. Int. Konf. Inst. trop. Vet. med. Orlando, Florida, USA.

Zhang Zhong Hang (1987): Elimination of the Gametocytes of Theileria annulata of cattle by Primaquin Phosphate. Vet. Parasitol. **23**, 1.

2. Soil-borne Diseases

2.1. Introduction

Soil-borne diseases are caused by aerobic – *B. anthracis* and *B. cereus* – and anaerobic – *Clostridium* spp. – spore-bearing bacteria which belong to the family *Bacillaceae*. The vegetative forms of these pathogens produce the resistant spore under unfavourable environmental conditions. Soil contaminated with spores remains a potential source of infection for a long time.

The development of infection with *Bacillaceae* is bound to two basic properties of the pathogens, their capacity for

- invasion and
- toxin production.

Pathogens which cause soil-borne diseases develop their tissue damaging action only outside of the phagocytes; they are destroyed rather quickly when phagocytized. By producing a capsule (*B. anthracis*), they are able to protect themselves from phagocytosis. The enzymes produced by these bacilli – e.g. collagenase, which by destroying the collagene of the muscle contributes to the extension of the gas gangrene – enhance their virulence. Characteristically, the toxins develop their toxic effect not only at the site of their production but also systemically, being distributed via the blood and the lymph, thus causing alterations in the visceral organs (liver, kidney); they are basically exotoxins. The organism is able to neutralize the toxic effect of the bacilli producing antitoxic humoral antibodies. Toxins can be transformed *in vitro* to toxoids by adding chemicals, e.g. formalin. Toxoids or anatoxins still keep their antigenic property and therefore can be used for vaccine production. The term "toxin" is not well defined. The widely used term "toxic antigen" only defines serologically demonstrable proteins or protein containing compounds which are produced during the metabolism of *Clostridia* and either remain inside the cell and/or are released outside of, or are set free after the destruction of the cell. Using sophisticated methods, a large number of substances has been demonstrated, often in very low concentrations, which contradicts this description. The pathogenic effect on the organism of only a few toxins is known. It is supposed that often only the combined effect of several substances is able to develop particular tissue damage. The large number of antigenic metabolic products of *Clostridia* which have been found during the last years requires the depiction of all these in an antigenogram since a precise classification has become impossible using the classic "toxinology" (Böhnel 1988). The toxins of the *Bacillaceae* have a characteristic tissue- and celltropism, e.g. the toxins of *C. tetani* and *C. botulinum* for the CNS, the haemolysins produced by many species of *Clostridia* for the red-blood cells; the lecithinase produced by many *Clostridia* has a necrotizing effect on cells of the connecting tissue and muscles.

The pathogens which cause the complex of soil-borne diseases can almost never be transmitted between animals; they are released from the organism only when in agony and/or *post mortem.*

Bacillaceae are able to reproduce in the soil during a saprophytic phase or can get into the soil with biodegradable products, where they persist in their vegetative form or as spores. *B. anthracis* spores can survive in humous soil for decades or even longer, but only for a period of months in heavy soils which are relatively poor organically speaking. As an average, 1000–10000 *C. perfringens* spores per gram soil have been found in soil samples, and 100000 or more in contaminated soils, mostly as vegetative forms. The survival rate of the pathogens in the soil is influenced by

– the antagonistic effect of the soil and the flora in the soil on a large part of microorganisms foreign to the soil;
– antibiotics, which are produced by the flora in the soil;
– lack of suitable nutrients as well as unfavourable temperatures and pH values which are the most important factors of influence to the survival rate.

With the exception of thermophilic bacteria, the vegetative forms of bacilli have little resistance to heat. They can be destroyed in a few minutes at 100 °C. Spores in contrast are resistant to heat in different degrees depending on the species. The soils of the arid and semi-arid pasture ranges of the tropics which are poor in humous dry out extremely during the dry season. This means that an almost sterile state develops in the soil which allows *B. anthracis* and *Clostridia* spores, for example, to survive since no competing bacterial flora remains; they are more resistant under these conditions than in humid soil. Bacillus spores survive a pH range of 4–9; optimal conditions exist where the pH values are neutral to slightly alkaline. Radiation from the sun destroys the spores and vegetative forms of *Bacillaceae*; if brought to the surface by ploughing, the soil can be decontaminated by radiation from the sun. The destruction of the vegetative forms and spores of *Bacillaceae* by means of disinfectants depends on the concentration and time it takes to take effect, the pH within the environment of the reaction, the temperature as well as the species, amount and age of the bacteria (Böhnel 1986).

The importance of *Clostridia* as causal organisms of saprozoonoses depends on their capacity to exist not only in the soil but also in water and inside the system of living organisms – being perhaps only temporarily saprophytic – and to infect animals and humans from there (Table 36).

The grouping of soil-borne diseases into an epizootiological complex is justified especially in the tropical animal production system. Pathogens of soil-borne diseases are not only bound to the soil. There are also influencing factors from soil physics and chemistry which favour their development. Furthermore, environmental influences in the broadest sense are able to control the epizootiological development through their action on the vegetation and availability of water.

Epizootiological characteristics of soil-borne diseases are:

- the infection occurs through the ingestion of spores or vegetative forms of the pathogens. Botulism is an exception: it is an intoxication caused through the intake of toxins which have developed inside a carcass and/or in feed mostly of animal origin;
- an infection transmitted between animals is almost impossible;
- the occurrence of soil-borne diseases is bound to the infected pasture;
- the occurrence of varieties of pathogens and of their subtypes may be geographically different and may be very specific;
- influences of weather, conditions of the feed and the soil have a particular effect on the development of the infection;
- while the spores of the pathogens of soil-borne diseases can survive in the soil for a long time, a pasture once infected may remain so permanently (span. campo maldito = dammed field);
- the clinical features and the pathology of disease depend on the capacity of the pathogen for toxin production.

Soil-borne diseases are especially hazardous to animal production in the tropics and subtropics, because

- the semi-arid pasture ranges of the tropics on the one hand present excellent conditions for the survival of the spores, and on the other hand the conditions of the pasture are favourable to the infection of the animals during the dry season;
- the production systems of the tropics present manifold causes and circumstances which contribute to the distribution and concentration of the causal organisms of soil-borne diseases on certain pastures and/or sites, e.g. by concentrating animals around salt licks and water holes, overstocking, lack of water and minerals, and insufficient disposal of carcasses or total lack thereof;
- scavenger birds and carnivores distribute the pathogens with their faeces;
- typical mistakes of pasture management in the tropics, like overstocking, lack of water and minerals, a lack of feed reserves during the dry season as well as supplementary feeding with highly nutritious feed or grazing on irrigated pastures which may also favour the development of soil-borne disease.

Table 36. Aetiological grouping of the most important infections and toxications caused by *Bacillaceae* in domestic animals (Böhnel 1986, Seifert et al. 1986).

Growth conditions	Genus	Disease complex	Route of infection	Route of toxin absorption	Species of the causal organisms	Favourable environmental factors for the development of the disease
Family Bacillaceae: Gram-positive, labile rods, forming endospores						
Aerobe	**Bacillus**	Bacillus intoxication	per os	per os	*B. cereus*	Spoiled feed
		Anthrax	per os	per os	*B. anthracis*	Dry season, irrigated, flooded pastures
Anaerobe	**Clostridium**	Entero-toxaemia complex	per os	enteral	*C. perfringens* types A-E (F), *C. sordellii*, *C. difficile*	Mistakes in feeding and management, intensive pasture rotation, over-stocking
		Gas gangrene complex	per os, parenteral		*C. chauvoei*, *C. haemolyticum*, *C. histolyticum*, *C. novyi* types A-C, *C. perfringens* types A-F, *C. septicum*, *C. sordellii*, *C. chicamensis* Madagascar field strains 217, 335, 735, Mexico field strain 809 other field strains	Changes in the permeability of the intestine, skin or mucous membranes lesions, dry season, tough ligueous fodders, plants, lack of fodder, overstocking
		Toxication complex	per os, parenteral	per os	*C. botulinum* types A-F, *C. perfringens* types A-F	Mistakes in management, overstocking, mineral deficiency (P), spoiled feed
			parenteral		*C. tetani*	Deep, anaerobic lesions

The variety of pathogens which exists in the tropics in combination with the particular conditions for the development of soil-borne diseases caused by the tropical environment do not allow the continuation of the classical grouping of the complex of clostridioses. Instead, the infections should be classified according to their aetiological and epizootiological characteristics as demonstrated in Table 36. It is also shown here that beside the tropical environment, the influences of the production system, no matter whether it is extensive or intensive, are important for the emergence of the disease complex.

2.2. Soil-borne Diseases Caused by *Bacillaceae* (Aerobic Bacilli)

2.2.1. Anthrax
(Charbon bactéridien, Carbunclo carbonoso, Miltsiekte, Milzbrand)

Anthrax is one of the oldest recorded diseases in both humans and animals; the infection is mentioned in the Bible (Exodus, chapters 7–9) as one of the seven plagues of Egypt. Vergil (70–19 BC) also described outbreaks of anthrax in humans and animals in ancient Rome. The "Hippocratica", a collection of veterinary documents from the 10th century makes reference to anthrax. In 1877, Robert Koch discovered the causal organism of anthrax and the development of the infection. The name anthrax has its origin in the typical dark colour of the blood (Gk. anthrax = carbon).

Aetiology
Bacillus anthracis belongs to the family *Bacillaceae*; its taxonomic status as distinct from *B. cereus*, *B. mycoides* and *B. thuringiensis* was confirmed by DNA hybridization and G+C content determinations (Böhm and Späth 1990).

   B. anthracis organisms are aerobic, Gram-positive, typically rod-shaped, stout, non-motile, spore-forming rods which measure 1.0–1.5 by 3.0–10.0 µm appearing in body fluids as short chains, and in cultures as longer ones. The bacilli may also occur singularly or in pairs. The shape of the bacilli in body fluids is square like a brick, but has rounded edges when in cultures. In tissue fluids however, *B. anthracis* has a well defined capsule composed of poly-D-glutamic acid, which is not formed in a culture; it can be well demonstrated using Wright's, Giemsa and polychrome methylene blue stain, or by the application of the immuno-fluorescence technique (Fig. 52 on page 1 annex). The capsule is supposed to be responsible for the virulence of individual strains. The spores, which are never found in the animal's body while it is alive, are elliptical or oval and are formed under suitable environmental conditions equatorially without causing a swelling of the sporangium.

   B. anthracis grows on most of the ordinary culture media, forming colonies of 3–5 mm in diameter with a grey, frosted appearance on nutrient agar after 24 hours. The margins of the colonies are very irregular because of highly tangled outgrowths of bacterial filaments from the edges of the colonies which give the colonies the so-called "Medusa head" appearance. Typically, these outgrowths taper and curve back in the same direction so that the colony as a whole almost appears to be spinning. In contrast to *B. anthracis*, the recurvature of the outgrowths of non-pathogenic *B. cereus* and other *Bacillus* spp. is either non-existent or less noticeable, this being a diagnostic hint for the identification of *B. anthracis*. Rough margins of the colonies are typical for more pathogenic strains, while smooth colonies are less virulent.

Due to its reduced biochemical activity, *B. anthracis* cannot be differentiated from other *Bacillus* spp. because of its fermentation characteristics.

B. anthracis produces a tripartite extracellular toxin consisting of three distinct, relatively unstable protein components known as

- the oedema factor (EF) or factor I;
- the protective antigen (PA) or factor II;
- the lethal factor (LF) or factor III.

Depending on their concentration, these factors may cause the following effects synergistically

- I + II cause severe local oedemas and have a strong immunizing action;
- I + III are inactive and have little immunizing action;
- II + III have an immunizing effect and are lethal;
- I + II + III cause oedemas, have an immunizing effect and are lethal (Freer 1988, Hambleton et al. 1984).

Plasmids are involved in the production of these toxin components; without plasmid PX01, *B. anthracis* loses its toxigenicity and virulence. Virulent and some avirulent strains of *B. anthracis* also harbour a smaller plasmid (PX02), which is involved in the synthesis of the capsule. Without PX02, capsule formation is impossible, which explains why the avirulent Sterne vaccine strain which carries PX01 but not PX02 remains stable. With DNA probes, PCR and specific toxin antigen detection, it has been possible however, to show that strains which do not produce a capsule and are non-pathogenic in laboratory animals – thus having been discarded as inconsequential – are *B. anthracis* strains which lack the PX02 gene.

Mutants, or Pasteur strains, either arise spontaneously or can be induced in cultures and may differ considerably from a field strain of *B. anthracis*. Growth under unfavourable conditions will induce such mutations.

While vegetative cells of *B. anthracis* are not resistant to adverse environmental conditions, the spores are very resistant and can withstand the action of physical and chemical agents for considerable periods. Spores may persist in the soil for decades and possibly even longer. At room temperature they survive in blood for 270 days, in carcasses for 311 days, in slurry for 150 days, in skins for years, and in waste water for 2 years. Contaminated soil can still be infective after 60 years. The spores withstand humid heat at 120 °C for 10 minutes and dry heat at 160 °C for more than 1 hour (Mitscherlich and Marth 1984, Seifert 1992).

Occurrence

Anthrax is thought to have originated in early Mesepotamia and northern Africa, from where it spread with the process of domestication of wild ungulates which were subsequently disseminated into areas such as Eurasia, America and Australia. Nowadays, the disease occurs in all typical pasture ranges of the semi-arid and humid regions of the world. Anthrax, however, does not occur in the high steppes of the Andes. Newly developed pastures in the tropical forests, e.g. in Brazil, are primarily not infected but sometimes have become contaminated through vaccination with live virulent (Pasteur) vaccines. Rather limited but highly contaminated foci of infections are found around watering places, in traditional cattle markets, around salt licks etc., where animals are concentrated and/or infected animals have died. The traditional animal holder (nomads) are well aware of these infected sites ("campos malditos") and often prevent the animals from entering such areas because of the knowledge passed on by their forebears. In Europe, typical anthrax areas have been reforested over the centuries and have thus been excluded from the pasture ranges (Seifert 1992).

Epizootiology

All mammals, as well as some birds and reptiles are susceptible to the *B. anthracis* infection. Anthrax, however, is above all an infection of herbivores, especially cattle, horses and small ruminants which ingest the pathogens as spores or vegetative forms from the soil. Buffaloes and camels are also susceptible, but pigs rarely become infected, and carnivores only exceptionally. In recent years in the Luangwa Valley Game Park of Zambia, elephants, hippopotami, African buffaloes (*Syncerus caffer*), antelopes and also wild dogs have died in large numbers because of anthrax. The same has been reported from other game parks in South Africa, Namibia and Tanzania. Anthrax in poultry is as good as non-existent (De Vos 1990, Seifert 1995).

Discrepancies between susceptibility and prevalence of the disease can probably also be explained in terms of differences between and in the behaviour of the animals, such as feeding habits and in the routes of infection. In areas deficient in phosphorus, the likelihood of cattle contracting anthrax will increase. Age also affects the susceptibility of animals to anthrax, adults being generally more vulnerable than the young or sub-adults.

The initiation of an outbreak of anthrax depends on interrelated factors which include specific properties of the bacterium, environmental factors, factors affecting dissemination of the organism and certain human activities. In anthrax, the ability of the bacterium to survive outside its host, to enter and successfully infect its host, and to multiply *in vivo* are of particular importance.

The vegetative bacilli which are disseminated by sick animals in large amounts from their body openings or which are released from slaughtered or dead carcasses sporulate in and at the soil at temperatures of 20–32 °C. In the closed carcass, little or no sporulation takes place. If carcasses are buried on the pasture those sites can remain a source of infection for years to come since the spores regularly reach the surface of the soil through its capillary system and thus the fodder plants. Spores may also reach the surface with the ground water from deeply buried material which contains spores. Waste water from tanneries and slaughter houses may contaminate the pastures through flooding via their waste water channels.

B. anthracis spores apparently survive best in neutral to alkaline soil at a pH level not lower than 6.0, with enough calcium and a relatively high nitrogen content. Drought, lack of fodder, overstocking and intensive pasture rotation are causes and/or factors which favour an outbreak of anthrax. By grazing down to the last remaining parts of plants during the dry season which are covered with soil particles, the animals can be heavily infected with spores of *B. anthracis*. If the animals are crowded in an infected pen the contaminated dust which gets inhaled may lead to a pulmonary infection.

Under suitable conditions (humidity, temperature, pH and content of nutrients), the spores may grow out within the upper layer of the soil and be ingested by the animal with the fodder (incubator areas). The basis of this "incubator area" or "soil capability" concept is that *B. anthracis* survives in soil in a dynamic state in which it undergoes cycles of germination and sporulation dependent on fluctuating conditions in the micro-environment (Van Ness 1971). This can lead to an infection with vegetative forms which mostly shows a peracute course. A favourable environment for the development of such a mechanism of infection exists during the onset of the rainy season and also when fodder reserves for the dry season are produced on irrigated pastures.

The long survival time of enzootic anthrax areas especially under the conditions of nomadic pastoralism as well as on ranches and game parks in Africa may be maintained by a biotic-abiotic cycle. During the biotic phase, animals die from anthrax, the carcasses are opened by scavengers and the anthrax spores which then form contaminate the environment. During the abiotic phase, spores are washed down drainage channels during the rainy season to low-lying, poorly drained areas where they accumulate almost exclusively in the upper few centimetres of the soil. The spores become suspended in drinking water, particularly during droughts, when water levels are low and stagnant pools dry up. Animals are then infected when they use waterholes (De Vos 1990, 1994).

In intensive animal production systems, anthrax occurs either if feed of animal origin (carcass, bone and blood meal) is used, or through pulmonary infection by inhalation of dust from feed concentrates or from the contaminated soil of the pen. Whether insects can transmit cutaneous anthrax in animals seems doubtful, though *B. anthracis* has been found on insects and in their excreta. Reportedly, transmission of anthrax by insects is considered to be important in the CIS and on Java. Scavenger birds and carnivores are able to spread *B. anthracis* spores over wide areas without getting infected themselves.

Anthrax is one of the most important and most severe zoonoses of the tropics and occurs in humans only as a consequence of a *B. anthracis* infection in animals. The pathogens may enter the human organism aerogenically, cutaneously and enterically as vegetative forms or spores whenever man gets in contact with sick animals, diseased animals which have been slaughtered, animal products or carcasses of infected animals. Insects can also transmit the infection to humans from sick animals or carcasses of animals which died of anthrax (De Vos 1994, Seifert 1992).

Pathogenesis
After infection, the spores soon germinate and form actively dividing, encapsulated vegetative cells which produce toxin. After subcutaneous infection, oedema develops at the site of infection (humans). From there the bacilli are transported via lymphatics to regional lymph nodes where further multiplication takes place and from where vegetative bacilli continuously enter the blood stream. Initially, the RES, particularly the spleen, tries to control the infection until it is overwhelmed. During the final 10–14 hours of the host's life, the bacteraemia doubles every 95 minutes reaching 10^8 bacteria/mL in sheep (Jones and Klein 1967). The variation in the severity and the extent of lesions that develop depend on the virulence of the organism, the susceptibility of the animal, the infective dose and the route and site of infection.

While it used to be supposed that the pathogenic development of the *B. anthracis* infection is only a consequence of hypoxia because of the enormous multiplication of the *B. anthracis* organisms in the blood, the combined action of the toxins is now thought to be the responsible pathogenic mechanism. It used to be postulated that the toxin production is correlated directly to the development of the capsule which at the same time would determine the pathogenicity of the strain. Empirical observations in the field have shown that rough strains which possess a more developed capsule are found in cases of severe enzootics. It is now known that only those strains which possess a large plasmid (PX01) which does not withstand heat treatment produce toxin. The production of the capsule as a determining factor of virulence depends on a smaller plasmid (PX02). The three factors of the toxin have different effects as described above (aetiology) and react differently in different combinations. The precise synergistic effect of the toxin factors on the organism is still not fully understood; it has not been possible to demonstrate an influence on the CNS. Apparently, extreme hypoxia, hypoglycaemia and alkalosis seem to contribute to most of the clinical signs observed. Loss of body fluids into the tissues (oedema) and body cavities, such as the lungs, mediastinum, and peritoneal and pericardial cavities, causes haemoconcentration in diseased animals. Respiratory distress and severe hypoxia in animals *in extremis*, have been attributed to pulmonary oedema, utilization of available oxygen by the massive numbers of circulating bacilli and/or depression of the nervous system. The neuromuscular irritability and convulsions during the terminal stages of anthrax are probably the consequence of decreased serum calcium and serum potassium. Death is the consequence of apnoea caused probably by a combination of the effects of the toxin combinations on the CNS, lack of oxygen, damage to the capillaries, lung oedema and secondary shock (De Vos 1994, Seifert 1992).

Clinical Features

Anthrax mostly shows the course of a peracute or acute septicaemia. After pulmonary infection with contaminated pen dust, the incubation period for peracute anthrax may be less than 12 hours. Mostly the animals die apoplectically without presenting any symptoms of the disease. Only secretions of tar-like blood from mouth, nostril and anus hint at anthrax (Seifert 1960).

After an incubation period which probably ranges from 1–14 days, the course of acute anthrax does not last longer than 2–3 days. After a rise of body temperature up to 42 °C, the animals may remain standing with staring eyes and a hanging head; they become depressed and then lie down. Initial excitation is followed by listlessness and recumbency with severe disturbances of the circulation and respiration; scattered small haemorrhages occur in the visible mucous membranes, the mucosae are blue-violet in colour; the tongue, the throat and the abdomen as well as udder and/or preputium and perineum become oedematous. There is a loss of appetite, rumination is suppressed and acute digestive disturbances may set in. Some animals may develop diarrhoea, which is usually haemorrhagic. In lactating cows, milk production decreases and the small amount of milk still secreted is either blood-stained or yellow. Pregnant animals usually abort. Dark-red tar-like blood is secreted from the body openings. The bacilli are found in large amounts (up to 10^8 organisms/mL) in the circulating blood and can be recognized because of their typical capsule.

The course of the subacute or chronical form of the disease, usually extends for more than 3 days before either recovery or death occur. The most frequent sign is an oedematous swelling of the throat and neck, following primary infection of the pharynx, peripharyngeal tissues and regional lymph nodes. The swelling in the pharyngeal region may become so extensive that it interferes with respiration, and the ingestion of food and water. The infection may remain localized or it may progress to a septicaemia, which is usually fatal. Chronic anthrax often occurs as a consequence of accidents in vaccination. Especially after vaccinations with Pasteur vaccine, oedemas may appear along the throat and the whole underbelly which may persist for several weeks. Large numbers of characteristic bacilli can easily be demonstrated in the oedema liquid. Animals which show signs of chronical anthrax may recover without treatment. The chronical course is typical for anthrax in pigs which have a high level of resistance. Symptoms of disease in pigs are mostly only an oedema of the throat. Pigs which have recovered may harbour the infection for a rather long time (De Vos 1994, Seifert 1992).

Pathology

At necropsy it is noticeable that the tar-like blood either does not clot at all, or only poorly. The development of *rigor mortis* is also incomplete, and dark, tar-like blood-stained fluid oozes from the natural body openings, such as nose, mouth and anus. All serosae are stained dark-red and are covered with petechiae and ecchymoses; degenerative changes appear in the parenchymatous organs; extensive pulmonary oedema, excessive amounts of blood-tinged serous fluids in the peritoneal, pleural and pericardial cavities and oedema and haemorrhage are found in individual lymph nodes. The spleen is enlarged, the capsule appears tight, the pulpa is pulpy, soft and even semi-fluid, blackish-red in colour, this splenomegaly being the classical characteristic and pathognomonic finding at necropsy. The mucosae of the gastrointestinal tract show signs of inflammation which are both, catarrhal and haemorrhagic, free-flowing blood being found in the intestine; the congestion of subcutaneous blood vessels lends a "fiery appearance" to the carcass. Large numbers of *B. anthracis* bacilli are present in all tissues. The carcass of a ruminant shows marked bloating soon after death and putrefaction sets in quickly.

In pigs and carnivores, only a subacute affection of the intestinal tract may appear with haemorrhagic gastroenteritis and swelling of the lymph nodes as well as swelling in the mouth and pharynx. The clinical symptoms of anthrax in dromedaries are similar to those in cattle (De Vos 1994, Seifert 1992).

Diagnosis

In spite of some spectacular signs found during necropsy, it remains difficult to diagnose anthrax because of gross pathological features. In areas with high incidence of anthrax, however, the appearance of the typical symptoms and gross pathological lesions will be helpful in diagnosing the disease. Confirmation of diagnosis is easy by demonstrating the *B. anthracis* organisms in an animal which is diseased or has just died. In order to prevent the opening of the carcass, a blood smear is best taken from a vessel of the ear and stained with Giemsa which gives an excellent presentation of the typical capsule. In order not to endanger the environment and the investigator, a plastic bag should be put on the ear, and after having cut off part of the ear the closed bag should be sent to the laboratory. Under no circumstances should a carcass which is suspected of being infected with *B. anthracis* be opened. It must also be borne in mind that as soon as an animal dies, the capsule of the anthrax bacillus starts to disintegrate and the protoplasma starts to degenerate, making the stain more and more faint until only ghost-like bacilli are seen. Smears from oedematous fluids which surround localized lesions (pigs) may also be useful for diagnosing anthrax.

Material which has been obtained from organs, bones and also the blood of animals only after the animal has been dead for a certain time has to be investigated with cultural techniques. In anthrax, putrefaction sets in very soon after death and *B. cereus* and *B. subtilis* are found in the peripheral blood. They are microscopically difficult to distinguish from *B. anthracis*. On nutrient agar, *B. anthracis* develops the typical Medusa head (see aetiology) which is a helpful hint for identifying *B. anthracis*. Control of motility and pathogenicity in mice, guinea-pigs and rabbits is a further method of identification. The laboratory animals have to have died 2–3 days p.i. showing the typical features of septicaemia. *B. anthracis* has to be demonstrated microscopically in the blood of these laboratory animals. In our laboratory we have developed a method for distinguishing between *B. cereus* and *B. anthracis* by means of gaschromatographic analysis of cell wall components (Heitefuß 1991, Heitefuß et al. 1991). The lytic effect of bacteriophages (gamma, W and Y) which are specific for *B. anthracis* can also be used to identify cultures suspected to be *B. anthracis*. Another useful diagnostic test for the identification of *B. anthracis* is the culture of *B. anthracis* on nutrient agar with low concentrations of penicillin. Under these conditions, the bacilli swell and filaments appear as chains of spores, referred to as "strings of pearls". A highly sensitive and specific enzyme-linked lectin-sorbent assay can also be used to differentiate *B. anthracis* from other *Bacillus* spp.

Serological and immunofluorescence diagnostic methods so far have not been developed to a degree which guarantees a reliable diagnosis.

In Russia, an allergen test with "Anthraxin" is claimed to be highly specific and sensitive for the recognition of an actual infection and suitable for gauging post vaccination immunity. The antiquated Ascoli test for the detection of *B. anthracis* antigen in dried-out tissues (leather) of animals which have been dead for a long time is not reliable (Cole et al. 1984, De Vos 1994, Seifert 1992).

Treatment

If diagnosed on time, anthrax can be treated successfully. Hyperimmune sera which were used previously have been replaced by antibiotics. Most of modern antibiotics with a broad spectrum of action may be applied, especially those with long action. REVERIN, an oxytetracycline compound, which can be applied intravenously will be successful even in almost hopeless cases. Antibiotic treatment has to be combined with symptomatic treatment of the circulatory system, application of electrolytes and dextrose.

Control

Measures of Management

The most efficient control of anthrax will be achieved when measures of animal management, feeding, keeping and immunoprophylaxis are combined and well coordinated. At the same time, the local conditions of the animal production system, like weather, quality of the soil and vegetation have to be taken into consideration. It will often be enough to keep animals away from contaminated sites which can be done by fencing out known "campos malditos" and/or by reforesting such areas. Contaminated watering places should be closed and clean, hygienic drinking water should be provided. Appropriate herd management, i.e. adapted stocking rates, maintenance of fodder reserves for the dry season, reduction of the intensity of pasture rotation as well as provision of clean and uncontaminated drinking water are further means to reduce the incidence of anthrax. It may even be enough to reduce the stocking rate on the pasture only at the peak of the dry season in order to prevent outbreaks of anthrax.

In intensive production systems, care has to be taken not to use feed or feed-additives which might be contaminated as for instance blood, bone and carcass meal which have been sterilized insufficiently. If bone meal cannot be produced *lege artis*, the crude bones can be burnt instead and the bone ash used as feed additive.

A further important contribution for the prevention of anthrax is the appropriate disposal of carcasses. Carcasses of animals which have died of anthrax have to be disposed of immediately to prevent scavengers from spreading the pathogens. An anthrax carcass is best burnt at the site where the animal died; burying it will lead to permanent contamination of the pasture.

Immunoprophylaxis

The first efficient, attenuated, live anthrax vaccine was developed in 1887 by Pasteur and is still used today, for instance in Mexico, in a slightly modified form. Pasteur obtained his strain I by culturing *B. anthracis* at temperatures of 42–43 °C after 15–20 days and strain II after 10–12 days, which was less attenuated. The organisms had lost their ability to sporulate: type I was only pathogenic for mice and young guinea-pigs, but II was pathogenic for rabbits as well.

It was only in 1983 that Mickesel et al. (Hambleton 1984) found out that the plasmid which is responsible for the production of the lethal toxin was destroyed by culturing the organism at a raised temperature. The disadvantage of the Pasteur vaccine is that though it is able to induce a massive immunity, it can at the same time still provoke chronical oedematous anthrax, since component I of the toxin which triggers anthrax oedema has apparently not been destroyed or perhaps mistakes have been made in the laboratory, when attenuating the vaccine strain. The Pasteur vaccine, even in an improved form with the addition of saponin as an adjuvant, should not be applied in the tropics without having previous confirmation that the disease really occurs. Anthrax has been brought to regions not previously infected through vaccination with the Pasteur vaccine, for instance in the Andes of northern Peru, the Matto Grosso in Brazil, and it has been spread in Madagascar.

Ever since Sterne obtained a capsule-free *B. anthracis* strain (34 F_2) in 1937 by culturing it under CO_2, this Sterne spore-vaccine has become the most successfully applied vaccine for the prevention of anthrax in animals. With slight variations, the vaccine is produced with a spore content of 5×10^6 spores/mL with the addition of saponin. Saponin acts as an adjuvant and enhances the immunizing effect. For the most part, the vaccine is produced by culturing the Sterne *B. anthracis* strain in Roux bottles on solid agar. When sporulation is attained, the material is removed and the vaccine diluted in order to obtain the required spore concentration. This method is prone to contamination and, especially under the conditions of the vaccine laboratory in developing countries, may lead to vaccination accidents. In our laboratory we have developed a method for cultivating the Sterne strain in a bioreactor and are also able to obtain sporulation in suspension. Thus a high spore density can be obtained and a vaccine produced which guarantees an immunoprotection for several years. Since the culture medium is removed during the process of sporulation and the spores are resuspended, a highly concentrated vaccine can be produced. 0.2 mL applied intracutaneously with a special intra-dermal needle as it is used for tuberculinization or with a high pressure needleless dermal injector into the caudal fold produce a long-lasting immunoprotection. Calves which have been vaccinated after weaning and given a booster injection 6 weeks later receive a lifelong immunity. It is prerequisite that the animals remain in the infected areas thus getting constantly boosted through regular field infection (Böhnel and Seifert 1992, Seifert 1960).

Sterne spore vaccines are also produced as aluminium-hydroxide-adsorbed vaccines. In this case, the aluminium hydroxide replaces saponin as the adjuvant and is supposed to reduce the potential danger of vaccination with a live *B. anthracis* strain.

Often *B. anthracis* spore vaccines do not contain the minimal number of spores of 5×10^6 spores/mL. This may be because the spores have died but it may also be because the production laboratory does not have enough expertise in the technology and is scared of producing vaccination accidents with its vaccines.

In the meantime, vaccines have been developed which are supposed to be applied especially to humans which only contain the immunizing fraction of the toxin. From the Sterne strain, the immunizing fraction of the toxin has been isolated through cross-flow filtration, and afterwards, the concentrate is precipitated with alum (Hambleton et al. 1984). For the application in animals in the tropics, this vaccine is inappropriate because it is inferior to the Sterne spore vaccine and too expensive to produce.

Since anthrax is a major threat to wildlife, a number of trials vaccinating wild living animals has been made. In South Africa, game has been vaccinated either after having been captured, or from the air with a syringe fired from a dart gun. We have carried out preliminary trials to vaccinate game animals orally with the Sterne strain distributed in a bait (Rengel 1993).

2.2.2. *Bacillus cereus* Intoxication

Aetiology and Occurrence
Bacillus cereus is a rather big Gram-positive rod which is motile because of its peritrichous flagella. The cells have a diameter of 1.2 by 3.0–5.0 µm. The rods may appear in chains, and the stability of the chains determines the shape of the colony which may be quite different in some isolates. The endospores appear in central or in a paracentral position and do not inflate the sporangium. The organism sporulates in many media whenever enough oxygen is available, though the vegetative cells may also grow anaerobically. *B. cereus* has determined biochemical properties which show slight variations in different strains.

During its exponential phase of growth, *B. cereus* produces a number of extracellular metabolites to which belong several toxins:

- diarrhoea enterotoxin: a thermolabile protein, MW approx. 38000–46000, of complex structure which is inactivated through proteolytic enzymes. It has a necrotizing lethal action and causes food-poisoning and infections;
- emetic toxin: it is highly stable, has the character of a peptide, MW <10000, is not produced at temperatures above 40 °C, is resistant to proteolytic enzymes and is thermostable at 126 °C for 90 minutes. It is probably released during sporulation;
- primary haemolysin: a thermolabile protein, MW approx. 49000–59000 with

cytolytic effect, neutralized by cholesterol and antistreptolysin 0, has a lethal necrotizing action and causes extra-intestinal infections;
– secondary haemolysin: a thermolabile protein, MW approx. 29000–34000, susceptible to protease, toxicity unknown;
– phospholipase C or lecithinase: composed of two phosphatidylcholine-hydrolases, MW 23000 and 29000, and sphingomyelinase, MW 24000, the latter with a haemolytic action;
– exoenterotoxin: a thermolabile protein, MW 57000, lethal to rodents, probably similar to diarrhoea toxin;
– toxin isolated by Ezepchuk: protein, trypsin-stable, lethal (Walz 1993).

B. cereus intoxication caused by feeding *B. cereus*-contaminated alfalfa in Dubai has been described by Wernery et al. (1992). The disease occurs in racing camels in the U.A.E. but is probably more widespread in ruminants in the tropics under similar intensive feeding conditions. It can also occur in cats and dogs and *B. cereus* has been found to cause severe haemorrhagic mastitis in cows with a fatal generalized infection.

Pathogenicity, Clinical and Pathological Features
B. cereus multiplies easily in pressed alfalfa bales kept under humid tropical conditions where it can produce its toxins which are in part highly resistant to environmental influences. Little is known about the pathogenic action of *B. cereus* toxins. Apparently the necrotizing component is an important factor in causing tissue-damaging processes. The mode of cell destruction is still unknown and it is doubtful whether the cell-destroying factor which is also active against erythrocytes can be separated from the haemolytic activities. Enterotoxin-producing strains enhance permeability of blood vessels and diathesis in the intestinal loop and the skin of rabbits. It is supposed that *B. cereus* enterotoxin damages the blood vessels locally and systemically leading to a release of protein-containing exudate and of erythrocytes (Walz 1993).

According to clinical and pathological findings, the disease among racing dromedaries in the U.A.E. resembles that of a haemorrhagic diathesis or haemorrhagic disease (HD), as the disease was called. It occurs primarily during the summer months with outside temperatures of 40 °C and high humidity. Only young racing camels are affected.

The initial stage of the disease is characterized by severe agranulocytosis, fever as high as 41 °C and by inappetence. Some animals develop a cough and a marked uni- or bilateral enlargement of the submandibular lymph nodes. Additionally, complete atonia of the first compartment of the rumen, abdominal pain and regurgitation have been observed. The faeces were covered with fresh or tar-like blood.

Infected dromedaries die between the 3rd and 7th day, becoming recumbent 2 or 3 days before death; in some cases, CNS disturbances, lacrimation and hypersalivation develop. A sharp rise in serum enzymes is observed in the final stage of the disease.

At necropsy, ecchymotic haemorrhages of varying severity are seen in the pharynx, trachea, epicardium, subepicardium, abomasum, intestinal tract, but primarily in the ascending colon and renal pelvis.

Histologically, an intermediate to severe loss of lymphocytes in the lymphatic tissues, including the spleen and tonsils are found as well as haemorrhaging, necrosis and karyorrhexis in the follicular centres. A severe haemorrhage is also seen in the abomasum, intestinal tract, and the myocardial interstitium, the latter extending into the endocardium. Pronounced necroses appear in the renal tubules; the liver exhibits a pan-lobular fatty degeneration as well as pronounced necrosis of the sinusoidal epithelium (Wernery and Kaaden 1995).

Treatment and Control
Only symptomatic treatment is possible. To control the disease it has to be guaranteed that the animals are only provided with feed which has been processed appropriately and does not provide the conditions in which *B. cereus* can multiply.

2.3. Clostridioses

Clostridioses are caused by *Bacillaceae* which belong to the genus *Clostridium*. It was back in 1935 that Bergey gave a precise definition of this group of pathogens: "*Clostridia* are rod-shaped bacteria which are normally motile because of their peritrichous flagella but also may be non-motile. They produce ovoid to ball-shaped spores which normally do not inflate the bacilli. Under normal conditions at least in the early phase of growth, they are Gram-positive and do not reduce sulphate. Many strains are anaerobic, though some may grow under normal atmospheric pressure. Generally *Clostridia* occur in the soil, in the sediment of fresh- and sea-water and in the intestinal tract of humans and animals."

Since the classification of *Clostridia* by means of their serological, biochemical and morphological properties does not produce unequivocal results, a large number of techniques has been proposed for a precise determination. At the same time the taxonomy of *Clostridia* remains equivocal. Constant attempts are made with modern methods to improve the classification of *Clostridia* (Giercke-Sygusch 1987, Heitefuß 1991, Seifert 1990, 1995). Böhnel (1986) has presented an extensive overall view regarding the problem of taxonomy and classification of *Clostridia*.

The aetiological grouping of clostridial infections in humans and animals as it has been presented by Zeissler et al. (1958) for the conditions of Europe cannot be maintained at present. An aetiological connection between disease syndromes and a specific pathogen as it has been claimed there – and is also usually still so presented in the international literature – cannot be maintained, at least under the conditions of tropical animal production systems. For instance we now know that *C. perfringens* can also cause an infection which resembles blackleg. Modern methods of identification of *Clostridia* by means of gaschromatography have allowed the diagnosis of the disease complex aetiologically and it can be shown that a species-specific grouping of clostridioses in connection with singular syndrome complexes should not continued any more. In contrast, it seems logical to divide the disease complex of clostridioses according to its epizootiological mechanisms into three groups:

– gas gangrene,
– enterotoxaemia and
– toxication.

Though the division in diseases which are caused by singular pathogens is still maintained in modern text books, it does withstand the results of intensive research in many parts of the world. More and more pathogens are isolated which do not fit into the classical pattern (Böhnel 1986, Seifert 1990, 1995).

The disease syndromes as they were well defined previously, especially in Europe, are much more heterogeneous and blurred under the different ecological influences of the tropical animal production system. But the patterns of clostridioses are changing even in industrialized countries. The disposal of organic waste, the application of antibiotics as feed additives and the indiscriminate oral application of antibiotics to humans are, for example, causes for a different pathogenic behaviour of *Clostridia*. Even a classical disease like blackleg has changed its epizootiology since cattle breeds have been produced which have such a low immune competence that they almost do not react to vaccinations against *C. chauvoei*. The appearance of locality-specific pathogens is only observed as soon as reliable vaccines no longer give protection as has been reported from many parts of the world (Böhnel 1986, Salinas 1991, Seifert 1960, 1975, 1987, Seifert et al. 1983). Unknown factors become relevant where identical pathogens are apathogenic in one place while they cause infections in another.

Regarding the systematology of the following presentation of the complexes of clostridioses, please refer back to Table 36.

2.3.1. Gas Gangrene Complex

Aetiology
The following clostridioses which are usually described separately are summarized below as the gas gangrene complex:

– Blackleg (Blackquarter, gas gangrene, Charbon symptomatique, Carbunclo sintomático, Rauschbrand);
– Malignant oedema (Swollen head, Braxy, Edema maligno, Malignes Ödem);
– Bacillary haemoglobinuria (Bazilläre Hämoglobinurie);
– Infectious necrotic hepatitis (Black disease, Infektiöse nekrotische Hepatitis).

It is difficult to distinguish between blackleg and malignant oedema when they occur in a tropical animal production system. Though bacillary haemoglobinuria and necrotic hepatitis present distinct disease syndromes, they can be considered as being special forms of gas gangrene, and the *Clostridia* species considered to be specific for these infections can also occur in cases which present the picture of classical blackleg. In a number of investigations we found that, beside the classical causal organisms of blackleg, a number of locality-specific *Clostridia* which do not show any similarity with *C. chauvoei* were responsible for outbreaks of gas gangrene showing the typical symptoms of blackleg, for example *C. chicamensis* which was isolated in the northern Andean ranges of Peru (Seifert 1975), in Madagascar the strains 217, 335 and 735 (Seifert et al. 1983), and in Mexico 5 distinct groups of strains, strain 809 being especially pathogenic (Gonzales 1995, Salinas 1991). *C. sordellii* can also cause gas gangrene. So far, the following pathogens have been identified as causal organisms of gas gangrene:

- *C. chauvoei, C. septicum, C. chicamensis* and field strains which have been characterized precisely (335 and 735 Madagascar, 5 groups of strains in Mexico) (Gonzales 1995, Salinas 1991, Seifert 1992);
- *C. novyi* types A–C, *C. sordellii, C. histolyticum, C. haemolyticum, C. carnis* and *C. fallax*;
- *C. perfringens* types A–E and field strain (217 Madagascar) (Seifert 1992, 1995).

C. chauvoei, s. feseri. The classical causal organism of gas gangrene was first described in 1865 by Feser and Bollinger studied its pathogenic effect in 1875; it received its present name *C. chauvoei* from a French bacteriologist of the 19th century.

   C. chauvoei is a rod-shaped bacterium with rounded poles, about 0.5–0.7 by 2.0–6.0 µm, which stains from Gram-positiv to Gram-labile, older cultures are mostly Gram-negative. The organism is typically pleomorphous and under varying cultural conditions produces involutional forms, so for example it can look like a lemon, a cigar, a balloon, a drumstick or other things. The cells are motile and have peritrichous flagella. Spores which are formed on solid media and in broth are oval, occur in central or subterminal positions and distort the shape of the cell. They are resistant to the effects of boiling in water as well as to phenolic and quaternary disinfectants.

   C. chauvoei has high requirements regarding culture media and culture conditions: under anaerobic conditions irregular vine leave-shaped, translucent or opaque, granulated, glossy or matt, slightly raised or low convex, pale grey colonies grow on blood agar after 24–48 hours with varying types of filaments and β-haemolysis. The cultural properties vary depending on the cultural conditions and may be also specific for several strains. Cultures in PYG broth are turbid, produce a smooth sediment and reach a pH of 5.0–5.4 after 4 days' incubation. The optimal incubation temperature is 37 °C; the growth is stimulated through addition of different C-sources. *C. septicum* can be distinguished from *C. chauvoei* because of fermentation of saccharose, and the non-fermentation of cellobiose and trehalose. In PYG broth acetic acid, butter acid, formic acid, butanol, CO_2 and H_2 as well as few amounts of lactic-, succinylic- and pyruvate acid are all produced. *C. chauvoei* is sensitive to chloramphenicol, clindamycin, erythromycin, penicillin G and tetracycline.

The supernatant of the culture is i.v. non-toxic for mice, but following an s.c. or i.m. injection it is pathogenic to mice, guinea-pigs and hamsters. The infectivity is enhanced when $CaCl_2$, which leads to destruction of the tissues is added to the culture broth. Soil is the natural habitat of *C. chauvoei*.

The pathogenicity of *C. chauvoei* is caused by a complex of toxins which have different effects. They are characterized as follows:

– alpha toxin: lethal, necrotizing and haemolyzing, MW 27 kDa, as part of a larger 53.5-kDa complex, which is called the soluble immunizing component;
– beta toxin: desoxyribonuclease;
– gamma toxin: hyaluronidase;
– delta toxin: haemolysin which is sensitive to oxygen;
– oedema factor: thermolabile, no antigenic action;
– soluble immunizing component: thermolabile, protective antigen;
– non-soluble immunizing component: thermolabile, protective antigen.

C. chauvoei continues to multiply in the carcass for several hours after the death of the animal, sporulates later on and survives in the decomposing carcass for 180 days, in dry muscles for 8 years, and in the soil of contaminated pastures ("campos malditos") for 25–30 years. The spores survive at 100 °C for 30 minutes (Mitscherlich and Marth 1984, Seifert 1995).

C. septicum, one of the causal organisms of classic malignant oedema, is a rod 0.6–1.9 by 1.9–35.0 µm in size. Because of peritrichous flagella, it is usually motile, and is a Gram-positive straight or bent rod, which appears singularly or in pairs and stains in older cultures irregularly. In PYG broth the organisms are Gram-positive in young cultures and become Gram-negative in older cultures, the staining being often irregularly, and rods which are stained intensively Gram-positive appear with unstained spots. In broth, the bacilli are straight or bent; under distinct cultural conditions they can become extremely pleomorphous. The spores are oval, subterminal and inflate the cell.

The cell wall contains L-lysin instead of DAP, glutaminic acid and alanine; the sugars of the cell wall are glucose, galactose, rhamnose and mannose; the cell wall is sensitive to lysozyme.

Circular colonies with a diameter of 1–5 mm with distinct irregular or rhizoid margins develop on blood agar; they are slightly raised, translucent, grey, glossy and produce β-haemolysis.

Cultures in PYG broth are turbid, produce a smooth sediment and reach a pH of 4.7–5.3 after incubation for 5 days. The optimal temperature for incubation is 37–40 °C. The presence of CO_2 is not a requirement for growth but the best growth is obtained in an atmosphere up to 100% CO_2. Fermentable carbohydrates, serum or peptic digested blood contained in the medium stimulate growth.

The pattern of toxins of *C. septicum* is almost identical to that of *C. chauvoei*. Antisera neutralize the toxins of both species: alpha toxin, gamma toxin and delta toxin. A cross-immunity exists to beta toxin; neuraminidase is only produced by *C. septicum*. Also because of these properties it is doubtful whether *C. chauvoei* and *C. septicum* can be considered as a different species (Seifert et al. 1990, Seifert 1992, 1995).

It is diagnostically important that in contrast to other pathogenic *Clostridia, C. septicum* can grow into long filaments on the surface of the liver of infected guinea-pigs.

At room temperature, sporulated cultures of *C. septicum* have remained virulent for 24 years (Mitscherlich and Marth 1984). *C. septicum* is also a dangerous pathogen in humans if it infects wounds. The pathogenesis of the human wound infection is similar to that in animals. The organism occurs in the soil and in the intestine of humans and animals (Seifert 1995).

C. chicamensis grows in PYG broth as a slender (0.3 by 5.0 μm) usually singular Gram-labile rod which is morphologically similar to *C. chauvoei* but does not produce involutional forms. Under strict anaerobic conditions, very small button-shaped colonies develop on blood agar with short filaments which rise in the centre and which typically are sunk into the agar surface and turn from a greyish to a greenish colour as soon as they are brought into an O_2 atmosphere; they produce β-haemolysis. In broth the organisms are Gram-negative. Contrary to *C. chauvoei*, the *C. chicamensis* infection causes a pale reddish slushy non-gaseous gangrene in guinea-pigs which is characterized by a pronounced histolysis. In contrast to *C. chauvoei* and *C. septicum, C. chicamensis* does not ferment lactose. By means of gaschromatography, butter acid has been found as a fermentation product which is not produced by the organisms mentioned before. The pathogen has been isolated from carcasses of cattle which present lesions typical for blackleg in the mountain ranges of the northern Peruvian Andes (Seifert 1975). Through determining the metabolic products and cell wall components by means of gaschromatography, it has been possible to identify *C. chicamensis* as a singular species (Seifert et al. 1990).

The Malagasy field strains **335** and **735** are morphologically similar to *C. chauvoei* but have different cultural and metabolic properties (Seifert et al. 1987). Strain **809** isolated in northern Mexico (Tamaulipas) also differs in its metabolic and morphological properties from *C. chauvoei*. Because of the pattern of metabolic short-chained and the long-chained fatty acids contained in the cell determined by means of gaschromatography, it could not be found to be comparable to any known causal organism of blackleg or malignant oedema, though it produces typical lesions of blackleg. Strain 809 grows as a slender (0.3 by 5.0–8.0 μm) Gram-positive rod with a tendency to produce filaments. After incubation for 48 hours, subterminal or terminal spores appear which inflate the cell. Round, convex, smooth colonies grow on blood agar, having a smooth margin and a diameter of about 3 mm which produce strong β-haemolysis

(Salinas 1991). In total 5 groups of distinct pathogenic clostridia have been isolated in northern Mexico which produce symptoms of gas gangrene in guinea-pigs but are different from known reference strains as far as their metabolic properties and cell wall components (long-chained fatty acids) are concerned (Gonzales 1995).

C. haemolyticum is, because of peritrichous flagella, a motile rod, 0.6–1.6 by 1.9–17.3 μm which appears singularly or in pairs and stains being Gram-positive in young cultures, and Gram-negative in older ones. The oval spores appear subterminally and inflate the cell. Surface colonies on blood agar have a diameter of 1–3 mm, are round, slightly raised or convex, translucent, grey with a glossy, granulated or mosaic-like surface and a fringed scale-like margin. β-haemolysis is produced on blood agar by the beta, gamma or delta toxins; opaque zones appear around the colonies on egg-yolk agar because of the activity of lecithinase. The spores survive for 5 minutes at 100 °C, and thus specimens can be heated in order to obtain pure cultures. The cell wall contains meso-DAP, alanine and glutaminic acid. PYG broth becomes turbid with a granulated fluffy sediment; the pH is reduced to 5.0–5.5 after 24 hours incubation, the optimal incubation temperature being 37.5 °C. *C. haemolyticum* is one of the most fastidious for growing *Clostridia* and is extremely susceptible to oxygen. Products of fermentation are propionic-, butter- and formic acid. *C. haemolyticum* is sensitive to chloramphenicol, clindamycin, erythromycin, penicillin G and tetracycline (Bergey 1986, Hatheway 1990, Seifert 1995).

The supernatant of *C. haemolyticum* broth cultures is toxic for mice. Whole cultures are pathogenic for cattle, sheep and laboratory animals. The relevant lethal toxin is phospholipase C which is identical to *C. novyi* beta toxin. It hydrolyses lecithin and sphingomyelin and haemolyses erythrocytes. Because of the close relationship of the toxins and also the bacteriological properties of *C. haemolyticum* and *C. novyi*, *C. haemolyticum* is considered by a number of authors as *C. novyi* type D (Hatheway 1990). *C. haemolyticum* can only be distinguished from *C. novyi* B by means of toxin neutralization.

C. histolyticum is a motile straight Gram-positive rod, 0.5–0.9 by 1.3–9.2 μm because of its peritrichous flagella and appears in pairs or in short filaments. The spores appear oval, are situated centrally or subterminally and inflate the cell slightly. Surface colonies on blood agar have a diameter of 0.5–2.0 mm, are between circular and irregular in form, as well as between flat and slightly convex, translucent and semi-opaque, greyish-white, glossy with a granulated or mosaic-like surface and a straight to undulated margin; they produce β-haemolysis. The cell wall contains meso-DAP, glutaminic acid and alanine. The optimal incubation temperature is 37 °C, the growth being stimulated by carbohydrates. *C. histolyticum* is characterized by its strong proteolytic activity; gelatine, beef, milk, casein, collagen, haemoglobin, fibrin, elastin, protein, coagulated serum, muscle and liver, brain and Achilles tendon are digested; acetate, traces of formeate, lactate and succinate are produced. The lecithinase and lipase reaction are negative; therefore, *C. histolyticum* is difficult to diagnose if not tested with protein reactions.

C. histolyticum is sensitive to chloramphenicol, clindamycin, erythromycin, penicillin G and tetracycline.

The supernatant of broth cultures is toxic for mice; the toxicity becomes lost in old cultures (protease activity). *C. histolyticum* is pathogenic for all laboratory animals, the pathogenicity being caused by 5 toxins:

- alpha toxin: necrotizing, non-haemolytic, cross-immunity to *C. septicum* toxin, inactivated through proteolytic enzymes;
- beta toxin: collagenase, consists of 7 components with a MW of 68, 115, 79, 100, 110, 125 and 130 kDa;
- gamma toxin: thiol-activated proteinase which digests skin, gelatine and casein but is inactive to collagen; MW 50 kDa;
- delta toxin: proteolytic enzyme, elastase, MW 50 kDa;
- epsilon toxin: oxygen-labile haemolysin, serologically identical to haemolysins from *C. tetani*, *C. septicum* and *C. novyi* (Hatheway 1990).

Liver broth cultures kept for 25 years at room temperature bear spores which are still alive; the same is true for spores inside dried meat (Mitscherlich and Marth 1984).

C. novyi is divided by Bergey (1986) into types A, B and C.

Type A: In PYG broth, Gram-positive rods 0.6–1.4 by 1.6–17.0 μm grow singularly or in pairs. They are mostly motile and carry peritrichous flagella. The oval spores are situated centrally or subterminally and may inflate the cell wall which contains meso-DAP, glutaminic acid and alanine. Surface cultures on blood agar have a diameter of 1.0–5.0 mm, can be circular or irregular, flat or raised, translucent or opaque, grey, matt or glossy, with a crystalline or mosaic-like internal structure and an indented, undulated, frayed or rhizoid margin. The culture may grow like a film over the whole surface. In PYG broth, a turbid, smooth or fluffy sediment with a pH of 5.1–5.8 is produced after a week of incubation. The optimal incubation temperature is 45 °C, though most strains also grow at 37 °C; strict anaerobiosis is prerequisite. In PYG broth, butter- and propionic acid as well as small amounts of acetic and valerian acid and propanol are produced.

C. novyi A is sensitive to chloramphenicol, clindamycin, erythromycin and penicillin G; a resistance to tetracycline may exist. The supernatant of the culture of some strains is toxic for mice, and the whole culture is pathogenic for guinea-pigs, rabbits, mice, rats and pigeons.

The main lethal toxin of type A as well as type B is a necrotizing alpha toxin. Type A also produces gamma(phospholipase C)- and epsilon(lipase)-toxin; some strains produce delta(haemolysin)-toxin. Type A and B can be distinguished because of the production of epsilon toxin by type A. The type A lecithinase (gamma toxin) is antigenetically different from the lecithinase (beta toxin) of type B and C from *C. haemolyticum*. The toxicity can be transferred between the strains through bacteriophages. *C. histolyticum* A has been isolated from soil, marine sediments, human and animal wounds including gas gangrene (Bergey 1986, Hatheway 1990, Seifert 1995).

Type B: The cells are larger than those of type A, 1.1–2.5 by 3.3–22.5 µm; mannose is fermented; milk and meat are digested; lipase is not produced on egg-yolk agar and the electrophoretic patterns of the cell proteins differ from one another. Like type A, type B produces the lethal necrotizing *C. novyi* alpha toxin. The lecithinase which is produced by type B is the same beta toxin which produces *C. haemolyticum*. The beta toxin can be demonstrated through haemolysis and haemolysis inhibition with antitoxin. Type B also produces zeta(haemolysin)- and eta(tropomyosinase)-toxin. Some strains produce small amounts of theta toxin (lipase) (Bergey 1986).

Type C: This pathogen, originally isolated from water buffaloes which have suffered from osteomyelitis, is almost indistinguishable from type A; probably because of losing its phages, it is not toxigenic (Bergey 1986). The spores of *C. novyi* C withstand 105 °C for 6 minutes, and the cultures survive at room temperature for 25 years.

C. sordellii forms Gram-positive, straight rods, 0.5–1.7 by 1.6–20.6 µm in PYG broth which appear singularly or in pairs and are regularly motile and carry peritrichous flagella. The spores are oval and are situated centrally or subterminally and are often found as free spores. They inflate the cell slightly. The cell wall contains meso-DAP, glucose and traces of galactose, glutaminic acid and alanine. Circular colonies (1–4 mm diameter) grow on blood agar which can be round or irregular, flat or raised, translucent or opaque, grey or chalk-white with matt or glossy surface, a granulated or spotted internal structure and a scaled, indented or smooth margin. The haemolysis varies, most strains showing slight β-haemolysis. The optimal incubation temperature varies between 30–37 °C. *C. sordellii* digests milk, casein and meat; acetic- and formic acid as well as small amounts of isocapron-, propionic-, isobutter-, butter- and isovalerian acid are produced.

The supernatant of broth cultures is only occasionally toxic for mice, *C. sordellii* is pathogenic for humans, cattle, sheep, guinea-pigs and mice. Though the antitoxin of *C. sordellii* neutralizes the toxins of *C. difficile*, the toxins of both species are different. *C. sordellii* produces three toxins which are identical to those of the apathogenic *C. bifermentans*:

– a lecithinase, a phospholipase C, serologically related to alpha toxin of *C. perfringens*;
– an oxygen-labile haemolysin;
– a fibrinolysin.

Furthermore, a lethal factor is produced, called beta toxin, which is responsible for the pathogenicity of *C. sordellii* and distinguishes it from *C. bifermentans*. The beta toxin can be divided into two toxic factors which both have dermonecrotic and haemolytic properties.

C. sordellii has been isolated from soil, normal human faeces, human clinical tissue samples, blood cultures and numerous tissues from animals (Bergey 1986, Hatheway 1990, Seifert 1995).

***C. perfringens*.** Though *C. perfringens* is mainly responsible for causing enterotoxaemia, it is already discussed here because it can also be found as a causal organism of gas gangrene.

C. perfringens produces a number of toxic effects either *in vitro* and/or *in vivo*, and because of this it has been divided into types A, B, C, D and E. The five types cannot be distinguished because of their bacteriological and biochemical properties. And as well, a reliable differentiation of the types cannot be accomplished by means of gaschromatographic determination of fermentation products and/or cell wall components (Heitefuß 1991).

C. perfringens is a non-motile straight 2.4 by 1.3–19.0 μm stout rod with blunt ends, which carries no flagella and appears singularly or in pairs. The organism rarely sporulates under normal conditions *in vivo* and *in vitro*; if spores appear, they are large, oval, placed centrally or subterminally and inflate the cell. Surface colonies on blood agar normally have a diameter of 2.0–5.0 mm, are circular with a smooth margin, cupola-shaped, grey or greyish-yellow and translucent with a shiny surface; they vary significantly under different cultural conditions.

The cell wall of *C. perfringens* contains LL-DAP; lactose, glucose and rhamnose may appear. Some strains possess a polysaccharide capsule which has a variable composition according to the strain.

The type and degree of haemolysis depend on the type of the organism and on the type of blood contained in the culture medium. PYG broth becomes turbid and forms a smooth sediment; the pH is reduced to 4.8–5.6 after one week of incubation. Type A, D and E grow best at 45 °C, B and C at 37 and 45 °C. The range of temperature for growth can vary from 20–50 °C; there are also psychotropic strains which grow at 6 °C. Carbohydrates stimulate growth. Metabolic products are deoxyribonuclease, acid phosphatase, ribonuclease, elastase, hyaluronidase, amylase, neuraminidase, haemagglutinin, galactosidase, reductase and dismutase; bacteriocins are also produced which are active against other strains of the same species. All 5 types produce large amounts of acetic-, butter- and lactic acid and sometimes small amounts of propionic-, formic- and succinylic acid in PYG broth.

C. perfringens bacteriophages have been known to the scientific world since 1949. Regularly S- and R-strains are susceptible to phages while mucoid strains are resistant.

Chloramphenicol, clindamycin and metronidazole are effective against most isolates, but less so erythromycin and tetracycline (Bergey 1986).

Cultures of *C. perfringens* survive at room temperature for only a few days; the spores resist 100 °C for 80–90 minutes. Beef remains infective at 15 °C for 330 days, and in dried whale meat the spores have survived 21 years (Mitscherlich and Marth 1984).

C. perfringens types are distinguished because of their pattern of toxin production, either *in vitro* and/or *in vivo*; the following toxins are produced:

- type A alpha toxin,
- type B alpha, beta and epsilon toxin,
- type C alpha and beta toxin,

- type D alpha and epsilon toxin,
- type E alpha and iota toxin.

Alpha toxin: a phospholipase C with a MW of 43 kDa which hydrolyses lecithin. It is produced in large amounts by all 5 types and has a lethal effect by lysing the membrane of the cell and by hydrolizing the lecithin of the cell wall; it causes myonecrosis in gas gangrene of humans and animals and can also cause intravascular haemolysis. Furthermore, it may cause hypertension and brady-cardia and eventually shock which is the most frequent cause of death with gas gangrene (Hatheway 1990).

    Beta toxin: produced by types B and C. It is a polypeptide sensitive to trypsin with a MW of 40 kDa which enhances the permeability of the capillaries. It causes increased blood pressure and bradycardia and gastrointestinal symptoms in humans and animals (Hatheway 1990).

    Epsilon toxin: is produced by types B and D as a prototoxin and is only transformed through proteolytic enzymes, especially trypsin into a potent thermolabile toxin of a MW of 34.25 kDa. It increases the permeability of vessels and causes necrosis in tissues. In the intestine, the permeability for proteins is enhanced and thus the reabsorption of toxins. Oedemas are produced by this toxin in the kidneys, lung, pericardium and especially the CNS, the latter explaining the apoplectic death during enterotoxaemia. Epsilon toxin can be demonstrated by means of an ELISA which is an alternative to toxin-neutralization in mice (Naylor et al. 1987).

    Iota toxin: is the only toxin produced by type E. It is a prototoxin which has to be activated by the proteolytic enzymes of the organism; it increases the permeability of vessels, causes necrosis, also of the skin and is lethal after i.v. application. It consists of two immunologically and biochemically different components, iota A and iota B with an MW of 47.5 and 71.5 kDa (Stiles and Wilkens 1986).

    Beside the lethal toxins mentioned above, further **factors of virulence** of *C. perfringens* have been described. They are

- **gamma**: a lethal toxin of mainly diagnostic importance which apparently does not play a part in the pathogenesis of the disease;
- **delta**: a lethal haemolysin which is produced by some B- and C-strains *in vitro*;
- **eta**: a lethal toxin with the same effect as gamma;
- **theta**: a haemolysin with lethal and necrotizing effect in laboratory animals; it causes haemolysis on blood agar;
- **kappa**: a collagenase which is the cause of gas gangrene during natural infection of humans and animals;
- **lambda**: a protease which is produced by B- and E- strains as well as some D-strains;
- **my**: a hyaluronidase, apparently of little importance in the pathogenicity;
- **ny**: a desoxyribonuclease which is produced by all strains but apparently of little pathogenetic importance (Bergey 1986).

Additionally, the subtypes of *C. perfringens* produce a number of other substances which have been called toxins. How far they influence the virulence is unknown. An **enterotoxin** has also been described as a protein with a MW of 35 kDa, which is composed of a peptide of 309 amino acids; it contains a free sulfhydryl group. The effect of the enterotoxin is directed towards the epithelial cells of the intestine; it changes the cell metabolism and the macro-molecular synthesis. By increasing the concentration of intracellular calcium ions, morphological damage, disorders of permeability and loss of tissue liquid and ions occur (Hatheway 1990).

Occurrence

All types of gas gangrene occur worldwide. Though the disease does not usually cause sudden losses in great numbers during extremely dry years, it may result in high mortality rates, especially in calves. The local importance depends on the local ecological conditions, the structure of the production system and perhaps also the effected animal species and breed. Characteristically the different species of pathogens appear strictly localized and pastures, once contaminated, remain so for a long time. The gas gangrene complex is of special importance in the semi-arid belt of pasture ranges of the world since the ecological conditions in this zone favour the development of the infection. On a particular pasture range in different altitudes or ecological regions, the disease complex may be caused by different species of pathogenic clostridia (Seifert 1975). Here it is unimportant whether the infection occurs orally or through wounds. It has not been possible to confirm that a correlation exists between geological-chemical factors (minimum content of calcium of the soil) and the occurrence of gas gangrene. While blackleg occurs worldwide, a prerequisite for the appearance of *Hepatitis necroticans infectiosa* is the infestation of the pasture with liver flukes.

Epizootiology

C. chauvoei is primarily a member of the normal intestinal flora of animals but can also be found in the organs and muscles of healthy slaughtered animals. At the same time, *C. chauvoei*, *C. septicum* and the other causal organisms of gas gangrene are components of the normal bacterial flora of the soil where they can persist as vegetative forms or spores for decades. In soil which has a high content of organic matter as well as in clay and chalk soil, the pathogens are supposed to remain infective for a long time. Gas gangrene is enzootic both on the low humid grass lands of the tropics as well as on the middle hilly, sandy arid and mountainous pasture ranges.

Gas gangrene occurs mainly in ruminants; cattle are the most susceptible followed by sheep, goats and sometimes pigs. Antelopes, buffaloes and horses may also be infected. *C. chauvoei* has been isolated from hippopotami and fresh-water fish. Guinea-pigs, hamsters and mice are suitable laboratory animals.

The infection of the soil is the consequence of the occurrence of gas gangrene in animals which release the pathogens during the disease or *post mortem*. The pathogens may, however, also be scattered as natural inhabitants of the intestine by animals which have died because of other diseases. It is also supposed that pastures can be contaminated by healthy animals through defecation.

Gas gangrene, especially blackleg, is a soil-borne disease which is often restricted to areas which are known to have been infected for long periods. Within these regions, there may however be pasture ranges which are especially dangerous ("campos malditos"). The infection usually occurs on pastures. In stabled animals, the infection may appear only as an exception when feed (hay) from infected pastures is fed. The disease outbreak on the pasture is mostly concentrated during the peak of the dry season especially when the grass has been grazed down to the ground and parts of soil are ingested with the feed.

Pathogenesis

The infection with clostridia causing the gas gangrene complex is mainly oral. When the animals are forced to graze the remaining tough and ligneous pasture down to the soil during the peak of the dry season, they ingest large amounts of spores of clostridia which occur at the site. The ligneous feed causes erosions of the mucosae of the gastrointestinal tract which allow the spores to enter the blood stream; they are also supposed to be able to penetrate the intact mucosae. The further path of the infection is doubtful. The spores are distributed via the blood stream through the whole organism; they appear in the muscles or in the visceral organs and may be eliminated by the host defence system. External influences, like blows to the animal, as well as injections with irritating drugs, e.g. prostaglandines or compounds which reduce the resistance (corticosteroides), favour the replication of the pathogens in the tissue. Experience has shown that the infection may appear after animals have been rounded up or vaccinated as well as after having walked long distances in order to seek feed or water. Apparently the muscle activity consumes oxygen which leads to a low redox potential in the muscle – a condition which is optimal for the development of the clostridial infection. The toxins released during the multiplication of clostridia, still improve the environment in the tissues for the further development of the clostridia by reducing the redox potential further and altering and destroying the surrounding tissue. Histolysis appears which can be seen as a blackish-red change of colour in the tissue and black-red stained oedema is present. The muscles become spongy; accumulation of gas in the muscle – the metabolic product of the clostridia – leads to the characteristic gas gangrene, which has the typical smell of butter acid because of the short-chained fatty acids produced metabolically by the clostridia (Fig. 53 on page 1 annex).

Typically, well-fed animals fall ill most frequently. The age group of cattle which is affected most is between 8 and 18 months. Only rarely do older animals become affected, and if so, they are usually bulls. If animals are affected which are younger than 6 months, it is because the dams have been immunized insufficiently and thus have not been able to immunize their calves passively. It has not been confirmed that there is a natural resistance in older animals, but one may suppose that older animals have received continuous natural boosters which has led to the development of a protective antibody level which is also transferred passively to the calf and can protect it during the first months of life. Another reason why younger animals are more susceptible may be that while changing from milk teeth to permanent ones, the lesions which appear in the mucosa during this process may be foci of infection. Gas gangrene mostly occurs in beef cattle and sheep in extensive pasture systems at the peak of the dry season, it only appears exceptionally while feeding conditions are good (Seifert 1960, 1975, 1995).

In sheep, the infection can often start from a wound, for instance a dog bite. The infection can also appear frequently after sheering, tail docking, castration and during parturition. Under those circumstances, a high disease incidence can cause considerable economic losses. Even under these conditions a herd immunity can develop, but is not as well expressed as after oral infection. In contrast to sheep, parenteral infection is unimportant in cattle. Whether or not tabanides are able to infect cattle with a contaminated hypostome seems doubtful. After an i.m. injection of prostaglandines, however, numerous infections in cattle occurred in Germany showing the typical picture of blackleg. We were able to reproduce this experimentally in guinea-pigs injecting prostaglandines i.m. with a needle previously contaminated with *C. septicum*. Apparently the local enzymatic action of prostaglandines produces ideal conditions for the development of pathogenic clostridia inside the muscle tissue.

In English literature, the so-called "**big head**" is described which is supposed to be caused by *C. novyi*, *C. sordellii* and *C. chauvoei*, and is characterized by an oedema on the head, face and neck. No development of gas gangrene is observed. The disease appears only in young rams which provoke the infection when they knock their heads against each other during fights. *C. perfringens* may also lead to gas gangrene in ruminants as a contaminant of wounds.

A characteristic disease caused by *C. novyi* type B, the **black disease**, occurs especially in sheep. Through migrating larvae of *Distoma hepatitis*, necrotic processes are caused in the parenchym of the liver from where the infection with *C. novyi* type B originates. The spores may be either already present in the liver or have been introduced by the larvae. The toxin produced by *C. novyi* type B leads to the sudden death of the animals.

A further special form of gas gangrene is the **bacillary haemoglobinuria** caused by *C. haemolyticum* (*C. novyi* type D). It is an infectious toxaemic disease which again is difficult to distinguish *post mortem* from the gas gangrene caused by other clostridia. The disease occurs especially in cattle on the American continent from Texas down to Chile along the West Coast, but is also supposed to appear in sheep. The pathogen has also been found in Great Britain and in Turkey. The infection also starts in the liver. Pathognomonic is haemoglobinuria which is always present (Seifert 1992, 1995).

Clinical Features

A disease outbreak of gas gangrene mostly occurs suddenly at the peak of the dry season, the course being so rapid, generally less than 24 hours, that clinical signs are not often observed in affected animals prior to death. Thus the animals are usually found dead on the pasture. During the further course of the infection in an affected herd, the animals will rarely die in large numbers although cases appear day by day. Sick animals can only be identified if the herd is observed closely.

Most affected cattle do not recover, but a small proportion of sheep and other animals which contract the disease as a result of an infected wound may survive.

In cattle, the first clinical signs are a rise in the body temperature, loss of appetite, ruminal stasis, and if the primary lesion is situated in one of the larger muscle groups of a limb, as is most often the case, there is initial stiffness followed by lameness which rapidly increases in severity in the limb concerned. The muscles or muscle groups most commonly affected are those situated in the shoulder, buttock or loin regions, but those in the chest and neck regions and, more rarely, muscles such as the intrinsic and extrinsic muscles of the tongue, and the diaphragmatic and sublumbar muscles or myocardium may be primarily involved. The clinical signs, in the few cases in which they are evident, vary according to the site of the lesion(s). If the muscles affected are superficial, there is subcutaneous swelling and crepitation at the site. The swelling increases rapidly in severity and size, and initially is hot and painful, but terminally it becomes cold and painless. Lesions situated deeper in the body are not clinically detectable. Occasionally, lesions may develop simultaneously in more than one location. Affected animals soon become recumbent, manifest signs of dyspnoea and have an accelerated pulse rate.

The clinical signs of localized gas gangrene which develop after a wound becomes infected by *C. chauvoei* are very similar to those of blackquarter and are also dependent on the site of the infection. In sheep and occasionally in cattle and goats, the vagina, cervix and uterus may be infected following parturition. In such cases, the perineal and adjacent tissues are severely swollen and emphysematous, and blood-stained droplets of serous fluid may ooze from the surface of the affected parts. In the case of contaminated injection sites, lesions are localized to the site of injection and are oedematous and crepitant. This localized reaction expands rapidly, causing swelling of the tissues and the exudation of drops of blood-stained serous fluid through the skin, the exudate having a rancid smell. Affected animals may die within 18 hours of being injected with a contaminated needle. The course of the infection in pigs is usually less acute, the infection being localized.

Localized gas gangrene as a consequence of a wound infected with *C. chauvoei* rarely occurs in horses. However, in cases in which this does occur, the infection may spread and involve large areas adjacent to the initial lesion. Affected horses either die without being obviously ill, or are severely ill, in which case they manifest depression, signs of shock, and an unwillingness to move the affected part if it is a limb or the neck. The swelling, which is initially hot and painful, becomes cold, oedematous and crepitant. Most affected horses die, but the affected tissues may slough off in those that do recover, following intensive antibiotic treatment, leaving large cavities a few centimetres deep which heal slowly.

Typical for bacillary haemoglobinuria is the high mortality rate of up to 95%. Characteristic clinical features are icterus and haemoglobinuria as well as oedemas of the brisket (Kriek and Odendaal 1994).

Pathology

The characteristic *post mortem* finding is a bloated carcass which lies sideways with the limbs stretched out or bent at an awkward angle (Fig. 54 on page 1 annex) and which putrefies very quickly. The typical crepitating swellings of the muscles are extended *post mortem* from the limbs onto the entire body, and frothy fluid may ooze from the nose and anus. The subcutaneous, intramuscular and other intestinal tissues in the vicinity of the affected musculature are distended by an either yellow or blood-stained or even dark red oedematous fluid that contains bubbles of gas. The colour of the oedematous fluid and its content of gas depends on the clostridia species that caused the infection. On cutting into an affected muscle, a reddish fluid exudes and there is a characteristic rancid butter smell caused by the metabolites of the respective clostridia. Lesions in the muscle are well-circumscribed, spongy due to the accumulation of gas bubbles between muscle fibres, friable, and either uniformly dark red or have alternating streaks varying in colour from a mottled greyish-red to pale yellow or black. These features again depending on the species of the causal *Clostridium* (Fig. 53). These lesions can also appear in the tongue and the diaphragm and myocardium, the latter being accompanied by a fibrinoid to fibrino-haemorrhagic pericarditis. Affected muscles tend to be dryish towards the centre of the lesions. The spleen is usually not enlarged, but regional lymph nodes are hyperaemic and oedematous. All serosae are of a blackish-red colour, the body cavities are filled with a blackish red liquid which can also be amber in colour depending on the causal clostridia species. In the intestinal tract, signs of a catarrhalic gastroenteritis are found. The liver, spleen and kidney soon decay *post mortem*.

Histologically the muscle lesions are characterized by extensive necroses caused by the toxins of the bacteria, and numerous bacilli may be found.

The lesion of **gas gangrene in sheep** following the infection of a wound resemble those of cattle which have died of typical blackleg.

Black disease in sheep shows characteristic grey-yellowish necrotic foci in the liver which are often found along the route of migration of the young liver flukes. An enlarged pericardium is filled with amber liquid which is also found in the body cavities. The capillaries of the subcutis are ruptured which because of the fast decay of the carcass leads to a black colour of the skin (black disease). Because of the fast onset of putrefaction of the carcass in the tropics, a few hours after death the disease cannot be distinguished from the classic gas gangrene (blackleg).

In **bacillary haemoglobinuria** it is observed that *rigor mortis* sets in rather quickly; the characteristics of the gross pathological lesions are a subcutaneous oedema which is altered *post mortem* in such a way that it is difficult to distinguish from the gas gangrene of blackleg. Typically, the body cavities are filled with blood-stained liquid. As a consequence of an enteritis which is either catarrhal or haemorrhagic the contents of the intestine are mixed with clotted blood. Anaemic infarcts in the liver are pathognomonic; they are slightly raised, clear in colour and surrounded by a bluish-red zone of congestion. The kidneys are dark, friable and covered with petechiae; burgundy-red urine fills the bladder. The lesions have to be distinguished from those of babesiosis and fern poisoning.

The **histopathological** lesions of gas gangrene can be divided into three stages of development.

First stage: swelling and oedematization; the muscle fibres lose their transverse striation, their nuclei show pyknosis, karyorrhexis and karyolysis. The perivascular tissue becomes infiltrated by serous fluid and haemorrhages.

Second stage: the infiltration increases, cells start to immigrate and the production of gas begins.

Third stage: the degenerated muscle cells and fibres are lysed, cell infiltration increases and the intramuscular connecting tissues also finally become affected. Its fibres degenerate and become anything from grainy to squamous. Many degenerated red-blood cells and numerous clostridia can be found. A haemorrhagic emphysematous necrosis appears (Seifert 1995).

Diagnosis
In typical cases of gas gangrene a presumptive diagnosis can be made if

- the incidence of the disease is concentrated in pastures which are known to be infected;
- only weaned calves are affected;
- typical gross pathological lesions are observed on the carcass, such as crepitating gas gangrene, *rigor mortis* with the characteristical position of the limbs, bloating and blue-violet colour of the mucosae as well as of the skin on the muzzle, udder and vagina.

It is important to distinguish the gross pathological lesions from those of other infections. Similar swellings can appear on the animal with anthrax and pasteurellosis, but in contrast to gas gangrene they are permanently hot and painful and never show the characteristic crepitation of gas gangrene. With gas gangrene, the blood-stained faeces (anthrax) and the pneumonic symptoms of pasteurellosis are missing.

Because of the heterogeneous origin of gas gangrene it is paramount to isolate, demonstrate and differentiate precisely the causal *Clostridium*, otherwise no effective and sensible preventive measure can be initiated. The laboratory diagnosis, however, only seems to be easy to the inexperienced investigator. The microscopic demonstration of clostridia in the oedema liquid is irrelevant. Immediately *post mortem*, saprophytic, anaerobic and also sporulating organisms multiply which cannot be distinguished from pathogenic clostridia under the microscope. The demonstration of the *Clostridium* using direct fluorescence serology often produces unreliable results in the tropics where the commercial conjugates do not react with the specific local organisms. Serological differences also exist between subtypes of one species. The direct immunofluorescence, however, is a valuable diagnostic tool where homologous and site-specific conjugates are available (Seifert 1975).

To isolate the causal *Clostridium* by means of bacteriological cultures a well equipped laboratory and personnel with experience in diagnosis of anaerobic organisms are required. A standard technique for sampling, transport of the sample, isolation, identification and differentiation of pathogenic clostridia has been developed in Goettingen. In order to get precise information about the identity of the pathogen, the metabolically produced alcohols and short-chained fatty acids produced in PYG broth as well as long-chained fatty acids which compose the cell wall are identified by means of gaschromatography and are compared biometrically with the pattern of fatty acids of reference strains. Last but not least it has been possible to demonstrate with this method that the aetiology of gas gangrene in the tropics is highly varied and site-specific (Gonzales 1995, Giercke-Sygusch 1987, Heitefuß 1991, Salinas 1991, Seifert 1987, Seifert et al. 1990, Seifert 1995).

Treatment
Whenever it is possible to recognize outbreaks of gas gangrene in time, diseased animals can be treated by applying high doses of oxytetracycline (10 mg/kg b.w.) intravenously. The animals, however, cannot be slaughtered for months after having recovered because the destruction of the muscles results in scars which only disappear slowly. The animals do not usually recover their previous conditions.

Localized gas gangrene can be treated by applying high doses of antibiotics directly into the swelling. Penicillin has to be applied in doses of at least 10000 I. U./kg b.w., but a better effect is obtained with oxytetracycline (15–20 mg/kg b.w.). The application of antisera is of little effect and usually impossible at present since appropriate immune sera are not available. Surgical treatment through incisions into the emphysema as it is still practiced in human medicine by applying simultaneously hydrogen-peroxide has proven to be useless.

During an outbreak of gas gangrene in a herd, it is economically justifiable to apply LA-oxytetracycline, i.m. (5 mg/kg b.w.) to all affected animals. Thus further cases of gas gangrene and losses are prevented (Seifert 1992, 1995).

Control
The prevention of the gas gangrene complex in the animal production system of the tropics always has to be a combination of measures of management, feeding and breeding together with immunoprophylaxis. At the same time the particular ecological conditions of the site and the socio-economic situation of the animal holder also have to be taken into consideration.

Management, feeding and breeding. Experience has shown that not only immunization but also the adaptation of management and feeding to the local epizootiological situation may prevent the gas gangrene complex. The best vaccine becomes inefficient wherever the seasonal peak of the disease complex has not been taken into consideration during pasture rotation. As already described above, the seasonal peaks of the disease depend on the specificity of the species and strain to the site, the type, condition and availability of the fodder and the stocking rate on the pasture. Thus only by reducing the stocking rate during the dry season by opening the rotating pasture system on the particular pasture can the incidence of the disease be reduced considerably. If supplementary feeding is applied, only smooth roughage should be used and enough drinking water provided in order to reduce the walking distance for the animals. Furthermore, minerals have to be provided in order to prevent the animals from licking soil. Perhaps those spots on the pasture which have been identified as extremely infected ("campos malditos") could be fenced off and reforested. Another alternative would be to graze an infected pasture only during the rainy season when high and rich pasture is available. Cultivation of fodder crops may also reduce the incidence of disease.

Castration, ear tagging and vaccination should not be carried out on the peak of disease incidence, that is during the dry season in the tropics. Wherever gas gangrene is present when carrying out such interventions, extreme care should be taken in using only clean and sterilized instruments and changing the injection needle after each application. An alternative is the use of a needleless pressure injector (dermojet) (see anthrax II/2.2).

Immunoprophylaxis. With immunoprophylaxis the protection of the calf or young animal is paramount. If calves have been immunized properly they will later on, if they remain on the same site, receive a natural booster through natural infection and thus obtain a long-lasting immunoprotection. Wherever calves fall ill at an age less than 6 months, the dams have to be immunized before calving in order to provide their calves with an efficient passive colostrum-derived immunity. To vaccinate calves which are younger than 6 months is contraindicated since the passive antibodies from the mother and the actively applied antigen would neutralize each other. The vaccination of young animals should be carried out at an age of 6–8 months with an interval of 6 weeks as a booster vaccination. The young animals should be given a booster again in the following year. If site-specific vaccines are applied and the collateral measures carried out which have been characterized above, lifelong immunoprotection can be obtained and the yearly repetition of vaccination be omitted (Seifert et al. 1983, Seifert 1992, 1995).

To prevent gas gangrene through vaccination, several types of vaccines are applied:

- multi-component vaccines, mostly combinations of whole cell cultures which have been inactivated with formalin which may contain small amounts of formalin toxoids. *C. chauvoei* + *C. septicum* + *Pasteurella multocida* are often combined. Mixtures with up to 8 components are marketed;
- cell component vaccines which contain components of the cell wall and/or flagella of the *Clostridium*;
- toxoids, usually with the addition of formalin toxoidized metabolic products of the *Clostridium* which are mostly crude and have rarely been purified;
- live vaccines from attenuated strains, which so far have not become popular. They have been produced in the former Soviet Union and it has been claimed that they have had a better protective effect than formalized whole cell cultures. Because of the potential danger which the injection of attenuated

clostridia could present such a vaccine is not an alternative for application in the tropics.

The vaccines are often produced from laboratory strains which have no relationship to the pathogenic strain relevant to the site. Severe allergic reactions may be triggered by the vaccination especially through whole cell vaccines which also contain lower molecular metabolites of the *Clostridium*. Wherever vaccination is carried out in short intervals (3–4 months) because of the insufficient immunoprotection of the vaccine, such vaccination accidents are frequent. In order to provide a sufficient immunoprotection against the respective site-specific pathogen, the combination of different strains and species of clostridia of the gas gangrene group in a vaccine requires a multiplication of the amount of each antigen which again enhances the danger of producing an allergic reaction. The scientific discussion about whether whole cell vaccines, purified toxoid vaccines, purified cell vaccines or a mixture of purified toxoids and cells can prevent the gas gangrene most efficiently is extensive. We have found that highly purified and concentrated toxoids, or a standardized mixture of purified and concentrated toxoid with washed cells give the best protection. Calves which have received an intracutaneous injection of the vaccine concentrate into the caudal fold and have received a booster vaccination 4–6 weeks later obtain a lifelong immunity. Monovalent homologous site-specific vaccines in combination with a precise diagnosis guarantee the most efficient protection against gas gangrene (Böhnel 1986, Salinas 1991, Schaper 1991, Seifert et al. 1983, Seifert 1992, 1995).

2.3.2. Enterotoxaemia Complex

Aetiology
Because of similar aetiology, epizootiology, pathogenesis, diagnosis, treatment and control, the following infections so far described as singular diseases are summarized as enterotoxaemia:

- *Clostridium perfringens* type A infections, s. sudden death, haemorrhagic enteritis, enterotoxaemia, necrotic enteritis, equinal intestinal clostridiosis, yellow lamb disease;
- *Clostridium perfringens* type B enterotoxaemia, s. lamb dysentery, *Dysenteria neonatorum infectiosa*, scours, lamb diarrhoea, dysenterie, anaérobie des agneaux;
- *Clostridium perfringens* type C enterotoxaemia, s. in sheep: struck, haemorrhagic enterotoxaemia, necrotic enteritis, Braxy-like disease; haemorrhagic enteritis or enterotoxaemia in calves; haemorrhagic necrotizing enteritis, necrohaemorrhagic enterocolitis in foals; and haemorrhagic or necrotizing infectious enteritis, enterotoxaemia, *Enteritis necroticans toxica infectiosa*, bloody scours in piglets;
- *Clostridium perfringens* type D enterotoxaemia, s. pulpy kidney disease, enterotoxaemia, overeating disease.

The enterotoxaemia complex is caused singularly or in combination by the different types of *C. perfringens*. All types of the pathogen as well as *C. sordellii* and *C. spiroforme* may be causal organisms for this disease complex. The classification of specific syndromes for different animal species and also in humans into specific types as mentioned above which is still continued in relevant text books, is difficult to sustain. This is also because of the more or less similar epizootiology of the enterotoxaemia complex.

The bacteriological characteristics of *C. perfringens* and *C. sordellii* have been described above (II/2.3.1). *C. spiroforme* is a Gram-positive anaerobic rod which often draws attention to itself because it appears curled up like a snail shell. Its toxicogenic properties are similar to *C. perfringens* type E. *C. spiroforme* toxin can be neutralized by *C. perfringens* type E antitoxin. The helic structure is a result of the singular semi-circular cells being lined up in a row. The singular organisms are non-motile and measure about 0.3–0.5 by 2.0–10.0 μm. The iota toxin produced by *C. spiroforme* is a binary toxin which is composed out of the components iota-a and iota-b which are similar to the analog components of *C. perfringens* iota toxin. Using immunoelectrophoresis, *C. spiroforme* antiserum precipitates both components of both species; *C. perfringens* antiserum, however, precipitates only the b-components of *C. spiroforme*. The molecular weight of iota toxin is 43–47 kDa (Hatheway 1990).

Occurrence
Enterotoxaemia caused by *C. perfringens* and *C. sordellii* occurs worldwide in all domestic animals. *C. spiroforme* is only responsible for a similar disease in rabbits. Enterotoxaemia is triggered by mistakes in feed management and the influences of weather and may occur as well in both intensive and extensive animal production systems of temperate and tropical climates. The special climatic conditions of the tropics are an important influence factor for the development of the disease. The local specificity of enterotoxaemia depends on the locally restricted occurrence of the organism in the soil and sometimes in the feed.

Epizootiology
Enterotoxaemia is the consequence of an increased influx of high proportions of C-containing nutrients as well as feed with a high protein content into the duodenum. *C. perfringens* or *C. sordellii* which can already be present in the intestine multiply massively because of this sudden increase in nutrients and produce their characteristic pattern of toxins at this point and/or later when they sporulate because the excessive provision of nutrients is discontinued. The production of toxins in the cells takes place during the vegetative stage and can already be demonstrated 3–4 hours after the beginning of the multiplication of cells. As soon as the organism has adapted itself to the increased supply of nutrients which stimulated the excessive multiplication of clostridia (C-sources, proteins), these nutrients are reduced and the vegetative forms of the clostridia sporulate. Through sporulation, large amounts of toxins are released together with the spores from the cell. The free toxin is activated through proteolysis and

enhanced in its effect, i.e. enterotoxin is bound to the membrane of the epithelium of the intestine within one minute whereby ions play an important part. Magnesium ions may reduce the interaction between toxin and receptor. The *C. perfringens* enterotoxin changes its beta-structure into a mainly alpha-helix-structure and reacts with the membrane of the intestine in such a way that the transportation of amino acids is inhibited. This leads to the formation of pores in the membrane of the intestine thus making it permeable for molecules up to a MW of 200000.

Mistakes in feed management and/or other anthropogenous influences which may trigger enterotoxaemia are

- pasture rotation in such a way that the animals are suddenly brought onto a pasture with fresh young grass especially at the beginning of the vegetation time, e.g. the onset of the rainy season;
- grazing on frozen fodder (tropical highlands) in which the starch has been transformed into readily available sugars ("struck");
- a sudden change of pasture when animals are brought from poor pastures to reserved pastures with fresh fodder rich in nutrients, for instance at the end of the dry season;
- an irregular supply of supplementary feed, especially molasses during the dry season or in a feed-lot;
- a disproportionately high percentage of dry matter and a low proportion of crude fibre in the diet as it occurs in intensive feeding using industrial by-products;
- mistakes in the use of milk exchangers or excessive supplementary feeding in young animals as well as mistakes in management which allow calves to take a disproportionately high amount of milk from the mother (Seifert 1995).

The disease is always triggered by a disproportionately excessive feed intake which is rich in nutrients. Contrary to former suggestions that it is mainly the content of protein which is responsible for the toxin production, we have been able to show that C-sources and other so far non-identified toxicogenic components in the feed are responsible for the toxin production (Böhnel et al. 1989, Seifert 1975). Under such conditions in ruminants, the pH in the rumen will be reduced and the complete metabolism of carbohydrates in the rumen is impaired. Feed especially with a low content of crude fibre and structure leads to the degeneration of the rumen villus and in the long run to a reduction of the pH in the rumen. In such cases, the disease may show a protracted onset. The discussion is irrelevant about whether, when triggering the development of entero-toxaemia, the spores have to be present in the feed and have to be able to pass through the rumen, or if they are occasionally ingested, and whether or not the pathogens are already present in the intestine. However, the reduction of pH in the rumen apparently favours the passage of the spores through the rumen. The numerous metabolites which are produced by the clostridia in the intestine have a wide range of pathogenic effects on the organism as soon as the toxins are reabsorbed by the mucosa. Epsilon toxin in particular, which is produced by *C. perfringens* in the intestine, damages the mucosa of the intestine and thus allows proteins of higher molecular weight to pass through. Since at the same time large amounts of feed reduce the peristalsis of the intestine, the production of toxins and toxin reabsorption are aided. Furthermore, the toxins lead to the destruction of the endothel of the vessels, produce oedemas in all organs with capillary haemorrhage which appears in the brain stem as characteristic, focal, symmetric encephalomalacy (FSE) which is the cause of the often peracute course of the disease. Sudden death with lesions of FSE in feed-lot cattle observed in Latin America as well as in Africa, which has been attributed to molasses poisoning was probably caused by *C. perfringens* enterotoxaemia. Whether the activation of epsilon toxin, perhaps triggered off in the intestine through the varying trypsin-content of fodder plants, is responsible for the course of disease or not is still not known (Böhnel et al. 1989). The outbreak of the disease may also be favoured through unhygienic keeping conditions of the animals as well as stress and probably also supplement feed which is contaminated with clostridia. We have also been able to observe that those strong and well-fed animals in a feed-lot which had a high position in the hierarchy of the group are most often exposed to the disease.

In lamb dysentery and necrotizing infectious enteritis in suckling piglets, it is assumed that the colostrum inhibits the trypsin, thus preventing the destruction of beta toxin.

Clinical Features
The course of enterotoxaemia is mostly peracute. The animals die suddenly on the pasture without having shown previous symptoms of a disease. Diarrhoea and symptoms of abdominal pain are only occasionally observed. Signs of nervous disorders, like excitation, ataxia, blindness, tooth grinding and convulsions may appear; they are typical indications for lesions in the CNS. The animals may walk in circles. The temperature may rise or be reduced to subfebrile, the mucosae being grey-violet in colour. A typical symptom is the opisthotonus which is present in the animals when they are in agony; this can also be reproduced in laboratory animals with *C. perfringens* toxins. The prognosis is unfavourable, and the animals die with violent abdominal pain and tonic-clonic convulsions. The morbidity in a herd depends on the number of animals which have had access to the feed which caused the disease outbreak; it can be very high. Shortly before death, the clostridia penetrate the intestinal wall and are distributed throughout the organism where they multiply before the onset of *rigor mortis* which is as soon as the redox potential has been reduced. This leads to the histolysis with its characteristic rancid smell because of the metabolically produced fatty acids. The pulpy kidney in sheep is the result of this development.

Pathology
Necropsy findings vary according to the course of disease. The rumen is usually overfilled with feed, and the mucosa of the gastrointestinal tract is reddened,

showing mucosal petechiation. In peracute cases in which a necropsy has been performed soon after death, it is often difficult to recognize pathological lesions. With careful inspection, circumscribed small grey-yellow foci of necrosis can be found in ruminants on the surface of the kidney; the renal cortex is clear to mid-brown in colour and is of a friable consistency.

During the acute course, catarrhalic enteritis with ulceration of the mucosae is found in all animal species. The pericardial cavity contains an excessive amount of serous fluid especially in young lambs in which clots may form on exposure to air. In older animals, ecchymoses appear in the myocardium and petechiae or haemorrhages in the muscles of the belly and in the intestine. There is usually severe oedema and congestion of the lungs.

The most common lesion in subacute cases in lambs is the presence of necrotic foci and ulcers up to 10 mm in diameter in the mucosa of the small and large intestine. These lesions are surrounded by a zone of hyperaemia and a haemorrhage; some coalesce to form large lesions. Ulcers are usually visible through the intestinal serosa. In severely affected and more chronical cases the ulcers may perforate, resulting in peritonitis and copious peritoneal effusion that contains fibrin which causes adherence of adjacent intestinal loops. The intestinal contents vary from being normal to containing almost pure blood. In calves, the macroscopic lesions are located in the small intestine, particularly in the ileum, ranging from severe generalized enteritis to extensive necrosis of the mucous membrane. Small, yellow diphtheric patches occur in the necrotic areas and the intestinal contents are blood-stained. Other lesions include congestion of the mucous membrane of the caecum and colon, swelling and severe haemorrhaging of the mesenteric lymph nodes, as well as congestion of the liver, spleen and kidneys.

In foals which have died of the disease, the macroscopic lesions are restricted to the alimentary tract and are those of a haemorrhagic enteritis. The content of the entire intestinal tract is fluid and contains blood. Round, dark-red areas, about 10 mm in diameter, are consistently present in the mucosa of the ileum but are less commonly present elsewhere in the small intestine. Some of them may show evidence of ulceration (Kriek et al. 1994).

The post mortal lesions found in camels are similar to those in cattle.

The fast onset of putrefaction is characteristic for any intoxication caused by *C. perfringens*. A few hours after death, anaerobic decay of the internal organs is already present. It is typical that the kidney which decays inside its capsule turning into a pulpy-greasy mass smells sharply of butter acid (rancid).

Diagnosis

Clostridial enterotoxaemia has to be suspected whenever there is a connection between a change in feeding and sudden death. Because of the mostly peracute course of the disease and the unclear symptoms as well as necropsy findings, the disease has also been taken for plant poisoning, other intoxications or grass tetany. Molasses intoxication is often suspected. The outbreak of the disease is mostly connected to a severe mistake in the management of the animals. After a change of pasture or uncontrolled feeding of supplements, a lot of mostly well-fed animals suddenly die, these being mainly young weaned animals but it can also be cows and bulls. The same is true for sheep. Nervous symptoms and opisthotonus shortly before death give a hint to the cause of disease. If the onset of the disease can be stopped by changing the feed, an important lead regarding the cause has been found. In enterotoxaemia in young animals, sudden death is pathognomonic; connections between disease outbreak and management as well as feeding are rarely found here.

The laboratory diagnosis has to be accomplished by demonstrating the toxins in the intestinal contents of dead animals. For this purpose the intestinal contents have to be collected immediately *post mortem*, should be conserved and transported deep-frozen; they have to be purified by cross-flow filtration or centrifugation and have to be tested for pathogenicity in mice. The laboratory animals die in convulsions and with symptoms of the characteristic opisthotonus. By neutralization with type-specific antitoxins, the relevant type of *C. perfringens* can be determined. We have found in contrast to other investigators that the toxin quickly loses its activity if the samples of intestinal contents are not kept deep-frozen until being processed. In order to avoid trials in laboratory animals, the toxin neutralization can now be carried out on permanent cell lines (BHK21- , Vero-cells) (Schaper 1991). A DOT-ELISA has also proved to be reliable under practical conditions to demonstrate the toxin of *C. perfringens* type A (Mehta et al. 1989).

The microscopic demonstration of clostridia in the intestinal wall is irrelevant. The isolation of *C. perfringens* and *C. sordellii* from the internal organs can be important if the same type has been isolated from a number of animals affected by the same disease outbreak so perhaps it can be prevented with the same vaccine (Seifert 1992, Seifert et al. 1992, 1995).

Treatment

If enterotoxaemia is recognized on time, it can be treated. High doses of tetracyclines (oxytetracycline 10 mg/kg b.w.) are efficient. It is best to apply them i.v.. If the type of *C. perfringens* which causes the respective disease outbreak is known in valuable animals (race horses and racing camels) an efficient treatment can be accomplished by applying type-specific hyperimmune serum. Spectacular successes of treatment have been obtained with homologous hyperimmune sera.

Control

Appropriate management of animals can control enterotoxaemia. Preventing sudden changes of pasture, the provision of feed containing structurally crude fibre and the gradual adaptation of the animals to new feed which is rich in nutrients, and/or changing the rotating pasture system to open pasturing are efficient means of disease control. Furthermore, care should be taken that feed or supplement feed in intensive production systems or the feed which is used for young animals is free of contaminants. The equipment for mixing and transportation as well as application of the feed has to be kept clean.

If the types of pathogens which are relevant at the respective site are known, enterotoxaemia can be prevented with a homologous toxoid vaccine. A booster vaccination shortly before the onset of the rainy season will be an efficient prophylactic measure. Enterotoxaemia in lambs can be prevented by booster vaccination of the ewes. The simultaneous application of partially purified hyper-immune serum and oil adjuvant toxoid vaccine protected lambs against *C. perfringens* type D enterotoxaemia. Site- and type-specific toxoids prevent enterotoxaemia much better than polyvalent vaccines (Seifert et al. 1992). As well, young animals which are fattened in a feed-lot can be protected in this way.

An outbreak of enterotoxaemia in a herd can be controlled by applying oxy-tetracycline (LA) to all animals (10 mg/kg b.w.).

An efficient measure for preventing the disease in a feed-lot is the application of feed antibiotics provided they are withdrawn at least 4 weeks prior to slaughter. In the USA and in Mexico where rations which are extremely poor in crude fibre and rich in dry matter are fed because of economic reasons, losses through enterotoxaemia are prevented by adding 2 mg chlortetracycline/kg b.w./day to the feed (Seifert 1992).

A new perspective for the prevention of enteritides and enterotoxaemias in humans and domestic animals is the application of probiotic organisms. *"Probiotic"* is Greek and means "for life". Probiotics are supposed to balance the microbial flora of the gastrointestinal tract and to suppress pathogenic organisms. In the meantime it has been found that probiotic organisms which are daily applied with the feed in the gastrointestinal tract can

- suppress reactions which favour the development of toxic and eventually carcinogenic metabolites;
- stimulate enzymatic reactions which detoxify potentially toxic substances which either have been ingested or produced enterally;
- stimulate or produce enzymes which are required for the metabolism of the feed;
- produce vitamins and other essential substances which are not provided in adequate amounts with the feed.

In connection with the pathogenesis of enterotoxaemia it has been found that the normal healthy intestinal flora is able to inhibit the proliferation and toxin production of pathogens through bacterial antagonism, bacterial interference and competitive exclusion. Probiotica apparently support this bacterial competition and thus prevent the emergence of bacterial infections and intoxications, probably also by the production of bacteriocins.

Probiotic organisms which have so far been applied to humans and/or animals are, amongst others, *Bacillus licheniformes, B. subtilis, Clostridium butyricum, Enterococcus faecium, E. faecalis, Streptococcus lactis, S. cremonis, S. diacetilactis, S. intermedius, Lactobacillus* spp., *Leuconostoc* spp., *Pediococcus* spp., *Propionibacterium* spp., *Bacillus* spp., *Saccharomyces cerevisiae, S.* spp., *Candida pintolepsii, Aspergillus niger, A. oryzae.* The application of probiotica may replace antibiotics in preventing enterotoxaemia in intensive feeding systems in the long run (Seifert 1995).

2.3.3. Complex of Toxications

The following diseases are summarized as the complex of toxications

- Botulism, caused by the ingestion of *C. botulinum* toxins which results from the multiplication of the pathogens in organisms other than the target animal;
- Tetanus, caused by the toxin of *C. tetani* which is produced after parenteral infection of the target animal with the pathogen.

2.3.3.1. Botulism
(Lamziekte, Botulismus)

Botulism in livestock is a non-febrile, highly fatal intoxication and not an infection which mainly affects cattle, sheep, goats, horses, mules, donkeys and also camels but rarely pigs in tropical animal production systems. The disease is usually the consequence of the ingestion of toxins with the feed.

Aetiology
The causal organism of botulism, *C. botulinum*, belongs as an anaerobic spore-bearing rod to the genus *Clostridium*, family *Bacillaceae*. Because of their pattern of toxins 8 types and subtypes of *C. botulinum* can be identified: A, B, C-α, C-β, D, E, F and G. Types A, B and E mainly cause human botulism through food contaminated with toxin. C-α has been isolated from wild ducks, pheasants and broilers. C-β is found mostly in Europe in minks, cattle and horses but also occurs in the tropics, while type D occurs in cattle in tropical pasture ranges especially in southern Africa. Type G has been isolated from soil in Argentina but so far it has not been possible to find a connection to disease outbreaks in humans and animals. It has also been called *C. argentinense* and has recently been made responsible for botulism in humans (Böhnel 1995).

A proteolytic *C. botulinum* type is called *C. parabotulinum*; other organisms which are similar to *C. botulinum* but not proteolytic and difficult to distinguish from *C. sporogenes* are considered as apathogenic *C. botulinum*.

This classification seems to be obsolete because of the present knowledge about the toxin pattern of *C. botulinum* (Böhnel 1995).

Morphologically, *C. botulinum* is divided into 4 groups which differ in size: group a) 0.6–1.4 by 3.0–20.2 μm, group b) 0.8–1.6 by 1.7–15.7 μm, group c) 0.5–2.4 by 3.0–22.0 μm and group d) 1.3–1.9 by 1.6–9.4 μm. Due to their flagella the bacilli are motile. Unfavourable growth conditions produce polymorphisms with up to 70 μm long bacilli. The vegetative organisms are Gram-positive; they become Gram-negative in older cultures especially when the sporulation has started. On blood agar, the colonies grow with β-haemolysis, and they are usually irregular with a partly bow-shaped dented margin: group a) may also present root-like filaments, group b) frayed filaments, and group d) may also sometimes present lobular thread-like filaments, the colonies being translucent grey, often with a granulated centre.

Because of the large number of types and strains it is almost impossible to find common data regarding the metabolic behaviour of the *C. botulinum* strains. Four groups of similar metabolic properties have been considered which, however, also include other clostridia

- group a): type A, proteolytic strains of type B and *C. sporogenes*;
- group b): type E, saccharolytic strains of type B and F;
- group c): type C and D, and *C. novyi* A;
- group d): type G, and *C. subterminale*.

Much more than with other clostridia, the spore production of *C. botulinum* varies in the different types and within the strains of those types. Toxigenic strains usually produce less spores while proteolytic strains sporulate quickly. At first a prespore is produced from the vegetative cell which has to mature to the (permanent) spore. The spore production depends on the environmental conditions in which the organism is kept. In young cultures, dormant spores often appear which do not germinate even if kept under optimal growth conditions. Old spores germinate immediately. The spores are very resistant to environmental conditions: they withstand 100 °C for 130 minutes in a neutral environment; in dry heat the spores survive for 5 minutes at 180 °C, but germinated spores as well as vegetative bacilli are destroyed at 60 °C in a few minutes. At room temperature, sporulated cultures have remained alive for decades; in cheese spores have survived at +4 °C for 4–5 years, in canned meat for 115 years, and for 11 years in the swamps of Saskatchewan (Böhnel 1995, Mitscherlich and Marth 1984).

As they are the cause of the disease, the neurotoxins produced by *C. botulinum* are usually called botulinum toxins. Botulinum toxin is considered to be the most toxic biological substance. The neurotoxins, C2 and C3 are exotoxins which are released from the organism during the logarithmic phase of growth. An exception is type E toxin which can only be found after the cells have been lysed. Not all of the toxins emerge from the cell as definite molecules. Monomolecular protoxins have been described for all types. They are split through proteolytic activation ("nicking"). With the word "progenitortoxin", complexes are described which are composed of toxic and non-toxic components and which can be split into the toxic molecules (derivative toxins). The molecular weight of the toxins varies between 12 and 900 kDa in type A, to 150–238 kDa in type F. The toxinogenesis depends on the environmental conditions of the vegetative organism. Toxin production, however, is also possible at low temperatures. In carcasses toxin production depends on a number of factors which are so far unknown. *C. botulinum* toxin not only originates in material of animal origin. Characteristically for the disease in humans it is produced in vegetables, especially beans. It has also been found in silage and brewers' grain.

Occurrence

Botulism is a classical enzootic of the arid and semi-arid pasture ranges of the tropics and occurs wherever there is a low phosphorus content in the soil. The disease is also beginning to appear increasingly in the tropical humid pasture ranges which have been developed after deforestation. Rice is usually planted there for a couple of years after clearing the trees in order to cover the cost of clearing. The rice plant robs the soil of the phosphorus, and thus the pastures which are planted after the rice, produce fodder with poor phosphorus content. While South Africa and Namibia used to be known to be especially affected by botulism, this intoxication now occurs more and more in the Sahel and in South America as a consequence of overstocking and insufficient mineral supply. In intensive animal production systems, botulism appears where fish-, carcass- and blood meal are fed. It has also been observed that carcasses which lie in waterholes cause type C and D botulism which is passed on through drinking water.

In cattle, botulism is usually caused by type C1 and D. In South Africa, type D and C are the most important, type B being found after feeding with grass silage and brewers' grain. In Australia, type C is the most common cause of botulism. In Argentina, the types A and D have been found to cause botulism; type A has been made responsible for botulism in Cebu cattle in Brazil (Böhnel 1986, 1995, Seifert 1992).

Epizootiology

Tropical soil is poor in phosphorus which is freely available for plants. Phosphorus is bound to aluminium in tropical laterite soil which makes it unavailable to plants and thus to the animals. Cultivation of rice after tree clearing by burning leads to a further enormous reduction of the phosphorus content of the soil. Overstocking not only causes an absolute deficiency of fodder but also an extreme relative lack of minerals in the fodder plants. The appearance of botulism on pasture ranges where the disease has so far been unknown is a typical indicator for overstocking and management mistakes. Thus the disease has appeared in the Sahel only after the nomads were forced to become sedentary around wells which had been installed (Seifert 1991).

Whenever animals only have a diet poor in phosphorus available, they try instinctively to cover their mineral requirements by seeking phosphorus-containing matter of animal origin. On the tropical pasture, there is usually an opportunity to do this in the surroundings of waterholes where remains of carcasses and bones of dead animals can be found (osteophagism). Small animals, for instance rodents which have died on the pasture, may also mummify there and become the source of intoxication. In South Africa, tortoises, hares and birds along with rabbits in Australia are frequently supposed to be the foci of toxin production. The botulinum toxin remains active in the shell of a tortoise for 600 days.

The botulinum toxin is produced inside the carcass of an animal which has died because of another cause. As soon as the *C. botulinum* organisms which live in the intestine saprophytically have entered the surrounding tissue *post mortem*, they start to multiply after the aerobic putrefaction has produced anaerobic environmental conditions. In the arid tropical climate, optimal conditions exist because the carcass mummifies rather quickly due to the low moisture and an anaerobic environment is soon built up inside the carcass. In this way, botulinum toxin can already be produced in a carcass the size of a mouse. Because of the high toxicity of botulinum toxin, even small amounts of ingested tissue will cause intoxication. Only few animal species like dogs and cats have a natural resistance to the botulinum toxins. 1 g of dry muscle of a mummified carcass may contain enough toxin to kill a cow. During their passage through the gastrointestinal tract the toxins are not affected by the metabolic enzymes of the organism and are reabsorbed by the mucosa of the stomach and intestine.

The oral toxicity depends on the natural formation of complexes. The active toxin is composed of a heavy and a light chain, both chains being bound through disulfite-bridges. The larger the complex the higher the oral toxicity because the enzymatic inactivation is reduced by the size of the complex. The toxic action of the neurotoxins of *C. botulinum* results in three steps

1st: binding to the respective ganglioside receptors;
2nd: receptor-specific endocytosis;
3rd: intracellular lysis.

The toxin has the highest affinity to the nerve endings of the striated muscles. Thus the release and not the production of acetylcholine is blocked. The muscle-paralyzing effect may last for weeks. Because of their action, the toxins of the different types of *C. botulinum* cannot be distinguished from one another since they all produce paralysis by affecting the peripheral nerve system. They have no or only little effect on the CNS (Böhnel 1995, Seifert 1992).

Clinical Features
The clinical signs caused by intoxication with types C and D in cattle vary considerably and depend on the quantity of toxin which has been ingested. The incubation period varies from about 18 hours to 16 days, but death may occur for up to 17 days after animals have had access to toxic feed. Because of the destruction of the cholinergic nerve-muscle connections, the typical symptoms of botulism intoxication are starting with excitation, diarrhoea and visual impairment. The affected animals are afebrile and manifest partial or complete flaccid paralysis of the muscles used for locomotion, mastication and deglutition. The paresis which progresses to paralysis, usually commences in the hindquarters and then spreads progressively forwards affecting in turn the fore limbs, the neck and head. Hypersensitivity and aggressiveness may occasionally manifest themselves. The animals' appetite is usually unaffected and they will masticate food for as long as they are able. Ruminal tympany often develops in animals which lie in awkward positions in the terminal stage of disease.

Four stages of the disease have been observed: peracute, acute, subacute and chronical. Animals with the peracute disease die suddenly without any clinical signs being noticed. The acute course lasts for about 1–2 days. Walking stiffly, the animals have become dull: they walk with their forelegs splayed, they lie down and are just able to get up on their forelegs only to drop back down both rapidly and helplessly. Initially the head is held upright, but in the later stages, the neck is typically doubled back with the head resting on the flanks. Impairment of the muscles used for mastication and deglutition is usually pronounced, causing boluses of feed to accumulate in the back of the mouth. Paresis and paralysis of the tongue also set in, the tongue either lying on the incisors or protruding from the mouth. Paralysis of the tongue is a pathognomonic feature. The tongue can be pulled out of the mouth whereby the mouth can be opened manually with relative ease. During the subacute course which lasts about 3–7 days, the clinical signs differ from those of the acute disease only in degree and in the period required for their evolution. Affected animals walk with a stiff, sluggish gait and usually prefer to lie down becoming gradually recumbent. Chronic botulism is characterized through a long incubation time and a clinical course that lasts longer than 7 days. At first the animals are still able to eat, but gradually appear tucked up and have staring coats. Swallowing and chewing are not always affected; paralysis only develops to a certain extent, but the animals lose condition, and with progressing clinical signs, death usually takes over. The signs of type B botulism are different. The disease is characterized by anorrhexia, profuse salivation, regurgitation of feed and water after ingestion, and dehydration. Affected animals have a reduced appetite, reduced milk production and sometimes diarrhoea. Though the tonus of the skeletal muscles is decreased, the animals are not ataxic, nor are the tail or tongue paralysed. Severely affected animals usually die, but those which are mildly affected may recover after 3 to 6 weeks.

In sheep, symptoms of botulism are similar to those in cattle: animals are usually affected after the first change of teeth; pregnant ewes are usually affected, too. The first clinical symptom is usually a paralysed tail, the animal walking with a stilted gait and remaining separate from the herd. Prior to the paralysis of the limbs, the neck muscles are affected and the animal becomes unable to keep its head in a normal position. Salivation starts and the animal dies without signs of agony. Sheep are relatively resistant to type G toxin depending on the intensity of the intoxication, and up to 50% of affected animals may recover. In goats, the most prominent symptoms are a stilted gait and difficulties in getting up. Paralysis of the muscles used for mastication and deglutition is not as obvious. Often the disease takes a chronical course.

Horses are very sensitive to toxins of all types of *C. botulinum*. The symptoms of type B and C intoxication vary to a certain degree but are similar in principle. Intoxicated animals do not usually live longer than 48–72 hours, and in the case of type B intoxication only 24–40 hours; for intoxicated foals it is even less. The clinical signs start with the horses walking with a stilted gait. They are initially lethargic and stand with slight or complete carpal flexion; no febrile reaction appears. After about 24 hours, depression sets in. The animals stand with a hanging head and neck and show signs of salivation; they have no facial expression and appear "dumb". They retain their appetite and when intoxicated with type C toxin have no difficulty in deglutition which is different from type B intoxication where swallowing is inhibited and feed which has been taken falls out of the mouth again. The horses retain their appetite and become constipated. Mydriasis and irregular reflexes of the pupilae appear from the beginning of the intoxication though it is also been observed that eye, anal, caudal and skin reflexes have not been affected. Within 24 hours these signs usually develop into severe muscular tremors of the fore and hind legs which disappear after the animals have been in the prone position for a short period, the animals becoming sternally recumbent and lying with an extended neck and the muzzle resting on the ground. Terminally, horses assume lateral recumbency while paddling with their legs (Böhnel 1995, Kriek and Odendaal 1994, Seifert 1992).

There are few reports about botulism in camels. During a catastrophic outbreak of type C botulism in dromedaries in Chad a peracute course of the toxication appeared. The sick animals had difficulty in standing, developed hind quarter paresis, then they collapsed and died within a few hours (Provost et al. 1975).

Diagnosis

The diagnosis of botulism has to determine if the intoxication has been caused by the *C. botulinum* toxin and at the same time it should provide information about the source of the toxin.

Under the conditions of a tropical animal production system, it is difficult to find animals which show clinical symptoms. If clinical symptoms are observed, a presumptive diagnosis may be possible if the epizootiological development of the disease is taken into account. Here especially the paralysis of the tongue and a lack of resistance in the animal when its ears are touched are pathognomonic for botulism.

The toxin can be demonstrated in the blood, the contents of the gastrointestinal tract, the tissues and body liquids of affected and dead animals, but only if the samples are taken from carcasses before putrefaction has set in. Because of its affinity to the tissue, the toxin disappears from the body liquids rather quickly. The toxin can be demonstrated in the supernatant of centrifuged samples either through toxin neutralization with the respective antitoxin in mice or with the CFT developed by Weiss and Weiss (1988) which is claimed to be able to demonstrate even small traces of *C. botulinum* toxin. A problem, however, is that all these tests require the availability of the respective *C. botulinum* antisera which at present is difficult because they are rarely commercially available.

The isolation of *C. botulinum* organisms from carcasses is irrelevant.

In order to identify the source of intoxication the suspected feed can be analysed for its toxin content. Böhnel (1995) has summarized the techniques applied at present. The principle is to incubate the material in order to induce toxin production and later on to demonstrate the presence of toxin as already described above either with toxin neutralization in mice or the CFT. The isolation of the *C. botulinum* organisms would of course be diagnostically important provided they represent the same type as the toxin identified in the affected animal (Böhnel 1995, Seifert 1992).

Treatment
Under the conditions of animal production systems in the tropics, treatment of botulism is virtually impossible. Valuable animals can be treated with hyperimmune sera of the homologous *C. botulinum* type – this is only a theoretical possibility because the animal will usually die before the relevant *C. botulinum* type has been identified and the respective hyperimmune serum acquired. The hyperimmune serum has to be applied intravenously; because of the difficulty in knowing which type is required, a mixture of types C and D is used in South Africa (Kriek and Odendaal 1994, Seifert 1992).

Control
Prerequisite for the prevention of botulism is appropriate management and feeding of the animals. Adapted pasture management and adequate provision of phosphorus to the animals can prevent the disease. Mineral licks, perhaps also the application of technical phosphorus acid mixed with drinking water or added to a molasses lick are practical means for disease prevention (Seifert 1971). With cultivated pastures, fertilizing should already provide the fodder plants with an appropriate level of phosphorus.

Another important prerequisite for the prevention of botulism is to carefully remove all carcasses and parts of carcasses from the pasture and dispose of them preferably by burning.

Vaccination against botulism is an effective way of preventing the disease where the disease still occurs in spite of appropriate mineral supply. This may be necessary especially for horses and when cattle are fed with poultry litter or similar unconventional feed stuffs. Bivalent toxoid vaccines of types C and D are available which contain aluminium hydroxide gel as an adjuvant. An oil adjuvant has been tried experimentally, and this type of vaccine has produced a better immunity. Some vaccines are prepared by mixing an antigen from *C. chauvoei* with a *C. botulinum* toxoid (Kriek and Odendaal 1994, Seifert 1992).

2.3.3.2. Tetanus

Symptoms of tetanus were described in the *Corpus Hippocraticum* during the 3rd and 4th century A.D. The connection between tetanus and an infection with soil bacteria was recognized back in 1884 by Nicolaier in Goettingen. Tetanus is a toxaemia which is caused by a specific neurotoxin of *C. tetani*. The organism is found in the soil and the intestinal contents of humans and animals.

Aetiology
C. tetani belongs to the genus *Clostridium* of the family *Bacillaceae*, and it is a mostly motile rod because of its peritrichous flagella, 0.5–1.7 by 2.0–18.0 μm in

size. It appears singularly or in pairs and stains in young cultures as Gram-positive and after 24 hours incubation as Gram-negative. The spores are mostly round and terminal (drum-stick shape); sometimes they may be ovoid and appear subterminally. When incubated under strict anaerobic conditions flat colonies with a diameter of 4–6 mm are formed on blood agar; they are translucent, grey with a matt surface, have irregular and rhizoid margins and produce β-haemolysis. On humid agar they have a tendency to form themselves into swarms. The optimal incubation temperature varies, depending on the strain, between 32 and 37 °C; *C. tetani* grows in most enriched culture media with neutral or slightly alkaline pH.

The biochemical properties of *C. tetani* may vary extremely depending on the strain which makes a differentiation based on biochemical parameters impossible.

C. tetani produces a complex toxin composed of tetanospasmin, haemolysin, tetanolysin and a non-spasmogenic toxin. The toxin production is controlled by a large plasmid; whether or not phages which have been found with *C. tetani* are involved in the toxin production is doubtful. Tetanospasmin, the real tetanus toxin, is produced intracellularly as a protoxin of little activity, its amount and toxicity depending on the availability of specific toxinogenic factors in the environment, the toxinogenic capacity of the strain and the type of cultivation (*in vitro*). The tetanus toxin is composed of 1315 amino acids with an MW of 150700 Da, (fragment BC heavy chain 98300 Da, fragment A light chain 25288 Da). The sequences of amino acids are very similar to those of the toxins of *C. botulinum* types A, B and E. The intracellular toxin (ITT) is split by proteases into the tetanus toxin (TT) which consist of a light chain, fragment A or alpha L, and a heavy chain fragment BC or beta H. The pharmacological effect of tetanospasmin depends on both parts of the molecule. Fragment BC causes the penetration into the neurones by binding its amino-end to the specific receptor of the membrane. The fragment A inhibits the release of the transmitter substance acetylcholine. Fragment C is of immunological interest since it is non-toxic, but the antibodies targeted against it protect against the toxic effect of the whole toxin. The specific toxicity of tetanus toxin varies depending on the animal species between 5 by 10^7 and 2 by 10^8 MLD/mg N. At the motor endplates, the effect of the toxin is about 100 times less than botulinum toxin A or B; inside the cell the effect is similar. The toxic effect of tetanospasmin can be inhibited, i.e. toxoidized through chemical influences (formalin) whereby the antigenic action is maintained. Through formalin bridges, tetanus toxin may connect to amino acids which are contained in the culture medium and thus produce allergenic substances. Therefore, toxoid vaccines should be produced from toxin which has been purified previously through toxoidation (Böhnel 1995, Seifert 1992).

Epizootiology

C. tetani is a soil bacterium which also appears in virgin soil. It can multiply within the soil in symbiosis with other anaerobes and/or aerobes; therefore tetanus must be considered as a *Saprozoonosis*. *C. tetani* appears in all soils worldwide. No agreement exists about whether the presence of *C. tetani* in the soil depends on the soil type or not. C. tetani also appears in the intestinal contents of humans and animals. With an increasing population of humans and animals and intensification of land use, the amount of organisms is mounting; they are more frequent in ploughed fields than on pasture. The infection with *C. tetani* usually occurs through a deep wound, though superficial abrasions which are covered with a crust also provide conditions suitable for the development of *C. tetani*. In a tropical animal production system, injuries caused by tough thorns are the usual foci of infection. Whenever lambs are castrated or their tails docked with elastic rubber rings, anaerobic conditions appear around the constriction and through the abrasions caused by the ring, spores of the pathogen can penetrate. In this way, heavy losses can appear after such interventions. Infections of the navel and lesions due to difficulties during parturition (twins) also act as ports of entry for the *C. tetani* infection in sheep. Cattle are relatively resistant to the infection with *C. tetani*; apparently they possess normal antibodies and/or produce protective antibodies early on as a consequence of a minimal infection.

The spores of *C. tetani* are unable to multiply in healthy tissue which contains oxygen. The same is true for wounds which are open and bleeding and/or are well supplied with blood. The wound in which the *C. tetani* infection develops is a deep lesion caused by a puncture which is contaminated with small amounts of soil or dung. In the tropical production system animals are often rounded up and penned, so deep injuries at the fence, in the chute or on the transport vehicle can easily occur. After aerobic organisms have consumed the oxygen inside the wound, the strictly anaerobic *C. tetani* can multiply. The causal lesions are only rarely found, e.g. after careful inspection of the hoof. The pathogens remain localized inside the necrotic tissue of the primary site of infection. If *C. tetani* spores manage to join up with destroyed tissue in the blood, or get into the inner organs via the lymph, they will be distributed and settle, especially inside the liver and spleen. Under appropriate anaerobic conditions, idiopathic or cryptogenic tetanus will be caused from there. In such cases no connection between a wound and the intoxication will be found. As soon as the invaded *C. tetani* organisms have ceased to multiply, they decay autolytically and release a highly effective neurotoxin. The neurotoxin is reabsorbed by the motor nerve endings in the surrounding of the site of infection and reaches the spinal cord and further the CNS by migrating along the nerve sheaths. This leads to the development of the typical ascending symptoms of tetanus, which is a spasmic-tonic contraction of the striated muscles. In ascending tetanus, a migration speed of the toxin of 7.5 mm/hour has been calculated. In cases of a massive infection and toxin production, the toxin may reach the CNS directly via the blood and the lymph and then from there cause descending symptoms which are the usual pathogenesis of the tetanus infection in animals (Böhnel 1995, Seifert 1992).

Clinical Features
The incubation period varies between 1–3 weeks; on average it is 10–14 days but can also be much longer. In sheep the first symptoms can already appear 3–10

days after castration or docking; navel infections can already show symptoms after 48 hours. The incubation time depends on the amount of spores which has been implanted, the site of infection, the type of lesion in which the infection develops and the species of affected animals. The symptoms of tetanus are similar in all species of animals. At first, typical difficulties in chewing and stiffness of the neck and the hindquarters appear, the affected muscles appearing hard. 1–3 days later, generalized convulsions of the entire muscular system set in. The head and neck are kept in a straight position, the tail is raised and the limbs are kept in a sawhorse-like position. The muscles are rigid and tonic spasms and hyperaesthesy are apparent. Reflex sensibility increases and convulsions are triggered through the smallest stimuli like noises, light or contact. Typical symptoms which appear in horses are difficulties in chewing, stiffly erect ears, an extended tail, wide open nostrils and protrusion of the nictitating membrane, the latter being an extremely pathognomonic feature (Fig. 55 on page 1 annex). The animals are only able to walk with stiff limbs and usually remain standing like a sawhorse. The slightest excitement will trigger the typical spasms of the muscular system. Horses fall down with the characteristic torticollis and wide open eyes, remaining stiffly so for minutes with typically bent fore legs and extended hind legs (Fig. 55). They sweat profusely, and both circulatory and respiratory disorders appear along with tachycardia and an increased frequency of respiration and hypostasis in the mucosae; defecation and urination are impaired. No disorders of consciousness are observed during the course of disease. The temperature may be slightly elevated at the beginning of the clinical symptoms, but later on it may be subnormal but also febrile. In horses the mortality rate is mostly high, often over 80%. Animals which recover do not acquire a challengable immunity (Böhnel 1995, Merck 1991, Seifert 1992).

Pathology
During necropsy, no pathognomonic macroscopic or histological alterations can be found.

Diagnosis
The clinical picture of tetanus is rather pathognomonic. It is, however, necessary to carry out a differential diagnostic investigation in order to assure that the disease is not confused with other diseases like hypocalcaemia, other affections of the CNS, botulism and perhaps BSE. If the site where the infection started can be found, the causal organism may be isolated from tissue obtained by biopsy.

Treatment
The treatment of tetanus has several starting points:

– toxin neutralization,
– antibiotic application,
– treatment of the wound,
– symptomatic treatment.

The duration of the treatment may require up to 4 weeks; the prognosis in horses and cattle is rather bad, and the mortality rate will be about 80%. Cattle recover relatively often. If animals do recover, they will do so completely.

The first toxin neutralization was described by Behring and Kitasato (1890) [cited from Böhnel 1995]. Antitoxin only neutralizes free toxin. Toxin which has already been bound to the nerve endings cannot be influenced any more. The effect of the applied hyperimmune serum depends on the dose, the duration of the disease and the method of injection. The most efficient way of application is an injection into the subacharachnoidal space, but an intravenous injection is also successful. Using a subacharachnoidal injection, toxin which already has been bound to the spinal cord can be neutralized. Only antitoxin which has been purified should be applied and if possible it should originate from the homologous animal species in order to avoid allergic reactions. Applying tetanus toxoid simultaneously is also recommended as it is supposed to support the antitoxin application.

The application of high doses of antibiotics (penicillin, chloramphenicol, clindamycin, erythromycin and tetracycline) is supposed to inhibit the further multiplication of *C. tetani* at the site of infection and thus the further toxin production.

Symptomatic treatment of convulsions can be achieved by applying chlor-promazine (0.4 mg/kg b.w. i.v. or 1.0 mg/kg b.w. i.m.) and acetylpromazine (0.05 mg/kg b.w.) twice daily for 8–10 days until the symptoms have disappeared. Curare-like compounds, tranquillizers or sedatives are especially efficient at the beginning of the clinical symptoms (Böhnel 1995).

Control

Tetanus can be prevented efficiently through active immunization. At the same time, care should however be taken to prevent circumstances in animal manage-ment which favour injuries in animals appropriate for the development of the infection. At present, vaccines are produced as formalin- and/or glutaraldehyd-toxoids. The vaccine is available as a whole vaccine (diluted whole toxoid) or as an adsorbed vaccine (toxoid-bound to aluminium hydroxide, aluminium phosphate or calcium phosphate). Combined vaccines which also contain other toxoids from clostridia are commercially available. As a standard, 150 IU tetanus toxoids are applied to all animal species and boosted after 4 weeks. A challengable immunity will have been achieved 10–14 days after the booster vaccination. The immunoprotection lasts for about a year. It is not influenced by the simultaneous application of antitoxin. Young animals should be vaccinated with 3 and 5 months of age, the vaccination being repeated after 12 months. The immuno-protection will last 2–4 years and has only to be repeated every 2–4 years. Regular active immunization is required wherever the incidence of tetanus is high.

Recombined vaccines containing fragments C of *C. tetani* have been produced experimentally on *E. coli*. Synthetic peptides are also being experimented with. So far there is no practical application of these vaccines (Böhnel 1995).

References:

Bergey, D. H. (1986): Manual of Systematic Bacteriology. 2, Williams & Wilkins, Baltimore.

Böhm, R. and G. Späth (1990): The taxonomy of *Bacillus anthracis* according to DNA- DNA hybridization and G + C content. Salisb. Med. Bull. **68**, 99–101.

Böhnel, H. (1986): Methodische Untersuchungen zur Herstellung von Toxoid-Vakzine gegen Clostridiosen der Wiederkäuer. Habil. FB Agrarwiss., Göttingen.

Böhnel, H. (1988): Die Toxine der Clostridien. J. Vet. Med. B **35**, 29–47.

Böhnel, H. (1995): Botulismus und Tetanus, vol. II/4. In: Blobel and Schließer – Handbuch der bakteriellen Infektionen bei Tieren. 89–179, Fischer, Jena, Stuttgart.

Böhnel, H., B. Pawelzik and R. Schaper (1989): Untersuchungen zur saisonalen Toxigenität verschiedener Kulturpflanzen für Clostridien im Zusammenhang mit dem sog. Wildsterben. Berl. Münch. Tierärztl. Wschr. **102**, 310–317.

Cole, H. B., J. W. Ezzell, K. F. Keller and R. J. Doyle (1984): Differentiation of *Bacillus anthracis* and other *Bacillus* species by lectins. J. Clin. Microb. **19**, 48–53.

De Vos, V. (1990): The ecology of anthrax in the Kruger National Park, South Africa. Salisb. Med. Bull. **68**, 19–23.

De Vos, V. (1994): Anthrax. In: Infectious diseases of livestock (eds: Coetzer, Thomson and Tustin), 2, 1263–1289, Oxford Univ. Press, Cape Town/S.A.

Freer, J. H. (1988): Toxins as virulence factors of gram-positiv pathogenic bacteria of veterinary importance. Virulence mechanisms of bacterial pathogens. Amer. Soc. Microbiol. Washington, 265–267.

Giercke-Sygusch, S. (1987): Untersuchungen zur Erstellung eines Altas anaerober Bakterien. Göttinger Beitr. Land. Forstwirt. Trop. Subtrop. 20.

Gonzales-Salinas, J. (1995): Untersuchungen zur Ätiologie von Clostridiosen in Nordost-Mexiko. Diss. FB Agrarwiss., Göttingen.

Hambleton, P., J. Anthony Carman and J. Melling (1984): Anthrax: The disease in relation to vaccines. Vacc. **2**, 125–132.

Hatheway, C. (1990): Toxigenic clostridia. Clin. Microbiol. Rev. **3**, 66–98.

Heitefuß, S. (1991): Untersuchungen zur Identifizierung von aeroben, anaeroben und fakultativ anaeroben Bakterien mit gaschromatographischen Methoden. Göttinger Beitr. Land. Forstwirt. Trop. Subtrop. 57.

Heitefuß, S., D. Lawrence and H. S. H. Seifert (1991): Differentiation of *Bacillus anthracis* from *Bacillus cereus* by gas chromatographic whole-cell fatty acid analysis. J. Clinic. Microbiol. 7.

Jones, W. I. and F. Klein (1967): *In vivo* growth and distribution of anthrax bacilli in resistant, susceptible and immunized hosts. J. Bact. **94**, 600–608.

Kriek, N. P. J. and M. W. Odendaal (1994): Botulism. In: Infectious diseases of livestock (eds: Coetzer, Thomson and Tustin), 2, 1354–1371, Oxford Univ. Press, Cape Town/S.A.

Kriek, N. P. J. and M. W. Odendaal (1994): *Clostridium chauvoei* infections. In: Infectious diseases of livestock (eds: Coetzer, Thomson and Tustin), 2, 1325–1331, Oxford Univ. Press, Cape Town/S.A.

Mayr, A., G. Eißner and B. Mayr-Bibrach (1984): Handbuch der Schutzimpfung in der Tiermedizin. Parey, Berlin.

Mehta, R., K. G. Narayan and S. Notermans (1989): DOT-enzyme linked immunosorbent assay for detection of *Clostridium perfringens* type A enterotoxin. Int. J. Food Microbiol. **9**, 45–50.

Merck & Co. (1991): The Merck veterinary manual, 7th ed., Merck, Rahway/USA.

Mitscherlich, E. and E. H. Marth (1984): Microbial Survival in the Environment. Springer, Berlin.

Naylor, R. D., P. K. Martin and R. T. Sharpe (1987): Detection of *Clostridium perfringens* epsilon toxin by ELISA. Res. vet. Sci. **42**, 255–256.

Provost, A., P. Haas and M. Dembelle (1975): Premiers cas au Tschad de botulisme animal (type C): intoxication des dromadaires par l'eau d'un puit. Rev. Elev. Med. vet. Pays trop. **28**, 9–12.

Rengel, J. (1993): Untersuchungen zur oralen Immunisierung mit dem *Bacillus anthracis* Stamm 34 F$_2$ (Sterne). Göttinger Beitr. Land. Forstwirt. Trop. Subtr. 77.

Salinas, A. (1991): Die Entwicklung einer für Nordost-Mexiko standortspezifischen Vakzine zur Verhütung des Gasödems. Diss. FB Agrarwiss., Göttingen.

Schaper, R. (1991): Methodische Untersuchungen zur Produktions- und Wirksamkeitskontrolle von Rauschbrandvakzinen. Göttinger Beitr. Land. Forstwirt. Trop. Subtrop. 61.

Seifert, H. S. H. (1960): Die Bekämpfung eines "Rindersterbens" in der Cordillere Nord-Perus. Zbl. Vet.-Med. **10**, 991–1015.

Seifert, H. S. H. (1975): Untersuchungen zur Ätiologie von Anaerobenseuchen bei Rindern und Schafen in Lateinamerika. Zbl. Vet.-Med. B, **22**, 60–86 and 177–195.

Seifert, H. S. H. (1987): Identification of pathogenic bacteria by head-space gaschromatography. Anim. Res. Dev. **26**, 100–111.

Seifert, H. S. H. (1987): Problematik der Verhütung von Bodenseuchen in extensiver Wiederkäuerhaltung der ariden Tropen. Göttinger Beitr. Land. Forstwirt. Trop. Subtrop. **28**, 102–112.

Seifert, H. S. H. (1991): Bodenseuchen als Indikator für Überstockung und Umweltzerstörung (eds: Mensching and Seifert), Göttinger Forsch. Proj. Sudan. Akad. Wissenschaften, Göttingen.

Seifert, H. S. H. (1992): Tropentierhygiene. Fischer, Jena.

Seifert, H. S. H. (1995): Clostridiosen, vol. II/4. In: Blobel and Schließer – Handbuch der bakteriellen Infektionen bei Tieren. 5–88, Fischer, Jena, Stuttgart.

Seifert, H. S. H., H. Böhnel and A. Ranaivoson (1983): Verhütung von Anaerobeninfektionen bei Wiederkäuern in Madagaskar durch intradermale Applikation von ultrafiltrierten Toxoiden standoretspezifischer Clostridia. Dtsch. tierärztl. Wschr. **90**, 274–279.

Seifert, H. S. H., H. Böhnel, A. Depping, S. Giercke-Sygusch, A. Heine, A. Ranaivoson, F. Roth and U. Sukop (1987): Ätiologie und Inzidenz von Bodenseuchen in Madagaskar. Dtsch. Tierärztl. Wschr. **1**, 22–27.

Seifert, H. S. H., S. Giercke-Sygusch and H. Böhnel (1990): Identification of pathogenic bacteria by headspace gaschromatography. Anal. Microbiol. Meth., Plenum Press, New York.

Stiles, B. G. and T. D. Wilkens (1986): *Clostridium perfringens* iota toxin: synergism between two proteins. Toxicon 24, 767–773.

Van Ness, G. B. (1971): Ecology of anthrax. Sc. **172**, 1303–1307.

Walz, A. (1993): Nachweis und Eigenschaften von *Bacillus cereus*- Stämmen, isoliert von arabischen Kamelen (*Camelus dromedarius*). Vet. Med. Fak. München.

Weiss, H. E. and H. Weiss (1988): Nachweis von Clostridium-botulinum-Toxin mittels Mikro-Wärmekomplementbindungsreaktion. Tierärztl. Umsch. **43**, 117–126.

Wernery, U. and O. R. Kaaden (1995): Infectious diseases of camelids, 19–27. Blackwell Wissenschafts-Verlag, Berlin.

Wernery, U., H. H. Schimmelpfennig, H. S. H. Seifert and J. Pohlenz (1992): *Bacillus cereus* as a possible cause of haemorrhagic diseases in dromedary camels (*Camelus dromedarius*). In: Proc. 1st Int. Camel Conf., Dubai, 51–58, R & W Public. Ltd., Newmarket, England.

Zeissler, J., C. Krauspe and L. Rassfeld-Sterneberg (1958): Die Gasödeme des Menschen, Bd. 1: Geschichte, Beziehungen zur Veterinärmedizin, Bakteriologie und allgemeine Pathologie. Steinkopf, Darmstadt.

3. Contact Diseases

3.1. Introduction

Contact diseases are transmitted by direct or indirect contact from an infected to a susceptible organism. The most intensive contact occurs during mating when infection of the genital system takes place; even congenital transmission of the infectious agent can occur. The pathogen may also be transferred to the susceptible organism by licking or cutaneous contact of the animals, e.g. at the feed- or water trough. Even the transmission of aerosols, as for example the transmission of *Mycobacterium bovis* after coughing and inhalation, is an indirect transmission, in which the dispersed infected particles are carried through the air which then is the vehicle for the infectious agents. Many different vehicles are involved in the transmission of contact infections. *Mycoplasma m. mycoides* can become air-borne and be carried over limited distances, but salmonellas and viruses can be carried much farther with water and can also be ingested by drinking. It is because of these mechanisms that water-borne diseases are considered as a specific epidemiological group in tropical medicine (human). For the transmission of endoparasitoses, water is a vehicle of special importance. Considering epizootiological mechanisms in animal diseases in contrast to soil-borne diseases, water-borne diseases do not represent an independent disease complex. Viral infections are often carried with non-living vehicles, for instance motor vehicles, as well as tools used in animal production and feed. Last but not least there are the living vehicles which are able to carry infectious agents externally or after ingestion with feed inside their intestinal tract to transmit the pathogen with their faeces. All kind of rodents which can carry pathogens without getting into immunological contact with the agents are vehicles of this kind. A rat which carries the FMD virus from one cowshed to another for instance is a *vehicle* and not a *vector*. The concept of vector should be limited strictly to the function of true vectors which transmit vector-borne diseases when ingesting the pathogen actively, which they then harbour and eventually multiply in order to transfer it later on actively. Whenever arthropods carry causal organisms of contact diseases either actively or passively, they are vehicles and not vectors. Nevertheless, vehicles similar to vectors can become reservoirs of infection for the causal organisms of contact diseases. This is true not only with non-living but also with living vehicles. The importance of humans as vehicles of causal organisms of contact diseases is not to be underestimated. Because of his very varied activities, last but not least travelling worldwide, man can spread exotic pathogens internationally with his clothes as well as with other objects, especially products of animal origin.

The possibilities for the transmission of causal organisms of contact diseases increases with the tenacity of the organisms against environmental influences. The more resistant the agent is to the influences of environmental conditions, as well as to autolysis during decay of a carcass, the longer a site remains infectious, and the better and farther the pathogen can be carried indirectly. The tenacity of causal organisms of vector-borne diseases against environmental influences is of rather little relevance; it is only relevant in so far as they may survive for a certain time when they are attached externally to the vector. This is quite different with causal organisms of contact diseases. Measures for disease control depend to a high degree on the resistance of the pathogens to environmental influences.

It is typical for contact diseases that after recovery from infection with most causal organisms of contact diseases, the organism has produced a more or less lasting sterile immunity (I/2.2.3.2). The same is true after vaccination with live vaccines. Efficient vaccines are available for controlling most of the contact diseases occurring in the tropics. If these vaccines are prepared carefully and if an adequate veterinary infrastructure is present which provides a vaccination scheme adapted to the respective animal production system, contact diseases also can be controlled in the tropics. There are, however, important dependencies of the organism regarding its capacity to produce a lasting protective immunity in relation to its genetically determined resistance as well as to factors of productivity, management, feeding and environment.

Most of the contact diseases which are discussed in the following chapter also used to occur in countries with a temperate climate. Some of them have been almost classical diseases which decimated the animal population in Europe for centuries and have been brought to Africa only during colonization at the end of the last century. It is not the tropical environment which limits the occurrence of these epizootics on the tropical location at present. Rather it is the under-development, the missing infrastructure, the socio-economic as well as the political situation and in some countries also the geographic conditions which are responsible for the persistence of contact diseases in underdeveloped countries. Furthermore, disease outbreaks are favoured through reservoirs of infectious agents in latently infected populations of game. In many regions of Africa, it is not possible to eliminate contact diseases for the time being since a reinfection of domestic animals can continuously occur from neighbouring game population. The interchange of the pathogen from the domestic to the game animal and back can also contribute to increasing the pathogenicity of the pathogen. This knowledge has to be taken into account if novel production systems, for instance game ranching, are planned and if cattle production in the region is supposed to continue. An important influence factor which is favourable to the outbreak of contact diseases in the tropics is the climatic stress which always appears during the often sudden seasonal changes. The stress on the animal at the onset of the hot, humid weather at the beginning of the rainy season often leads to an outbreak of contact diseases as for instance of FMD in South America.

In the industrialized countries, economically important contact diseases have been eliminated with the principle "test and slaughter" (T+S). A prerequisite for such a policy is the availability of diagnostic techniques to detect antibodies in the herds from which the disease has to be eradicated, and financial resources to recompensate the animal holder for those animals which have to be slaughtered because of being carriers. Problems with the reliability of the diagnostic techniques, as well as for example in Germany (leukosis and swine fever), have caused unjustified financial burdens to the animal holder in the past and thus have led to a loss of confidence in such control measures organized by the government. In a tropical animal production system, there usually exists a wide heterogenicity of animal breeds, of the condition of the animals and of environmental influences on the animals, factors which cause the immune response to an infection to become heterogeneous as well. This in many cases may produce results of testing which are difficult to standardize and to judge. The appearance of non-agglutinating, so-called incomplete antibodies in brucellosis infection makes it difficult to apply the T+S scheme to the extensive tropical animal production system in programmes for eradicating brucellosis. A lack in veterinary infrastructure and financial resources of the government for the organization of programmes to carry out T+S are further obstacles for the eradication of contact diseases. On the other hand the acute or latent incidence of contact diseases (FMD, Rinderpest) in developing countries is used by some industrial countries especially the USA and the EC as an important trade barrier. The justified policy of the industrial countries to protect their own animal industry against exotic pathogens can also be used to maintain a stability in prices for their own animal products and to guarantee the home market. In spite of its excellent quality, Argentinian beef is much less expensive on the international market than the product of much poorer quality which is produced inside the EC. Contact diseases are also an insurmountable obstacle for the export of breeding animals. Even when the infection has disappeared after systematic vaccination, the animals cannot be exported because they still are carriers of antibodies although these are of sterile nature. The potential buyer demands negative reactors. Thus Latin American countries for instance are unable to upgrade their Cebu herds by importing breeding stock from India because Rinderpest and FMD are enzootic there.

The causal organisms of contact diseases belong to the *Bacteria* and viruses. In Table 37, the pathogens are summarized which are discussed in this book. Because of the continuously changing systematic grouping (Bergey 1984, Rolle and Mayr 1993) a pragmatic listing has been chosen in order not to complicate matters for applied work in the tropics. The family and genus names have, however, been taken from Bergey (1984) while the numeric order results from the grouping chosen for this presentation.

Table 37. Aetiological classification of causal organisms of contact diseases relevant in animal production systems of the tropics

Fungi:	Viruses:
Genus *Trichophyton*	*Herpetoviridae*
Genus *Histoplasma*	Malignant catarrhal fever
Bacteria:	*Iridoviridae*
	African swine fever
Mycoplasma	
Genus *Mycoplasma*	*Poxviridae*
	Pseudocowpox virus
Actinomycetes and related organisms	Buffalo pox virus
Order *Actinomycetales*	Camel pox virus
Family *Mycobacteriaceae*	Pox in small ruminants virus
Genus *Mycobacterium*	Lumpy skin disease virus
	Contagious ecthyma virus
Family *Dermatophilaceae*	
Genus *Dermatophilus*	*Paramyxoviridae*
	Rinderpest virus
Aerobic rods	Pest of small ruminants virus
Gram-negative rods	
Genus *Brucella*	*Picornaviridae*
Genus *Salmonella*	Foot and mouth disease virus
Genus *Escherichia*	
	Retroviridae
Vibrionaceae	Ovine pulmonary adenomatosis virus
Genus *Pasteurella*	Maedi-visna virus
Lactobacillaceae	*Rhabdoviridae*
Genus *Listeria*	Stomatitis vesicularis virus
Genus *Erysipelothrix*	
Spirochaeta	
Genus *Leptospira*	

3.2. Contact Diseases Caused by Fungi/Mycoses

3.2.1. Dermatophytoses
(Ringworm, Dermatomycoses, Tiña, Glatzflechte)

Aetiology and Occurrence
Dermatophytoses are diseases of the superficial, keratinized layers of the skin and appendages, such as hair and nails, and are caused by closely related myceliae fungi known as dermatophytes (Scott 1994). Some of the important mycoses in livestock are

Dermatophytoses caused by:	in:
Microsporum gypseum	horses
M. nanum	pigs
Trichophyton mentagrophytes	horses
T. verrucosum var. *autotrophicum*	sheep
T. verrucosum var. *verrucosum*	cattle

Sporotrichosis caused by:	
Sporothrix schenckii	equines

Epizootic lymphangitis caused by:	
Histoplasma farciminosum	equines

Mycetoma caused by:	
Curvularia geniculata	horses
Madurella mycetomatis	horses

The dermatophytes are specialized keratinophytic fungi, classified as geophilic, zoophilic or anthropophilic species.

Trichophyton verrucosum is primarily a pathogen of cattle but can be transmitted to other animals and humans and is thus also a causal organism of a zoonosis. The dermatophyte causing infections in sheep and goats resembles *T. verrucosum* morphologically but has other cultural requirements and is named *T. verrucosum autotrophicum*. The former practice of naming a pathogen according to the animal species in which the infection appeared cannot be maintained. Generally, the trichophytes are divided because of morphological and cultural characteristics into

- **endothrixe**, in which the mycelium grows inside the hairs and
- **exothrixe**, in which the mycelium grows inside as well as on the hair.

The latter are mainly those causing dermatophytoses in animals. They are divided into types forming large and small spores. Those which form large spores are *T. verrucosum* and *T. gallinae* which possess large round spores and an extrapilary sheath; those which produce small spores (*T. mentagrophytes* and *T. rubrum*) form a sheath with much smaller spores which also appear intrapilarily together with filaments of mycelium.

The dermatophytosis of cattle and especially calves (ringworm) is mainly caused by *T. verrucosum*. The fungus is easily transmitted to humans. *T. equinum* occurs mainly in horses. Small carnivores are attacked by a large number of *Trichophytum* spp. The disease is relatively rare in small ruminants (*T. verrucosum* var. *autotrophicum*). The disease called "favus" in mammals in which the skin becomes scale-like with penny-sized, round bowl-like colonies of fungi is caused by *T. quinckeanum*.

Dermatophytosis caused by *Trichophyton* spp. occurs worldwide. In the intensive animal production systems of the tropics, it is important wherever calves are raised with powdered milk or milk exchangers. Last but not least it is relevant as a zoonosis under such conditions.

Epizootiology, Pathogenesis, Clinical and Pathological Features
Trichophyton spp. find excellent conditions for multiplication in the hot, humid climate of the tropics. Mostly the pathogens are transmitted by the material of the box in which the calves are kept. Wherever calves are kept in closed sheds in the tropics, the relative humidity may rise to 95% (Gulf States). Under such conditions all the animals often become infected.

Though animals of all age groups may contract the disease, it is most common in calves at an age of 6–8 months. With good keeping and feeding, spontaneous recovery may follow. After an incubation time of 2–4 weeks, the hair breaks off at the infected parts of the skin and/or is lost completely and the characteristic round sharply circumscribed thick, elephant skin-like plaques with a diameter of up to 10 cm appear. It is mostly parts of the skin around the eyes, the ears, the mouth and in the neck which become infected. The infection may spread over the back and other parts of the body.

The appearance of trichophytosis is favoured by a hot, humid climate, bad management, early weaning and unbalanced feeding especially if vitamins are missing in the diet. The spores of the fungi are extraordinary resistant; they may survive inside the crust of the infected skin but also on the wood of the box for years.

Diagnosis
The typical lesions of the skin are pathognomonic. Microscopically, the fungi can be recognized on the infected hairs if the samples are treated previously with KOH. The mycelium and the filaments of spores, the latter with a diameter of 4–6 µm, can easily be identified. A precise differentiation of the relevant species can only be achieved by culturing the organism.

Treatment
Improvement of maintenance and feeding of the animals, application of vitamin A in the feed and keeping animals outdoors are important prerequisites for a successful treatment. Fungicides applied topically are effective. Phosphoric acid esters (dichlorphos) and thiabendazole have also been applied with good results. The treatment has to be repeated frequently.

Control
The disease can be prevented by improving the keeping and feeding of the calves and by carefully cleaning and disinfecting the sheds in which the calves are kept. In Argentina, calves are reared outdoors in portable individual small sheds which are placed on the pasture at such a distance apart that the animals do not have any contact with each other. At the end of the 70's in East Germany, a Russian trichophyton-vaccine (LTF 130) was applied in large calf rearing farms which

not only prevented the disease but also was claimed to have a therapeutic effect (Rotermund 1980). At present in Germany, a life vaccine is on the market which contains the attenuated avirulent *T. verrucosum* strain TV-M-310. Calves are supposed to become definitively protected after a booster vaccination.

3.2.2. Epizootic Lymphangitis
(Lymphangite épizootique, Linfangitis epizoótica, Lymphangitis epizootica)

Aetiology and Occurrence
Lymphangitis epizootica is caused by *Histoplasma farciminosum*, a fungus which belongs to the family *Moniliaceae*. A common feature of histoplasmas is that they appear on a warm-blooded animal at body temperature in the yeast-like phase while changing into mycelium when kept at room temperature. During the yeast-like phase they spread like yeast, and during the mycelium phase they build conidia. Thus they take an intermediate position between mould and yeast. The systematic classification of the organism has been discussed in the past and is still not yet clear (Rolle and Mayr 1993).

The histoplasmas are found in large amounts in the pus of an opened abscess, they completely fill out the leukocytes leading to degeneration of the leukocytes. Depending on the duration of the infection they appear also as free organisms. With increasing age they multiply by budding; plasma is released at the pointed end of the cell from which a bud develops and a new double-contured membrane is built. Subsequently the budded cell separates from its mother cell.

Histoplasmas may remain virulent for 6 months in the pus of an abscess; they withstand sun radiation for 5 days. They even survive the effect of 20% $CaCO$, but the organisms can be killed in one hour with 1% formalin or phenol (Mitscherlich and Wagener 1970).

The disease occurs in the Mediterranean, in Asia, Africa, and it has been reported from Columbia, Peru and Uruguay.

Epizootiology, Pathogenesis, Clinical and Pathological Features
Equines are susceptible to the infection, but less so donkeys. The disease has, however, also been observed in highly productive dairy animals in an intensive production system in Peru (Seifert 1995). The pathogens enter the organism through skin lesions. The vehicle for this are tools for grooming, harnesses and saddlery and the walls of the stable. Haematophagous flies are also supposed to transmit the infection.

The incubation time may be rather long; starting from small wounds ulcers appear at first on the limbs, the withers, the saddle area, the brisket at the neck and also at the head. Subsequently, granulations which bleed slightly arise at the margin of the ulcers and blood-stained pus is secreted. After healing in the surroundings of the ulcer, painless nodes appear between the size of a hazelnut and walnut which gradually become soft and erupt, releasing thick, yellow honey-like pus, often mixed with blood. Neighbouring ulcers coalesce to become extended surface ulcers. The ulcers extend via the lymph vessels becoming enlarged and extending over the surface of the body forming tough, painless tracts which are closely attached to the skin. Node-like swellings build on these, appearing rosary-like. When these erupt, new ulcers appear continuously; erosions can also show up on the nasal septum. If the pathogens reach the eye via the lymph, the consequence will be exophthalmus at first and later on the ulcerative destruction of the eye (Fig. 56 on page 2 annex).

The disease usually has a chronic course which may take months. Dry, warm weather can favour recovery, but with the onset of the next rainy season the disease will erupt again.

Diagnosis

With Gram-staining, the histoplasms can be found as yeast-like organisms of a size of 3.5 by 2.0–3.0 µm intracellulary in the pus of the ulcera. The pathogens grow in a CO_2 atmosphere on Sabouraud-glucose-agar with 10% horse blood. After 2–8 weeks incubation, tiny grey or yellow floccular colonies appear (Scott 1994).

Treatment

As a therapeutic measure, the altered nodules, ulcers and lymph vessels can be excised surgically and treated with antibiotics topically. The application of hyperimmune serum has also been recommended.

Control

Strict quarantine measures and the detection and elimination of infected animals are the best scheme of prevention. Infected animals by all means have to be separated. The stables and cowsheds have to be cleaned and disinfected carefully. The application of a formolized aluminium hydroxide vaccine has been claimed to have prophylactic effect (Mitscherlich and Wagener 1970).

Other mycoses, like sporotrichosis caused by *Sporothrix schenckii*, mycetoma caused by *Actinomyces*, *Streptomyces*, *Actinomadura* and *Nocardia* spp., Rhinosporidiosis caused by *Rhinosporidium seeberi*, aspergillosis caused by *Aspergillus* spp. and particularly *A. fumigatus*, zygomycosis caused by mucorales such as *Mucor*, *Absidia*, *Mortierella* and *Rhizopus* spp., candidiasis caused by *Candida* spp., cryptococcosis caused by *Cryptococcus neoformans* and mycotic abortion caused by a variety of fungi are infections in humans and animals which also occur in the tropics. Because of their local and economically limited importance in productive domestic animals they are not discussed in the context of this book. For information on these mycoses see Scott (1994).

3.3. Contact Diseases Caused by Bacteria

3.3.1. Mycoplasmoses

Mycoplasmas, formerly called PPLO (pleuropneumonia-like organisms), are non-sporulating, Gram-negative, non-motile bacteria which do not possess a determined shape of the cell. Their polymorphism is the consequence of the missing cell wall. Due to their small size (0.1–0.3 μm) and their polymorphism, they are able to pass through the usual bacteriological filters (0.1–0.3 μm). Their characteristic growth is such that from a round or slightly oval cell (central body of 125–220 nm size), a thin homogeneous filament develops. Its end is butt-shaped and swollen at first, but later on it branches out like a fork into new filaments. Chain-like spherical corpuscles (coccoid forms) develop in these filaments which are set free by constriction and grow out to new filaments. On special agar mycoplasmas develop characteristic tiny colonies of a size of 0.1–0.6 mm diameter which look like a fried egg.

Of the 36 recognized species of the genus *Mycoplasma*, *M. mycoides* ssp. *mycoides*, *M. mycoides* ssp. *caprae* and *M. agalactiae* ssp. *agalactiae* are relevant for the presentation in this book.

3.3.1.1. Contagious Bovine Pleuropneumonia (CBPP)

(Pleuropneumonia contagiosa bovum, Lung plague, Lungsickness, Longsiekte, Péripneumonie contagieuse des bovidés, Peripneumonia contagiosa bovina, Lungenseuche)

Aetiology
CBPP is caused by *Mycoplasma mycoides* ssp. *mycoides*. The organism grows best *in vitro* in an atmosphere with 5% CO_2 and on nutrient medium which contains 10–30% horse serum and 0.2% glucose. Growth is relatively slow; the mycoid stage which is a characteristic of strains is caused by a matrix of galactan surrounding the colonies and is found after several days of growth in liquid medium. Colonies on solid media have an average diameter of 1.0 mm, a dense central core, a granular structure thinning towards the edge, and a finely serrated border. The elementary body is the basic reproductive unit which gives rise in broth cultures to homogeneous filaments which, after developing spherical bodies lead to the formation of coccoid forms.

Two types of *M. m.* ssp. *mycoides* are recognized: large colony (LC) and small colony (SC). They cannot be differentiated serologically but are different morphologically, culturally and in their pathogenicity, and can be distinguished through mouse protection tests. LC types occur almost exclusively in goats, rarely in sheep while SC types cause CBPP in cattle. *M. m. mycoides* LC also causes mastitis, arthritis, and, occasionally, contagious caprine pleuropneumonia and a fatal systemic disease in goats.

The serial passage of *M. m.* ssp. *mycoides* in culture and in animals alters both its virulence and pathogenicity (Schneider et al. 1994, Seifert 1992).

The tenacity of *M. m.* ssp. *mycoides* is rather low: sun radiation and usual disinfectants destroy it quickly; it is inactivated in 60 minutes at 50 °C, and in 2 minutes at 60 °C. Dried on clothing it survives for 15 days and can also remain vital for at least 1 year in frozen infected lung tissue.

Occurrence

CBPP was first described in 1550 by Gallo and, with the exception of South America and Madagascar, has occurred throughout the world at some time or another. It was eradicated from Australia, North America and from Europe at the beginning of the 60's. It reached Africa during the colonialization period where it continues to occur with changing extension in a belt which reaches from 18° north down to about the Equator where it spreads in the hot, humid regions of the continent (Fig. 57). CBPP is enzootic farther south in Angola. In the Near East, CBPP occurs in the Gulf States and Jordan. CBPP probably reached China in 1919/20 through Hong Kong and Shanghai with Australian dairy cattle. Australia became infected around the middle of the last century, but as has been stated, it has been eradicated in the meantime (Schneider et al. 1994, Seifert 1992).

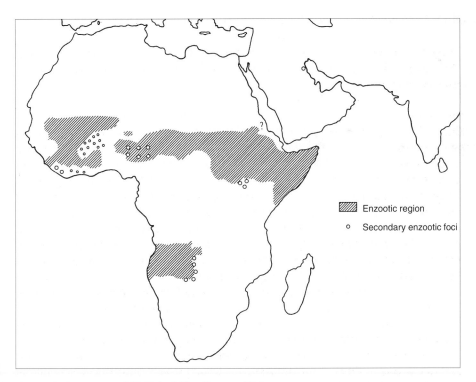

Figure 57. Occurrence of CBPP in Africa (Provost 1981).

Epizootiology

Under natural conditions, breeds of domestic cattle, buffaloes and antelopes are susceptible to *M. m. mycoides*, cattle of the type *B. taurus indicus* being relatively more resistant to the infection than animals of the type *B. taurus typicus*. West African taurine autochthonous cattle are extremely susceptible; with this type of cattle, the disease usually has an acute course which almost always leads to a lethal outcome. The relative resistance of the cebuines apparently depends on their origin since they have already been confronted with CBPP for millenia in the regions of their origin in Central Asia. The difference of resistance between the Sanga cebus of the Sahel and the West African taurine cattle is of great practical importance, the latter being mostly kept in the Sudan region. During the peak of the dry season in West Africa the nomads enter agriculture areas with their Fulani cebus in order to take advantage of the crop residues as feed. When the nomadic Cebus get in contact with the taurine cattle of the sedentary farmers, they infect these animals and thus regularly cause severe outbreaks of disease. Antibodies to *M. m.* ssp. *mycoides* have also been found in African buffalo (*Sincerus caffer*), impala (*Aepyceros melampus*) and in camels. Whether the infection with *M. m. ssp. mycoides* in camels has any significance is disputed (Wernery and Kaaden 1995)

Natural transmission of *M. m.* ssp. *mycoides* occurs through droplet infection either from cattle with the clinical disease or from subclinical carriers ("lungers") which pass the organism on to susceptible animals in close contact. The aspiration of infected aerosols which are discharged from infected animals during coughing or exhalation may occur over a distance of up to 20 m and more. It is facilitated by crowding the animals around watering places or at night in a pen or paddock. Of course, in cowsheds the infection will spread most easily, the direct contact of susceptible with diseased animals, however, being mandatory for transmission. Neither ingestion of infected fodder nor direct exposure to diseased organs will cause transmission of *M. m.* ssp. *mycoides*. Since the organisms are excreted with the urine of affected animals a "urinary tract to nose" route of transmission is suspected; this is also assumed in the case of transmission with infected milk and placental fluid which are, however, modes of transmission which are of no or little epizootiological relevance. The same is true for a supposed transmission by ticks. Since *M. m.* ssp. *mycoides* remains virulent for years in the sequestered lung tissue of latently infected animals, the disease can be carried through nomadic pastoralism over wide distances and also be spread through cattle markets.

The intensity of infection is related to the concentration of infective organisms in a herd at any given time. This is at its highest during acute outbreaks while low intensities occur during the early stage of an outbreak and result in slow-spreading, long incubation periods and a slow build-up of infection. Under these circumstances, many months may pass before the enzootic becomes apparent. The level of susceptibility depends considerably on the level of resistance and immunity of the individual animal, since animals which have recovered from CBPP are less susceptible. New-born animals and animals brought into the herd from other places mostly constitute the nucleus of a new outbreak (Schneider et al. 1994, Seifert 1992).

Pathogenesis

The pathogenesis of CBPP is still not clear. It is assumed that a diffusible toxin produced by *M. m. mycoides* stimulates fibrous granulation tissue and proliferation resulting in capsule formation around infected necrotic tissue (Fig. 58 on page 2 annex). A carbohydrate, galactan, the major antigen of *M. m.* ssp. *mycoides* increases subsequent infection with live organisms and has physiological effects similar to those of the endotoxins of Gram-negative bacteria. Apparently, immunologically induced cell damage and autoimmune hypersensitivity reactions are also involved in the development of lesions, including agglutinating antibodies which probably cause local lesions in the lung (Schneider et al. 1994).

Clinical Features

The incubation time is uncertain and can last from a few days up to several months. Depending on the resistance level of the animal, and the intensity of exposure, the disease takes an acute, subacute to chronic course, or the acute course is followed by a chronic stage which may last for years (lunger) as a latent phase of the disease. The majority of cases are subclinical. The clinical symptoms start with the characteristic short, dry cough which becomes more and more painful. This is a consequence of multiple serous cellular bronchopneumonic foci in the lung. With a moderate temperature (40 °C), the respiration becomes more frequent and aggravated, feed intake and milk production are reduced, the animals are listless with a staring, lustreless coat and they become emaciated. As the disease progresses, the stage of hepatization of the lung sets in with lobular extension of the disease also reaching the pleura. The febrile reaction increases to 42 °C, tachycardia and dyspnoea become evident, the coughing increases, both in frequency and intensity, the animal is reluctant to move and stands with its head extended, mouth open, tongue protruding, and the carpal joints turned out. Contraction of the muscle of the abdominal wall occurs after each inspiration, while expiration is frequently followed by a characteristic grunt or groan. The animals aspirate with flared nostrils. There is a bad smelling mucoid discharge from the nostrils, and frothy saliva accumulates around the mouth which becomes mucopurulent in the later stages of disease when subcutaneous oedema of the lower parts of the chest and abdomen as well as emaciation also become evident. The locality of the affected areas in the lungs is revealed by the presence of dull sounds during percussion. Death usually occurs after 2–4 weeks with profuse diarrhoea, and in West African taurine cattle it is usually after a week. With Fulani cebus, the disease takes a comparatively mild course, and after a convalescence of several weeks, the animals evidently seem to be healthy. The traditional herdsmen are aware of this particular course of the disease and prevent their animals from getting in contact with animals of neighbouring nomads, but most of all they avoid officials of the veterinary service. These animals mostly harbour capsulated sequesters which excrete the pathogens remittently.

In subacute cases, the lung lesions are more localized and an infrequent cough may be the only sign. In chronic cases, the only usual clinical symptoms are emaciation and a cough which usually occurs when the animal stands up. In calves up to the age of 6 months only arthritis and no pulmonary lesions will develop.

Whether a sterile state of immunity is reached after recovery is unclear. It is a fact, however, that animals which have recovered from the disease obtain a challengable immunity for at least 2 years. Little is known about the mechanisms of immunity and about the question of whether humoral or cellular systems are involved. The mortality rate varies between 30 and 70% depending on the level of resistance of the respective breed (Schneider et al. 1994, Seifert 1992).

Pathology

The characteristic *post mortem* findings in CBPP are localized in the lung. In the early stage, signs from a severe fibrinous pleuropneumonia are localized on portions of the lobes of the lung, showing marbling which is the consequence of varying stages of grey or red hepatization. The marbling is accentuated by the detention of interlobular septa, the interstitium, and the vessels and air passages by a serofibrinous, sometimes blood-stained exudate. A clear, yellowish fluid exudes from the cut surface of the lung (Fig. 58 on page 2 annex).

Interstitial oedemas and a typical slushy gelatine-like extension of the interstitium, lymphangitis and thrombosis of the lymph vessels are further characteristic features. Infarcts, varying in size from about 10–300 mm, are frequently present in the affected lung tissue, which are the result from thrombosis of inter- or intralobular arteries and lymph vessels. The infarcts subsequently become sequestered from the adjacent parenchyma by granulation tissue, and thereafter they are referred to as sequestra.

In chronic cases, the sequestra which may be very large and even may involve a whole lobe are the prominent feature of the *post mortem* findings. Smaller sequestra become replaced by fibriotic scars; those recently formed are surrounded by a thin fibrous capsule, the necrotic contents are firm, pinkish to yellowish-grey, and the original pulmonary architecture may still be discernible to some extent.

The sequestra are organized with a capsule of granulation tissue in the course of the convalescence which may take a year or longer leading to the pathognomonic picture of a "brick-like" lung as it is called in German. In this stage a thick layer of proliferative tissue appears on the pleura which can be pulled off. The lymph nodes are swollen and oedematous, the thoracic cavity is filled with a yellowish-grey, clear or turbid liquid which contains pieces of fibrin.

In acute cases of CBPP, lesions in other organs can also appear, such as serofibrinous pericarditis, infarcts in the kidneys, and serofibrinous polyarthritis and tendosynovitis of the carpus and tarsus in calves.

"Occult" cases of CBPP show only oedematous lymph nodes in the mediastinum and no discernible lesions in the lungs.

Microscopically, pneumonic processes of the alveoli and hyperaemia of the alveoli capillaries are found, the alveoli being flooded with serofibrinous exudate. The interlobular septa are widened by the accumulation of serofibrinous exudate, the latter being found in and around the bronchi and bronchioles. Vasculitis and thrombosis of intralobular and interlobular arteries and the lymphatics are frequently present in pneumonic parts, culminating in necrosis of a lobule or part of a lobule, or in infarction of multiple lobules and intervening septa. Typical features of lesions in the lymph nodes are hyperplasia of lymphocytes in the follicles and medullary cords, the accumulation of oedematous fluid, fibrin and neutrophils in the subcapsular, cortical and medullary sinuses and trabeculae (Schneider et al. 1994, Seifert 1992).

Diagnosis

Because of the epizootiological situation, taking into account the affected breed, a diagnosis based on the clinical picture and necropsy findings is rather easy. The causal organisms can be isolated culturally from animals during the febrile phase or shortly *post mortem* from blood, pleural exudate and/or affected lung tissue. The specimens can be sampled in tubes or sterile filter paper, but the material has to be kept on ice.

To detect latently or chronically infected animals, almost all serological tests are suitable. Galactan, the major antigenic component of *M. m.* ssp. *mycoides* can be found in all body fluids during and for some time after the acute stage of the disease. Since it is extremely stable it can also be detected in putrefied material as well as in formalin-fixed samples of lung tissue. Suitable tests for the detection of galactan are the Ouchterlony double-immunodiffusion test, radial immuno-diffusion and counter-immunoelectrophoresis. For the detection of antibodies to *M. m.* ssp. *mycoides*, the CFT (recommended by the O.I.E.), the passive haemag-glutination (PHA) and the slide agglutination test (SAT) can be applied. The PHA can detect antibodies against *M. m. mycoides* at lower levels and for a longer time than the CFT can in animals suffering from the subclinical or chronic disease after vaccination – an asset when monitoring vaccinated cattle, nearly all of which are CFT-negative 2 months after vaccination. Antibodies detectable with the SAT disappear earlier which renders this test less suitable than the other two. The ELISA detects antibodies for as long as 2 years after vaccination; its reliability, however, is influenced by the occurrence of non-specific reactions. An intra-cutaneous allergic reaction has also been developed to detect routine carriers of CBPP, a test which may be suitable for controlling animals before they are brought into an intensive production system (feed-lot). It has been found that all *M. m. mycoides* strains which have been isolated so far do not present immunolo-gical differences. It has also to be mentioned that the level of antibodies against *M. m.* ssp. *mycoides* which can be detected serologically does not necessarily also give information about the level of protective antibodies being present (Schneider et al. 1994, Seifert 1992).

Treatment

Under practical field conditions, treatment is not applicable and not indicated because of reasons of disease prevention. Treated animals are highly likely to be chronic carriers. Also, in order to recognize a disease outbreak on time, sick animals should be slaughtered immediately. Highly valued breeding animals, if quarantined appropriately could perhaps be treated with tylosin (7.5 mg/kg b.w. during 5 days); tetracyclines and chloramphenicol are also effective.

Control

Eradicating infected herds by slaughter as it has been carried out in the industrialized countries and also in South Africa is a control method which is difficult or impossible to impose in traditional African animal production systems, though the veterinary code of most of these countries requires this. As mentioned above, the nomads are well aware of the relatively mild course of disease in their animals. Therefore, they will hide sick animals until they recover. Socio-economic reality also prohibits such a scheme which used to be tried out during the colonial period as well. It is intended, however, to establish, in Kenya for instance, disease-free zones separate from infected areas, schemes to which CAT and CFT have been applied.

CBPP can only be prevented efficiently by regular yearly vaccination of all susceptible animals, which has to be combined with veterinary control measures. In Tanzania, Kenya, the Central African Republic and Senegal, the disease has been eradicated through regular vaccination campaigns. Provost (1981) believes that if 100% of a cattle population older than 6 months are vaccinated continuously for 3–5 years the disease will disappear. With this scheme it is however necessary to stamp-out all clinically and chronically infected animals. Provost asks international organizations to provide the funds for recompensation of the animal holder.

Vaccination against CBPP is never harmless because inactivated vaccines do not produce a challengable immunity. The first vaccines applied against CBPP contained dried pleural exudate from infected animals which was injected into the tip of the tail. Severe local reactions, loss of the tail and up to 15% total losses of vaccinated animals were the consequence. The application of formolized inactivated vaccines produced from broth cultures only provides a rather low protection and even enhances the susceptibility to the infection, though nowadays attempts are being made to produce vaccines which have been inactivated by ultrasound. The application of live vaccines produced in embryonated hens' eggs did not prove their worth. Severe general and also anaphylactic reactions did result after a booster vaccine.

Since mid of the 60's, a bi- or trivalent vaccine (*M. m. mycoides* + rinderpest virus resp. + *B. anthracis*) called *Bisec* or *Neobisec* has been applied especially in francophone West Africa. The application of combined vaccines has a significant disadvantage. It has been proven that the required concentration of antigen has to be increased in proportion to the components of the antigen contained. Consequently this vaccine should contain three times the amount of antigen of each single component. This requirement is limited by the desirable volume of the dose and possible side reactions. Furthermore, interferences of the single components of antigen have been found for the bivalent type of vaccine. Finally it should be taken into account that the protective time obtained by the different antigens varies considerably between the different pathogens (rinderpest!) (Häusser 1989).

At present, only live vaccines provide an efficient immunogenic effect. Vaccine strains of *M. m. mycoides* are KH3J, T1 and V5. Since 1970, only T1 and KH3J have been recommended by the FAO; T1 is generally preferred. T1 was isolated in 1951 in Tanzania and is now used in the 44th passage; a streptomycin-resistant variant T1-SR (Maison-Alfort) is also applied. KH3J which has been isolated in Juba/Sudan is cultivated in the 88th passage. It is less immunogenic than T1 but in contrast to T1 does not cause side reactions in taurine cattle. Therefore, highly susceptible animals (West African *B. t. typicus*) which are vaccinated against CBPP for the first time, after an initial vaccination with KH3J can be boosted with T1 and then continued to be vaccinated yearly with T1. The highly immunogenic Australian strain V5 causes severe side reactions in Africa, therefore it cannot be applied on this continent. Comparative vaccination trials brought the following results regarding the relationship of virulence: T1 > V5 >> KH3J; the same is true for the immunogenic capacity. The appearance of local reactions after vaccination is partly considered to be a prerequisite for good immunogenicity. Therefore, the vaccine strains have been attenuated by passages. Attenuation not only reduces the Willems-reaction which varies between local oedemas up to the total loss of the tail but also the immunogenicity as with KH3J. The appearance of "lungers", i.e. chronically infected animals, is also a problem which can be the consequence of the application of live vaccines.

The minimal vaccine dose is 10^7 organisms/dose; however, a higher dose is usually recommended. The protective effect of the vaccine varies between 60 and 70%. It depends on the production technique, the storage time and the transport conditions. Often fresh broth cultures are applied which have to be used as soon as possible. As a rule, not more than 4 weeks should pass between the time of production and application, a time during which the concentration of live antigen can drop by several logarithms. Many failures of vaccination against CBPP are caused by the above mentioned influence factors. Freeze-drying is the best way to prevent these problems; freeze-dried vaccines, however, can only be stored for one year cooled or frozen. We have developed a novel technology to improve the method of vaccine production (Häusser 1989). Through continuous culture in a bioreactor system on the one hand, the output of antigen can be enhanced and on the other hand, the survival rate of the antigen is enhanced considerably through subsequent microfiltration and resuspension.

3.3.1.2. Contagious Caprine Pleuropneumonia (CCPP)

(Pleuropneumonia contagiosa caprae, Contagious pleuropneumonia of goats, Pleuropneumonie des chèvres, Lungenseuche der Ziege)

Aetiology and Occurrence
Mycoplasma mycoides ssp. *caprae*/*M. m. capri* (type strain PG3) used to be considered the causal organism of CCPP. In 1976, MacOwan and Minette isolated a mycoplasma in Kenya, and designated it F-38 which since then has been shown to cause most African outbreaks of the highly contagious CCPP in Africa. *M. m.*

ssp. *mycoides* LC which has also been isolated from goats with pneumonia produces septicaemia, arthritis, mastitis, encephalitis, conjunctivitis, hepatitis and pneumonia in goats but not CCPP. Currently both *M. m.* ssp. *caprae* and mycoplasma F-38 should be considered causative agents of CCPP, the latter being the principle cause of the syndrome as it occurs in Africa.

M. m. ssp. *caprae* can easily be cultivated in embryonated hens' eggs but also in broth cultures. F-38 is much more fastidious, but it can be grown in Newing's tryptose medium and in medium B. *M. m.* ssp. *mycoides* produces almond- to ball-shaped bodies in broth with a diameter of 75–250 nm which themselves grow branched filaments. *M. m.* ssp. *caprae* grows better and more quickly than *M. m.* ssp. *mycoides* on solid media.

All of the above mentioned mycoplasmas are very labile outside their hosts, and are easily deactivated by heat, dehydration, sunlight and disinfectants.

CCPP occurs in many African and Near East countries, especially where goat production is important. It has been known in Arabia for a long time as "Boufrida". The disease has also been described in eastern Europe, the CIS and the Far East. In Mediterranean countries, CCPP has occurred since the middle of the 60's. A description of CCPP outbreaks in North America in 1978 was erroneous.

Epizootiology and Clinical Features
CCPP only affects goats. Sheep and deer can be infected artificially. Since the aetiology was doubtful until the importance of mycoplasma F-38 was recognized, the epizootiology of CCPP is not well defined. The natural infection occurs aerogenically by direct contact, and an indirect transmission is impossible.

After an incubation time of 8–20 days, the animals may show three different courses of the disease

- peracute: during this, dyspnoea develops quickly and the animal may die in a few hours, mostly, however, 3–5 days after the appearance of the first clinical symptoms;
- acute: after a sudden rise of temperature up to 42 °C, mild, pneumonic symptoms appear at first, the animals are listless, exhausted and lag behind the herd seeking shade where they try to stay, and may recover after 10–15 days. In severe cases, the pulmonary symptoms increase. The affected goats' breathing is laboured; later they may grunt in obvious pain or bleed. Frothy nasal discharge and stringy salivation are often seen shortly before death. Sometimes vesicular exanthema appears on the lips and on the udder. Death occurs 4–15 days after the onset of clinical signs. In the main, highly susceptible breeds are affected by the acute course of the disease whereby morbidity may reach 100% and the mortality rate 70%;
- chronic: with few apparent clinical symptoms, only sporadic coughing, emaciation and diarrhoea may be present. Characteristically the animals recover from this form of the disease almost completely and do not show symptoms or lesions after recovery. Sequestra like with CBPP do not appear, but many of the animals may remain carriers.

Seasonal peaks of disease outbreaks are the rainy season and in the Mediterranean the winter (Maré 1994, Seifert 1992).

Pathology
The lesions of CCPP caused by mycoplasma F-38 are confined to the thoracic cavity and can involve one or both lungs, showing nodules of up to 7.5 mm in diameter which present congestion around the lungs in advanced cases. Chronic pleuritis causes visceral, pleural thickening and adhesions to the chest wall. In contrast, *M. mycoides* ssp. *caprae* affects a wide variety of organs, causing lesions that include encephalitis, meningitis, lymphadenitis, splenitis, and urogenital tract and intestinal lesions (Maré 1994).

Diagnosis
In enzootic areas, the disease is recognized because of its epizootiology, clinical and pathological features. *M. m. caprae* can be isolated during the febrile phase of the disease from lung tissue or blood on suitable nutrient media. Mycoplasma F-38 is more fastidious but grows either on medium B or Newing's tryptose medium. Isolated organisms can be identified by immunofluorescence, metabolic inhibition or most specifically with the growth inhibition test, applying monoclonal antibodies. Latex agglutination and ELISA are suitable tools for detecting serum antibodies. The CFT should not be applied since goat serum reacts anti-complementarily, and a cross-reaction exists between *M. m.* ssp. *mycoides* and *M. m.* ssp. *caprae* with the CFT (Maré 1994, Seifert 1992).

Treatment and Control
Treatment with tylosin at a dosage of 10 mg/kg b.w./day over 3 days is effective. Tetracyclines and tiamulin can also be applied. Because goats are usually kept by small-holders, it would be difficult to carry out schemes of disease eradication on a basis herd by herd. The treatment of individual animals therefore should be taken into consideration.

No real proven vaccines for controlling the F-38 infection are available yet. Only recently an F-38 vaccine incorporating saponin is on trial in Kenya. In the former Soviet Union and in China, aluminium-hydroxide-formalin vaccines have been applied as it is claimed that they protected the animals for 14 months.

3.3.1.3. Contagious Agalactia of Goats and Sheep

(Agalactia contagiosa, Agalaxie contagieuse, Infektiöse Agalaktie der Ziegen und Schafe)

Aetiology and Occurrence
Contagious agalactia which mostly occurs in goats is caused by *Mycoplasma agalactiae* ssp. *agalactiae*. This mycoplasma species has the morphological features typical for this genus: the filaments grow out to a length of 10–40 µm and branch out frequently. It is easy to isolate the organism from infected specimens

by making cultures on appropriate media. The pathogens can also been grown on embryonated hens' eggs and tissue culture. *M. a.* ssp. *agalactiae* can be attenuated with passages on embryonated hens' eggs and in tissue culture. No known relationship exists of the antigens of *M. a.* ssp. *agalactiae* to those of *M. m.* ssp. *mycoides* and *M. m.* ssp. *caprae*. The pathogens do not survive for long under environmental conditions and they are destroyed quickly, especially in a dry environment.

Contagious agalactia was recognized in Italy more than 170 years ago and has spread from there over the entire Mediterranean, including North Africa, Switzerland, as well as Hungary, the CIS and the Near East.

Epizootiology, Pathogenesis, Clinical and Pathological Features
It is mainly lactating goats which are susceptible to *M. a.* ssp. *agalactia*, sheep are less so, and other ruminants can only be infected experimentally. Exotic breeds are highly susceptible. The natural infection only occurs through direct contact and through the ingestion of contaminated feed. Straw bedding infected with contaminated milk seems to be an important vehicle of infection.

After an incubation time of 3 days up to 2 months, a febrile reaction develops. Subsequently, the pathogens become localized in the udder, in the eyes and in the joints whereupon the disease may lead to a fatal outcome with a highly febrile course or enter a chronic stage. In lactating goats, the course of the disease is characteristic; the milk becomes grey-white, sticky and forms a grey pus-containing sediment which is separated from the supernatant by a narrow layer of red blood cells. The milk has a salty or bitter taste and gradually milk production dries up, the affected half of the udder becoming atrophic and often shrinking to the size of a juvenile udder. Cold abscesses may appear in the parenchym of the udder which contain whitish-yellow pus. The abscesses may produce metastases, for example around the saliva gland or the ear. Strangely enough, an udder which has been severely affected may recover after the next parturition completely and produce as much milk as before. Often, however, subsequent to the infection of the udder an affection of both eyes may occur, but this can also be an independent development. The conjunctivae are thickened, reddened and watery; sensitivity to light and itching appear. At first it may seem that the condition is improving and only a slight cataract remains on the cornea. Often, however, a severe affection of the eye develops, the eye being destroyed by multiple ulcers. The goat becomes completely blind because the keratitis or the ulcers are organized by granulation tissue. Characteristic affections of the joints appear either simultaneously or later on, often remaining the only symptoms in non-lactating animals. The joints become hot, swollen and painful developing bony outgrowths and synovitis; the animals die because of decubitus. In some animals, pulmonary affections with coughing, tracheitis, secretion from the nostrils and finally pneumonia can appear. Pregnant goats abort; orchitis is a typical feature in male goats. The disease can last for weeks; morbidity and mortality vary widely. Recovery from the disease leads to a long-lasting immunity.

Diagnosis

A clinical diagnosis of contagious agalactia is unreliable. Pulmonary forms caused by *M. a. agalactiae* are difficult to differentiate from pneumopathias caused by *M. ovipneumoniae, M. arginini, M. capricolum* and *M. m. mycoides*. The latter usually being acute or peracute with severe lung pathology.

The causal organism can be isolated from affected tissue, especially from the swollen joints, by culturing on appropriate media. In order to detect antibodies in the serum of affected animals, the CFT is the best suited method; agglutination and precipitation produce unspecific reactions.

Treatment

Treatment is successful by repeated application of high doses of tylosin and oxytetracycline (10 mg/kg b.w.) as long as no severe lesions have appeared. Recovered animals will usually remain carriers of mycoplasmae because the application of antibiotics will not achieve a *sterilisatio magna*.

Control

Because goats are mostly kept by small-holders, it is difficult to carry out schemes for eradicating the disease on a herd by herd basis. However, standard sanitary procedures, such as isolation, quarantine, restriction of animal movement and disinfection of infected premises are effective. Infected animals also have to be separated and kept apart during treatment. Milk from sick animals has to be destroyed. The colostrum should be heated prior to feeding it to the kid.

Both inactivated and live vaccines are applied for immunoprophylaxis. In Rumania, an aluminium-hydroxide formalin- and/or phenol vaccine has been used successfully. It causes a local side reaction and reduces the milk production but protects the animals for one lactation period. Furthermore, an attenuated live vaccine is available which is prepared from the attenuated strain AG1. This vaccine also produces side reactions (Mitscherlich and Wagener 1970, Seifert 1992).

3.3.2. Contact Diseases Caused by Actinomyceta

3.3.2.1. *Tuberculosis*

Aetiology

Tuberculoses in humans and domestic animals are caused by

- *Mycobacterium tuberculosis*, which is pathogenic for humans, carnivores and parrots; it causes only latent infections in pigs and cattle;
- *Mycobacterium bovis,* which is pathogenic for cattle, humans, dogs, pigs, goats and horses;
- *Mycobacterium avium*, which is pathogenic for poultry and pigs, but only as an exception for humans, cattle and horses.

Beside the above mentioned mycobacteria, there are the pathogenic species *M. paratuberculosis* and *M. leprae*. *M. paratuberculosis* is discussed later on. Furthermore, there exists a number of mycobacteria which are mostly saprophytic.

Mycobacteria belong to the genus *Mycobacterium* of the family *Mycobacteriaceae*. These are singular, slender, acid-fast lightly curved rods of different length (0.2–0.6 by 1.0–10.0 μm) which characteristically appear in specimens of infected material in small clusters, the bacteria themselves being joined by one pole of the rod, thus forming an acute angle. Because of their wax coating they are extraordinary resistant (acid-fast). Mycobacteria can be demonstrated with special stains (Ziehl-Neelsen) by taking advantage of their acid-resistance. All members of the mycobacteria complex are slow growers and require enriched media at an incubation temperature of 37 °C. For successful culturing, the most frequently used solid media are the Löwenstein-Jensen (L-J)- and the Ogawa-medium, both containing eggs, phosphate buffer, and magnesium salts; the L-J medium also contains asparagine. Visible growth of *M. bovis* may occur on solid media after 3–4 weeks or may take longer to appear. Colonies of *M. bovis* are white, flat, and moist and are easily emulsified. The different species of mycobacteria can be differentiated by immunological techniques, gaschromatographic analysis of the cell wall components, comparison of homologous enzyme sequences, DNA/DNA homology, plasmid profiles, restriction endonuclease analysis and rRNA sequencing.

M. bovis can survive on a shady pasture for weeks, in the soil for months, but exposed to sun-radiation it dies in a few hours. It survives in acid milk for 15 days and in milk products such as cheese and butter for weeks. *M. bovis* is destroyed when milk is heated for 30 minutes at 65 °C; cream has to be heated at 85 °C for 2 minutes. The organism can survive for over 2 years in frozen carcasses. As disinfectants, formalin, cresol-sulfuric acid, as well as chlorine-containing compounds have to be applied in a 4% solution in order to be effective.

• Bovine Tuberculosis
(Tuberculosis bovum, Tuberculosis bovina, Tuberkulose)

Occurrence and Significance
Bovine tuberculosis occurs in extensive as well as in intensive production systems of the tropics. The disease is of significant economic importance in two different production systems where it can appear in two different forms:

– pulmonary tuberculosis in highly productive cattle in intensive dairy production systems;
– gastrointestinal tuberculosis in the extensive as well as nomadic animal production systems. In Madagascar for instance, cattle of the traditional pastoralists are affected up to 60%.

The economic losses which are caused by this enzootic, to which rather little attention is paid in tropical developing countries, are enormous. They depend on the shorter lifetime of the animals, sterility, reduced beef and milk production as well as macro-economic problems caused through restrictions in the export of beef (Madagascar). Wherever milk and dairy products are consumed without previous heat treatment, last but not least in nomadic production systems (Sudan), bovine tuberculosis is an important zoonosis.

Epizootiology, Pathogenesis, Clinical and Pathological Features

Bovine tuberculosis is mainly caused by *M. bovis* of which cattle are the maintenance hosts. Since game ruminants and water-buffalo can also be infected, they may play a restricted role in the transmission of *M. bovis* to cattle (Huchzermeyer et al. 1994). Whether or not this is also true for camels is unknown. *M. avium* and *M. bovis* only cause local lesions and usually do not result in progressive disease, infected animals not being contagious. The infection only constitutes a problem in false positive diagnosis of the intradermal tuberculin test. The primary mode of spread of bovine tuberculosis between herds is by the introduction of infected animals in non-infected herds. Possible routes of infection are respiratory, alimentary, congenital, cutaneous, venereal and via the teat canal. Infection via damaged skin is possible. In the tropical animal production systems, evidently two important routes of infection are relevant:

- aerogenic transmission in animals of intensive dairy production systems; the infection is acquired aerogenously by the inhalation of infected droplets from a coughing or sneezing animal with open tuberculosis or infected dust particles from the floor of the paddock or cowshed; the primary focus of infection is localized in the lung;
- oral alimentary infection, typical for the extensive production systems of beef cattle; in this case, bacteria excreted with faeces, urine, milk, lochia from cows with tuberculous endometritis (some authors consider this as high as 5%), and mycobacteria-laden exudate from the lung are ingested with contaminated feed and water. The primary focus with this route of infection is in the lymph nodes of the intestinal system.

The transplacental transmission, whereby the primary focus will appear in the liver of the new-born animal, is of little importance in tropical production systems.

The period of infectiousness may be extremely long, and excreting carriers may transmit the infection for months or even years.

Depending on the route of infection, the pathogens usually at first affect the lymph nodes of the organ which has been the first target of infection. Such a primary focus can be macroscopically inapparent, lympho- and haematogenous, and the infection can remain dormant for a length of time. Independent of the level of resistance and immunity of the animal, the pathogens reach the surrounding tissues with the further progress of infection where small tubercles gradually appear around the supplying vessels. Now the regional lymph nodes of the respective organ become distinctly affected. With a relatively little dose of infection, focal tubercles appear after several weeks which consist of round or longish epithelial-like epithelioid cells as well as of multinucleoid Langhans' giant cells which are surrounded by lymphocytes. After massive infection an exudative process develops. A necrotizing mass composed of epithelial cells and invading lymphocytes as well as polynuclear leukocytes is a result of this development which, according to its localization, appears as a caseous pneumonia or a caseous characteristic alteration of the intestinal lymph nodes. Coalescence and expansion of pneumonic foci result in the development of large areas of caseating broncho-pneumonia. Bronchial and tracheal ulceration may result in intrapulmonary spread. The miliary tubercles can develop into larger granuloma with progressive expansion. Sometimes masses of new tissue develop on the serous mucosae of the body cavities. The granuloma become necrotic in their centre with increasing growth, gradually turning caseous with a tendency to calcify. The tubercles may become surrounded by dense fibrous tissue and the further development of the pathological lesions come to a halt. On the pleura or mesentery, spread is aided by respiratory or intestinal peristalsis movements which cause nodular lesions which are diffuse, caseous, plaque-like or clustered, respectively. The latter lesions are referred to as TB-grapes, and TB-pearls after calcification (Huchzermeyer et al. 1994, Seifert 1992).

The process of chronic tuberculosis localized on a particular organ can last for weeks and months. Depending on the condition, the challenge of stress and environment and perhaps productivity, and in accordance with the individual level of resistance and immunity of the animal, the infection may extend to generalized or miliary tuberculosis. During this initial stage, the disease process may also become encapsulated by granulation tissue and enter a chronic stage which can last for years. The final generalized stage of tuberculosis – the break-down to the infection – appears as soon as the host defence mechanisms and the environmental influences to the animal give way to the further expansion of the pathogens. Generalized tuberculosis is characterized by the presence of miliary or larger lesions in organs and lymph nodes throughout the body. Miliary lesions are small and are translucent in the early stage, but become caseous and calcified as they age. In advanced cases of tuberculosis with widespread dissemination, the large peripheral lymph nodes, such as submaxillary, prescapular, precrural, and supramammary, may be easily palpable or visible. Suppressed moist coughs may be heard, especially early in the morning, after exercise, and when infected animals are exposed to low environmental temperatures. Dyspnoea may be apparent and persistent ruminal bloating can occur. Hepatic involvement may also be part of generalization, the liver presenting miliary or coarse, nodular lesions. In the kidney, miliary lesions of the cortex may be found. Tuberculous osteomyelitis of the vertebrae, ribs and flat bones of the pelvis may be found in young animals. Finally, meningitis may also be the consequence of generalization. In the gastrointestinal tract, ulcers in the mucosa of the upper alimentary tract, abomasum, or small or large intestine are the consequence of oral infection. Tuberculous mastitis is a further consequence of generalization, the udder and the supramammary lymph nodes being firm and enlarged. Tuberculous mastitis usually appears as chronic organ tuberculosis in which most of the lobules are replaced by granulation tissue. Disseminated miliary tuberculous mastitis is less common. Caseous mastitis may also develop in which large, irregular areas of dry, yellowish, caseous necrotic tissue develop, surrounded by a hyperaemic reaction zone. The fourth manifestation of tuberculous mastitis is tuberculous galactophoritis in which the affected lactiferous ducts are dilated and filled with exudate, while the acini are relatively uninvolved (Huchzermeyer et al. 1994). During all stages of the disease, the pathogens may be excreted from the affected organ which, however, becomes especially massive during the final break-down of the animal.

About 1% of new-born calves from tuberculous cows are born suffering from congenital tuberculosis. Infection of foetuses is generally via the haematogenous route, secondary to tuberculous endometritis.

In the intensive dairy production system of the tropics with exotic high pro-duction breeds, the course of the disease is usually relatively short. The expansion of the infection in the lung is favoured by the environmental conditions, especially in the humid tropics. The animals are mostly kept close together which again is favourable to the transmission of the pathogens from an infected animal to a susceptible one. The productive age of a cow will seldom exceed three calvings.

The tuberculosis in the extensive beef cattle production systems of the semi-arid savannahs usually shows an inapparent course. The animal holders often will not even notice that their animals are infected. The disease only becomes apparent when the animals are slaughtered and the extensive tuberculous lesions especially on the lymph nodes of the gastrointestinal tract become evident in the carcass. Animals with affections of the lungs are rarely found.

The necropsy findings depend on the affected organ and the intensity of the disease process. Slight alterations of the lymph nodes are difficult to recognize. Therefore, positive reacting animals will not necessarily present recognizable gross pathological lesions. In comparison, it is rather easy to diagnose the tuberculous alterations in the lung and its regional lymph nodes. Along the gastrointestinal tract, lymph nodes are evident which are enlarged and partly caseous on the cut surface. When cut into, they crunch as though sand were being cut into. Furthermore, tuberculous processes can be found as already mentioned above in all organs of the animal and in the mammary gland as well.

Diagnosis
Bovine tuberculosis cannot be diagnosed simply because of its clinical symptoms. If appearing enzootically, however, the inspection of the carcass of the slaughtered animal will give enough indications to ensure the diagnosis. It is easy to demonstrate the characteristic acid-fast, slender rods in their typical position microscopically with Ziehl-Neelsen staining on impression smears from caseous tubercles of affected organs or lymph nodes. The fastidious cultural isolation and identification of the pathogen or the transmission of tissue from affected organs to guinea-pigs will only be required in exceptional cases.

For eradication of tuberculosis on a herd by herd basis, the tuberculin test is applied. As an antigen, an extract of *M. bovis* is used which has been cultivated on synthetic media, subsequently being broken down and purified (PPD = purified protein derivate). Through this procedure, unspecific reactions are largely excluded. In contrast to the tuberculin injection on the shoulder as it is required in Germany and other developed countries, the intracutaneous application of 0.1 mL (1.0–3.0 mg antigen/mL) into the caudal fold has proven its worth under the conditions of tropical production systems. In order to prevent the transmissions of causal organisms of vector-borne diseases, the injection can be carried out with an airpressure injector, for instance the dermojet. Swelling, warming and pain at the injection site, diagnosed after 24 and 48 hours, are indicators for the presence of an infection with *M. bovis*. Because in the small-holder production systems of the tropics, cattle often are latently infected with *M. avium*, the simultaneous application of tuberculin produced from *M. avium* can be helpful as a measure of differential diagnosis. When comparing the reaction of *M. avium* tuberculin with *M. bovis* tuberculin to a *M. avium* infection, the reaction of *M. avium* tuberculin will be twice as strong as that of *M. bovis* tuberculin.

Treatment
In South Africa, bovine tuberculosis has been treated with isonicotinic-hydracide (INH), a therapeutic effect of 78% being claimed. The application of 10 mg INH/kg b.w. daily for 8 weeks is recommended (Huchzermeyer et al. 1994). The treatment is considered economical under the conditions of animal production in South Africa. The big disadvantage of this scheme is that in many cases the pathogens will not be eliminated from the organisms: the tuberculous processes will only be encapsulated and the chronic stage of disease stabilized. Thus treated animals will remain carriers and excrete the pathogen. The drug is furthermore excreted with the milk and after termination of treatment the disease may recur.

Control
As it has been shown in the industrialized countries, the scheme of "test and slaughter" (T+S) is the most efficient and practical method for the control of bovine tuberculosis. By detecting infected animals and separating them temporarily, this policy can be carried out during a protracted length of time in order to make it economically feasible for the animal holder. Wherever no financial resources are available to recompensate the affected farmer – which is gradual in most cases in developing countries – a healthy herd of young animals can be built up. Tuberculin tests every 3–6 months have to ensure that any newly infected animal can be eliminated immediately from the disease-free herd. Since human tuberculosis occurs quite frequently in the tropics, the healthy herd has to be taken care of by healthy personnel. In contrast, false positive reactions will appear during control testing when cattle have become infected with *M. tuberculosis*, e.g. from human sputum.

Efforts to prevent bovine tuberculosis through vaccination with the BCG (Bacille Calmette-Guérin) vaccine have been unsuccessful. Though the course of disease in cattle has been positively influenced in some cases, the infection could not be prevented. Since, however, the animals have become positive reactors through vaccination the subsequent chance of elimination of bovine tuberculosis by T+S has become impracticable.

• Tuberculosis of Dromedaries

Tuberculosis in dromedaries has been diagnosed especially in Egypt since the turn of the century. It is caused by *M. tuberculosis* and *M. bovis* but atypical myco-bacteria have also been found to be causal organisms (*M. kanasii, M. aquae, M. aquae* var. *ureolyticum, M. fortuitum* and *M. smegmatis*) (Elmos-Salami et al. 1971, cited from Wernery and Kaaden 1995).

The studies in particular by Mason (1912, 1917, 1918) on Egyptian dromeda-ries have provided information as to the epizootiology and the pathological changes found in the disease. The organs most frequently affected in the drome-dary are the lungs, bronchial and mediastinal lymph nodes, pleura and liver. The trachea, kidney and spleen can also be affected. Miliary nodes on the surface of the lung and deeper in the tissue have been observed. Mycobacteria have been isolated from these lesions that cause typical tuberculous lesions in the guinea-pig and rabbit. Similar changes in the organs of dromedaries have been described in India by Leese (1918) and in Somalia by Pellegrini (1942) (both cited from Wernery and Kaaden 1995). The lesions occurred primarily in the lungs whereas the disseminated form of tuberculosis was rarely observed. The alimentary form has not yet been reported in the camel.

Difficulties have arisen in estimating the prevalence of tuberculosis in camels, since inconsistencies in the interpretation of the intradermal tuberculin test occur. Positive tuberculin test results (*M. avium* and *M. bovis*) were seen in 10–22% of Australian dromedaries whereby no indicative lesions were found after the animals were slaughtered. A program to control tuberculosis in camels based on intradermal tuberculin tests would therefore exhibit some inherent deficiencies (Wernery and Kaaden 1995).

• Paratuberculosis
(Enteritis paratuberculosa bovum, John's disease, Paratuberkulose)

Aetiology and Occurrence
Paratuberculosis is caused by *Mycobacterium paratuberculosis*, an acid-fast, non-motile, short, straight Gram-positive rod, which is similar to *M. bovis* but only 1.0–1.5 μm long. In smears from cultures, the organisms appear different from *M. bovis* in tangled clumps, this being caused by intracellular filaments. The cultural characteristics of *M. paratuberculosis* are similar to *M. bovis* but may need a longer time for growth. Furthermore, mycobactin has to be added to the culture media. In biochemical tests, *M. paratuberculosis* shows similar characteristics to *M. avium* and therefore has been grouped with the *M. avium* complex of mycobacteria. Recently, restriction endonuclease and DNA hybridization

analysis have identified a cattle group and a sheep group of *M. paratuberculosis*. *M. paratuberculosis* can be differentiated from other mycobacteria by its myco-bactin-dependence and by thin-layer gas-liquid chromatography.

The environmental resistance of *M. paratuberculosis* is similar to that of *M. bovis*, the same is true for its sensitivity to disinfectants.

Paratuberculosis occurs throughout the world and is considered to be the most important infectious disease threatening the cattle and sheep industries in some countries such as New Zealand, Australia, the U.K. and Mediterranean countries. The infection causes serious economic losses in intensive dairy and beef produc-tion in the tropics (Huchzermeyer et al. 1994, Seifert 1992.)

Epizootiology
Under natural conditions, young cattle, older animals as well to a limited degree, small ruminants, water-buffaloes, camels and pigs are susceptible to infection from *M. paratuberculosis*. It is considered as one of the most important and wide-spread diseases of the Bactrian camel in the CIS, but is less prevalent in drome-daries (Wernery and Kaaden 1995). A number of wild animals, such as deer and buffaloes can also be infected. Apparently the susceptibility in cattle depends also on the breed, exotic beef breeds being more sensitive. There is furthermore an age-related resistance to infection, new-born calves being the most susceptible while old animals are quite resistant. It is supposed that the rumen barrier protects the adult animal while the increased intestinal acidity favours the infection with the acid-fast mycobacteria in suckling pre-ruminant calves (Huchzermeyer et al. 1994, Seifert 1992). An important factor for the development of the infection is the environmental influence, this for example being a reason why dromedaries, kept in the desert, rarely become infected (Wernery and Kaaden 1995).

Pathogenesis
The pathogens excreted with the faeces of infected animals are ingested with feed and drinking water. Calves are mostly infected during the first weeks of life but can also become infected congenitally. After an extremely long incubation period of up to 18 months, it is mostly a subclinical stage of disease which develops, the animals however already being able to shed pathogens. A single exposure to the infection is not sufficient to cause the disease, and repeated infection over a period of time is required for the development of clinical signs.

The pathogens ingested with the feed enter the lymphatic system through the tonsils and the intestinal mucosa being taken up by the regional mediastinal lymph nodes where they cause chronic inflammation. It is exceptional that the bacteria spread haematogenously through the organism; they mostly multiply and persist in the intestinal mucosa. Being contained intracellularly in macrophages, they resist the digestion by lysosomal enzymes and multiply uninhibitedly. They do not produce toxins, but monokines are released from the invaded macro-phages. The monokines enhance the inflammatory reaction and the development of granulomatous lesions. Because of the granulomatous ileitis, villous atrophy appears in the affected areas resulting in decreased absorption of nutrients. Loss of plasma proteins into the intestinal lumen, and probably histamine which is a result of antigen antibody reactions are involved in the development of diarrhoea. A catarrhal enteritis of the ileum precedes the development of the typical granulomatous lesions and may be the initial cause of the wasting syndrome in infected cattle. While cell-mediated immune response occurs at the beginning of the infection, cattle may become anergic during the final stage of disease which may allow infected macrophages to enter the blood stream resulting in bacteraemia and the spread of organisms to most organs and tissues (Huchzermeyer et al. 1994, Seifert 1992).

Clinical Features
The incubation period can last from 6 months up to 15 years, most animals developing clinical symptoms at an age of 3 to 5 years. The development of the disease process is very slow; animals infected as calves develop clinical evidence mostly only after the first or second parturition. The characteristic symptom is chronic but intermittent diarrhoea which does not respond to treatment, dehydration, and emaciation being the most obvious clinical signs. The diarrhoea is profuse, homogeneous, semi-fluid or watery, and greenish. Periods of remission occur which may last for weeks or even months. Fever is uncommon and the appetite is usually maintained despite the progressive weight loss. Submandibular oedema, unthriftiness, a rough coat and a dry skin may develop. Animals eventually become recumbent and die in a state of cachexia. The morbidity and mortality rate in infected herds are rather low, around 1–2%, though up to 100% may be infected.

The course of disease in small ruminants and camels is similar, though the diarrhoea is not as spectacular: the faeces may be normal in consistency throughout most of the course of the disease. Some sheep may shed wool, but no submandibular oedema can be seen (Huchzermeyer et al. 1994, Seifert 1992).

Pathology
At necropsy, the extreme cachexia as well as the inflammatory lesions of the mucosa of the intestine especially the ileum are apparent; the most spectacular *post mortem* findings are thick, irregular, transverse folds, like convolutions of the brain. The submucosa is mostly unaffected; the regional lymph nodes are enlarged, humid and infiltrated with white-yellowish foci which may, as with bovine tuberculosis, be caseous. Similar tissue of granulation is also found histologically in the folds of the mucosa like with tuberculosis. The bacteriae can be found in clusters inside the epithelioid cells. The macro- and microscopical lesions in small ruminants and camels are similar to those in cattle.

Diagnosis
The necropsy findings on the intestinal tract are pathognomonic. The disease can be confirmed by microscopic inspection of biopsy specimens from the intestinal mucosa. The organisms can be directly demonstrated microscopically with Ziehl-Neelsen staining. The pathogens can also be cultured from faecal samples of infected animals.

For schemes of disease eradication, serological tests are available such as the CFT, ELISA, AGID and gamma-interferon assay. For field work three allergy tests are used. The intradermal test with avian tuberculin can be applied comparing the result of a simultaneous test with bovine tuberculin. It will give conclusive results if obvious symptoms of paratuberculosis are present. An intravenous injection of Johnin prepared from *M. paratuberculosis* or also with purified avian tuberculin (PPD) is used to confirm the disease. A rise in body temperature of at least 0.8 °C 6–8 hours after injection is regarded as a positive indication of the disease though it only occurs in 50% of clinical cases. Johnin can also be used for intradermal testing (Huchzermeyer et al. 1994, Seifert 1992).

Treatment
No efficient therapeutics are available.

Control
Wherever the disease is enzootic, careful diagnostic measures ought to prevent the introduction of the infection into a herd. Care should be taken to prevent calves from getting infected by rearing them separately from the cows. Schemes for disease eradication on a herd by herd basis are difficult to carry out because of the unreliability of the available field tests. One can, however, try to combine the allergenic tests with a serological test in order to initiate a T+S scheme.

3.3.2.2. Dermatophilosis

(Dermatophilus congolensis infection, Streptothrichosis, Bovine cutaneous streptothrichosis, Lumpy wool of sheep, Strawberry foot rot of sheep, Dermatose contagieuse, Estreptothricosis, Klontwol, Streptothrichose)

Aetiology and Occurrence
The causal organism of dermatophilosis, *Dermatophylus congolensis*, is a pleomorph bacterium which belongs to the family *Dermatophilaceae* of the order *Actinomycetales*. "Lumpy wool of sheep" (dermatitis of sheep) and "strawberry foot rot of sheep" are also caused by *D. congolensis*. The organism is Gram-positive, non-acid-fast and stains well with ordinary aniline dyes. *D. congolensis* is aerobic or facultatively anaerobic and grows at temperatures between 20–37 °C on ordinary peptone nutrient media but it grows better when these are enriched with blood or serum, looking like a fungus in the form of branching filamentous mycelium with a diameter of 0.5–1.0 μm in grey-whitish colonies which later on turn yellow or orange, and are of a viscous consistency on the surface of the agar; a thick floccular sediment develops in broth after a week. During growth, the filaments become transversely and longitudinally septate, the diameter of the hyphae increasing to 5 μm; the tip of the hyph remains unseptated. The hyphae are surrounded by a capsule and form a chain of small, coccoid zoospores like a rosary which are dormant and can survive in the dried crusts of lesions for several months. Under favourable conditions of moisture and temperature, the zoospores

become flagellated at the poles, which makes them highly motile and thus able to infect new sites on the same or on other susceptible animals. After shedding the flagellae, the zoospores grow into new hyphae. With reduced supply of nutrients and decreasing temperature the formation of spores is enhanced.

Dermatophilosis can occur worldwide but is a typical enzootic of the hot, humid tropics. It is widespread especially in Africa including Madagascar, as well as in Australia and New Guinea. It has appeared in Argentina, Canada and the USA. Sporadic cases have also been reported from Europe. Camels also present characteristic lesions of dermatophilosis, the infection being reported from Mauritania, Kenya and the U.A.E (Seifert 1992, Wernery and Kaaden 1995).

Epizootiology
Under natural conditions cattle, small ruminants, equines, camels and humans are susceptible to *D. congolensis*. Furthermore, certain species of game like zebra and deer and lambs are extremely sensitive to the infection. There are significant genetically determined differences in resistance in cattle. The epidermis of animals is protected against infections by three barriers: the coat or fleece, the sebum which impregnates the hair or wool and forms a layer on the skin, and the *Stratum corneum*. The degree of development of each barrier, and hence the protection it provides, varies widely between animals and at different sites on the body (Van Tonder and Horner 1994). Highly productive European cattle are highly susceptible, less so African Sanga cebus and the West African taurine breeds only a little, especially N'Damas which are well adapted to the environment of the African humid savannah. While Malagasy cebus put up with the infection relatively well, imported exotic breeds and combination crosses (Renitelo = Limousin × Cebus × Africaander) succumb to the disease with serious symptoms which in young animals may lead to a fatal outcome. Apparently this also depends on breed-specific differences in the layer of lower fatty acids which protects the skin of the animals as well as the specific activity of phagocytosis of the different breeds. The intensity of the clinical symptoms varies significantly depending on the season: serious clinical symptoms which appear during the rainy season may disappear almost completely during the dry season. Fulani cebu which enter the Sudan zone in order to feed on crop residues during the dry season and remain there during the rainy season fall ill with severe clinical symptoms (Fig. 59 on page 2 annex) which may subside as soon as they return to the dry climate of the Sahel. There is no special resistance to dermatophilosis of any age group of cattle.

The zoospores are released from active clinical processes especially during the rainy season and may be transmitted by direct contact or through a vehicle, e.g. ticks (*Amblyomma variegatum*) and flies (*Musca domestica* and *Stomoxys calcitrans*) as well as other insects which feed on dermal secretions. *A. variegatum* is considered to be of special importance in transmitting the infection, the lesions being found particularly in areas of the body most infected with ticks (Van Tonder and Horner 1994). Thorns from brushes and awns from pasture grasses can also take up the spores and infect susceptible animals. Dipping probably also contributes to spreading the disease. Rainwater as well as dip- and spray liquid on the body surface favour the spread of the pathogens. This is especially true for sheep.

Because of their positive chemotrophism to CO_2, the zoospores are directed towards small lesions on the skin. They require humidity in order to be able to move around. The hyphae which develop from the spores in the epidermis mainly invade the sheaths of the hairs. The inflammation which develops subsequently produces an exudate which is enriched with granulocytes, the epidermis which is gradually keratinizing being lifted off from the corium, on which a new layer of epidermal cells develops. The second layer again is invaded by zoospores from the hair sheaths and then is lifted off. Subsequently a third epidermis grows, the infected epidermis being again separated from the uninfected epidermis by a layer of exudate and so the process continues in cycles until a thick, laminated crust is formed, composed of alternating layers of para- and orthokeratotic hyperkeratosis and exudate. The new layers of epidermis are formed mainly as outgrowths from the sheaths of the follicles, and are also infected by bacterial filaments originating from these sites (Van Tonder and Horner 1994). During the course of infection, an allergic reaction of the delayed type develops which results, after reinfection, in an accumulation of granulocytes which enhances the growth of the hyphae but at the same time stimulates the healing of the lesions. Specific antibodies are also produced which favour phagocytosis. The failure of trials to vaccinate against dermatophilosis may depend on the fact that the granulocytes do not come in contact with the zoospores which germinate in the superficial cell layers of the epidermis.

It is thought that *D. congolensis* and certain pox viruses such as orf in sheep and the lumpy skin disease virus in cattle may have a synergistic action.

Clinical and Pathological Features

The first lesions in cattle mostly appear on the ventral areas of the body, such as axillae, brisket, inguinal area, scrotum or udder; in small ruminants they appear in the inguinal region and on the back; in goats they appear especially on the face and on the ears, probably because goats get infected when feeding on contaminated brushes, at the same time being hurt by the thorns. At first the typical "paint brushes" appear: these are tufts of hair which are glued together with exudates from the inflammation process. They coalesce to form a large mosaic-like area on the skin of treebark-like crusts which stick to the skin (Fig. 59). With the continuing disease process, the crusts can be removed leaving moist reddish wounds which are covered with a grey exudate and which bleed slightly. In animals with a low level of resistance, the infection spreads over the entire body surface. The lesions can also appear on the cheeks and even spread to the mucosa of the oesophagus and rumen where they cause ulcers. The animals become emaciated and may die with a febrile general reaction. In Madagascar it has been reported that the infection spread to the lungs in imported Brahman cebus.

In sheep, dermatophilosis appears as "lumpy wool" during the rainy season. Dipping especially provides favourable conditions for the appearance of the disease. The wool gets glued together by the exudate produced from the infection, and beneath the wool, the typical wart-like proliferations of the skin are found which can also appear on hairless parts of the skin and around the joints. When the lower legs are affected, the condition is generally referred to as "strawberry foot rot". Here strawberry-coloured, wart-like proliferations appear which have a surface which heavily secretes fluid and smells repulsive. Myiasis may complicate the course of the disease.

Diagnosis

The appearance of the characteristic skin lesions during the rainy season is pathognomonic. The disease can be confused with lumpy skin disease. To confirm the diagnosis, impression smears from the concave base of freshly removed scabs or of suspensions of recently formed scabs in distilled water are stained with Giemsa. The pathogens can be recognized as typically branched filaments which are both transversely and longitudinally septate, and show the characteristic parallel rosary of coccoid elements. In older lesions or moist specimens only spores are usually found. The diagnosis can be confirmed further by applying the immunofluorescence technique; conjugates from monoclonal antibodies are now available. The presence of antibodies against *D. congolensis* in the serum of animals is not diagnostically relevant and also not a criterion for programmes of disease eradication (O.I.E. 1990).

Treatment

Severe symptoms of the disease can be treated through topic application of anti-septic drugs. Formalin used to be injected intravenously. Treatment with tetra-cycline or other antibiotics has also been recommended. 20 mg/kg b.w. oxytetra-cycline LA was claimed to have definitely cured the disease in exotic cattle in Nigeria (Wilson and Amakiri 1989), though it has mostly been observed that the disease has recurred in the rainy season even following successful treatment.

Control

Preventive measures should aim to interrupt the chain of epizootiological events required for the occurrence of outbreaks of dermatophilosis. Animals with skin lesions due to dermatophilosis have to be separated from non-infected animals or they have to be culled. It can be helpful to add 0.03% copper sulphate to cattle and sheep dips in order to destroy the zoospores in the dip liquid but it can also have a therapeutic effect on the pathogen on the skin of the animals. Regular control of ticks and flies can contribute to limiting the spread of the infection within the herds. In sheep, the replacement of the dip with dusting (I/4.3.2.1) will be helpful in controlling the infection. Trials to prevent dermatophilosis by vaccination have so far not presented convincing results, though it has been possible to show that an increased titre of antibodies was produced after repeated artificial infection (Makinde and Ezeh 1981).

3.3.3. Contact Diseases Caused by Aerobic Rods

3.3.3.1. Brucelloses

Brucellosis as a febrile disease of humans already described by Hippocrates 450 A.D. is one of the most important zoonoses in the tropics. Not only does brucellosis present a serious hazard to human health but it is also a disease in domestic animals with important economic consequences. The causal organism was at first isolated by Bruce in 1870 from the liver of a patient which had died with undulated fever (Malta fever), he called it *Micrococcus melitensis*.

The different species of the genus *Brucella* are able to infect a broad range of hosts and cause abortion, infertility and reduced productivity especially in large and small ruminants as well as pigs.

Aetiology
At present, six species of pathogens are classified as *Brucella*:

- *B. abortus*, the causal organism of bovine brucellosis and Bang's disease in humans;
- *B. melitensis*, the causal organism of brucellosis in small ruminants and undulating or Malta fever in humans;
- *B. ovis*, the causal organism of brucellosis in sheep;
- *B. suis*, the causal organism of brucellosis in pigs which also can be transmitted to humans;
- *B. canis*, the causal organism of brucellosis in dogs;
- *B. neotomae*, which occurs in desert-rats in the USA.

Brucellae are small cocci, cocco bacillae or short rods; they are 0.5–0.7 by 0.6–1.5 μm, non-sporulating and non-encapsulated and not acid-fast bacteria which stain Gram-negative. The different species cannot be distinguished from each other morphologically. For microscopic demonstration in or outside of tissues selective staining methods are applied which are able to show the tiny bacteria (Stamp or Hansen' staining).

For culturing, brucellae require complex media; they grow best if special peptones, like tryptose and trypticase-soya-peptone, are added to the medium at a neutral pH; a 3–10% CO_2 atmosphere with an incubation temperature of 37 °C is required. Delicate translucent colonies of 2–3 mm in diameter grow on blood- or glucose-agar. *B. ovis* grows always in the M-(mucoid)-form, *B. abortus* and *B. melitensis* grow at the beginning in the S(smooth)-form and later dissociate into the R(rough)- and the M-form.

The species can be differentiated because of their sensibility to phages, their metabolism and their biochemical structure as far as it correlates with their antigenic characteristics. The biotypes are differentiated serologically (agglutination) applying specific monosera (A, M and R). There is no single test by which a species may be identified with absolute certainty, but a combination of growth characteristics, colonial and cellular morphology, staining properties, agglutinating antisera and biochemical reactions will allow an accurate identification (Bishop et al. 1994).

The genera *Bordatella*, *Campylobacter*, *Moraxella* and *Actinetobacter* are morphologically related to the genus *Brucella*. *Yersinia enterocolitica* (YE 09) is antigenically closely related to *Brucella* and is agglutinated by *Brucella* antisera. It can be differentiated from *Brucella* because of its motility (Table 38).

Pasteurization and the usual disinfectants (chalk solution, caustic soda, formalin 2%, lysol 1%) destroy the brucellae. The brucellae survive the production process of soft cheese (curing up to 6 months); in contrast, the preparation of hard cheese destroys the organisms. Brucellae survive in butter for up to 4 months, for up to 6 weeks in milk, in cooled meat for 14 days, and in ice cream up to 30 days. In milk, the organisms are found mainly in the fatty components. The environmental resistance of the pathogens depends on whether they are protected against the radiation of sunlight and against high temperatures. Neutral soil pH and a moist environment, which is rich in organic material, are favourable elements for survival. In liquid manure, the brucellae survive for months, for 22 weeks in humid faeces, up to 4 months in aborted foetuses and afterbirth, 44 days in the dust of streets, in tap water for 30 days, 51 days in sterile water, for 2–5 weeks in the soil of paddocks, up to 2 months in desert soil, and up to 2 years in frozen soil. Contaminated straw remains infectious for longer than a month (Weidmann 1991).

Occurrence

Brucelloses occur worldwide in domestic and game animals. They create a serious economic problem for the intensive and extensive animal production systems of the tropics. While the disease has been eradicated in most industrial countries, especially in Europe, through intensive schemes of control and eradication, its occurrence is increasing in developing countries in an even aggravating epizootiological situation. This depends on the policy of many developing countries of importing exotic high production breeds without having the required veterinary infrastructure and the appropriate level of development of the socioeconomic situation of the animal holder. Furthermore, the increasing international animal trade with increasing movements of animals and the trend towards intensification of animal production favour the spread and transmission of the infection. Even highly developed countries like the USA and France have so far not been able to eradicate brucellosis completely. In intensive dairy production systems of the tropics, an incidence of infection of up to 80% can be found. In the extensive animal production systems of the Sahel an average disease incidence of 25–30% has to be calculated (Seifert 1990). In eastern Sudan (Butana) we found an infection rate in cattle of almost 22% and 13.6% in sheep (Weiser 1995). In Central Africa, an incidence of infection in cattle of above 30% has been found; economic losses of the yearly income of the animal holder have been calculated of up to 6% (Domenech et al. 1982).

Brucellosis in camels can be caused by *B. abortus*, *B. ovis* and *B. melitensis*. The infection has been reported from Egypt, Sudan, Somalia, Ethiopia, Kenya, Chad, Tunisia, Nigeria, Niger, Russia, Mongolia, Libya, India, Iran, Saudi Arabia, Kuwait and the U.A.E. A serological prevalence of 2–23% was found (Wernery and Kaaden 1994). In two studies carried out in the Sudan, we found 37% (Seifert 1990), respectively 27.4% (Coombs test) and 27.0% (SAT) of camels infected with *B. abortus.* (Butana) (Weiser 1995).

Epizootiology
Large and small ruminants and humans are susceptible to *Brucella* spp., whereby the species may infect their main host but may also mutually infect other animal species as well as game. Camels, horses and buffaloes are also susceptible. Apparently *B. ovis* is only pathogenic for sheep, while *B. melitensis* and especially *B. abortus* can mutually infect their respective main host. They also mostly occur in wild ruminants.

The infection is usually introduced into a herd through latently or acutely infected animals. The infection occurs mostly by the ingestion of material which has been contaminated with the excretions of a female which has aborted but may also have had a normal birth, though brucellosis also has to be considered an important venereal disease. Contaminated feed (hay) can spread the infection from infected pastures over large distances when it is traded. Insects and game animals may also carry the infection over long distances.

Since the period of lambing is limited in sheep, infection occurs mostly during the lambing season. The sheep dog may become a vehicle of infection at the same time.

Pathogenesis
The susceptibility of the animal depends significantly on their natural resistance, their age, their level of immunity and on environmental stress. If the infection is introduced into a non-infected herd in which all animals are immunologically naive to brucellosis, so called abortion storms may occur and almost all pregnant cows will abort.

The infection with *Brucella* takes place through the mucosae or the injured and/or intact skin, through the oral ingestion of contaminated feed and the invasion of the pathogen via the upper parts of the intestinal tract; but infection through the mucosae of the respiratory system or the eyes occurs also frequently. The transmission during mating is important but perhaps not as much as often supposed.

After infection of the regional lymph nodes, bacteriaemia occurs which can last for 1–3 weeks and distribute the organisms to the lymphatic system, the large parenchyms and other organs and tissues. The facultative-intracellular organisms may infect all organs and tissues. In pregnant animals, the uterus is a preferred site of infection where it leads to a necrotizing placentitis. In non-pregnant animals, the first infection often occurs in the udder followed by the infection of the uterus later after the onset of pregnancy. In cattle, the uterus is the central site of multiplication of the pathogen; the enhanced virulence of the brucellae inside the reproductive system is supposed to be the consequence of the increased level of the sugar erythrol which is maintained in the reproductive system. A characteristic exudative and proliferative process develops in the gravid uterus starting from the epithelium of the villus of the chorion.

The variable facets of clinical symptoms which are typical for brucellosis are the consequence of the individual level of host defence which is specific for each breed, but also for each individual and which is the result of the sum of influences of genetically determined resistance, level of immunity, age of the animal, productivity, condition, environmental influences as well as virulence of the pathogen. The following complexes of symptoms are typical consequences of these mechanisms:

- embryonal early death and thus symptomless infection;
- abortion in the first third of pregnancy;
- abortion after the seventh month of pregnancy;
- birth of weak calves – a typical feature with the infection of game;
- *retentio secundinarum*;
- inflammation of the seminal vesicle and vesicular gland in bulls;
- epididymitis in rams;
- chronic inflammation of the epididymis in males, of the joints, tendon sheaths, synovial bursae, especially at the carpus in cows; the regional lymph nodes being enlarged and containing loads of brucellae, a typical feature in autochthonous animals.

Wherever the disease is enzootic in relatively resistant autochthonous animals as it is mostly the case in the tropical savannahs, abortion is rare but the infection causes typical signs which lead to a significant reduction in productivity. These are:

- late first calving age;
- long intercalving time;
- herd fertility below 60%;
- comparatively low milk production.

After recovery from an apparent or inapparent abortion, females are protected against a renewed infection because of the development of a massive immunity. They may even become fertile again and but if not, it is because of permanent lesions in the reproductive system which appear due to metritis as a consequence of a *retentio secundinarum*. The male animal may suffer of orchitis and show symptoms of oligo- and aspermia. This, to certain degree, leads to a "self cleaning" of the herds. Since traditional herdsmen are more or less aware of these mechanisms, they will be reluctant to cooperate in schemes for eradicating brucellosis. There is, however, an important problem with the immunity that develops after recovery from an acute infection. It mostly will remain an unsterile immunity in females and thus the animals may remain lifelong carriers and can excrete the pathogens. This is especially true of those animals which have developed chronic processes (hygroma).

Clinical and Pathological Features
The clinical and pathological features are different for the brucelloses of ruminants.

With **bovine brucellosis** the incubation period varies between 14 and 120 days. If an infection appears in a herd which has so far been immunologically naive, all pregnant animals will abort. Where the infection is enzootic, usually only first calving animals abort. A few days before abortion, a slimy pus-like, grey-whitish to reddish secretion appears in the vagina. The aborted foetuses are covered by a yellowish, slimy layer; the foetuses may be macerated. The afterbirth is oedematous, slushy, the affected cotyledons, or parts of them are covered by a sticky, odourless, brownish exudate, and are yellowish-grey as a result of necrosis. Parts of the intercotyledonary placenta are thickened, oedematous, yellowish-grey and may contain exudate on the surface. Large amounts of pathogens are excreted with the evil-smelling, dirty-grey lochiae. *Retentio secundinarum* is usually the consequence of abortion and can lead to permanent sterility. Microscopically, the stroma of the chorion is infiltrated by numerous mononuclear cells and some neutrophils. Many of the chorionic epithelial cells are packed with numerous intracytoplasmic bacteria.

Though many foetuses may show no gross changes, petechiae are often to be found in the abomasum and on the mucosa of the bladder of the foetus. Spleen, liver and lymph nodes are enlarged, pneumonia with greyish-white foci may be present, sometimes accompanied by pleuritis. Multifocal bronchopneumonia, bronchitis and bronchiolitis and necrotic foci or microgranulomas in the liver, the lymph nodes, spleen and kidneys can be found microscopically.

Gross lesions are not evident on the udder, though the supramammary lymph nodes may be enlarged. Microscopically, lymphoplasmacytic and histiocytic interstitial mastitis is evident.

Infected bulls at first show an acute febrile general reaction with a heavily swollen and painful scrotum. The animals refuse feed and are depressed; the febrile phase may last for weeks. The testes and epididymis are enlarged. Acute orchitis is characterized by multifocal or diffuse necrosis of the testicular parenchyma, and a focal, necrotizing epididymitis. Microscopically, the seminal epithelial cells are necrotic and desquamative, and large numbers of organisms are present. In the chronic stage, spermatic granulomas develop in the testicular parenchyma and epididymitis in response to dead sperm. Brucellae are excreted with the sperm.

Hygromas, in particular on the carpal joints (Fig. 60 on page 2 annex) are a characteristic feature of a chronic infection as it occurs especially with the autochthonous cattle of the nomads of the Sahel, but also the taurine animals of the farmer of the Sudan zone. These are sometimes found on the tarsus as well. Furthermore, the regional lymph nodes and lymph vessels are usually enlarged.

Brucellosis in small ruminants is characterized by lesions which appear mainly in the male animal (orchitis, epididymitis) as well as inflammations of the joints and bursae. But abortion may also occur in the female presenting the typical yellowish, sticky layers on the placenta. The consequences of brucellosis in small ruminants are infertility, a high mortality rate in lambs/kids, mastitis and reduced milk production. The lesions on the placenta are more prominent with the *B. melitensis* infection than when *B. abortus* is the causal agent. In small ruminants, *B. abortus* infection often may be inapparent, both in males and females. After a *B. ovis* infection, epididymitis and inflammation of the vesicular gland are the most prominent symptoms in the male animal, this also having a significant influence on the herd fertility.

Diagnosis

The diagnostic alternatives for the recognition of brucellosis are more varied than with any other infectious disease. This depends on the reaction of the organisms on the chronic infection which partly occurs intracellularly and which can be detected only incompletely with the usual serological techniques. In order to carry out eradication schemes within infected herds it is, however, essential to identify all chronically and/or latently infected animals. This is last but not least the reason why the diagnostic techniques are continuously being improved. There are difficulties in recognizing latently infected animals especially in autochthonous breeds of the tropics. Apparently the autochthonous breeds of the tropics react to a certain level only with so-called incomplete, non-agglutinating antibodies in the infection, a phenomenon which cannot be identified with the classical serological techniques (I/2.2.2). Because of this problem the diagnostic methods for detecting brucellosis which are available at present are discussed in detail as follows.

The **direct demonstration** of the causal organism can be achieved

- microscopically in specimens from foetal membranes with staining according to Stamp or Köster;
- fluorescence-serologically species-specifically of specimens of foetal membranes or content of the abomasum of aborted foetuses;
- culturally on special nutrients, preferably from the contents of abomasum from aborted foetuses or samples obtained by puncturing altered joints or bursae;
- in animal experiments with guinea-pigs, injecting material from placenta, abomasum contents or altered joints or bursae.

Other organs from which specimens may be collected are lymph nodes, udder tissue, male sex glands, as well as milk and seminal plasma. Even contaminated samples can still be examined culturally by adding antibiotics (Bacitracin).

By the **indirect demonstration** of the pathogen by means of serological techniques, only the question about whether the animal has produced an immunological response to the agent can be answered. It is not possible to determine whether this reaction is a sterile or unsterile immunity; nor are the varied serological techniques able to detect all stages of infection. Especially a weak in-

Table 38. Serological test methods for diagnosing brucellosis by means of their special characteristics and commentaries for their application

Name of test	(–) Characteristics and (≫)remarks
TAT (tube agglutination test)	– detects presence of agglutinating antibodies – inexpensive – can soon be used again after 45/20 vaccination – demonstrates infection at early stage – can be doubtful or negative with chronic infections – can be very strongly influenced by vaccination or other infections – prozone phenomenon possible – based on international standardization ≫ should be reduced in favour of simpler survey tests and replaced by more conclusive tests
PAT (plate agglutination test)	– the same applies here as for TAT, but PAT is simpler and quicker to conduct, even in the field, and displays no zone phenomena ≫ can therefore be used on the farm or the market in general as a quick test for survey studies
MET (mercaptoethanol test)	– simple modification of TAT, in which unspecific IgM antibodies are eliminated – detects chronic infections – fewer false positives as with TAT, PAT, or Card Test (or other tests with reduced pH) – slow – repugnant smell – serum quality can influence the test ≫ useful for detecting advanced infection stage
RBPT (rose bengal plate test) or CARD (card test)	– qualitative quick agglutination test – few false negative reactions – good for discovering early infection stages – may produce false positive reactions by vaccination or other antigen contacts – simple and quick – relatively sensitive – can be automated ≫ useful survey test, can be used on farm and at market
CFT (complement fixation test)	– fewest false positive reactions – detects chronic infections – does not detect reagents with incomplete antibodies – sometimes negative in early infection stages – rather expensive and complicated ≫ most desirable corroboration test

Table 38. (Continued)

Name of test	(–) Characteristics and (≫)remarks

ELISA, polyclonal antibodies (enzyme-linked immunosorbent assay)
- very sensitive, good for detecting latent carriers and incomplete antibodies
- severely hampered by vaccination
- relatively simple and easily to be automated
- still too little standardized
≫ very good as control test in free areas and as survey test in areas where no vaccinations have been performed; complicated, cannot be carried out everywhere

ELISA, monoclonal antibodies (enzyme-linked immunosorbent assay)
- like ELISA with polyclonal antibodies
- but more complicated
- vaccine titres detectable
≫ thus, useful like polyclonal ELISA, also applicable after vaccinations

AGT (antiglobulin test) or Coombs test
- very good for detecting chronic infections and incomplete antibodies
- few false negative reactions
- perhaps false positive reactions after vaccination
- relatively complicated
- still little standardized
≫ suitable in areas where brucellosis incidence is chronic and vaccinations have not yet been performed

MRT (milk ring test)
- detects milk-brucellae antibodies
- tests only possible on lactating animals
- very uncertain at individual-animal level
≫ only applicable on entire herd, yields a rough picture of the status of infection

RIV (rivanol agglutination test)
- elimination of serum proteins with following standard agglutination test
- detects chronic infections
- suitable for detecting vaccine titres
- few false positive reactions
- occasionally negative with early infections
- simple and quick
≫ good for complementing CFT after S 19 vaccination

IHLT (indirect haemolysis test)
- antigen titration after removal of haemolysins
- not very specific
- better suited than CFT after previous S 19 vaccination
- no prozone phenomena
- relatively sensitive
≫ suitable particularly as complementary test, e.g. to CFT

Table 38. (Continued)

Name of test	(–) Characteristics and (≫)remarks

AT (anamnestic test)
- antibody analysis following 45/20 vaccination
- helps detect false negative reactors
- may have a high false-positive rate
- ≫ complementary test; for use when routine tests fail

RIA (radio-immunoassay) Less widely used tests
- works with radioactive marking
- is sensitive already in early infection stages
- can distinguish quite well between vaccination and infection
- less advisable where eradication with traditional means is making good progress
- ≫ complementary test, particularly in the final stages of an eradication programme, to lower the rate of false-negative reagents

HIG (haemolysis in gel test)
- simple, precise, and repeatable
- no standardization required, thus interlab comparisons possible
- must be further studied as to applicability
- ≫ possibly applicable as mass survey test

CIEP (counter immunoelectrophoresis)
- method based on immunoelectrophoresis
- can use haemolysed serums or such with anticomplementary activity
- yields quick results
- ≫ feasible as complementary test for otherwise spoiled serums

TRACK (TRACK XI system)
- antigen detection with marked antibodies
- simple and quick
- economical
- hitherto only available from manufacturer without antigen, but is said to be easily adapted for brucella antigen
- ≫ if proven as useful test in practice, could be used as main test

RID (radial immunodiffusion test)
- produces precipitate bands by radial immunodiffusion
- good for distinguishing between vaccination and infection in sheep
- simple
- better in early detection than late detection (chronic brucellosis)
- ≫ suitable for control of infection-free sheep

AGID (agar-gel immunodiffusion test)
- simple, similar to RID
- gel precipitation of brucellae sera
- ≫ as with RID

fection which is limited to small foci in the lymphatic system will not be recognized serologically. Generalized brucellosis will however be detected, but only if it has lasted for some time. Since generalized processes may develop from localized infectious foci, a repetition of the examination has to be carried out after a certain waiting time if doubtful results have been found during the first investigation. A special problem is the recognition of antibody titres following vaccination with inactivated or live vaccines; the origin of the serological reaction can be detected by combining several test methods. Unspecific reactions which for instance are caused through infection with *Yersinia enterocolitica* can cause a further complication.

In Table 38, the test methods for the recognition of brucellosis which are used at present are compiled and commented on. The commentaries are supposed to help choosing a technique which is best suited to develop a control scheme adapted to the respective socio-economic situation of the animal production system (Weidmann 1991). For the interpretation of the results of the serological investigation and to judge the relevance of the infection for the fertility of the animals on a farm, within a herd or a population, the following formulas can be applied, which have been presented by Domenech et al. (1982):

- abortions caused by brucellosis are calculated as follows: (a : b − c : d) × 100: e × f. Here, a = number of abortions of positive reactors, b = total number of positive cows, c = number of abortions of negative reactors, d = total number of negative cows, e = calving rate, f = total number of positive reactors in the herd;
- abortions caused by other influences than brucellosis are calculated as follows: c : d × 100 : e.

Treatment
Treatment of brucellosis in domestic animals is not indicated. Humans are treated with antibiotics (doxicycline with rifampicine). Relapses are, however, possible.

Control
The strategies for preventing brucellosis have to be adapted to the animal production system (III). Failures of disease control are mostly due to the application of a scheme for which neither the veterinary infrastructure exists, nor the required reliable serological laboratories and the animal holder does not have the socio-economic prerequisites. Principally two alternatives exist:

- test and slaughter (T+S), i.e. recognition of all animals which have responded immunologically to a *Brucella* infection and subsequent culling of the reactors. Part of the scheme has to be a careful control of all animals which will be newly added to the herd as well as a production system which prevents contact with infected neighbouring farms and/or contaminated feed or pastures;
- vaccination of exposed herds with inactivated or live vaccines. It may be

temporarily accepted that carrier animals remain in the herd or in the neighbourhood. Under these conditions the socio-economic situation of the animal holder and the situation of the veterinary infrastructure do not yet allow an elimination of carrier animals.

Live vaccines give the best protection against brucellosis. Since a permanent immune response of the organism against the intracellular bacteria must be attained, this is best stimulated by living, but attenuated bacteria which are thus permanently present. However, in handling these vaccines there is a risk of accidental self-vaccination of the personnel and thus of subsequent infection. Male animals cannot be vaccinated with a live vaccine because it may cause inflammation of the sexual glands. The following live vaccines are available at present:

– **strain 19 (Buck 19, S 19)** has been attenuated by Buck through culture on a potato culture medium. The vaccine has been applied for a long time worldwide for the immunoprophylaxis of brucellosis to weaned calves at an age of 4–7 months; normally a protection of about 70% is obtained. Though adult animals also respond well immunologically, cows cannot be vaccinated with S 19 since they may abort. An advantage of S 19 is that it produces mainly non-agglutinating incomplete antibodies, and thus in the long run the alternative is to eradicate the brucellosis through serological selection. Persisting agglutinating antibodies will not be present at the time of the first calving especially whenever only weaned calves are vaccinated. Wherever an unclear epizootiological situation exists, last but not least in the extensive animal production systems in Africa, vaccination with S 19 ought to be the best alternative. A number of experiments involving application of a reduced dosage via the conjunctiva have significantly expanded the possibilities of use of this vaccine. Titre formation is then less a problem and vaccination of all animals of a herd can be performed without producing carrier animals. Conjunctival vaccination requires a booster after 6 weeks producing an immune response comparable to subcutaneous application.
– **strain Rev. 1** is an attenuated *B. melitensis* strain which is applied for the vaccination of small ruminants and also for beef cattle. Only young animals should be vaccinated; conjunctival application has also recently been applied. The protective effect lasts for 4–5 years. Systematic vaccination with Rev. 1 may prevent brucellosis in extensive sheep production systems efficiently. Wherever sheep brucellosis occurs close to a cattle ranch cattle, can also be protected with Rev. 1. Extreme precautions have to be taken when applying the vaccine in order to protect the personnel against a possible infection with this *Brucella* strain which is highly pathogenic to humans.
– *B. suis* S 2 was developed during the 70's in China as a vaccine to be applied orally for the immunoprophylaxis of brucellosis in large and small ruminants. With sheep, goats, beef cattle, and pigs rates of protection of 83, 82, 75 and 72% respectively have been attained with a single application. Booster vaccination scarcely increases the level of immunity obtained. The special

feature of oral administration also allows it to be administered via drinking water, as there is a broad overdose latitude (over 15 times the normal dose) without negative side reactions in pregnant animals. Within a few months after immunization, the titres detected with most immunological tests fall to a level allowing T+S policies. Recently, *B. suis* S 2 has also been applied parenterally without causing side reactions in pregnant animals. The vaccine seems to be ideally suited to vaccinate small ruminants but it should also be applicable in extensive cattle production (nomadic pastoralism).

Inactivated vaccines with *Brucella* antigen do only provide a short-living and comparatively low immunoprotection which has to be enhanced through booster application. The main advantage of inactivated vaccines is that they may provide a temporary protection until the infrastructure and other prerequisites are installed which allow a T+S control scheme. Inactivated vaccines can be applied to adult cows and also bulls without producing a side reaction. An indication for the application of inactivated vaccines exists where a systematic brucellosis control scheme is planned for infected herds (ranch). Under those conditions one would start to vaccinate the weaned calves regularly with S 19 and protect the adult animals with inactivated vaccine until these can be replaced by young immune animals. Dead vaccines are also indicated where brucellosis has been newly introduced into a basically clean herd until the infected animals can be eliminated. An important advantage of the inactivated vaccine is its innocuity to humans. The following inactivated vaccines are applied:

- **45/20 (Dyphavac)** has to be repeated yearly preferably as a booster vaccination in order to provide an efficient protection. The antigen is also marketed as a combination vaccine, e.g. with *Clostridia* antigens. The advantage of the vaccine is that it produces almost no agglutinogenic antibodies. In Australia, the vaccine is applied for the anamnestic test in order to identify carrier animals which with simple test methods which do not detect incomplete antibodies cannot be recognized. After application of 45/20, such hidden reactors produce agglutinating antibodies and can be culled.
- **H 38 vaccine** is produced from an attenuated *B. melitensis* strain containing aluminium hydroxide as an adjuvant and is mainly applied to small ruminants. After a single application a higher protective antibody titre is produced as with 45/20. Since this titre includes all types of immune globulins, it is serologically detectable. All age-groups of animals can be vaccinated; the immune response is immediate.
- **B 112-H 105** is a mixture of the strains *B. abortus* B 112 and *B. melitensis* H 105 which have been inactivated by adding formalin. The vaccine is applied to small ruminants in herds where symptoms of acute infection are present. All animals of the herd can be vaccinated; the vaccine is agglutinogenic.
- *B. melitensis* vaccine is produced culturing *B. melitensis* with addition of *Brucella* antiserum and attenuated with an oil-adjuvant. It is applied only in small ruminants and is not agglutinogenic.

To develop an efficient scheme of vaccination against brucellosis it is necessary to analyse exactly the conditions of production and the economic perspectives of the animal production system (III).

3.3.3.2. Salmonelloses

Aetiology and Occurrence
The genus *Salmonella* which belongs to the family *Enterobacteriaceae* consists of a single species, *S. enterica*, which comprises 7 subspecies (*S. enterica, salmae, arizonae, diarizonae, houtenae, bongori, indica*). *S. enterica* ssp. *enterica* contains all the serovars found in warm-blooded animals formerly assigned to subgenus I which are differentiated from each other by the combination of their somatic (O) and flagellar (H) antigens and, to a lesser extent, by their biochemical reactions. The serovars, of which over 1700 have been identified, do not have a species status (Venter et al. 1994). Salmonellas are exclusively pathogenic, Gram-negative rods which with a few exceptions (*S. gallinarum*) are motile because of their flagellae; there is no morphological difference between the subspecies/serovars. The serological classification is achieved by the determination of the following antigens

- O-antigen: the somatic antigen is composed of a lipopolysaccharide-protein complex (LPS) of the cell wall which contains lipid A and a core portion. The polysaccharides determine the serological specifity. 67 O-antigens are known so far which are identified by Arabic numerals 1–67. The complete O-antigen is not only the mature immunogen of the salmonella; it also possesses virulence properties. It elicits both humoral and cellular immune responses from salmonella infections;
- H-antigen: the flagellae antigen is part of the flagellae of the bacterium and is proteinaceous. The specifity is determined by the pattern of amino acids and it is thermolabile. The H-antigens are designated by a combination of letters of the alphabet and numerals;
- K-antigen: the hull-antigen is as a microcapsule set on the bacterial cell wall. Because of its own specifity it may interfere with the determination of the O-antigen.

For identification purposes, the results from the agglutination of the O- and H-antigens are compared with known antigenic formulae contained in the Kauffmann-White diagnostic scheme, which is updated annually because new serovars of salmonellas are identified continuously.

Salmonellas have simple nutrient requirements and grow on ordinary salt-glucose media over a temperature range of 10–49 °C under aerobic conditions. For isolation, selective enrichment media are used (tetrathionate medium, Kauffmann-Müller tetrathionate medium, selenite broth). Through subsequent culture on discriminating agar (MacConkey, deoxycholate agar, *Salmonella-Shigella* agar), the genus *Salmonella* can be differentiated from other *Enterobacteriaceae* by lactose fermentation and other biochemical reactions, since the genus *Salmonella* does not ferment lactose.

The resistance of salmonellas to heating depends on the medium in which they are kept; a broth culture is killed after only 10 minutes at 80 °C. In contrast, they resist cooking in sausage pulp for 2 hours; they are supposed to survive for years in dried faeces; they perish after 3 weeks in water. In salted meat, the organisms survive for 75 days, in acid milieu, however, they are destroyed rather quickly. The usual disinfectants (creolin 3%, chalk milk 5%, caustic soda 2%) inactivate the pathogens in a few minutes. Attention should be paid to the fact that salmonellas are not destroyed when feed pellets are manufactured. The process heat which is produced when the feed is pressed through the pelletizer is not enough to destroy the organisms (Rolle and Mayr 1993).

Salmonellas occur worldwide. A certain site-specifity of some species used to exist, but the pathogens have been distributed worldwide through international animal trade, especially by trading feed internationally, in particular fish meal. This is of particular epizootiological importance to the existing balance between pathogen and host, which is due to the site-specific immunity developed by the animals, and it is disturbed significantly when new salmonella serovars are introduced into premises.

Epizootiology, Pathogenesis, Clinical and Pathological Features
The *Salmonella* infection occurs mostly through the ingestion of infected feed and water or through contact with the contaminated excretions of latently infected animals; shedders always playing a major role in the dissemination of the organisms. Calves aged between 3 and 12 weeks are the most frequent victims of salmonellosis, but mature cattle may also suffer from the clinical disease. *S. dublin* is regarded as a serovar specific to cattle and only rarely infects other species of animals and humans. Other serovars are not host-specific and direct or indirect transmission between cattle as well as between cattle and other domestic or wild animals may occur. No specific *Salmonella* serovars have been identified in small ruminants. The same is true for pigs. In horses, at least 40 of the known serovars have been isolated. *S. equi* causes abortion in pregnant mares.

Virulence factors produced by salmonellas are responsible for enteric and systemic clinical signs and lesions of salmonellosis. The LPS of the O-antigen is associated with the invasiveness and enterotoxin production. Several *Salmonella* serovars produce an enterotoxin similar to the thermolabile toxin of *Escherichia coli* and the cholera toxin produced by *Vibrio cholerae*, which leads to the secretion of Cl^-, HCO_3^- and Na^+ and water into the intestinal lumen, resulting in diarrhoea. Furthermore some serovars of *Salmonella* excrete a cytotoxin which causes increased permeability in intestinal epithelial cell membranes. With their type 1 heli, the salmonellas adhere to the epithelial cells and because of serovar-specific plasmids, some of the most common species may enhance their virulence and their resistance to antimicrobial drugs with R-plasmids (Venter et al. 1994).

Salmonellas produce septicaemic diseases which are mostly febrile through the toxins which are released from the cell. The pathogens may invade all internal organs from the intestinal tract and cause symptoms which in some cases are specific for an animal species. It is characteristic for the *Salmonella* infection that after recovery, a latent stage of infection appears during which the pathogens persist in the intestine and the animal may become a shedder, perhaps even for months. According to the specifity of the *Salmonella* species to an animal species, a distinction must be made between primary and secondary infections. Primary intoxications caused only by the ingestion of toxins are unimportant in animals.

Primary salmonelloses are specific for an animal species and often also for the site of the animal production system. All primary salmonelloses may show an acute and/or chronic course, the affected animals usually becoming typical shedders of the pathogen during the latter.

Because of the effect of its toxins on horses, *S. abortus equi* causes periodic fever, enteritis and colic. The animals become emaciated and anaemic. Due to the affinity of the *S. abortus equi* toxin to the cells of the placenta of the gravid uterus, abortion occurs mostly between the 4th to 8th months of pregnancy. The pathogens are excreted with the lochia; the mares may remain latently infected for more than a year after recovery. *S. abortus equi* may also cause local inflammations in the intestine and joints in foals which is a typical symptom of foal paresis.

S. abortus bovis and *S. abortus ovis* cause enzootic abortion in cattle and sheep. In cattle, other *Salmonella* species may also be associated with abortion.

S. gallinarum (s. *S. pullorum*) is the cause of enteritis in chickens. Chickens hatched from infected eggs fall sick during the first days of life or may even die already before hatching. Animals which recover from the infection remain shedders. Cocks may transmit the infection with the sperm.

Secondary salmonelloses are caused by different species of salmonellas and may also occur as an opportunistic infection during other diseases. Enteritides are a typical form of secondary salmonelloses in cattle, especially calves (*S. typhimurium* and *S. dublin*) as well as in piglets (*S. cholerae suis* and others). These infections which mainly appear in young animals mostly originate from infected adult animals. The disease usually appears already in the first weeks of life because the passive immunity transferred by the mother has been weak or has already decreased. With a febrile reaction, diarrhoea, inflammation of the joints and emaciation appear. A typical feature is a bent back (calf paresis). The mortality rate may reach 30%.

Enteritis, splenomegaly and necrotic foci in the liver, spleen and kidneys are to be found during necropsy. In piglets, diffuse caseation or circumscribed ulcers appears in the solitary lymph follicles.

Diagnosis
Mostly because of non-specific clinical symptoms and necropsy findings, a presumptive diagnosis has to be confirmed by bacteriological examination of faeces or specimens from the affected animal, last but not least because of its relevance to human health. The salmonellas can be cultured at first in special enrichment media (see aetiology) and subsequently be characterized through their metabolic properties. Their metabolic pattern in response to different C-sources permits a rough differentiation. The final typing has to be done by determining the O- and H-antigens according to the Kauffmann-White scheme.

Treatment
Salmonellas are rather sensitive to a number of antibiotics, for instance ampicillin, amoxycillin, chloramphenicol, gentamicin, trimethoprim-sulphonamide combinations, fluoroquinolones and nitrofuran derivates. Bacterial resistance has been found towards several antimicrobials, e.g. neomycin, tetracyclines and some sulphonamides. Oral administration of antimicrobial drugs may lead to metabolic disorders and eventually can be overcome by feeding probiotic organisms. In addition to the causal treatment, supportive therapy and good nursing are important.

Control
Hygienic premises, cleanliness, provision of non-contaminated feed and drinking water as well as appropriate feeding are important prerequisites for the prevention of salmonellosis. In the tropics, fish meal is often highly contaminated and may especially lead to the infection of poultry and pigs. When unloading fish from fishing vessels in Peru, they are suspended in water from the port which has been found to carry heavy loads of salmonellas (Seifert 1995). The process of production is uncontrolled and does not guarantee the complete destruction of salmonellas during cooking and drying. In 1984, 34% of the fish meal imported to Germany was contaminated with salmonellas.

Infected and/or latently diseased animals have to be separated and treated. They can only return into the herd if they are found to be negative through bacteriological control.

Infections of young animals caused by salmonellas are best prevented by type-specific active immunization of the mothers with formalin-adsorbed vaccines which has to be carried out in time before parturition. The young animals themselves can also be protected through active immunization. In Germany, a live vaccine prepared from salmonellas attenuated by genetic engineering is now available which is applied orally to pigs.

All measures to control salmonellosis have to take into consideration that an insufficient treatment may produce shedders which remain in the herd. Such carriers not only are a zoosanitary hazard for the farm but also may become a menace to human health through contaminated animal products.

3.3.3.3. Escherichia coli Infection

(Colibacillosis, Colienteritis, Colisepticemia, Colibacillose)

Aetiology and Occurrence
Only the species *E. coli* belongs to the genus *Escherichia* which is a natural inhabitant of both, the human and animal caecum and becomes pathogenic only under specific conditions. *E. coli* is a middle-sized stout rod with rounded ends which is mostly motile because of its flagella but occasionally may be non-motile; it stains Gram-negative. The organism may form a capsule from mucopolysaccharides which causes a slimy appearance of colonies on the agar surface. *E. coli* grows on

simple nutrient media and forms colonies on the surface of solid media which can be smooth or rough, flat or curved, round or irregular, and dry or slimy. Haemolysis is occasionally produced. Because of its biochemical behaviour, *E. coli* is easy to distinguish from salmonellas: like salmonellas, it has O-, H- and K-antigens and if it is encapsulated, M-antigen. The practical serodiagnosis is limited to the demonstration of O- and K-antigen. The respective antigen formula follows the scheme of *Salmonella* classification; 49 of the H-antigens, and 91 of the K-antigens are recognized internationally (Rolle and Mayr 1993).

   E. coli survives heating at 60 °C for only 15 minutes; in dried material the organism may survive for several months. *E. coli* is sensitive to ordinary disinfectants.

   E. coli infections can become important especially in intensive production systems of the tropics in particular where the hygienic conditions of the premises of the animals are inadequate.

Epizootiology, Pathogenesis, Clinical and Pathological Features
E. coli causes diseases in young animals, especially in calves, lambs and piglets, which manifest themselves as septicaemia or enteritis. Inadequate keeping and feeding are usually the cause for the outbreak of the disease. Badly ventilated dirty sheds as well as overfeeding may stimulate *E. coli* as a normal inhabitant of the intestine to become pathogenic. The *E. coli* infection is often associated with other facultative pathogenic bacteria (*Clostridia*, *Pasteurella*) or viral infections. Different serovars may cause different symptoms. In contrast to other enteropathies (colidiarrhoea) colisepticaemia in calves is caused by exogenous infection with certain serovars, especially of the O-groups. In adult animals, *E. coli* may produce localized infections of organs e.g. the urogenitalial tract and the udder. With the enteritic forms (oedema disease in piglets, colidiarrhoea in calves), the pathogens multiply mainly in the jejunum and bacteriaemia does not occur. The enterotoxin produced by the pathogens causes dehydration of the body and enteritis through the release of liquid into the intestine.

   The septicaemic findings at necropsy are serofibrinous polyarthritis, meningitis or meningoencephalitis, sometimes interstitial nephritis; with enteritis, enlargement of intestinal lymph nodes, inflammation of the intestine and signs of dehydration will be evident.

Diagnosis
Whether the infection with *E. coli* has been the relevant cause of the disease of the animal is difficult to know because of the clinical and pathological features of the disease, neither is the cultural isolation of *E. coli* from specimens of the infected animal alone relevant. The enterotoxin produced by pathogenic *E. coli* can be demonstrated with the intestinal loop test in suckling mice or in tissue cultures. Serologically it has to be determined whether the isolated organism belongs to a pathogenic serovar. Other infections especially virus infections have to be excluded through differential diagnostics since even the demonstration of toxins and the differentiation of serovars may not always produce conclusive results (Rolle and Mayr 1993).

Treatment
In principle, the *E. coli* infection can be treated with the same antimicrobial agents as are effective against salmonelloses. Chloramphenicol, streptomycin and tetracycline as well as most of sulphonamides are especially effective.

Control
Diseases of young animals with the causal participation of *E. coli*, even if only opportunistic, can be prevented efficiently by organizing appropriate keeping, hygiene and balanced feeding. Special attention has to be paid to the early and sufficient provision of the new-born with maternal colostrum (I/2.2.3.1). Immunoprophylaxis is most efficient if herd-specific adsorbed vaccines are applied in time before parturition to the female animal. Inactivated vaccines may also be applied orally during the first 10 days after birth to the new-born.

3.3.3.4. Pasteurelloses

Pasteurelloses which represent a problem to animal health in the tropics are classified in German text books as:

- haemorrhagic septicaemia (HS) (Septicémie hémorrhagique, Septicémia hemorrhágica, Büffelseuche) caused by *P. multocida*, *P. haemolytica* and *P. pneumotropica*;
- "Wild- und Rinderseuche" caused by *P. multocida*;
- shipping fever (Transportkrankheit) caused by *P. multocida* and *P. haemolytica*;
- septicaemia haemorrhagica ovis (Schafrotz) caused by *P. multocida*;
- pasteurellosis of goats caused by *P. haemolytica*,
- fowl cholera (*Cholera avium*) caused by *P. multocida*

(Mitscherlich and Wagener 1970, Rolle and Mayr 1993). This classification takes a historical approach on the one hand, and classifies the diseases according to the affected animal species on the other hand. In English-language literature (Adlam and Rutter 1989, Nesbit and van Amstel 1994), the classification is done referring to the symptoms caused by the affected animal species. The pasteurelloses are divided as follows:

- haemorrhagic septicaemia (HS), the primary infection of cattle and buffaloes with *P. multocida*, serovar B and E, serovar 6:E being the most common in Africa;
- pneumonic pasteurellosis in cattle (pasteurellosis in cattle, shipping fever), which is usually a pneumonia which is caused by *P. multocida* (A) and *P. haemolytica* (A1, A2) and can be only an opportunistic infection associated with viral infections;
- ovine and caprine pasteurellosis (pasteurellosis of sheep) caused by *P. multocida* and *P. haemolytica* serovar A and T which epizootiologically is similar to

HS. In South Africa, a pneumonic pasteurellosis caused by *P. haemolytica* is distinguished from this disease. Furthermore, a mastitic pasteurellosis in post-parturient ewes and a septicaemic pasteurellosis in lambs are recognized;
– porcine pasteurellosis together with septicaemic pasteurellosis in piglets and pneumonic pasteurellosis in pigs, caused by *P. multocida* and *P. haemolytica*;
– fowl cholera, a primary septicaemic infection of poultry and ducks, caused by *P. multocida* and others; *P. anatipestifer* infection and perhaps primary or secondary infection in ducks, geese, turkeys pheasants and quails.

P. pneumotropica which also is associated with pulmonary infection now has been designated *P. dagmatis* but is not supposed to be involved in causing HS.

Depending on their adaptation to the host, pasteurelloses cause primary or secondary (opportunistic) infections, the pathogenesis of these infections more or less being the same. Since the epizootiology and the clinical and pathological features of pasteurelloses also only differ slightly, it is thought to be a more prag-matic approach to present the pasteurelloses as a single disease complex, and this last but not least because treatment and control of all pasteurelloses are identical, the efficiency of immunoprophylaxis only depending on the application of the causal, locality-specific species and strain as an antigen.

Aetiology and Occurrence
Pasteurella spp. are found as saprophytes or parasites of the mucosae especially of the upper air passages of the respiratory tract in mammals and birds. Causal organisms of pasteurelloses of domestic animals are *P. multocida* with its different serotypes, *P. haemolytica* and *P. anatipestifer*; *P. haemolytica* may occur associated with *P. multocida*.

Pasteurella spp. are non-motile, ovoid, Gram-negative rods of 0.15–1.25 by 0.3–1.25 µm which are usually surrounded by a capsule and characteristically stain more intensively at their poles (bipolars). They usually appear singularly, in pairs or in short chains; they only produce filaments and become pleomorphous after several passages in cultures. Pasteurellas grow readily on ordinary nutrient media with the addition of serum and form small, round dewdrop-like, grey iridizing colonies with a smooth, rough or also slimy surface which coalesce after longer-lasting incubation and produce a characteristic smell. Only *P. haemolytica* produces a β-haemolysis; it changes the colour of blood agar to brown.

The *Pasteurella* species can be classified because of their biochemical metabolic pattern and by using serological techniques. Because of the passive haemag-glutination of erythrocytes which have been sensitized with the capsule antigen of *P. multocida*, 5 serovars (A, B, D, E, F) of *P. multocida* can be distinguished at present; the previously recognized serovar C has been abandoned in the meantime. *P. haemolytica* has been divided into 2 biotypes (A and T). By means of DNA hybridization, the following subspecies of *P. multocida* are recognized at present:

– *P. multocida* ssp. *multocida,*
– *P. multocida* ssp. *septica,*

– *P. multocida* ssp. *gallicida*
– and others which are unimportant as pathogens of diseases for domestic animals in the tropics (Mutters et al. 1989).

Pasteurella spp. produce endotoxin and a protein toxin similar to other Gram-negative bacteria. Which chemical determinants are responsible for the pathogenicity of the organisms is however almost unknown. The pathogenicity is probably associated with properties of the capsule, though avirulent strains may also possess a capsule.

Pasteurelloses occur worldwide but are a particular problem in the tropics, especially the hot, humid tropics where environmental stress is an important trigger mechanism of this disease complex. Amongst the pasteurelloses which affect the large ruminants, HS in buffalo in Asia is of particular importance. It is caused by the serotype 6:B. In South and East Asia, HS is one of the economically most important diseases of cattle and buffaloes. In India it is thought that more than 40000 animals die every year; the incidence of the disease peaks in the rainy season. Pasteurelloses of cattle and small ruminants also occur however in other parts of the tropics, but probably are mostly secondary opportunistic infections. In Africa, *P. multocida* serovar 6:E is generally supposed to be the causal agent, serovar 6:B having been isolated in the Sudan (El Bashir 1992). Fowl cholera, caused by *P. multocida*, is a serious sanitary problem everywhere, but especially in the small-holder poultry production of the tropics.

Epizootiology and Pathogenesis

Environmental and other stress-causing influences are prerequisite for the development of pasteurelloses. Pasteurellas which exist saprophytically on the mucosa of the upper air passages of the respiratory tract become pathogenic under those conditions and can be transmitted to susceptible animals which have a low level of resistance because of external influences. How far mechanisms of immunity control the pathogenesis of pasteurellosis is unclear. In contrast, the introduction of the infection into a herd by carrier animals seems to be unimportant. Acutely infected animals excrete large amounts of pathogens which have in the meantime been enhanced in their virulence with saliva, faeces, urine and milk which subsequently can be ingested by susceptible animals. The infection probably starts in the tonsils, with factors of keeping, stocking rates, feeding, production and climate not only being responsible for the onset of the infection, but also for the course of the disease and the spread of the infection within a herd. With shipping fever, it is the stress which is exerted on the organism through the conditions of the transportation which triggers the disease. The spread of the infection by arthropods does not seem to be important.

In ruminants, the infection with *P. haemolytica* mainly causes an affection of the respiratory tract and septicaemia, in sheep as well. The pathogenic action of *P. haemolytica* in the lung depends on the cytotoxicity of the pathogen for phagocytes. During its multiplication, *P. haemolytica* produces an exotoxin. This cytotoxin, a glycoprotein, can be split into several subunits of which two can be distinguished serologically. *P. haemolytica* also produces a neuraminidase and a neutral protease which are supposed to have a pathogenic effect. Furthermore, the lipopolysaccharide of *P. haemolytica* is involved in the development of the pathogenesis of the disease. It influences the haemodynamics and changes the surfactan-layer of the alveoles of the lung. In addition it can stimulate the migration, the phagocytic activity and the mitosis of the leukocytes. The disease is triggered by environmental factors and/or viral infections (Bötcher 1988).

Clinical Features
The incubation period of pasteurellosis is usually 2–3 days but may be as short as 30 hours. The disease may run a peracute, acute or subacute course. The peracute disease is characterized by sudden death in cattle, generally with no premonitory signs being noticed, but some animals may develop dyspnoea, grunting and protraction for between a few and 24 hours prior to death. Acutely and sub-acutely infected animals are inappetent, pyretic and show anorexia, depression, profuse salivation and nasal discharge, a rapid respiratory and pulse rate, sensory disorders and rumen paresis before death. The subacute disease is characterized by the development of a firm, subcutaneous, painful swelling of the submandibular region which may extend to the neck, brisket and fore legs. While at first defaecation becomes reduced subsequently profuse diarrhoea with evil-smelling and blood-containing faeces appears, the anus and the vagina being swollen. Pregnant animals abort. Some animals may show circling movements and incoordination. Death can already appear 3–24 hours after the onset of the febrile reaction, but clinical signs can be present for periods ranging from 8 to 10 days. The disease is generally fatal if animals are not treated; the mortality rate can reach up to 98% (Bastianello and Nesbit 1994, Mitscherlich and Wagener 1970).

The course of pneumonic pasteurellosis as it is defined by Nesbit and van Amstel (1994) is similar to HS, with serous nasal discharge, lacrimation and photophobia, partial closure of the eyes due to swelling of the eye lids and adventitious sounds audible on auscultation of the cranio-ventral lung-fields being specific clinical signs. Respiration is painful, the nostrils are wide open, and the animals grunt and cough.

In pasteurellosis of small ruminants and in fowl cholera, the course of disease is usually peracute or acute as well; the birds may drop down dead from their perches.

Pathology
Characteristic are the oedematous lesions along the mandibular space, the pharynx, around the parotis, at the neck, the brisket and shoulder, the peracute and acute HS usually being characterized by severe generalized congestion and widespread petechiae and ecchymoses of the serosal surfaces and subcutis. The presence of suggillations in the abomasum, mild to moderate ascites, hydrothorax and hydropericardium, focal oedema at the thoracic inlet, and congestion, oedema and petechiation of lymph nodes are also characteristic. In cases with particular affection of the lungs, which is mostly caused by *P. haemolytica*, severe

oedema or a focal fibrino-purulent pneumonia, oedema and haemorrhages of the tracheal mucosa, adventitia and peritracheal connective tissue, and localized serofibrinous myositis of one or more muscles may be present. The lobules of the lung are irregular in colour, tough in consistency and often covered with a fibrinous layer. On the cut surface, reddish-brown and grey-red spots are scattered and mixed with one another. The interlobular septae are oedematous infiltrated, the lymph vessels are enlarged, and grey and red hepatization is evident. Typically, the body cavities including the pericardium are filled with liquid, and petechiae are to be found on the epicardium. The mucosa of the respiratory air passages shows catarrhal lesions; pointed and striped petechiae are present. Microscopically, the acute disease is characterized by hyperaemia and leukostasis throughout the body, the presence of bacteria in lymph vessels in the lungs and occasionally in blood vessels in other organs in tissues, as well as myositis marked by necrosis of the muscle fibres, interstitial oedema, haemorrhage and serofibrinous exudation (Mitscherlich and Wagener 1970, Nesbit and van Amstel 1994).

Diagnosis
The clinical and pathological features of pasteurellosis are pathognomonic. If pasteurellas are found in a blood smear of febrile animals or fresh carcasses stained either with Giemsa or with methylene blue in the Löffler manner, the diagnosis will be confirmed. With methylene blue staining, the pasteurella appear as bipolar bacteria. The cultural isolation of pasteurellas, as well as the serotype 6:B from a cadaver without an unequivocal case history, may be meaningless. The same is true if the pathogens are isolated from decaying specimens through an animal experiment with rabbits or mice, the latter mostly dying 24 hours p.i. In the blood of the laboratory animal and in impression smears of the organs, the characteristic organism can be found in large numbers. Serological titres in serum of animals which have recovered from the infection are irrelevant because such a test cannot prove whether the immunological reaction between pathogen and organism has been due to an actual disease. Specific identification of the isolated organisms as to species and serotype can be done by using ELISA. With this assay, serotypes responsible for causing HS are readily distinguished from those *P. multocida* strains which are non-pathogenic.

Treatment
The infection with *Pasteurella* spp. can be treated efficiently with chloramphenicol, tetracyclines, penicillin, ampicillin and sulphonamides (sulphamidine or sulphamethazine). It is mostly difficult, however, under the conditions of a tropical animal production system to treat animals because of the acute or peracute course of the disease.

Control

Chemoprophylaxis

Chemoprophylactic measures for preventing pasteurellosis are, as with almost no other infectious disease, useful for preventing the outbreak of the disease, especially when disease-provoking stress is consciously put up with. Application of oxytetracycline LA before shipping animals over a long distance will protect the animals efficiently against shipping fever. In the USA, feeder calves which are shipped for fattening from Texas to the Midwest are treated prophylactically in this way. During international sea- and air transport of breeding stock, the preventive application of tetracycline LA can also be an efficient protective measure. The antibiotic chemoprophylaxis of pasteurellosis is the only way to stop the infection immediately during a sudden outbreak and prevents its spreading to other animals or herds. 10 mg oxytetracycline LA/kg b.w. applied to all animals of an exposed herd will do. In such cases, the chemoprophylaxis replaces the application of hyperimmune serum which used to be applied.

Immunoprophylaxis

Vaccines prepared from *Pasteurella* spp. organisms, mostly from the *P. multocida* serovars B and E adding *P. haemolytica* inactivated with formalin and various adjuvants (saponin, aluminium hydroxide, alum, mineral oil) are used worldwide in the tropics to prevent pasteurellosis. Where non-purified formolized whole cultures are applied, vaccination with those products often results in anaphylactic shock caused by the metabolites of the pathogens. In order to prevent this phenomenon, a low dosage is often applied, though the minimum vaccine dose per strain per adult cattle should contain at least 2 mg of bacterial cells. The literature regarding the production of *Pasteurella* vaccines is uncommonly voluminous, which is an indication that so far it has not been possible to develop a vaccine which fulfils all expectations. Vaccines which contain mineral oil as an adjuvant are supposed to protect the animals for 12 months against infection. With the Göttingen continuous bioreactor system combined with a subsequent tangential filtration cascade, a highly purified *P. multocida* serovar B vaccine has been prepared which does not produce side reaction and is highly efficient (El Bashir 1993). It is now used in the Sudan. Prerequisite for the efficiency of a *Pasteurella* vaccine is that it is prepared from the site-specific homologous pathogen and serovar. The application of a homologous or even herd-specific vaccine may often produce spectacular success, especially with opportunistic *Pasteurella* infections.

The efficiency of vaccination schemes in preventing pasteurellosis depends on choosing the time of vaccination in such a way that the animals are in an optimal condition to produce an active immunity. This is done best when the vaccination is carried out before the onset of the rainy season. Furthermore, appropriate keeping and feeding of the animals should minimize the emergence of stress.

3.3.3.5. *Listeriosis of Ruminants*

Aetiology and Occurrence

Listeriosis in ruminants is caused by *Listeria monocytogenes*. The relatively small (0.4–0.5 by 0.5–2.0 μm), Gram-positive, non-acid-fast, non-spore-forming rods, have rounded poles and are motile because of their peritrichous flagella; this means that they show typical reeling movements. In young cultures they tend to produce filaments since they are mostly Gram-labile. *L. monocytogenes* is facultatively anaerobic and produces small to medium-sized colonies with β-haemolysis on solid media which are flat, moist and shiny, at first translucent but later grey-white and opaque. The incubation temperature varies between 20 and 37 °C; the optimal pH is in the neutral range. Listerias produce acid from carbohydrates but not gas; they are catalase-positive and metabolize aesculin. The organisms can be classified because of their biochemical-cultural properties and can also be classified serologically because of their H- and O-antigens. 5 serovars, designated 1 to 5, and a number of subtypes have been identified. Only *L. mono-cytogenes* is of major importance as a pathogen for ruminants in the tropics. Serotype 5 (now called *Listeria ivanovii*) is commonly isolated from cases of listerial abortion in sheep, but is of low pathogenicity.

Listerias are highly resistant to environmental influences and are able to multiply in the environment under appropriate conditions. They can still be found after 790 days in pond water at a temperature of 2–8 °C and are able to multiply at temperatures from 18–20 °C. In humid soil they can survive for 11 months; wintery cold preserves the pathogens (Rolle and Mayr 1993).

Listeriosis occurs worldwide; it is not peculiar to the tropics, but apparently it is important in the animal production systems of the tropical highlands.

Epizootiology, Pathogenesis and Clinical Features

L. monocytogenes is widely distributed in nature and has been isolated from a wide variety of healthy and diseased mammals and bird species as well as from soil, water, sewage, silage, vegetables and fruit; it has been possible to isolate the pathogens from at least 42 domestic and wild mammals, 22 species of birds, fish, crustacea and insects. The natural reservoir of infection has not been recognized yet. Silage is supposed to be a source of infection though it is unclear whether the listeria are ingested with the silage or whether the change of diet to silage is a disease-triggering stress. Listeriosis has been associated with feeding large quantities of poor-quality silage with a pH in excess of 5.5. Spoilage and high pH of silage result in growth of the bacteria, particular in the top and side layers of the silage (Schneider 1994). The transmission of the pathogen can also occur through infected dust and dirt as well as by contact with infected animals. The excreta (lochia, faeces, urine, nasal secrete and milk) of infected animals are contaminated. Arthropods are also supposed to be vehicles of infection. It is also suspected that, apart from nutritional stress, reduced resistance of the animal because of adverse environmental factors may predispose the organism to disease (Schneider 1994).

Characteristically, listeriosis occurs in the same herd in yearly intervals during the same season. Low night temperatures on the pastures of the tropical highlands have been associated with this. Usually only 10–30% of the animals of a herd become infected. Whether the infection of domestic animals with *L. monocytogenes* has any importance as a zoonosis is not yet clear. By all means the excreta and products of infected animals should be handled with great care.

Listeriosis in ruminants causes three clinical features:

– meningoencephalitis,
– septicaemia,
– organlisteriosis, mastitis, and especially abortion which seldom occur together in an outbreak.

In cattle listeriosis, CNS disorders are the prominent feature of the disease. Signs presented by the affected animals are incoordination, depression, walking in circles, tilting of the head, and evidence of unilateral facial nerve paralysis characterized by drooping lips, ears and eyelids, and paralysis of the muscles of the jaw and pharynx which interferes with mastication and swallowing. The animals show increased irritability and often stand for long periods with food and drooling saliva hanging from their mouths. Lethargy frequently progresses to somnolence which is followed by generalized paralysis and death. Occasionally convulsions and paddling movements with the fore limbs may be seen in terminal cases, the animals dying in a final comatose state. Abortion occurs after early infection between the 4th and 7th months of pregnancy; after a later infection, stillbirth and/or the birth of weak calves are typical features. Mastitis may appear subsequent to abortion, the infected part being hard and nodular.

Listeriosis in small ruminants mainly appears as a meningoencephalitis. The typical symptoms are apathetic behaviour, a tilted head, paralysis of one or both ears and paralysis of the muscles of the jaw and pharynx, refusal to feed and drink, turning movements leaning on solid objects, recumbency, paddling movements as well as tonic-clonic convulsions. The infection nearly always has a fatal outcome. Pregnant ewes abort, give birth prematurely or have still-born calves, and CNS affections only sometimes become evident.

Septicaemic listeriosis is the most common form encountered in ruminant neonates, and is sometimes found in monogastric animals. In these cases, many organs and tissues are affected (Rolle and Mayr 1993, Schneider 1994).

Pathology
Gross pathological lesions caused by listeriosis in ruminants are not normally noticeable. Examination of aspirated cerebrospinal fluid reveals the presence of inflammatory cells, the fluid being grossly turbid. Microscopic changes are confined to the white and/or grey matter of the brain stem, presenting signs of meningoencephalitis with focal lymphocytic leptomeningitis and lymphocytic perivascular infiltration in the grey and white matter. In foetuses and neonates, a granulomatosis is found in all organs, the liver being infiltrated with miliar,

pinpoint sized white necrotic foci. Foetuses are usually severely autolysed when expelled (Schneider 1994, Seifert 1992).

Diagnosis
Only a provisional diagnosis can be made based on clinical signs; intoxication especially through poisonous plants can cause similar clinical signs in the tropics. Confirmation of the diagnosis has to be made by isolating the organisms from the brain or the internal organs as well as faeces, urine, blood and milk of affected animals. Since the pathogens can be very irregularly distributed in the body, specimens should be collected from various organs. The exposure of tissues to a temperature of 4 °C for up to 2 months, known as cold enrichment, is advisable in order to promote the isolation of *L. monocytogenes* from tissue specimens (Schneider 1994, Seifert 1992).

Treatment
The early stage of listeriosis can be treated with high doses of tetracyclines, neo-mycin, furazolidone, sulphonamide-trime-thoprim mixtures, penicillin fluoroqui-nolones, erythromycin and rifampicin. As soon as signs of the affection of CNS have appeared, the prognosis becomes unfavourable (Schneider 1994, Seifert 1992).

Control
Immunity to listeriosis is based on the cell-mediated immune response. So far, no convincing results of immunoprophylaxis with formolized suspensions of cultures have been presented, though reports of successful vaccinations are also available (Schneider 1994, Seifert 1992).

3.3.3.6. Swine erysipela

(Erysipela, Erysipelothrix rhusiopathiae infection)

Aetiology and Occurrence
Swine erysipela is caused by *Erysipelothrix rhusiopathiae*, the only species within the genus *Erysipelothrix*. The Gram-positive organisms are short rods, straight or slightly curved, 0.2–0.4 μm wide and 0.5–2.5 μm long. They may, however, be filamentous and 60 μm and more in length. *E. rhusiopathiae* is facultative-anaerobic and grows on agar in fine dewdrop-like colonies, in gelatine-stab in fine, brush-like filaments. Pathogens of little virulence grow rough (R-form), the more virulent smooth (S-form). The organisms have a poor capacity to ferment glucose and lactose and produce hydrogen sulphide in triple sugar iron agar. *E. rhusiopathiae* grows readily on most of the standard laboratory media, but cultivation is enhanced by the addition of glucose and, to a lesser extent, blood and serum to the medium in a slightly alkaline pH.

The serovars A, B and N can be differentiated by precipitation, the virulence becoming decreased from A via B to N; the differences are due to different thermolabile antigens.

E. rhusiopathiae is resistant to environmental influences: the organism can resist sunlight for 10–12 days, and it survives cooking for 1½ hours in meat; salting and pickling do not destroy the pathogen. *E. rhusiopathiae* is not in-activated through drying and it may even multiply in alkaline animal waste. The usual disinfectants (1%) destroy the organism. *E. rhusiopathiae* also exists outside of living organisms and is associated with the decay of organic matter.

Though *E. rhusiopathiae* is a cosmopolitan pathogen it occurs widespread in the tropics and outbreaks of erysipelas in pigs occur frequently during the hot, humid season.

Epizootiology, Pathogenesis, Clinical and Pathological Features
Animals other than pigs which are susceptible to *E. rhusiopathiae* are cattle, sheep, horses, white mice, pigeons, turkeys and several other species of birds as well as humans, but pigs between 2 months and 1 year are the most susceptible. Most infections are acquired from food contaminated by *E. rhusiopathiae*, soil and faeces. Transmission may also occur via bites form infected flies or by contamination of skin wounds (Spencer 1994). Animals which have recovered from an acute infection as well as those which are chronically infected excrete pathogens which will lead to an outbreak of the disease after being ingested by susceptible animals, especially under the influence of stress caused by environ-mental factors. Hyperthermia of the affected animal may be an important trigger mechanism – a factor of particular importance in tropical countries. The intro-duction of the pathogen into a stock is apparently rather unimportant. The organisms can also be found on the tonsils of healthy animals. The infection can occur as

- acute or peracute septicaemia,
- subacute skin erysipela or diamond skin disease,
- chronic arthritis and/or endocarditis.

During the peracute septicaemia, the animals die suddenly after an incubation period of 3–5 days without any noticeable clinical signs. This course is typical for suckling piglets older than 3 weeks or in the age group for pigs weighing 45–90 kg.

Acutely ill animals show a febrile reaction of up to 42° C, anorexia, thirst, somnolence, vomiting and conjunctivitis. They walk with a stiff gait on the tips of their toes and lie upright on their sternums separate from the other animals. They squeal when forced to stand up and shift their weight from one leg to the other. Discolouration of the skin usually appears within one day of illness; dark-red or purplish patches develop in the skin of the ear, snout, neck, ventral parts of the thorax, abdomen, groin, perineum and axillae. These erythematous areas are not sharply circumscribed and coalesce, affecting large parts of the body. At first the faeces are hard and may be streaked with blood or mucus, but diarrhoea may subsequently set in. The course of the disease will not last longer than 2–4 days before death supervenes. The mortality rate varies from 50–100%.

The milder subacute disease may develop from the septicaemic form and is characterized by the appearance of numerous well-circumscribed purplish-red, intensely hyperaemic and slightly raised patches, with a diamond rectangular or rhomboid shape ("bricks" in German). The animals have fever, show malaise, anorexia, a reluctance to stand and the hairs on the skin bristle where lesions are going to develop. The skin erysipela can develop into the chronic erysipela with affections of the joints and endocarditis, the latter always having a fatal outcome and can lead to sudden death in apparently healthy pigs.

Splenomegaly and enlarged lymph nodes are typical of necropsy findings after acute septicaemia. Haemorrhages on the serous membranes, and occasionally inflammation of the intestine are also to be found. The chronic form is characterized by cauliflower-like proliferations on the heart valves and perhaps the endocardium; stasis in the internal organs and transudates in the body cavities are other typical features (Seifert 1992, Spencer 1994).

Diagnosis

The lesions on the skin together with the febrile reaction are pathognomonic clinical features of erysipelas. The cultural isolation of the pathogens from blood, kidney and spleen as well as heart valves and joints has to confirm the diagnosis. In the tropics, swine fever and anthrax have to be excluded through differential diagnosis.

Treatment

E. rhusiopathiae is sensitive to most of the usual antibiotics, penicillin being the drug of choice, and the response to the administration of a single long-acting preparation at 10000–20000 units/kg is often rapid and dramatic in acutely and subacutely affected pigs. The application of antibiotics has replaced the use of hyperimmune sera completely.

Control

Appropriate management of pigs in sheds which are adapted to the tropical climate is the most important measure for preventing erysipela especially when fattening pigs in the tropics. Shade and ventilation in open sheds will contribute to prevent the disease. For the immunoprophylaxis, inactivated vaccines (adsorbed or lysed) are available which provide a 12-month-long protection after booster injections have been given. In contrast to that, the immunity produced by live vaccines containing attenuated antigens only lasts for 3–6 months. Recently in the USA, a live vaccine which is applied orally was introduced onto the market. Because of the local variants of *E. rhusiopathiae* which may occur in the tropics, the protective value of some of the vaccines produced from laboratory strains may be inadequate. The efficiency of the immunoprotection under the conditions of the tropical production system can be enhanced if the animals are vaccinated twice before the onset of the hot, humid season.

3.3.4. Contact Diseases Caused by *Leptospiraceae*

(Leptospiroses of ruminants, Leptospirosis, Redwater, Icterohaemoglobinuria infectiosa bovum, Leptospirose)

Aetiology and Occurrence
Leptospira spp. belong to the family *Leptospiraceae*. Unlike *Spirochaetaceae* they have hooked ends, are obligate aerobes and use fatty acids or alcohols and carbon as energy sources. Leptospires are helical, motile bacteria, consisting of a cell body or protoplasmic cylinder enclosed by a membrane or outer envelope. With a cell width of 0.1 μm their length ranges from 3–30 μm, depending on the number of whorls. A helix of protoplasma is wound around the central axis thread and only the central part is helical, while both ends are curved like a hook or are swollen like a button. Thus the pathogens take the shape of a coat hanger or a walking stick. The organism's unique method of locomotion is effected by rotation about its longitudinal axis and by a flexing action. This is made possible by two periplasmic flagella wound around the length of the cell, one of which is inserted at the end of the helix. The organisms can move forwards and backwards. The pathogenic leptospires can be cultivated on solid, semi-solid and liquid media with the addition of serum and vitamins; they multiply by simple longitudinal division.

At present only two species are classified with the genus *Leptospira*: *L. biflexa*, a non-pathogenic saprophyte in moist environments especially in water fowl, and *L. interrogans* the serovars of which are pathogens. The two species are morphologically indistinguishable but can be differentiated biochemically and by their growth characteristics. Based on DNA homology, three genetic groups are recognized. Of the about 180 different serovars being isolated so far, antigenically related serovars have been grouped in 14 serogroups.

The serovars *pomona*, *grippotyphosa*, *hardjo* and *icterohaemorrhagiae* of *L. interrogans* are pathogens of domestic animals in the tropics.

In a moist environment, leptospires are extremely resistant: in water they can survive for weeks. Animal waste and decaying organic material with acid or neutral pH are detrimental to their survival, and the same is true for a dry environment (Hunter and Herr 1994, Seifert 1992).

Epizootiology, Pathogenesis, Clinical and Pathological Features
All domestic animals, game, especially rodents as well as humans are susceptible to *L. interrogans* and its serovars. The organisms have also been isolated from poikilothermic vertebrates, such as amphibians and reptiles. Rats and dogs are the most important reservoirs of infection in the intensive animal production systems of the tropics. Humans which have been in close contact with animals can become infected and fall ill to leptospirosis. The usual path of infection is through contact with contaminated urine or the ingestion of feed or water which have been infected with urine. The infection can also occur through the skin if for instance infected dogs implant the pathogens with their saliva. Whenever carrier animals

which excrete leptospires are brought into a herd the disease will spread rather quickly and manifest itself in that numerous animals will suddenly abort (abortion storm). The appearance of the infection may however also be mild or inapparent. The infection leads to a fast development of humoral, protective anti-bodies, a mechanism which causes the self-control of the disease in an affected herd.

Amongst the animals which are ruminants, calves in particular are exposed; sheep which only become infected if they are kept close to infected cattle show a milder course of the disease. After an incubation period which varies between 1–2 weeks, a febrile reaction with inappetence, dyspnoea, icterus, haemoglobinuria and anaemia develops. After a sudden rise of temperature up to 41 °C, combined with leukocytosis and albuminuria the haemoglobinuria which appears subsequently will usually only last for 48–72 hours. The icterus also subsides and is followed by a pronounced anaemia which again only lasts for 10 days. In dairy cattle which often are infected with serovar *hardjo*, the symptoms are difficult to detect because the haemolytic crisis seldom appears. Abortions occur about 2–5 weeks after the outbreak of the disease in the herd mostly in the 7th month of pregnancy. If the initial symptoms of the infection of the stock are overlooked, the sudden abortion of numerous animals may be the first diagnostic indication. The disease may often only cause a reduction of the milk production, the milk showing an altered consistency (yellowish with traces of blood) without clinical symptoms being found at the udder. Calves receive a massive passive immune protection with the colostrum and later on are capable of producing a good active immunity.

The most prominent *post mortem* findings with leptospirosis are anaemia and icterus. The urine is clear red or stained like port. The kidneys show the most typical signs which are red-brown patches on the kidney cortex which often are so prominent that they can be seen through the capsule of the kidney. The liver is enlarged and covered with numerous small miliar foci of necrosis. After the acute course of leptospirosis, petechiae are found on the epicardium and the lymph nodes (Hunter and Herr 1994, Seifert 1992).

Diagnosis
If other causes of enzootic abortion can be excluded and neither changes in the composition of milk nor clinical symptoms of the udder are present, leptospirosis should be suspected. The diagnosis has to be confirmed by the isolation of the pathogens from the urine of the patients and through serological tests with the determination of the relevant serovar. Agglutinines and lysins can be found in the serum of infected animals starting from the 10th day p.i. with the agglutination lysis reaction. A further important diagnostical tool is the fluorescence antibody test which can be applied to specimens from infected organs especially from foetuses in order to demonstrate the pathogens. An immunoperoxidase staining technique employing rabbit antileptospiral antibodies can be used to analyse tissues fixed in formalin-embedded in paraffin. The technique is serovar-specific. PCR assays are also now available which may detect even only a few leptospires. Silver impregnation techniques for the demonstration of spirochaetes in tissue

sections only are of value in acute cases, it not being possible to identify the serovar which causes infection. Intradermal testing with leptospiral antigen has only been performed with variable results in pigs (Hunter and Herr 1994, Seifert 1992).

Treatment
Leptospirosis can be treated successfully with a number of antibiotics especially tetracyclines at a dose of 10 mg/kg b.w. which has to be applied for several days. Dihydrostreptomycin can be used the same way in a dosage of 25 mg/kg b.w.

Control
Wherever leptospirosis is enzootic under intensive production conditions one should try to eliminate or keep away rats and other rodents from the animals. Susceptible animals should be separated from potential carriers of infection, like dogs and pigs. Furthermore, the access of the animals to potential sources of infection such as contaminated water places, creeks etc. should be restricted. The provision of clean drinking water is an important factor in the prevention of leptospirosis.

Abortion storms in cows and sows, as well as the renal carrier state in these species, may be arrested by a single treatment of 25 mg/kg b.w. dihydrostreptomycin when the infection is caused by *L. pomona*. Medication of the feed with chlortetracycline at 400–800 g/1000 kg for 10 days reduces the number of carriers on infected premises, but does not necessarily eliminate all carriers. New stock brought onto clean premises must be held in isolation for 2 weeks and should be given a single parenteral treatment with dihydrostreptomycin in order to eliminate a possible renal carrier state (Hunter and Herr 1994, Seifert 1992).

The regular yearly vaccination with formolized homologous serovar-specific or multivalent vaccines is applied especially in the USA. The vaccination protects the animals against clinical symptoms, especially abortion. It has, however, not been possible to prove that chronic affections of the kidneys can be prevented. Vaccinated animals probably remain as carriers and can excrete the agent.

3.4. Contact Diseases Caused by Viruses

3.4.1. Introduction

Contact diseases caused by viruses can be transmitted through direct or indirect contact as has been discussed already for contact diseases in general. This mechanism becomes further complicated in the tropics because of the presence of game animals which can be a reservoir of infection for a number of viruses pathogenic to domestic animals. This is especially true for the panzootics, FMD and rinderpest which because of the virus reservoir in wildlife are difficult to control in Africa. In order to point out the importance of game as a reservoir of infection and to show the numerous species of wildlife which may be involved, in Table 39 the species of game animals which are susceptible to rinderpest virus have been compiled. Wherever the mentioned species of game are present, it has to be taken into account that after controlling the disease with the domestic animals, a reservoir of infection can remain in the population of game animals. This scientific fact in particular has to be paid attention to where game ranching projects are supposed to be established in the environs of a cattle ranch or an extensive nomadic production system.

The contact infections caused by viruses discussed in the following chapter have been selected according to their prevalence and economic importance in tropical animal production systems. Not only those infections which are limited to the tropics have been included but also infections which under the conditions of the tropical production systems require other measures of control than those which are carried out in industrial countries. A typical example is FMD. On the other hand virus infections which have been brought only recently with exotic cattle from industrial countries to the tropics and which only appear in intensive animal production systems have been omitted. Infectious bovine rhinotracheitis/ infectious pustulous vulvovaginitis (IBR-IPV) are examples of this. The same is true for European swine fever and for virosis in poultry which have to be controlled with the same schemes as they are applied in industrial countries.

In accordance with current nomenclature and ordering systematology, the following contact diseases caused by viruses which occur in the tropics and subtropics will be discussed. They are caused by

- *Herpetoviridae*
  - malignant catarrhal fever (MCF)
- *Iridoviridae*
  - African swine fever (ASF)
- *Poxviridae*
  - pseudocowpox
  - other pox infections in ruminants
    - buffalo pox
    - camel pox
    - pox in small ruminants
    - lumpy skin disease (LSD)
  - contagious ecthyma (CE)
- *Paramyxoviridae*
  - rinderpest
  - peste des petits ruminants (PPR)
- *Picornaviridae*
  - foot and mouth disease (FMD)
- *Retroviridae*
  - ovine pulmonary adenomatosis (OPA)
  - maedi-visna (MV)
- *Rhabdoviridae*
  - stomatitis vesicularis (VS).

The arboviroses transmitted by vectors as well as rabies have already been discussed in II/1.5.1.

3.4.2. Malignant Catarrhal Fever (MCF)

(Coryza gangraenosa, Bovine epitheliosis, Snotsiekte, Bösartiges Katarrhalfieber des Rindes)

Aetiology and Occurrence
MCF is caused by a virus which is classified with the *Herpetoviridae*. This disease which occurs worldwide, but in Africa in particular (East and South Africa) as well as in the USA, is caused in Africa and in America by different virus types. The African wildebeest-derived MCF virus has been classified as *Alcelaphine herpesvirus 1* (AHV-1), because it occurs in the family *Alcelaphinae*, to which the wildebeest belong. AHV-2 virus, closely related to AHV-1, has been recovered from hartebeest and other game ungulates; it produced MCF in cattle after artificial infection. The American MCF virus which has not been characterized definitely has been isolated from sheep and is denominated as the sheep/goat associated form (SGA). It is supposed to be closely related to AHV-1. The virions of AHV-1 are approx. 109–220 nm in diameter and like other herpes viruses have icosahedral symmetry and comprise a DNA core enclosed in a capsid approx. 100 nm in diameter. The nucleocapsid matures in the nucleus of infected cells, and development occurs at the nuclear and cytoplasmic membranes. Virions emerge from cells by budding from the plasmalemma, or through channels of either the Golgi apparatus or the endoplasmatic reticulum. The virus grows on calf testis cells, calf kidney cells, foetal lamb kidney cells and Vero cells. It can be transmitted intracerebrally to rabbits (Barnard et al. 1994).

MCF virus is extremely sensitive to environmental influences and is difficult to preserve. It remains infective in blood at +5 °C.

Epizootiology, Pathogenesis, Clinical and Pathological Features
Cattle, sheep, buffaloes, bison, deer and especially gnus (*Gorgon taurinus*, wildebeest) are susceptible to the MCF virus. Wildebeest are involved in the epizootiology of MCF; though they do not manifest signs of clinical disease most, wild adult wildebeest are persistently infected with AHV-1. The route of natural infection for cattle is unknown. However, it has been shown that intranasal instillation of cell-free virus suspensions can produce infection in both cattle and rabbits. Henning (1956) already reported that the infection can be transmitted to cattle with blood of hunted, completely healthy, wildebeest. Maasai herdsman apparently believe that cattle contract MCF when they come into contact with wildebeest placentas or with the hair of juvenile wildebeest, shed when the calves are 3–4 months of age. In Kenya MCF usually occurs between April and July, following the wildebeest calving season. The same is true for South Africa but there is a further peak of incidence from September to November (Barnard et al.

1994). In cattle, diaplacental transmission has been proven. The "Voortrekker" of the Boers in South Africa observed in the last century that whenever their draught-oxen grazed where wildebeest had previously been on the pasture, the oxen died because of a dreadful disease. Because hunting gnus has been restricted in East Africa, MCF has become an increasing problem for the Maasai pastoralists. Even their cattle which have been adapted to the local environment for centuries and have been in contact with the disease continuously cannot resist the infection. Henning (1956) impressively describes that the white settlers already found MCF when they penetrated the continent with their draught-oxen. Cattle are probably the final link in a chain of infection in the game population. In the USA, sheep are apparently the reservoir of infection because MCF appears in cattle only when they have grazed together with sheep. Like gnus in Africa, sheep do not show symptoms of infection. Since MCF also occurs in the USA where no sheep are kept, other reservoirs of infection must be present. Though a seasonal peak of MCF occurs in the USA during the summer, it has not been proven that vectors may transmit the disease.

The pathogenesis of MCF is clearly unique, but at the same time poorly understood, and the process therefore remains speculative. Plowright (cited from Barnard et al. 1994) states three essential components in the pathogenesis of MCF:

- destruction of smaller lymphocytes, particularly in the germinal follicles of lymph and haemolymph nodes and the thymus;
- proliferation and infiltration in many tissues of large, granular lymphoblastoid cells, particularly around blood vessels and in T-cells-dependent areas in lymph nodes and the spleen;
- an irregular segmental angiitis, predominantly of medium-sized arteries, which affects all components of the walls of arteries and veins.

Incubation periods of a few up to 300 days have been reported. In the infected animal, the virus is thought to be maintained in the leukocytes for months.

The clinical course of MCF is variable. The mostly sporadic cases described earlier in Europe and the USA presented a much milder course than MCF in Africa at present. Because of its different clinical picture, MCF can be divided into four syndromes which can sometimes appear in combination with and/or follow each other:

- The peracute course has a high febrile (up to 42 °C) onset, the animals refuse water and feed, present muscle tremors, shivering and stupor, as well as evil-smelling diarrhoea which contains blood. With a rapidly worsening general condition, death supervenes after 1–3 days.
- The intestinal form of MCF also presents a severe course which has a lethal outcome after 4–9 days. Characteristic features are watery, evil-smelling diarrhoea which is mixed with blood, high fever, muscle tremors, reddened conjunctivae and lacrimation, sensitivity to light and a serous slimy excretion from the nostrils.

- The head-eye form represents the classical picture of MCF with high fever and the characteristic ocular lesions. Congestion and sometimes oedema of the conjunctivae and sclerae are accompanied by serous or seromucoid discharges which quickly become mucopurulent. Bilateral corneal opacity caused by oedema, starts at the limbus and progresses centripetally to eventually cause impaired vision or total blindness, affected animals often being photophobic. This form of MCF usually has a fatal outcome.
- The abortive form with slightly elevated temperature and only slight catarrhal symptoms on the mucosae of the head often only shows signs which are suspicious for MCF. In the course of this form an exudative dermatitis characterized by congestion and sloughing of the superficial layers of the skin occurs sometimes at the base of the horns and the dew claws, in the interdigital space, and on the flanks.

Nervous signs such as muscle tremors, incoordination, twitching of the ears, torticollis, and nystagmus, as well as aggressive behaviour and finally tonic-clonic convulsions have been reported from all forms of MCF. The way the animals support their head on fixed objects and keep it in an unphysiological position is characteristic (Barnard et al. 1994, Rolle and Mayr 1993).

The necropsy findings correlate with the clinical courses as described above. Enlargement of the liver, splenomegaly, *myodegeneratio cordis*, catarrhalic inflammation of the mucosae of the head and the intestine as well as of the reproductive tract are present. With the head-eye form, bronchopneumonia will be found aside from the typical lesions of the head mucosae. The CNS signs are caused by a distinct disseminated encephalitis which affects the grey as well as the white mass of the *mesencephalon*, the *pons*, the *medulla oblongata* and the *cerebrum*. Histopathologically the disease is characterized through a lymphocytical vasculitis with a distinct tendency to the arteries of the muscles, a lymphoid necrosis and reticulo-endothelial hyperplasia. The lymphopenia is a consequence of the extensive destruction of lymphocytes. Lymphopenia appears in the late stage of the disease together with an extension of the necrosis of the tissues. These lesions appear in all organs but especially in the intestinal tract, the eyes, the meninges, the epiphysis, the kidneys and in the skin. They are the basis for the focal necrosis and ulcera which appear on the surface of the organs.

Diagnosis

A presumptive diagnosis can be made on the epizootiology of the disease and because of the clinical and pathological features observed and if there has been evidence found of recent contact of the affected cattle with wildebeest. The diagnosis has to be confirmed by virus isolation from tissues of affected animals and/or demonstration of specific antibodies in the serum. The virus can be isolated from washed leukocytes on tissue cultures. AHV-1 virus can be identified with virus neutralization, also indirect immunofluorescence, ELISA and the immunoperoxidase test can be applied (O.I.E. 1990). The detection of antibodies in affected cattle is of limited value because only a small proportion of cattle develop a humoral antibody response late in the course of the disease (Barnard et al. 1994). FMD and rinderpest have to be especially considered for differential diagnosis.

Treatment

No treatment for MCF is known. Because of the unfavourable prognosis of the disease, affected animals should be put down.

Control

Restriction of the movement of wildebeest and sheep would be helpful in reducing the incidence of MCF.

A passive immunity based on cellular mechanisms has been demonstrated in calves. It is unclear whether it has a protective effect (Rolle and Mayr 1993). All attempts to protect exposed animals with attenuated or inactivated vaccines have so far been unsuccessful though virus neutralizing antibodies have been found in vaccinated cattle and rabbits serologically (Barnard et al. 1994).

3.4.3. African Swine Fever (ASF)

Pestis africana suum, African swine fever, Wart hog disease, Peste porcine africaine, Peste porcina africana, Afrikaanse varkpes, Afrikanische Schweinepest)

Aetiology and Occurrence

The causal organism of ASF was thought to belong to the *Iridoviridae* (Rolle and Mayr 1993). It is now thought justified to classify it into a group by itself (ICDV) (Plowright et al. 1994). The virus is an icosahedral cytoplasmic deoxyribovirus (ICDV), being considered an arbovirus, the only one which contains DNA, though it is also transmitted, which is important to note, by contact which is the reason why ASF has been grouped with the contact diseases for this presentation. The major virion has a diameter of 175–215 nm, the capsid consisting of hexagonal prisms about 13 nm by 5–6 nm with a central hole, being arranged in triangular facets on the surface of an icosahedron. The multiplication of the virus takes place in the cell cytoplasm, infected cell nuclei showing characteristic early clumping and condensation of chromatin on the nuclear membrane, while the nucleoli are vacuolated and fragmented. The adsorption and penetration of ASF virus into susceptible cells probably occur by endocytosis. About 50 virus-induced polypeptides have been found in infected cells of which at least 17 induce antibodies in natural infections. All strains of the virus possess a soluble precipitating and a haemadsorbing antigen which are strain-specific. Because of this, 8 ASF strains have been identified so far. The ASF virus has no relationship to the causal organism of European swine fever which is a RNA virus which belongs to the genus *Pestivirus*.

Adaptation of ASF virus to growth in cell lines leads rapidly to genomic changes, and the virus can be adapted to growth on pig kidney and Vero cells.

The ASF virus is relatively resistant to environmental influences and can be inactivated in serum at 56 °C only incompletely after 1 hour and at 60 °C only after 20 minutes. In decaying blood, the virus survives for 4 months, exposed to direct sun-radiation for 70 days, in faeces for 11 days, in ham 5–6 months, in bone marrow 6–7 months, and in frozen meat 104 days; it remains stable within a pH range of 2.0–13.4. 2% caustic soda and 2.5% formalin destroy the virus (Mitscherlich and Wagener 1970, Rolle and Mayr 1993).

Since being described for the first time in East Africa in 1909, ASF has been restricted with several severe outbreaks in the following years to Africa, mainly eastern and southern Africa. In 1982 the first epizootic occurred in West Africa and killed, for instance, about 80% of all pigs in Cameroon. In 1957, disease outbreaks with high mortality appeared in Portugal. Further outbreaks occurred in 1960 in Spain, in 1964 in France, in 1967 in Italy and in 1985 in Belgium. The disease was probably carried from Spain to Cuba in 1971 from where it spread throughout the Caribbean and appeared in 1978 in Brazil.

Epizootiology
Under natural conditions, domestic and wild suids are susceptible to ASF. The original vertebrate hosts of ASF are the wild suids of Africa, especially the warthog and to a lesser extent the bushpig. The wild boar (*Sus scrofa ferus*) of Europe and North Africa are highly susceptible but probably are unimportant as reservoirs of infection because they succumb to the disease. African wild suids, the warthog (*Phacochoerus africanus*), the bushpig (*Potamochoerus porucs*) and the giant forest hog (*Hylochoerus* sp.) only become infected latently and are the reservoir of infection for the disease in Africa. In the USA, the collared peccary (*Tayassu tajacu*) and in South and Central America the white-lip peccary (*Tayassu albirostris*) are supposed to become latently infected carriers.

In Africa a conception of the epizootiology of ASF existed such that it could be divided into an old silvatic cycle dependent on inapparent maintenance in warthogs or other wild suids and a new cycle based on virus maintained in domestic pigs. Wild suids are already infected as piglets by *Ornithodorus moubata porcinus* ticks which transmit the virus transstadially and transovarially. The infection is then also carried from the population of wild suids to domestic pigs by *O. moubata*, this being a vector-borne disease up to this stage. The contact between warthogs and domestic pigs usually results in a disease outbreak with domestic pigs. This connection was observed when the disease occurred the first time in domestic pigs in Africa and it could be confirmed by De Tray in 1957. The virus may also be spread between the domestic pigs by ticks which may harbour the virus for months. In Spain, the disease appeared in premises which had been kept free of pigs for months but in which *O. moubata* had survived. While no excretion of virus could be demonstrated with warthogs, infected domestic pigs excrete the virus over a longer period in nasal secretion, faeces and urine; the infection being a contact disease in domestic pigs is transferred orally and nasally. In Europe, the wild boar is not a reservoir of infection because it is killed by the disease as is the domestic pig. Due to the high resistance of the virus to environmental influences, the oral infection with contaminated feed is very important in domestic pigs.

The ASF virus can be carried by acutely or latently infected domestic pigs, by human and animal vehicles, by infected feed, infected water as well as virus-containing products of pigs and especially by feeding insufficiently heated waste from kitchens. Through the transport of hard sausages, smoked bacon and ham, the virus can also be carried over long distances thus having the characteristics of a true contact disease.

Clinical Features

An incubation period of 2–19 days leads to either a peracute or acute course of the disease which is characterized by high fever, thirst, inappetence, apathy, a staggering gait and muscle tremors as well as dyspnoea, and an increased respiration, the head being held in a characteristic position with an outstretched neck. The animals become recumbent with the fore legs spread in order to relieve respiration; a mucous, purulent excretion of the eyes and nostrils appears, and the faeces are hard, partly mixed with blood. Stasis develops at first on the ears and limbs, and on the whole body later on leading to haemorrhages on the limbs and on the belly. Shortly before death, the temperature drops to a subnormal level. During the peracute course, death may appear suddenly without clinical symptoms; the acute stage may last for 7 days, the subacute up to 70 days; the disease always having a fatal outcome in domestic pigs. A chronic course which sometimes appears may last for 2–15 months and is characterized by intermittent fever, emaciation, swelling of the joints and sheaths of the tendons, by keratitis, and atrophy of the bulbus. Chronically infected animals excrete the virus during the whole course of the disease.

Pathology

The gross pathological changes are similarly to European swine fever characterized by lesions of the circulatory system which are even more distinct with ASF. As is typical of the acute and subacute forms of the disease, blue or purplish cyanosis marks the skin of the snout, ear tips, tail and distal portions of the limbs in white-skinned pigs. Variable degrees of congestion together with sparsely scattered or densely distributed petechiae or ecchymoses frequently cover the ventral and lateral aspects of the neck, chest, abdomen and limbs. Enlarged and haemorrhagic lymph nodes, both superficial and visceral, are the most striking internal lesions in acute ASF. The gastrohepatic, mesenteric and renal nodes are often completely haemorrhagic and blackened in cross section. In the body cavities, a clear or turbid yellow-reddish liquid is found which congeals when exposed to air and becomes gelatinous. In the mucosa of the wall of the belly and in the mesenterium large reddish-black haemorrhages are evident. Haemorrhagic splenomegaly and lentil- to pea-sized haemorrhages appear under the capsule of the kidney; oedema of the wall of the bladder and gall bladder as well as petechial haemorrhages are further characteristics. The mucosae of the stomach and the intestine show haemorrhagic inflammation and congestion, button-like ulcers appear in the caecum, haemorrhages on the mucosa of pharynx and larynx as well as lung oedema with gelatinous or haemorrhagic enlargement of the interlobular connective tissue with foam in the trachea.

Microscopically, vascular changes occur throughout the body which are characterized by fibrinoid changes in vessel walls and thrombosis. Lesions in the upper layers of the dermis consist of marked congestion, perivascular oedema, haemorrhages of variable size and infiltrations of mononuclear leukocytes. In the lymph nodes, haemorrhages occur which may be anywhere from small and focal to extensive. In the spleen there is marked congestion and haemorrhages, depletion of lymphocytic cells, and pyknosis and karyorrhexis of cells belonging to the monocyte-macrophage and reticuloendothelial system in the white and red pulp are evident. Congestion, oedema, multiple small haemorrhages and fibrinous pneumonia may be present in the lungs. The brain and the meninges are often congested and oedematous and may contain multiple haemorrhages (Mitscherlich and Wagener 1970, Plowright et al. 1994).

Diagnosis

In regions where ASF is enzootic, a presumptive diagnosis can be made upon the characteristic *post mortem* gross pathological findings. By transmitting a suspension from dead animals' organs to domestic pigs which have been vaccinated against European swine fever and a non-immunized control group, it can be determined whether ASF is present. The diagnosis can be confirmed with the Malmquist and Hay haemadsorption:cytopathogenesis test (HAd-test). It is carried out by inoculating pig-leukocyte primary tissue cultures with suspected blood or tissue suspension. 16–48 hours p.i. a haemadsorption with pig-erythrocytes appears if ASF virus is present. Sera from recovered animals can be screened with the indirect immunofluorescence (IIF), the immuno-electroosmophoresis (IEOP), the contra-immuno-electrophoresis and the ELISA. Wherever ASF is enzootic, suspected herds should be tested with the ELISA or IEOP and the results controlled with the IIF. The ELISA is the most sensitive test to detect singular perhaps chronically infected animals (O.I.E. 1990, Rolle and Mayr 1993).

Treatment

No treatment for ASF is known; nor is it indicated.

Control

Though pigs which are infected with the ASF virus produce serologically detectable humoral antibodies, animals which have tested positive for ASF virus are fully susceptible to the infection. Sera from recovered animals also only give incomplete protection against the lethal infection. Whether or not this is due to the incomplete cross-immunity of the different virus strains is unknown. It has not been possible so far to produce an effective immunoprotection, either with inactivated or with attenuated live vaccines. The application of attenuated virus strains as vaccine antigen has produced outbreaks of chronic disease which contributes to the further distribution of the infection.

Control of ASF in Africa must rest on the rigourous prevention of contact between domestic pigs and wild reservoirs together with ruthless procedures for eradication and disinfection when the disease does occur (Plowright et al. 1994). Strict measures of quarantine and restriction of animal movements are required wherever the disease appears. Attention must be paid especially in the tropics to prevention of feeding insufficiently sterilized residues and carcass meal, last but not least of air-dried bone meal.

3.4.4. Pseudocowpox

(Paravaccinia, Milker's nodule (in humans), Kuhpocken)

Aetiology and Occurrence
Pseudocowpox is caused by a parapoxvirus which is very closely related and possibly identical to the bovine papular stomatitis virus and less closely related to orfvirus. It grows in cell cultures derived from bovine and sheep tissues (Munz and Dumbell 1994).

In contrast to the cowpox virus which appears to be confined to western Europe, pseudocowpox occurs worldwide, the incidence often being high. Wherever dairy production is intensified through management and breeding, especially in the tropics, pseudocowpox appears and can become a problem where mechanical milking is applied.

Epizootiology, Pathogenesis, Clinical and Pathological Features
The virus is usually introduced into susceptible herds by the purchase of infected animals, the infection being transmitted from animal to animal by contact especially during milking with the hands of the milker or the teat beakers of milking machines. Because of the short-lived immunity after infection, reinfection can occur during subsequent lactations.

After an incubation time of 3–7 days, localized ecthymatous lesions appear on the teats accompanied by slight fever. They turn into papules, vesicles and pustules within a few days, becoming covered with dark, horse-shoe shaped or ring-like scabs. The scabs are usually shed within a few weeks but occasionally they persist for months. The lesions mostly appear on the teats and on the scrotum in males. The affection may become complicated through secondary bacterial infection. Generally the disease does not adversely affect the general condition of the animal, but the milk production may be reduced perhaps also because the animals cannot be milked appropriately. Often the infection will be transmitted to the hands of the personnel who then may refuse to continue to handle the animal. The milkers may develop pox-like lesions on the hands, forearms or even on the face (milker's nodules).

Histopathologically the lesions are characterized by the appearance of cytoplasmic cell degeneration of the inclusion bodies along with vacuolization, hyperchromasia of the wall of the nucleus and pyknosis (Munz and Dumbell 1994, Seifert 1992.)

Diagnosis

A presumptive diagnosis can be made because of the clinical symptoms and the eventual appearance of typical lesions in the personnel. It can be confirmed by isolating the virus in tissue culture, numerous permanent line cells being useful. The diagnosis can also be verified by an electron-microscopic investigation of negatively stained preparations derived from scabs or vesicular material. HAH and ELISA can be applied as serological techniques (Munz and Dumbell 1994, Seifert 1992).

3.4.5. Other Pox Diseases in Ruminants

3.4.5.1. Buffalo Pox

A specific pox infection occurs in buffaloes (*Bubalus bubalis*) caused by *Orthopoxvirus* (*OPV*) *bubalis* which is related to the *Vaccinia* virus (*OPV commune*). Buffalo pox occurs in the Near East and in Asia and can cause serious economic problems. Lesions similar to those of pseudocowpox appear especially around the mouth and on the udder and scrotum. The infection shows a cyclic course and produces a solid immunity.

Humans who handle animals can likewise become infected by the pseudocowpox infection.

No effective vaccines to prevent the infection are available (Rolle and Mayr 1993).

3.4.5.2. Camelpox

Camelpox causes a generalized affection in dromedary and Bactrian camels, the infective agent being *OPV cameli*, which can be differentiated from other closely related *OPV* species through a number of tests which include the inoculation of embryonated hens' egg, CPE in tissue culture, the intracutaneous test in rabbits and the feather follicle test in chickens. Serological methods applied to characterize the virus are ELISA with monoclonal antibodies, DNA restriction enzyme analysis and a dot-blot assay using digoxigenin-labeled DNA probes. Apparently camelpox virus strains from different African countries have different genomes (Mahnel 1974, Munz 1992, Wernery and Kaaden 1995).

Camelpox occurs in Africa and Asia wherever camel husbandry is practised. The Australian dromedary population is so far free of the infection.

The camelpox virus causes a proliferative skin disease that primarily affects younger animals. Following an incubation period of 9–13 days, pustules develop in the nostrils, lips, eyelids as well as oral and nasal mucosae in mild cases. In more severe cases fever, lassitude, diarrhoea and anorexia appear, the eruptions being distributed over the entire body.

Mortality can reach 28% and the course of the disease be complicated by secondary bacterial and mycotic infections. Animals which have recovered from the infection develop a lifelong immunity. Enzootics occur in regular cycles dependent upon the rainy season in relation to the density of insect population and the number of immunized camels in the population. The virus does not seem to be pathogenic for humans (Wernery and Kaaden 1995).

Several attempts have been made to vaccinate camels against camelpox. At present vaccines are produced using an inactivated strain by Biopharma, Rabat/Maroc, and from attenuated strains in Saudi Arabia and the U.A.E. (Kaaden et al. 1992).

3.4.5.3. Pox of Small Ruminants

(Sheeppox and goatpox, Variola ovina and bovina, Clavelie, Scaappokke and Bokpokke)

Aetiology and Occurrence
Pox of small ruminants are caused by the sheeppox virus *Capripoxvirus* (*CPV*) *ovis* and the goatpox virus *CPV caprae*. Both viruses belong together with the lumpy skin disease (LSD) virus to the genus *Capripoxvirus* and are antigenically closely related. While *CPV caprae* protects against an infection with *CPV* ovis and *CPV caprae*, *CPV ovis* does not protect against *CPV caprae*. The viruses are not specific to their hosts but do prefer the homologous host-species to a certain degree. Recombined strains probably appear when the viruses occur in mixed flocks of sheep and goats. Differences exist regarding the virulence of the pathogens and the susceptibility of the respective animal species. Both viruses grow on tissue cultures from foetal organs of small ruminants and from cattle and may also be adapted to culture on embryonated hens' eggs (Munz and Dumbell 1994).

CPV viruses have little resistance to sun-radiation, day-light and putrefaction but can survive in dark, cool rooms up to 2 years and in the fleece of sheep for 2 months. Suitable disinfectants are formalin 2%, caustic soda 2% and hyperchlorite 0.1–1%.

Pox in small ruminants occur in Southeast Europe, the Near East, Asia and Africa. These infections especially cause serious economic problems in the flocks of small-holders in the Balkans and Turkey.

Epizootiology, Pathogenesis, Clinical and Pathological Features
The CPV virus infection is aerogenic, the virus being excreted by acutely or latently infected animals via expiration, saliva, milk and the pox pustules. The typical dark and cold sheds built of stone in Anatolia for example are important sources of infection in which the virus may survive for months. The infection can be spread with infected animals but also with skins and wool, *CPV caprae* being more infective than *CPV ovis*.

After an incubation period of 6–12 days and perhaps longer, a febrile phase with generalized viraemia develops. Secretion from the nostrils and the eyes is the first clinical sign together with swelling of the nostrils, lips and eyelids with subsequent conjunctivitis and mucopurulent rhinitis. Multiple skin lesions develop over the entire body and are easily seen on the muzzle and those areas free of wool or hair. As with pseudocowpox, pox lesions start as erythematous plaques and progress on to become reddish papules, the formation of umbilicated vesicles and pustules not always being present. In goatpox, the lesions are concentrated on the lips, around the eyes, on the scrotum or udder and on the medial aspects of the hind legs. During recovery, dark scabs are shed from the skin lesions as soon as the regeneration of the epithelium is complete, leaving glistening scars which persist for about 30 days (Munz and Dumbell 1994). At the same time, lesions appear in the respiratory and intestinal tract as well as mastitides. Depending on the severity on the disease the infection is characterized by different courses:

- the slight course: *variola ovina sine exanthema* and/or *variola ovina compressa*;
- the normal course: *variola ovina confluens*;
- the severe course: *variola ovina haemorrhagica-pustulosa* or *nigra* and *v.o. gangraenosa*

Depending on the course of the disease, the mortality rate can vary between 2 and 50%; in lambs it can be 80%.

Gross pathological findings are the typical foci of the pox disease on the skin which are reddish, solid, circular, and sometimes confluent papules covered by a pellicle. The lesions have congested borders. Papules, erosions and ulcers may occur in the mouth, pharynx, larynx and trachea. Grey-white nodules 5–20 mm in diameter appear in the lungs in about a third of cases, they may also be present in the kidneys and the gastrointestinal tract. Histopathological examination reveals hyperplasia of epidermal cells, local inflammation, oedema, cell degeneration and coagulative necrosis. Infected cells containing intracytoplasmic inclusion bodies and vacuolated nuclei are called "sheeppox cells" and are diagnostically relevant (Munz and Dumbell 1994, Seifert 1992).

Diagnosis
In areas where the disease is enzootic, a tentative diagnosis can be based on the clinical and pathological features. The virus can be isolated from skin and mouth lesions, blood and affected inner organs on sensitive cell cultures such as lamb kidney cells, and can be identified through direct or indirect immunofluorescence. Electron microscopy of negatively-stained preparations of lesions in the skin or mucuous membranes can be used for rapid diagnosis.

For the indirect demonstration of the virus, the indirect fluorescence antibody test, serum neutralization, agar precipitation (AGP) and counter-immunoelectrophoresis (CIE) can be applied (Munz and Dumbell 1994, Seifert 1992).

Control
Disposal of dead animals, contaminated feed and animal products as well as disinfection of premises are indispensable prerequisites for disease control.

Animals which have survived the infection acquire a solid immunity. Wherever the infection is enzootic because of permanent prevalence of the disease, a high percentage of the animals will gradually become immunized and thus the economic losses to the farmer can be kept low. Accordingly, the cooperation of the farmers with measures of control as they are required in Europe by statute will be very limited. Measures to carry out eradication schemes in order to control the disease are not feasible in tropical developing countries. Breeding animals which are supposed to be brought into enzootic areas should therefore receive a booster vaccination before shipment.

Effective cell culture-derived vaccines containing attenuated sheep- or goatpox virus are available in south-eastern Europe, the Near East, in Kenya and India. They will mostly provide an immunity for 12 months or longer. Since goatpox virus is less immunogenic, the immunoprophylaxis of goats is less effective (Munz and Dumbell 1994, Seifert 1992).

3.4.5.4. Lumpy Skin Disease (LSD)

(Dermatosis nodularis, Dermatose nodulaire, Dermatosis nodular, Knopvelsiekte)

Aetiology and Occurrence

LSD is caused by *Capripoxvirus bovis nodularis* type Neethling which is closely related to *CPV ovis* and *CPV caprae*. The average size of mature *CPV* virions is 360 × 260 nm. The virus grows in a wide variety of cell cultures and in embryonated hens' eggs, the replication being accompanied by the formation of intracytoplasmic inclusion bodies.

Apparently there is only one immunological type of LSD virus. A cross protection exists between LSD- and sheep- and goatpox viruses.

The LSD virus is remarkably stable: in skin scabs it survives at room temperature for at least 18 days, in necrotic skin nodules for up to 33 days and in tissue culture at +4° C for months. Kept at pH 6.6–8.6 no reduction of titre was observed after exposure to 37 °C for 5 days (Barnard et al. 1994, Rolle and Mayr 1993, Seifert 1992).

LSD only occurs in Africa including Madagascar. Since being encountered for the first time in Zambia in 1929 it has spread over the southern and eastern parts of the continent and at present is advancing via central Africa to the northwest. In 1989 the disease was reported for the first time outside Africa in southern Israel.

Epizootiology, Pathogenesis, Clinical and Pathological Features

All cattle breeds are susceptible to LSD; small ruminants, giraffes (*Giraffa camelopardalis*) and impala (*Aepyceros melampus*) are susceptible to artificial infection and antibodies have also been found in African buffaloes (*Syncerus caffer*) (Barnard et al. 1994). The morbidity varies between 5 and 50%, the mortality rate from 1–75%. There seems to be no breed-specific resistance, Cebus becoming infected just like other breeds, and exotic highly productive breeds possibly with more severe consequences. In South Africa, a reduction of pathogenicity of the LSD virus has been observed in the last decades. The biggest economical loss is caused by the reduction of the milk production, the loss of condition and permanent lesions of the skin. In Kenya, a pox-like disease has been observed in sheep which is supposed to be caused by LSD virus. Therefore, sheep are considered to be a reservoir of infection as well as African buffaloes.

LSD appears either sporadically or enzootically. The incidence of the disease is highest during the rainy season which is why it is suggested that biting insects play a major role in the transmission of the disease, though LSD may also spread in the absence of insects, but direct transmission by contact between animals is thought of as inefficient (Barnard et al. 1994). Since the infection can be transmitted with infected saliva and even with drinking water which has been in contact with infected animals, it is classified in the context of this presentation as a contact disease. Experimentally LSD can be transmitted with material from skin nodules or blood which has been collected during the febrile phase.

After subcutaneous or intradermal inoculation of cattle with the LSD virus, skin nodules erupt 7–19 days after inoculation. The virus is excreted 11 days after the development of fever in saliva, in semen for 22 days and in skin nodules for 33 days, but not in urine or faeces; viraemia lasts for 4 days. The virus replicates in pericytes, endothelial cells and probably other cells in blood and lymph vessel walls, causing vasculitis and lymphangitis. Immunity after recovery from natural infection is mostly lifelong; calves of immune cows are passively protected with a solid maternal immunity (Barnard et al. 1994).

After natural infection and an incubation period of 2–4 weeks, a biphasic febrile response appears which is followed after a few days by the development of the characteristic skin nodules which appear within a few hours. At the same time, the outer genitals and limbs may swell. The animals walk with a stiff gait and show symptoms of pain while walking; sensitivity to light, keratitis, lacrimation and salivation as well as mucoid or mucopurulent nasal discharge are evident. In the majority of animals, the superficial lymph nodes are enlarged.

Fairly uniform skin nodules with a diameter from 5–50 mm being well circumscribed, firm, round and raised, which are randomly distributed appear during the second rise in body temperature as the characteristic signs of the disease. Strings of nodules may be present in the perineum and on the vulva, sometimes forming large, irregularly circumscribed plaques. Skin lesions either resolve rapidly, or become either indurated or sequestrated leaving deep ulcers partly filled with granulation tissue that often suppurates. On the muzzle and the nasal and oral mucous membranes, round initially raised, plaques may be present which frequently ulcerate. After ulceration, respiratory problems and pneumonia can occur as a consequence of aspiration of purulent tissue. In bulls, lesions may be present on the scrotum, the preputial mucosa and on the glans penis, the animals becoming infertile because of orchitis; in cows, the udder and teats can be affected (Barnard et al. 1994).

LSD nodules can disappear completely and rapidly but remain endurated in some cases for some time, often for years as hard lumps ("sit fasts"), partly filled with granulation tissue which often suppurates. Bacterial infections and myiasis may complicate the course of disease.

LSD significantly affects the condition of the animal adversely and may lead to a fatal outcome in calves in particular. Pregnant animals can abort, and cows become anoestric.

The characteristic gross pathological findings are the skin nodules which sometimes are also found in the subcutaneous tissue. In cross-section, the nodules are greyish and the dermis and subcutaneous tissues oedematous, the nodules being found at the sites already described above. Histologically the nodules are conglomerates of fibroblasts, perivascularly accumulated histiocytes, lymphocytes, plasma cells and eosinophil granulocytes. Virus inclusion bodies, 2–10 μm in size, stain with haemalaun-phloxin as round or oval objects inside the histiocytes, epithelial cells and cells of the smooth muscles. Similar lesions as in the skin can be found on the mucosae of the mouth, the nose and pharynx, the trachea, vagina and forestomachs. Furthermore, circumscribed foci may appear in the kidney, liver, lung and the muscles. The lymph nodes are enlarged and moist on the cut surface (Mitscherlich and Wagener 1970).

Diagnosis
The skin lesions are pathognomonic but may however be easily confused with dermatophilosis and pseudo-lumpy skin disease (BHV-2 infection). Histological examination of the skin nodules will reveal the intracytoplasmic inclusion bodies. The virus can be demonstrated by electron microscopy and can be isolated from typical lesions on calf kidney cells and be identified through virus neutralization; in the same way, virus neutralizing antibodies can be demonstrated in serum of animals which have recovered.

Treatment
Only secondary infections of the nodules can be treated with antibiotics. General treatment has been tried by applying 5% formalin intravenously.

Control
So far, it has not been possible to prevent the introduction of the infection in non-infected areas by applying measures of restriction and quarantine. Nevertheless, attention should be paid to introducing only those animals which come from infection-free regions into areas where the disease has not so far been prevalent.

For the active immunization of cattle, vaccines are available which have been prepared from the original Neethling virus strain through attenuation on tissue cultures or passages on embryonated hens' eggs. They provide an immuno-protection which lasts for 3 years. After vaccination, local side reactions may appear in some of the animals which however disappear without complications. In Kenya, a modified sheep- or goatpox virus is used successfully for the vaccination of cattle against LSD.

3.4.6. Contagious Ecthyma (CE)/Orf

(Ecthyma contagiosum, Sore mouth, Scabby mouth, Contagious pustular dermatitis, Dermatitis pustulosa of small ruminants, Ulcerative dermatosis of sheep, Lip and leg ulceration, Venereal balanoposthitis and vulvitis, Lippengrind)

Aetiology and Occurrence
CE is caused by *Parapoxvirus ovis* which is morphologically and immunologically closely related to the parapoxviruses of cattle but not with other pox viruses. Cleavage of viral DNA with restriction endonucleasis reveals considerable genome heterogeneity between various strains of CE virus. The virus grows on cell cultures of tissues of sheep, bovine and human origin but not in embryonated hens' eggs. It is extremely resistant to environmental influences and can survive in scabs for years and survives even in putrefying tissue at room temperature for 17 days; at 64 °C, the virus is inactivated in 2 minutes. Appropriate disinfectants are creolin, chloramine and chlorinated lime in a 5% solution as well as other commercial products (Munz and Dumbell 1994, Rolle and Mayr 1993). Because of its high tenacity, the CE virus survives for a long time in sheds and sewer channels, even in sheep dips.

CE occurs in the traditional sheep production regions of the tropics and subtropics such as in South America and Australia as well as in South Africa, representing an economically important problem, even being considered one of the most important viral diseases in sheep and in some semi-arid countries in goats as well (Munz and Dumbell 1994, Seifert 1992). In US-american literature, CE is separated from balanoposthitis but it is unclear whether the latter is caused by a different virus (Merck 1991).

Epizootiology, Pathogenesis, Clinical and Pathological Features
Small ruminants, wild ruminants, as well as chamois, lamas and humans are susceptible to CE. It has not been possible to prove whether calves can be infected or not. The transmission occurs through direct and indirect contact as well as through drinking water and feed. Goats get hurt in and around the mouth when bruising themselves on thorny bushes; the same happens to sheep in the high Andes when they feed during the dry season on the remaining dried stems and spiny tough grass. Because of the high infectivity of the virus, the infection may also occur through the intact skin. A special feature of infection is the genital transmission of venereal balanoposthitis and vulvitis. Minor scabs on the prepuce are often overlooked when purchasing rams and thus the disease becomes introduced into the herd. Subsequently the infection is quickly spread by the rams throughout the flocks. Ewes usually show clinical symptoms at first; the morbidity may be up to 50%. The lambs become infected by ewes with lesions on the udder, the infection then spreading amongst lambs. Sheep are the natural reservoir of infection. Where sheep and lamas are kept together, lamas may also be infected. Humans become infected through small abrasions of the skin when handling infected animals. The typical scabs appear on the uncovered skin accom-

panied by a slight general reaction especially where contact with infected animals has occurred.

CE can be restricted to one region (face, genitals or claws) of the body or show a generalized clinical course. Mostly primary lesions will develop 2–6 days after infection at the point of entry of the virus to the body, the replication of virus taking place in the epithelial cells of the skin and mucosae of the body openings with proliferation of the keratinocytes and the development of reddish papules which within a few days change to pustules. The epithelial cells degenerate, become vacuolized and cytoplasmic inclusion bodies develop. The pustules become covered with dark brown scabs about 11 days after infection. The scabs slough within 2–4 weeks without leaving a scar.

With the labial form, after initial development of pustules, scabs develop sometimes as big as walnuts on the upper and lower lip, especially at the corner of the mouth, which slough after 2 weeks without leaving a scar. The lesions may also appear around the eye especially at the front eye corner and on the parts of the skin of the face which is not covered with wool as well as around the ears. Through secondary infections, the erosions may be complicated. The animals are unable to graze, become emaciated, and lambs may starve to death.

The pedal form appears regularly together with the labial erosions. It is characterized by lesions in the skin of the coronet and interdigital space, and may even extend up to the carpus and tarsus resulting in varying degrees of lameness and, in severe cases, sloughing of the hooves. Secondary infection may aggravate the condition.

CE often only appears in the genital form (venereal balanoposthitis and vulvitis); it may however also occur together with the other forms of the disease. The infection is noted mostly at first with the ram, an extensive scabby, often bleeding posthitis appearing on the prepuce (Fig. 61 on page 2 annex). The lesions may also occur on the scrotum, the perineum and ventral aspect of the tail. In chronic cases, often only very slight scabby erosions are found at the opening of the prepuce. The rams are affected adversely in their *potentia coeundi* and become unable to mate. Ewes sometimes show only a slight secretion from the vagina and scab-like layers can be found on the vulva. Lesions may also occur on the inner aspect of the hind legs, the udder and teats, thus not allowing the lambs to suckle; secondary infections and mastitis may be the consequence.

The generalized form, which is often fatal, is marked by severe and extensive lesions not only of the skin but also on the mucosae of the mouth, pharynx and oesophagus. Suppurative pleuropneumonia due to secondary bacterial infection is probably responsible for the high mortality rates.

Depending on the course of disease the infection can adversely affect the general condition of the animals and sometimes cause serious damage through secondary infection of the claws and reduced fertility.

Diagnosis
The clinical features of CE are pathognomonic especially if they only appear on one part of the body. The lesions may however be confused with those of blue-

tongue, sheeppox and FMD. The virus can be isolated through tissue cultures and the typical parapox virions can be demonstrated electron microscopically. Indirect immunofluorescence and other serological techniques can be used to detect antibodies in sera but are unreliable indicators of the immune status of the animal. A delayed hypersensitivity skin test may prove to be more useful in this respect (Munz and Dumbell 1994).

Treatment
The topic treatment of lesions in and around the mouth, on the claws and especially on the prepuce in the case of the ram will be very helpful in the recovery of infected animals. The application of antiseptic drugs with a back sprayer may be useful in treating large numbers of animals. Spraying the lesions on the prepuce with chloramphenicol quickly leads to healing of the posthitis.

Control
An important measure of control is the prevention of the introduction of the infection into the flock. Newly acquired animals must be quarantined and carefully examined for suspicious lesions.

Commercial vaccines for the prevention of the disease prepared from attenuated tissue culture virus are partly available where CE occurs, though these vaccines are not always effective. A multiplicity of incompletely attenuated virus strains is available which induce some degree of immunity about 3 weeks after inoculation. They are administered either by rubbing them into scarifications or by subcutaneous inoculation. A cell culture-propagated virus generally produces less effective vaccines than those propagated in sheep, the latter often causing a side reaction which may be serious in lambs. Lambs have to be vaccinated twice – at about 1 month and again 2 months later. Vaccination should never be attempted in flocks which are disease-free because it might introduce the infection. Farmers also use non-attenuated field strains as autogenous vaccines prepared from lesions or scabs from natural cases which are homogenized in glycero-saline solution (Munz and Dumbell 1994, Seifert 1992).

3.4.7. Rinderpest

(Pestis bovina, Cattle plague, Peste bovine, Peste bovina, Runderpes)

History and Occurrence
Rinderpest is one of the oldest known animal diseases and was first described on a papyrus found in Central Egypt from the 3rd millennium B.C. The centre of origin of rinderpest is Central Asia. The epizootic was probably brought to Europe by the Huns in the 4th century. At the time of Charlemagne, in the 10th, 11th, 13th, 14th and 16th century, and especially during the Thirty Years' War, disease outbreaks occurred repeatedly. The infection was probably always introduced each time by cattle which came from the eastern steppes and which

Table 39. Game species of artiodactilae susceptible to the rinderpest virus. (Provost 1981)

Sub-group	Family	Sub-family	English term	Scientific name	Sus-cepti-bility
Suiforme	*Hippopo-tamidae*		Hippopotamus	*Hippopotamus amphibius*	+
	Suidae	*Suinae*	Warthog	*Phacochoerus aethiopicus*	+++
			Bush pig	*Potamochoerus porcus*	++
			Giant forest hog	*Hylochoerus meinertzhageni*	++
Ruminants	*Tragulidae*		Water chevrotain	*Hyemoschus aquaticus*	++
	Bovidae	*Bovinae*	African buffalo	*Syncerus caffer*	+++
		Caprinae	Barbary sheep	*Ammotragus lervia*	+
		Antilopinae	Blackbuck	*Antilope cervicapra*	++
			Springbuck	*Antidorcas marsupialis*	+
			Gazelle	*Gazella sp.*	+
			Guerenouk	*Litocranius walleri .*	+
		Aepycerotinae.	Impala	*Aepyceros melampus*	+
		Reduncinae	Southern reedbuck	*Redunca arundinum*	+
			Bohor reedbuck	*Redunca redunca*	+
			Kob	*Adenota kob*	+
			Defassa waterbuck	*Kobus defassa*	+
			Common waterbuck	*Kobus ellipsiprymnus*	+
			Rhebuck	*Pelea capreolus*	+
			Oubi	*Ourebia ourebi*	+
		Oreotraguinae	Klipspringer	*Oreotragus oreotragus*	+
		Madoquinae	Dik-Dik	*Rhynchotragus sp.*	
				Madoqua saltiana	++
		Raphicerinae	Steinbok	*Raphicerus campestris*	+
		Cephalophinae	Duiker	*Cephalophus spp.*	++
			Duikerbuck	*Sylvicapra grimmia*	+
		Hippotraginae	Roan antelope	*Hippotragus equinus*	+
			Oryx	*Oryx sp.*	+
			Addax	*Addax nasomaculatus*	+
		Alcelaphinae	Hartebeest	*Alcephalus sp.*	+
			Tiang or Topi	*Damaliscus sp.*	+
			Blindled gnu	*Connochaetes (Gorgon) taurinus*	?
			White-tailed gnu	*Connochaetes gnou*	?
		Traguelaphinae	Bushbuck	*Tragelaphus scriptus*	+++
			Greater kuder	*Tragelaphus strepsiceros*	+++
			Derby Eland	*Taurotragus derbianus*	+++
			Oryx Eland	*Tragelaphus oryx*	+++
			Lesser kuder	*Tragelaphus imerbis*	+++
			Bongo	*Boocercus euryceros*	++
			Sitatunga	*Limnotragus spekei*	++
			4-horned antelope	*Tetracerus quadricornis*	+
	Giraffidae		Giraffe	*Giraffa camelopardalis*	+++
			Reticulated Giraffe	*Giraffa reticulata*	++
	Camelidae		Dromedary	*Camelus dromedarius*	+

tiological link and a hidden reservoir of infection, though they usually do not fall ill, but rather become carriers and excrete the virus.

Amongst laboratory animals, small ruminants, guinea-pigs, hamsters, mice, ferrets and Mongolian marmots are susceptible.

Because of the reduced tenacity of the virus in the environment, infected premises, pastures and other subjects are unimportant as vehicles of infection. The same is true for arthropods, birds and other animals which do not become infected themselves. The infection is a contact disease which is transmitted through direct contact between animals or is transmitted through the excretions of infected animals. Mitscherlich and Wagener (1970) have coined the concept of "migrating disease" because the infection is carried by the cattle on their migration.

Pathogenesis
Experimental infections can be established by all routes of parenteral inoculation and, more variably by intranasal or conjunctival installation. Natural infection usually occurs via the upper respiratory tract following inhalation of virus-containing aerosols, or via the oropharynx after ingestion of infected material. Within 24 hours of infection, the virus can be recovered from the pharyngeal lymph nodes and tonsils, the infection being associated with mononuclear leukocytes. Following primary multiplication in draining lymph nodes, viraemia enables the virus to infect and replicate in lymphoid tissues throughout the body. The virus has a predilection for T-lymphocytes; in the course of the disease it is also found in non-lymphoid organs, such as the lungs, liver and kidneys. The massive destruction of lymphocytes causes immunosuppression, a phenomenon which possibly needs to be taken into account when combining the application of live RV vaccine with other vaccines, such as *M. m. mycoides* vaccine. The virus is excreted from epithelial tissues 1 or 2 days before the appearance of fever or lesions, but the amount of excreted virus increases dramatically as the lesions develop and only starts to decline when the immune response becomes detectable some 4–6 days after the start of fever, the virus being excreted in copious large amounts with nasal secretions and faeces. The copious output of the virus explains why the disease can be so contagious despite the fragility of RV. Infected animals mount a vigorous response against the virus. Interferon is produced within 2 days of infection, enabling attenuated vaccines to protect cattle very rapidly against a challenge by the virulent virus. The early response consists predominantly of IgM antibodies, IgG antibodies being produced at the same time which usually persist for life. The severity of the cytopathology caused by the virus before the onset of antibody development influences the course of the disease. Virulent strains cause severe lesions before being restrained by the immune response. The persistence of immunity in recovered animals and those vaccinated with live virus vaccines contrasts with a short-lived immunity induced by inactivated vaccines, thus implying that recovered animals may be continually re-stimulated immunologically by RV antigen throughout their lives (Rossiter 1994).

Clinical Features

After an incubation period of 3–9 days, and in resistant animals up to 3 weeks, a sudden febrile reaction up to 41.5 °C appears during an initial prodromic phase. The animals are depressed, have a staring lustreless coat, reduced milk production, inappetence as well as reduced rumination and thirst; they grind their teeth, show dyspnoea, an increased pulse, and constipation with hard, dry faeces and reddened mucosae. The fever reaches its peak after 2–3 days. Lacrimation appears which is at first serous-mucous and later turns into a purulent conjunctivitis with petechiae on the conjunctivae and swollen lids. In contrast to MCF, the cornea does not become turbid. Leukopenia, especially of the lymphocytes, sets in.

In the continuing erosive phase, first spotty, and then later on diffuse reddening with petechiae covered with mucopurulent secretions appear on the mucosae of the nostrils which turn into crusts around the muzzle. The animals snort and throw their heads from one side to the other. Characteristic are the grey to grey-white semolina-like nodules on the mucosa of the mouth, on the inner aspect of the lips and cheeks, on the gums of the incisors, on the lower aspect of the tongue, on the palate and pharynx which emerge from an initially catarrhal inflammation and give the mucosae the appearance of being covered with a meal-like layer. Subsequently, caseous layers develop which can be sloughed revealing bleeding erosions. By extension and coalescence of these foci, the oral necrosis may worsen dramatically (Fig. 65). Similar lesions with nodules and pseudo-

Figure 65. Rinderpest. Inflammation of the oral mucosa (photo: Liess, from Mitscherlich and Wagener 1970).

membranes also appear on the mucosa of the vagina. The clinical diagnosis of rinderpest can only be established at this stage of the course of the disease.

During the late stage of the disease, the febrile reaction drops to subnormal. Diarrhoea with watery, evil smelling, greyish-yellow faeces, mixed with blood and pieces of epithelium is present. The mucosa of the rectum will prolapse. Polyuria with yellowish-red to coffee-brown urine, dyspnoea with an increased rate of respiration and pulse as well as evil-smelling breath, groaning and recumbency will finally be followed by death. Beforehand the diarrhoea may have ceased. In some cases, lentil-sized spots are found on the non-pigmented skin of the udder, the scrotum, the inner aspects of the thighs, but also on the neck, back, shoulder and on the lateral aspects of the thorax, which turn into blisters, these initially having been papules.

The peracute course of disease lasts for 1–2 days, the acute course for 4–7 days, and the mild course for 2–3 weeks. Latent infections with causal organisms of vector-borne diseases may become acute and complicate the course of the disease.

The morbidity varies in a wide range. Wherever the infection affects a cattle population which so far has been naive to infection (Amadowa Highland, Cameroon in 1960) it may be 100%, and the mortality rate may reach 90%. With Cebu breeds and those cattle which have been affected by the vaccination campaigns, the morbidity varies between 40 and 60% and the mortality rate between 10 and 20%.

The enzootic takes a characteristic course in an affected cattle population. At first only a few animals fall sick, and then after a week, numerous cases appear. Subsequently the disease spreads like an explosion in the herds because of the copious amounts of virus excreted by the animals which were the first to show symptoms. Most of the virus strains which occur in Africa present this picture. On the Arabian peninsula, especially where European high production dairy breeds have been imported, the losses have already appeared during the prodromal stage. On the other hand, in areas where the cattle population has obtained a relatively high level of immunity through repeated vaccination, a mild and subclinical course of the disease which is difficult to diagnose may be present (O.I.E. 1990).

Small ruminants can show severe typical symptoms of rinderpest depending on their genetically determined resistance, but in other cases only a mild clinical picture with symptoms of a bronchopneumonia.

Pigs of the European breeds present only an inapparent course of the disease. Asiatic breeds, however, after an incubation period of 5–14 days, show severe symptoms with high fever, inappetence, increased respiration, coughing, vomiting, erosions and layers on the oral mucosa, diarrhoea and there is a high mortality rate (Mitscherlich and Wagener 1970).

Pathology
During the *post mortem* inspection, dehydration, emaciation and rapid decay of the carcass are evident. The nose and face of the animal show traces of purulent excretions, the eyes are deeply sunk in and stasis of the conjunctivae will be noted. In the oral cavity, the necrotic epithelium has been sloughed in large patches and is distinctly demarcated from the surrounding healthy tissue. Characteristic are

the lesions on the visible mucous membranes, appearing similarly on the mucosae of the forestomachs. On the mucosa of the abomasum, diffuse, deeply dark patches are found, as well as petechia, oedematous infiltration of the submucosa, erosions, spotty necroses and pseudomembranes. The mucosa of the jejunum and caecum is of blackish-red colour and covered with meal-like layers, the mucosa of the rectum showing spotty haemorrhages or striped, almost blackish reddenings. The Peyer's patches are swollen and haemorrhagic, and focal necrosis is evident in the follicles of the lymph nodes. The mucosae of the respiratory tracts are reddened and swollen, show striped haemorrhages and mucopurulent exudate or are covered with croupous layers. The liver is enlarged, and the mucosa of the heavily filled gall bladder is covered with haemorrhages, erosions and pseudomembranes; nephritic lesions are evident, and circumscript reddenings and petechiae are found on the mucosae of the renal pelvis and the bladder. The spleen is negative.

Characteristics of the histopathological findings are acidophil granula, round or oval cell inclusions which contain the virus, as well as lipoides, histones and RNA. They are found in the cells of the reticulum of the lymphocytes and epithelial cells of the mucous membranes of the conjunctivae, the larynx, pharynx, abomasum, prepuce, vagina and in the tonsils (Mitscherlich and Wagener 1970).

Diagnosis

Epizootiology, clinical and pathological features of rinderpest are pathognomonic. The disease can be confused with FMD, MCF and MD. MD is less contagious, rarely showing lesions in the oral cavity, and if lesions appear in the mouth they are localized on the upper aspect of the tongue. The diagnosis during *post mortem* inspection should be concentrated especially on the lesions in the abomasum which presents an oedematous and greyish discoloured mucosa as well as the lymphoid necrosis of the Peyer's patches, and as well on the mucosae of the caecum, colon and rectum which have turned blackish.

The identification of the virus is accomplished with virus neutralization on tissue culture. Under field conditions, primary bovine kidney cells can be cultured on microtitre plates and can be infected with homogenisates from lymph nodes or spleen. After 24–48 hours p.i., the CPE and the virus can be detected by staining with immunoperoxidase (Wamwayi et al. 1991). The presence of the virus can also be verified by the demonstration of precipitating antigens with agar-gel immunodiffusion (AGID) or counter-immunoelectrophoresis in specimens from spleen, lymph nodes or nasal secretions of acutely infected animals. Hyperimmune serum from rabbits is used for antisera. Carrier animals which excrete the virus can be identified through virus detection in the leukocytes via tissue cultures (O.I.E. 1990, Rossiter and Wamwayi 1989). If cell cultures are unavailable, the specimens can be inoculated into known immune and susceptible cattle, as long as these are isolated from other susceptible animals (Rossiter 1994).

The indirect demonstration of RV from convalescent sera can be carried out with the virus neutralization test, ELISA, counter-immunoelectrophoresis or immunofluorescence (O.I.E. 1990, Rossiter and Wamwayi 1989).

Treatment
Wherever the local regulations allow it, the course of disease can be influenced favourably by applying broad spectrum antibiotics (tetracyclines) which also can be applied orally in order to treat the enteritis.

Control
As one of the most dangerous and economically important panzootics, rinderpest is controlled by stamping out in all countries where now the infection is exotic. Many international restrictions on the trade of animals and animal products are due to regulations which are supposed to prevent the introduction of rinderpest, though virtually all outbreaks of rinderpest in virgin areas have been caused by the importation of live animals (Rossiter 1994).

Not withstanding its occurrence, only one antigenic form of the rinderpest virus exists worldwide. Animals which have recovered from the disease acquire a long-lasting solid immunity. The precipitating, complement-fixating and virus neutralizing antibodies found at the beginning of the infection (3rd to 4th day p.i.) are IgM globulins which are replaced later on by those of the IgG class. Animals which recover from the natural infection have IgA globulins in their nasal secretion, which is not the case with vaccinated animals. The calf obtains a solid colostral immunity from its dam which is either naturally or actively immunized which, depending on the breed, may last for 3–8 months. If calves which are protected passively in this way are vaccinated, the passive antibodies and the active antigen neutralize each other, and thus the animals are completely un-protected, the phenomenon of "break-in immunity" being the consequence.

The immunoprophylaxis of rinderpest started with the application of inactivat-ed vaccines prepared from formolized suspensions of tissues from artificially infected cattle. It was only possible to obtain a short acting immunity. Live vaccines were at first prepared from RV strains which had been attenuated through passages in goats (600) (caprinized) and/or rabbits (800–1400) (lapinized). The virus has also been attenuated through culture in embryonated hens' eggs (avianized). Though the vaccines produced a good immunity, side reactions occurred in sensitive breeds (West African taurine cattle). In 1962, Plowright and Ferris were able to attenuate the RV through 90 passages in cultures on calf kidney cells in such a way that a long-lasting immunity with only negligible side reactions could be produced. This RV strain which is now used in the 90th to 120th passages as a so-called Muguga-modification of the Kabete-0-strain (RBOK), is now applied in all countries affected by rinderpest in order to produce lyophilized tissue culture vaccines. The multiplication of the antigen is usually carried out on calf kidney cells as monolayer or in MDBK cells in suspension. Animals vaccinated with this antigen acquire a lifelong immunity, the vaccination only producing occasionally negligible side reactions. The antibody production starts between the 7th to 17th day *post vaccinationem* and virus neutralizing antibodies persist for 6 years after a single vaccination. The vaccine also protects against the antigenically heterologous virus of the pest of small ruminants in goats. Field experience over many years, and careful serological control of the success of vaccination, as has been done with no other vaccination scheme, have shown that no other obstacle exists for the application of the vaccine but the thermolability of the final product. Under field conditions, especially in Africa, it is not always easy to maintain the required cold chain for the vaccine. In the meantime, a RBOK vaccine has been produced on Vero cells which is easier to produce and claimed to be more thermostable. It is supposed to have a biological half-life of 160 days at 37 °C (Mariner et al. 1990). Attempts have also been made to use the vaccinia virus which is very thermostable as a vector virus for the production of the immunogenic relevant antigen components of the RV. Two different recombined vaccines have been produced on the vaccinia virus: one contains the haemagglutinating, and the other the S-antigen of the RV. Both vaccines have protected cattle under experimental conditions against a dose of RV 1000 times the amount lethal for cattle (Yilma et al. 1988). It remains however doubtful whether a large scale spread of the vaccinia virus is acceptable, especially under the conditions of a tropical developing country while trying to control rinderpest.

The strategy to control/eradicate rinderpest so far has been to build up an immune animal population in the affected countries through systematic vaccination campaigns. With the so-called Joint Programme 15 (JP 15) from 1962–1966, the African rinderpest belt from Mauritania to Tanzania was vaccinated systematically supported by professional and material aid from the EC. Subsequently, rinderpest almost disappeared from Africa with the exception of sporadic outbreaks in the Sudan, Ethiopia and Somalia. With the reemergence of rinderpest 1979 in the western Sahel and on the Horn of Africa, an emergency campaign was carried out in 1980–1981 in most of the affected countries which was then continued, beginning in 1986, with the Pan American rinderpest campaign (PARC). This programme is supposed to help the affected African countries to rehabilitate their veterinary service in order to assist in carrying out regular vaccinations in enzootic areas and/or to maintain a sanitary cordon around regions where rinderpest is enzootic. At the same time the intention is to establish a vaccine bank and laboratories for quality control (Debre Zeit/Ethiopia and Dakar/Senegal) in order to guarantee the production of perfect vaccines which fulfil the requirements of the O.I.E. Additionally, investigations are intended to determine the shelf-life of the vaccine and to study the epizootiology of rinderpest especially as far as the role of game as a reservoir of infection is concerned. It is the aim of PARC to establish as high a level of immunity as possible in the exposed cattle population. Wherever the disease remains enzootic, 90% of all animals are supposed to be vaccinated. Furthermore, the intention is to control the movement and slaughter of animals in the enzootic areas and to carry out quarantine measures. Geographically, PARC has been divided into four blocs.

- The Chad-Cameroon-bloc: by building up a highly immunized animal population, the enzootic regions of West and East Africa were supposed to be divided up to which has happened in the meantime.
- The West African wall: by establishing a sanitary cordon from northern Nigeria through West Niger over Burkina Faso through South Mali up to the

border between Senegal and Mauritania, the West African coastal countries are protected against an invasion of the infection from the Sahel.
- The East African sanitary cordon: the spread of rinderpest towards the south along the African coast is prevented at the borders of Ethiopia and Kenya as well as in Kenya and Tanzania.
- The South Sudanese containment cordon: intensive vaccinations along the borders of the C.A.R., Uganda, Kenya and Ethiopia are supposed to contain the infection in those countries where it may appear.

The success of PARC is monitored by a programme of serological control with ELISA which has been established in the affected countries. It is also intended to combine the control of rinderpest with the control of CBPP. The disadvantages of a combined rinderpest CBPP vaccine have been discussed in II/3.3.1.1.

The FAO has developed the Global Rinderpest Eradication Programme (GREP) which aims to eradicate rinderpest worldwide. GREP is supposed to coordinate the strategies and professional expertise required for rinderpest eradication. This includes the development, production and application of efficient vaccines, reliable diagnostic techniques, a disease survey with serological methods and the processing of the data obtained in the field.

3.4.8. Peste des Petits Ruminants (PPR)

(Pest of small ruminants, Goat plague, Pseudorinderpest of goats and sheep, Stomatitis-pneumoenteritis complex of sheep and goats, Kata, Pest der kleinen Wiederkäuer)

Aetiology and Occurrence
Like rinderpest, PPR is caused by a virus covered by helicoids from the family *Paramyxoviridae* and the genus *Morbillivirus* which is antigenically closely related to RV. Recent molecular studies have clearly distinguished between PPR viruses isolated from Africa and those from outside Africa and it has therefore been suggested that PPR may have evolved from the adaptation of rinderpest virus to small ruminants on at least two separate occasions in Africa and Asia (Taylor et al. 1990). Through the determination of the proteins of the PPR virus and gene analysis, the PPR virus has been identified as distinctly different from RV. The PPR virus can also be distinguished from RV with serum virus neutralization.

Similar to RV, the PPR virus grows in tissue cultures on primary cells and cell lines; in lymphocyte cultures, both viruses prefer to grow on those of the homologous species. Vero cells are the most suitable for a first isolation of the PPR virus from specimens of an infected animal.

The PPR virus is thermolabile but relatively resistant in a cool environment and can survive for months in frozen animal tissue. It is inactivated at a pH of 3.0.

PPR was recognized for the first time in 1942 on the Ivory Coast and it was supposed that the disease was an infection limited to West Africa. In the meantime, the enzootic occurs within a belt in Africa between the tropic of the Cancer and the equator, on the Arabian peninsula and now also in India. The relatively fast spread of the disease during the last years has probably been caused not only by nomadic pastoralism but also by the exportation of living small ruminants from the Horn of Africa to the Arabian peninsula. Wherever PPR occurs, it is a considerable obstacle for anyone who intends increasing the productivity of small ruminants by the importation of exotic breeds (Diallo et al. 1989).

Epizootiology, Pathogenesis, Clinical and Pathological Features
Goats are more susceptible to the PPR virus than sheep; the autochthonous West African goats of the humid savannah are less resistant than the goats of the nomads of the Sahel. Cattle only become infected subclinically but do produce neutralizing antibodies. The PPR virus has also appeared in captive game (gazelle/*Gazella dorca*, nubic goats/*Capra ibex* and chamois/*Oryx gazella*). Recovered animals become immune to PPR as well as to rinderpest.

The epizootiology of PPR so far has been independent from that of rinderpest. Early during the febrile phase infected animals excrete the virus with the conjunctival secretion and saliva, later on also with the faeces. The transmission is aerogenic; ports of entry into the organisms are the nasal and pharyngeal mucous membranes. So far it has not been possible to prove that latently infected animals excrete the virus.

PPR has appeared as an epizootic with severe clinical symptoms and mortality rates of 70–80% in the agriculture regions of the West African coastal countries from Mauritania to Benin. In contrast, it is enzootic in the central African Sahel (Mali, Niger, Chad), where the disease only shows a subclinical course with low mortality, though 70% of the monitored animals reacted serologically positive to the PPR virus. Lefèvre (1991) considers this difference to be due to the stability of the PPR virus in the dry environment of the Sahel which allows the small ruminant population to become gradually and continuously infected and to develop a considerable level of immunity. Lefèvre also presumes that because of the same mechanisms of infection and immunity the small ruminants of the Sahel are unimportant as reservoir of infection for rinderpest (cross-immunity).

The course of disease and its clinical features are almost identical with and indistinguishable from those of rinderpest. During the acute course the temperature rises suddenly to 42 °C after an incubation period of about 2 days. The animals present severe general symptoms, are restless, have a dry nose, a staring lustreless coat and refuse feed. Stasis of the visible mucosae is apparent. Soon secretion of the nostrils appears which turns mucopurulent accompanied by a cadaverous-smelling expiration. A necrotic stomatitis develops in the oral cavity followed by a bronchopneumonia which is complicated by bacterial infection in its final stage. Profuse diarrhoea leads to progressive emaciation and after about 5–7 days, death supervenes during the acute course; the subacute form will also lead to a fatal outcome 14–21 days after the onset of the febrile phase. Morbidity and mortality are higher in young animals. The clinical picture of the disease may be altered through the activation of latent infections with causal organisms of vector-borne diseases.

A subclinical form of PPR which has an almost asymptomatic course appears especially in the dry areas of Central Africa. It is supposed to be a predisposing factor for other lung affections (Lefèvre 1991).

The gross pathological features of PPR are similar to those of rinderpest, especially those of erosive and necrotizing stomatitis, stasis and inflammation of the mucosae of the abomasum and enteritis. Typical lesions of a bronchopneumonia are always present.

Diagnosis

Epizootiology, clinical and pathological features allow a presumptive diagnosis. The typical zebra-marking of the inflamed intestinal mucous membranes is pathognomonic. Because of the similarity of PPR to rinderpest, the final aetiological diagnosis can only be established by isolation and identification of the virus (II/3.4.7). The virus can be isolated from swabs of the conjunctivae, the mucosae of the mouth and rectum as well as nasal secretion. The detection of virus-neutralizing antibodies in convalescent sera is diagnostically relevant. The same is true for antibodies which have been demonstrated with techniques as they have been described for the diagnosis of rinderpest (II/3.4.7).

Treatment

Broad spectrum antibiotics and sulphonamides can be used to control the secondary infections of enteritis and bronchopneumonia in order to influence the course of disease favourably.

Control

Strict control of animal movements being essential for controlling the disease however is almost unfeasible in production systems of small ruminants in Africa.

The immunoprophylaxis of PPR with RBOK rinderpest vaccine has been highly successful and provides immunity for at least 3 years, although its long term duration is uncertain (Rossiter and Taylor 1994). Diallo et al. (1989) developed a vaccine using the homologous PPR virus as an antigen attenuated through 55–61 passages in Vero cells. The virus has similar characteristics as the RBOK strain. Details of its efficacy in the field are being assessed (Rossiter and Taylor 1994).

3.4.9. Foot and Mouth Disease (FMD)

(Aphthae epizooticae, Aphthous fever, Fièvre aphtheuse, Fiebre aftosa, Aftosa, Maul- and Klauenseuche/MKS, Bek-en-klouseer)

Because of its worldwide occurrence, high contagiosity and the economic losses it causes to biungulates, FMD belongs to the most feared infectious diseases in domestic animals. It is not only because FMD adversely affects the health of the animals and their productivity but also because of international restrictions on animal trade which cause serious economic losses to the affected countries.

Aetiology and Occurrence

Fracastorius presented a precise description of a cattle disease in Latin in 1514 which exactly fits the complex of symptoms caused by FMD. In 1897, Löffler and Frosch proved that FMD is caused by a filterable virus.

The FMD virus belongs to the genus *Aphthovirus* from the family *Picornaviridae*. The complete virion has a diameter of 23 nm and its MW is 6.9 kDa, 69% being protein and 31% RNS. 32 capsomeres build a capsid in the form of a symmetric icosaeder, which surrounds the RNA nucleus. The virion contains a molecule of RNA in a single chain which is composed of 8450 bases and has a sedimentation coefficient of 35 S. The RNA is the infectious unit and acts as a messenger. The virion is composed of 60 identical copies of the structure proteins VP1, VP2, VP3 and VP4. The capsidale protein VP1 induces the production of neutralizing antibodies in cattle and pigs; the same can be accomplished with chemically broken down fragments of VP1. Three antigenic relevant regions have been described on the surface of type A 12. Two of these have also been found on the chains of polypeptides of VP1 and in subunits with a sedimentation constant of 12 S. Virus neutralization by these antigenic regions is induced through

- aggregation of virus units through which the virus loses its ability for penetration;
- blocking of the specific sites for adsorption;
- inhibition of the ability for penetration and decapsidation.

Regarding their function and structure, the antigenic components of FMD virus can be differentiated as follows:

- the complete virion, sedimentation constant 140 S with 60 copies of VP1–VP4;
- the empty capsid, sedimentation constant 75 S with 60 copies of VP0–VP3;
- the capsomeres, sedimentation constant 12 S, pentameres of VP1–VP3;
- the antigen VIA, sedimentation constant 3.8 S, a RNA polymerase (Schudel and Sadir 1986).

With the exception of RNA, all units also appear during virus replication within the infected tissue and have antigenic and/or allergenic properties. Because of their different antigenic structures the antibodies produced within the infected organism by these antigenic components are different. The same is true for the immunogenicity while the allergenetic activity is uniform. The immunizing component of the complete virion produces type-specific antibodies. This property also remains intact during gentle inactivation and inhibition of replication of the virus. Beside other antigenic qualities the complete virion also has an allergenic potency, the latter being localized especially in the capsomere-antigen. The RNA of FMD virus is antigenically unimportant.

FMD virus is antigenically not uniform but rather it occurs in serotypes which each may have subtypes or variants, which again are serologically different. The genetic differences are caused especially through a very wide variation in the protein VP1.

The worldwide occurrence of the FMD virus until the beginnings of the 90's is described in Fig. 66. In contrast to this presentation, the EC countries have now been declared FMD-free. In detail the situation is as follows:

– The American continent north of the Panama Canal including the Caribbean, Australia and New Zealand are FMD-free as well as the countries of the EC including the now adjoined Scandinavian countries since 1992.
– The so-called classical European types Vallée A, O and Waldmann C are enzootic worldwide with the exception of those countries mentioned above. By 1972, 29 subtypes of type A and 11 of type O had been described while type C only presents little variation, 5 serotypes having been isolated in Europe and South America (Table 40).
– The South African types SAT 1, 2 and 3 (Southern African Territories) are restricted to sub-Saharan Africa, although there have been incursions of SAT 1 into North Africa and the Middle East. By 1972, 7 subtypes of SAT 1, 3 subtypes of SAT 2 and 4 subtypes of SAT 3 were known (Table 40).
– The Asian type SAT 1 occurs in Asia as far as the border with Europe at the Bosporus; 3 serotypes have been identified (Table 40) (Schudel and Sadir 1986).

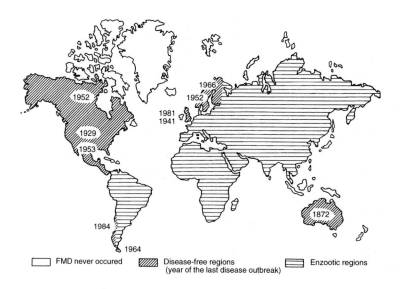

Figure 66. Worldwide occurence of FMD (Schudel and Sadir 1986)

Table 40. Subtypes of FMD virus recognized by the reference laboratory Pirbright until 1972 (Schudel and Sadir 1986)

Type O Vallée

01 Lombardy 46
    idt. Campos Brazil/58
02 Brescia/47
03 Venezuela/50
05 India/62
06 Pirbright-OVI/24
07 Italy/58
08 Brazil/60
09 Kenya/60
010 Philippines/58
011 Indonesia/62

Type A Vallée

A1 Bavaria/42
A2 España/43
A3 Mecklenburg/44
A4 Hesse/48
A5 Western Forest/47
A7 Greece/50
A8 Parma
A10 Kemron; Argentina/61
A11 Pirbright – AGB/Germany/29
A12 Pirbright – A119/32
A13 Brazil/59 – A Santos
A14 España/59
A15 Thailand/60
A16 Brazil/59 – A Belem
A17 Brazil/59 – A Guarulhos
A18 Venzuela/62 – A Zulia
A19 Argentina/63 – A Suipacha
A20 UdSSR/64 – Usbek/60
A21 Kenya/64 – Lumbwa
A22 Iraq/64 – MO-Variant
A23 Kenya/65
A24 Brazil/55 – A Cruzeiro
A25 A – Argentina/59
A26 A – Argentina/66
A27 A – Columbia/67
A28 Polatli, Turkey/69
A29 A – Peru/69
A30 A – Uruguay/68
A31 A – Columbia/69
A32 A – Venezuela/70

Type C Waldmann

C1 Field and vaccine strain
    Netherlands/62
    field strain GB/65
C2 997 field strain GB/53
    C. Pando Uruguay/45
C3 Brazil/55, C Resende
C4 Argentina/66-Tierra de Fuego
C5 Argentina/69

Type SAT 1

SAT/1 Bechuanaland 1
SAT1/2 Rhodesia/37
SAT1/3 South Africa/
    East Africa/49
SAT1/4 Southern Rhodesia/58
SAT1/5 South Africa/61
SAT1/6 South Africa/
    East Africa/61
SAT1/7 Israel/62 MO-Variant

Type SAT 2

SAT2/1 Rhodesia/48
SAT2/2 South Africa/59
SAT2/3 Kenya/60

Type SAT 3

SAT3/1 Rhodesia/34
SAT3/2 South Africa/59
SAT3/3 Bechuanaland/61
SAT3/4 Bechuanaland/65

Type Asia I

Pakistan/54
Israel/63
Kemron Asia I

The nomenclature of the subtypes which are differentiated as special serotypes is not uniform worldwide. Partly they have been marked in numerical order according to the date of their identification or, as in South America, with the place and year of their isolation together with the reference strain which identifies the subtype: for example, subtype A 81 Argentina/87 is a serotype which was isolated in Argentina 1987 and is related to a reference strain characterized in 1981.

The types and subtypes of the FMD virus are of varying importance to the epizootiology and immunoprophylaxis of FMD, especially where the infection is enzootic. An antigenic drift of the types occurs while schemes of control (vaccination) mostly lag behind. This is mostly the case where the vaccination is not carried out regularly. Consequently, continuously mild outbreaks of the disease occur. A disease outbreak can be caused by a type, a subtype or also a combination of types and subtypes or subtypes alone, the virus strains perhaps antagonizing each other.

The FMD virus is relatively resistant; it survives 5–10 weeks in protected places during cool weather, especially in tissues or other organic matter, a pH not below 6.5 being prerequisite. Drying, cold and a high concentration of salts do not influence the virus; it survives for years frozen. In the dirt of sheds it remains viable for 14 days, in animal refuse for 39 days, inside manure for 6 days, on the surface of manure during summer for 28 days, and in winter for 67 days. The virus can survive in feed for 15 weeks, on cattle hair for 4 weeks and in waste water up to 103 days. In animal products it is inactivated through the acidification of milk and meat, surviving however where no acidification occurs (lymph nodes, bone marrow, fat and blood) for months, the same being true for frozen meat (80 days) and salted meat (42 days). Heat inactivates FMD virus at 50 °C. In powdered milk it survives at 130 °C for 1–3 minutes. In an acid environment (pH 4.0) it is destroyed quickly. Appropriate disinfectants are caustic soda 1–2%, lime milk 5%, hot soda solution 5%, Na_2CO_3 4% or formalin 10%. Sensitive tools can be disinfected with citric and acetic acid (5%). Because of the tenacity of the virus, numerous ordinary disinfectants are inefficient.

Epizootiology

Susceptible to FMD virus are large and small cloven-hooved species including camels. Wild ruminants (African buffalo, wildebeest and kudus) can also become infected acutely and/or excrete the virus after subclinical infection. The same is true for wild suids. White-tail deer (*Odocoileus virginianus*) presented severe clinical symptoms often with a lethal outcome during an FMD type O outbreak in cattle in the Andean ranges of northern Peru. Since the infection remained for some time latently within the deer population, a reservoir of infection remained for the cattle of the ranches where the deer grazed on the pastures. The African game population including elephants as well as *Hippopotamidae* are an important reservoir of infection especially for SAT types. In the meantime, based on RNA hybridization techniques, genetic differences between original cattle viruses and viruses isolated from game have been found. Thus it has been established that the virus mutates to a particular sub-population in game. It has been possible to establish that an FMD virus infection leads to a long lasting carrier state in the oesophago-pharyngeal (OP) region in cattle, sheep, goats, African buffalo, wildebeest and kudus. Low titres of the virus in OP samples have been found after more than 2 years p.i. Animals vaccinated against FMD may also enter such a latent carrier state. In pigs, no carrier state has been detected so far. The epizootiological significance of such carriers so far remains unclear. It has not been possible to transmit the carrier virus experimentally through contact to cattle, sheep and goats. With buffaloes (*Syncerus caffer*) it has been possible to demonstrate that they transmit the carrier virus within their own species and possibly also to cattle. Carriers are most certainly a natural reservoir of FMD virus in enzootic areas and a potential source of antigenically mutated virus variants because variations of the virus with subsequent selection of virus mutants occur continuously in the animal during the carrier state (antigenic drift) (Wittmann 1990). Since the infection in domestic buffaloes (*Bos bubalus*) often remains inapparent, buffaloes kept in close contact with cattle are potential carriers of virus and sources of infection, especially in the difficult-to-control small-holder production systems in Asia.

Several reports deal with artificial infection of dromedaries with the FMD virus which yielded mostly only slight or clinically inapparent manifestations. Nor has there been any evidence of humoral antibody production (Nasser et al. 1980, Moussa 1988, Hafez et al. 1993) [all cited from Wernery and Kaaden 1995]. Only Richard (1986) [cited from Wernery and Kaaden 1995] was able to identify FMD virus types O, C and SAT 2 in 2.6% of sera from Nigerian dromedaries. Dromedaries kept for weeks in close contact with severe cases of FMD in cattle and small ruminants in various FMD epizootics in Ethiopia (Richard 1979), Oman (Hedger et al. 1980), Niger (Richard 1986), Saudia Arabia and Egypt (Hafez et al. 1993) [all cited from Wernery and Kaaden 1995] did not develop any signs of clinical symptoms. Though FMD is apparently not a problem for camels, the subclinical or latent infection acquired by camels which have been in contact with infected cattle or small ruminants may play an important role as a viral reservoir in the epizootiology of FMD (Wernery and Kaaden 1995).

The virulence of the virus in different animal species depends on the type of virus and the resistance and immunity of the affected animal population. European pig breeds for example can practically be wiped out by the South American O variants, but local breeds do not even show clinical symptoms. The same can occur with sheep and goats which are kept together with cattle affected by an acute outbreak, the small ruminants being spared from clinical symptoms. In the typical sheep-breeding regions of the tropics, small ruminants in general rarely show clinical symptoms and if so, they are only mild. Since the animal holders are aware of this, small ruminants are not often or almost never included into schemes of immunoprophylaxis, and as a consequence this establishes an important reservoir of infection for cattle. The same is true for local breeds of pigs kept by small-holders. It remains doubtful whether the observation in Europe that the virus adapted to pigs is less pathogenic for cattle and vice versa is valid for the tropics because of the above mentioned epizootiological mechanics.

The virus is carried and transmitted directly by latently or acutely infected animals and perhaps indirectly over large distances through numerous non-living objects (feed, drinking water, tools, animal products, as well as human clothing and vehicles) and living vehicles (humans who handle animals, rodents, straying dogs, wild animals and birds). Aerosols contaminated with virus and exhaled by infected animals may be carried windward over distances up to 60 km especially at night when the humidity exceeds 70%. Because of its high virulence, the virus spreads fast within a population especially if a new virus type has been brought into a so far immunologically naive population. Severe disease outbreaks occurred in South America when animals were imported from neighbouring countries carrying virus strains which were not present in the locally applied vaccines (Peru 1965).

In the tropics, disease outbreaks occur mostly with the onset of the hot humid season. The climatic stress suppresses the existing immunity in the cattle population which at first leads to sporadic and subsequently to severe and widespread disease outbreaks. Wherever FMD and the locally relevant virus type are enzootic, the course of disease is mostly rather mild. The lacking cooperation of many animal holders in South America, when schemes for immunological control of the disease are carried out, is last not least because of the mild course of the disease where it is enzootic.

Three factors determine the characteristics of FMD transmission: the quantity, duration and means by which the virus is liberated into the environment; the ability of the agent to survive outside the animal body; and the quantities of virus required to initiate infection at the primary infection sites. In viraemic animals, the FMD virus is present in most physiological fluids and hence potentially in nearly all secretions and excretions. Significant excretion occurs up to 4 days, or perhaps longer, before the appearance of lesions, so that apparently healthy animals can be an important source of infection, milk being particularly important. Domestic pigs are the most efficient excreters of the FMD virus. Since pigs are often intensively farmed, the disease in piggeries can result in an enormous multiplication of infectivity. Of all the secretions and excretions in acutely infected animals, saliva contains the highest concentration of virus, but nasal secretions, those of the oesophago-pharyngeal region, the urogenital tract and milk as well as semen also contain appreciable quantities of virus (Thomson 1994).

Pathogenesis
The respiratory tract is the usual route of infection in species other than pigs. In cattle and sheep initial viral replication occurs in the mucosa and the lymphoid tissues of the pharynx being predilection sites for FMD virus replication. The liberation of the virus into the efferent lymphatic system results in an initial viraemia whereby the virus reaches a wide variety of organs and tissues where further replication occurs. Development of characteristic vesicular lesions in FMD is dependent on infection of the squamous epithelium and persistent local irritation or friction. This explains why the mouth, feet and teats are predilection

sites for the development of lesions, and why pigs often develop lesions on the dorsum of the snout, as a result of snuffling (Thomson 1994).

Clinical Features

After an incubation period of 5–7 days in cattle, 1–6 days in small ruminants and 2–12 days in pigs a short febrile reaction and cessation of milk production are the first clinical symptoms. The febrile phase is when viraemia occurs and the virus has spread to the target organs where it multiplies. The animals salivate, the mucosa of the oral cavity becomes reddened whereupon the typical aphthae develop on the inner aspect of the lips, around the gums especially on the dental plate of the maxilla and at the tip and on the dorsal surface of the tongue. In severe cases, the entire mucosa of the tongue may slough off like the finger of a glove. The oral lesions mostly become complicated by secondary infection. Animals which graze on extensive pastures become unable to feed on the tough grass. Typical signs caused by these lesions are salivation, smacking of the lips and grinding of the teeth. Simultaneously or later, erosions appear on the feet, in the interdigital space, at the bulbs of the heel and around the coronet as well as occasionally on the udder, the animals becoming lame and showing a disinclination to stand. In severe cases, the hooves slough off which may be the cause for a fatal outcome of the disease especially in bulls because of the subsequent recumbency. Wherever FMD is enzootic, the infection often causes only a slight lameness due to an oedematous inflammation of the coronet, the animals occasionally not even presenting lesions in the mouth.

If the disease takes a favourable course, the lesions in the mouth may already heal after a week and those on the hooves after 2–4 weeks. Where, however, a new virus type has met with an immunologically naive population, not only severe erosions develop in the oral cavity and at the hooves but also the typical "tiger heart", a severe myocarditis, appears especially in young animals which may die suddenly without premonitory signs. Under those circumstances the morbidity may be up to 100%, and the mortality 50–70%. If a disease outbreak occurs during the calving season on a ranch in the tropics, it is all the worse if this takes place within the rainy season, and almost no calf will survive the infection; they mostly die without or only with general clinical symptoms in a few hours. In the extensive animal production systems of the tropics, the secondary infections in the mouth and on the hooves are of special interest. Because the animals are dependent on finding feed and water on their own, losses occur due to decubitus and emaciation since they become unable to move or graze on the tough grass.

In sheep, the lesions are mostly localized on the hooves, the disease manifesting itself in symptoms of lameness and the refusal of the animals to get up. Only slight lesions develop in the mouth about 1 week later. Young sheep are also especially affected; numerous lambs may die. Goats only show slight clinical symptoms which are restricted to the mouth. The often inapparent aphthae heal quickly and present only a neglectable discomfort to the animals. The clinical symptoms in pigs depend much on the respective virus type and the level of resistance of the affected porcine population. A disease outbreak in a piggery in the tropics stocked with highly productive exotic breeds may result in a few days in a morbidity of 100% and then the loss of all animals which die often without premonitory signs. The infection often takes its origin from an enzootic in cattle in the neighbourhood. The type O virus in particular is highly pathogenic for pigs. During a more protracted course, the typical disease symptoms in pigs are lameness with aphthae especially on the coronet, the cleavage of the claws and bulbs of the heel. After secondary infections which occur frequently, the hooves mostly slough off. Lesions also appear on the udder, the snout and occasionally on the mucosa of the oral cavity.

Pathology

The necropsy findings are characterized by the lesions which the replication of the virus causes to the cells of the *Stratum spinosum* of the epithelium. At first small cavities appear which finally develop into larger vesicles with a watery-clear content composed of destroyed epithelial cells and serous exudate; on the base of the vesicles, intact epidermis cells of the *Stratum spinosum* and *Stratum germinativum* remain which allow the complete and fast regeneration of the lesion. During necropsy, similar erosions are found on the mucosae of the oesophagus and the pillars of the rumen. The peracute course of disease in calves is the consequence of the grey-yellow or grey-white streaked or spotted, histologically hyaline and squamous vacuolous myocarditis.

Diagnosis

The diagnosis of FMD has to be confirmed by the direct demonstration of the virus and identification of its type. The clinical symptoms are easily confused with other infections like:

– *Stomatitis vesicularis*, which can be transmitted experimentally by contact (swab of oral saliva) to equines;
– mucosal disease, which shows focal ulcerations and not the typical aphthae of FMD and is much less contagious;
– bluetongue, which can be transmitted experimentally to susceptible sheep only parenterally;
– MCF, which is characterized by high fever and the typical inflammation of the mucous membranes of the head as well as nervous symptoms and cannot be transmitted experimentally by contact;
– rinderpest, which does not present neither lesions on the hooves nor the typical aphthae but it has a high febrile reaction.

The indirect demonstration of the causal virus through serology rarely is relevant since disease outbreaks only occur as an exception in populations which have neither been vaccinated nor been immunized through natural infection prior to infection. ELISA, CFT and virus neutralization (VN) are appropriate serological methods.

For the direct demonstration of the virus, the contents of the vesicles and epithelium of vesicles which have not yet erupted are the specimens of choice. If such material is unavailable, saliva and blood and/or swabs from oesophagus or pharynx from pigs from the larynx can be sent to the laboratory. The material has to be preserved cooled or frozen.

The determination of virus is still carried out with the traditional CFT but also the ELISA is now applied to FMD virus typing, the latter being more sensitive. If there is not enough virus present in the specimens, the virus can be multiplied in tissue cultures or suckling mice. Since FMD viruses usually grow quickly and produce rapid cytopathic effects in a wide variety of cell cultures, the most commonly used being primary or secondary calf thyroid cells and IB-RS-2 or BHK21 cell lines, the cell culture supernatant can be used for fast CF typing. Under field conditions in the tropics, suckling mice can be used to isolate the material from suspicious specimens; the dead and eviscerated mice may then be sent to the laboratory for virus determination.

For epizootiological studies and in order to establish the relationship of the isolated virus to vaccine strains and others known to have been previously present, newer techniques such as one- or two-dimensional electrophoresis of viral RNA fragments, isoelectric focusing of viral structural proteins or nucleotid sequencing of viral RNA are employed (Thomson 1994).

Treatment
Though no specific treatment for the FMD virus is available, especially in tropical extensive production systems, affected animals have to be treated. In industrialized countries, treatment is prohibited because of government regulations, and culling of infected and supposedly infected animals is required. Sick animals may be treated topically with mild disinfectants but also by applying broad spectrum antibiotics parenterally, tetracyclines in particular, in order to control the consequences of secondary bacterial infections. Breeding animals and bulls in particular may be saved this way. If, however, a disease outbreak occurs in suckling calves, no treatment will be possible.

Control
The strategies for controlling and prevent FMD in tropical developing countries are different in principle from those in the developed industrialized countries. There is often no functioning veterinary health service available in the tropics, nor do the economic and socio-economic prerequisites exist for carrying out schemes as they are applied in the industrialized countries. Beyond that, the geographic and climatic conditions as well as the breeds used in tropical animal production systems in the way they are kept and managed require a different approach (III). The intention to transfer concepts as they have been successful in Europe to the conditions of the South American countries may be one of the reasons why it has not been possible so far to control FMD successfully.

Basically, the control of FMD in tropical developing countries should attempt

– to prevent the introduction of the disease, especially of new FMD virus types

by applying strict measures of control and quarantine when trading and importing breeding animals as well as food of animal origin;
- to minimize the spread of the infection during a disease outbreak by limiting the traffic, especially of animals and feed;
- to apply a vaccination scheme using site-specific vaccines which is adapted to the local environment and management conditions, the type of animals and the socio-economic conditions of the animal holder.

As described in I/3.1.2 and III/3.1, the time of vaccination should not depend on the theoretically effective period of the applied vaccine. Instead, the animals should be vaccinated when they are in optimal condition and not while being of poor immunocompetence, as for instance after calving. Beyond that, the seasonal intensity of the infectious pressure should be taken into account. Thus it would be advisable to immunize pregnant cows at the end of the dry season with a booster vaccination thus providing them with an immunoprotection which outlasts the rainy season and the time of calving. Last but not least, practical points of view run contrary to carrying out vaccination campaigns during the rainy and calving season. Reduction of the milk yield caused through vaccination in dairy cattle requires a concept which is adapted to the conditions of intensive dairy production in the tropics in order to assure the cooperation of the animal holder (III/3.1 and III/3.2.

Because of reasons of opportunity, trivalent (type A, O, C) vaccines and perhaps one containing a fourth component (Asia 1) are mostly applied for vaccination against FMD. In Argentina, types O, C and two A types are contained in the vaccines. Due to the fast mutation of the A strains (Table 40), problems exist worldwide in actualizing the A component of the vaccine, this also being the case in India. It is a principle of vaccine production that the more components of antigen are contained in the vaccine, the more antigenic material of each antigenic fraction has to be included – a requirement which is not always followed, on the one hand because of the challenge to the animal, and on the other hand because of economic reasons. As a consequence, it can be assumed that monovalent vaccines provide the best and longest-lasting protection – a fact which has been proven under field conditions. Only bivalent vaccines of good quality provide a lasting immunoprotection of about 8 months' duration. With trivalent vaccines, the effective period is shorter. Furthermore, the protection provided by each single antigenic component is different; one type usually stimulates a shorter effective period. In Botswana, cattle were successfully vaccinated simultaneously against FMD-SAT 1 and rinderpest (Guillemin et al. 1987).

By applying properly produced vaccines, the animals are already protected one week after the application of the vaccine, the immunoprotection also depending on the breed, the age, the condition, the productivity and the environment in which the animals are kept. By no means should calves be vaccinated prior to weaning. The antigen applied with the vaccine is neutralized through the passive maternal antibodies acquired with the colostrum. This is especially true for autochthonous breeds which provide their calves with a solid maternal immunity which may last up to 8 months. Through a booster vaccination at an interval of about 6–12 weeks the titre of antibodies as well as their effective period can be enhanced. After a booster vaccination, virus neutralizing antibodies increase steeply. When polyvalent vaccines are applied, the weaker antigenic component of the vaccine is also stimulated. Through booster vaccination applied at the right moment, an immunoprotection may be obtained which lasts for a year. Because weaned calves for the most part only react poorly on the first vaccination, they must by all means be given a booster vaccination.

An effective immunoprotection of pigs through vaccination is difficult to accomplish. Pigs cannot be protected with the usual cattle vaccines which contain aluminium hydroxide as an adjuvant. Better success has been obtained with modern adjuvants such as MONTANIDE or DEAE. Nevertheless, the protective effect is unsatisfactory and only short lived, it may be enhanced by booster vaccinations up to 4 months.

It is especially difficult to protect cattle against SAT types. So far an efficient vaccine is only available against SAT 1.

In the tropics, vaccines of different quality are applied. This not only concerns the volume of injection, the effective period and the efficiency, but also eventual side reactions. This depends primarily on the technology of production, but also on the conditions of storage and transportation as well as the way the vaccine is applied. Vaccines available at present are produced with different technologies regarding the replication, purification and concentration of the antigen and the type of adjuvant applied.

The replication of antigen is carried out

- still in part with the technology developed in 1947 by Frenkel, e.g. in Argentina, but also in Botswana with SAT 1; epithelium of the tongue harvested from carcasses of healthy slaughtered animals is kept alive in a nutrient solution with the addition of antibiotics where it is infected with virus. The virus replicates in the cell suspension which is kept in stainless steel tanks and can be harvested after 24 hours. After homogenization, cells and cell rests are separated through centrifugation and filtration with Seitz filters;
- by means of a culture on BHK21 (baby hamster kidney) cells in monolayer or in roller-bottles;
- in suspension cultures of BHK21 cells, mostly in industrial fermenters with volumes up to 6000 L.

The purification and concentration of antigen is obtained

- with the traditional Waldmann method by homogenization, centrifugation and filtration in plate systems;
- with the concentration technique devised by Pyl. Here, treatment with chloroform replaces the filtration which was also previously required for decontamination. Since loss of the virus through filtration is prevented, lower volumina can be used for injection;

For the inactivation of the virus and in order to stimulate its antigenicity through adjuvants, the following alternatives are available:

- the traditional method of the Waldmann-Koebe-vaccine: the virus is inactivated through 0.5% formalin and adsorbed by aluminium hydroxide;
- inactivation with formalin or BEA (bromethylamine) = EI (ethyl-imin), addition of aluminium hydroxide and saponin as adjuvants;
- inactivation through BEA and the addition of a mineral oil + emulsifier or modern oily adjuvants (MONTANIDE). A distinction has to be made between simple and double oily emulsions.

In the meantime, experimental vaccines have been produced from subunits of the virus especially the protein VP1. By analysing the genome of VP1 of the FMD virus type A12 the sequence of the nucleotides and amino acids which comprise this antigenic component have been identified. Through recombination some of these sequences have been replicated in *E. coli* and *B.subtilis* but also with the *Vaccinia* virus as carrier. With some of those recombined vaccines specific antibodies against FMD virus have been produced in cattle and guinea-pigs (Schudel and Sadir 1986).

The efficiency of the vaccines in tropical developing countries depends a lot on the technology of production. The question always remains whether the titre of virus is high enough and especially whether the antigen used for vaccine production is still concurrent with the epizootiological situation in the field. The producers are often hesitant about replacing a virus strain which has been well adapted to the culture in suspension and which produces good titres even though a new variant has been relevant in the field for some time (drifting). An important factor is also the chosen adjuvant. Oily adjuvants are suitable for enhancing the antigenicity. On the other hand, they may produce swellings and abscesses at the site of injection under the conditions of extensive animal production in the tropics, a fact which may become an obstacle to the acceptance of the vaccine in the field. Especially where quality beef is produced, the use of oily adjuvants is not favourable for getting the cooperation of the animal holder in schemes for FMD immunoprophylaxis.

Side reactions caused by the application of FMD vaccines will hinder the motivation of the animal holder in carrying out vaccination campaigns for the control of FMD in the tropics considerably. Sudden vaccination alone in a dairy herd will reduce the milk yield of the affected herd considerably and result in noticeable economic losses for the animal holder. Local reactions at the site of injection, and perhaps with the development of abscesses may perceptibly reduce the quality of the carcass. Abortions subsequent to vaccination may not only be the consequence of poor quality vaccine but also because of bad treatment of pregnant animals – a problem which is not uncommon on a ranch. In the past, allergic reactions of the fast and delayed type after FMD vaccinations were a serious problem. For the most part they were caused by the incomplete removal from the virus concentrate of animal tissues or cells and cell rests which were used for virus replication. Apparently when saponin is used as an adjuvant this mechanism is more enhanced. The allergic reactions of the fast type may appear already after a matter of minutes but still up to 8 hours after the application of the vaccine especially following a booster vaccination. The consequence is restlessness, sweating, tremor, colic, coughing, dyspnoea, salivation, angineurotic oedema, acute circulatory disorders, and local or generalized exanthema of the skin. The animals often die abruptly with acute dyspnoea expelling frothy foam from the nostrils only a few minutes after getting an injection. At necropsy, extensive lung oedema and haemorrhages at the serosae of the body cavities as well as on the heart are apparent. Allergies of the delayed type appear about 3 weeks after the vaccination as urticaria with blisters of different size which are spread over the whole body; they may weep and be of tough consistency. Here, extensive oedema may appear in the intermandibular space, and on the neck, brisket and belly. The oedema may break open and release amber-coloured liquid. Extensive crusts on the weeping oedema appear which may persist and recur again for weeks after initial healing. The animals which have serious symptoms of general malaise become emaciated and may often not recover completely. Occasionally, thanks to an oedema on the hoof-corium, the hoof sloughs off and panaritium may also occur. Furthermore, proliferative eczema around the nostrils, the mouth, on the neck and on the back, around the vulva, on the udder and on the inner aspects of the hind legs can appear. To prevent allergic side reactions, new batches of vaccines should always be tested on a few animals in the field and observed for reactions. This recommendation should be followed especially if a booster vaccination is applied in the tropics in order to enhance the immunoprotection.

Side reactions which are caused by vaccines which have not been inactivated appropriately cannot be completely excluded. After an incubation period of about a week, a mild reaction may appear which perhaps only affects the hooves mostly in young animals. The connection between vaccination and outbreak of disease is confirmed if only vaccinated animals show similar symptoms as consequence of the vaccination. Animals which have been vaccinated with incompletely inactivated vaccines may excrete the virus for weeks (cattle 90 days p.i., sheep up to 9 months p.i.). In Brazil, the live virus was found in 52% of slaughtered cattle and 26% of exported carcasses. It was not possible, however, to transmit the isolates to susceptible animals.

3.4.10. Ovine Pulmonary Adenomatosis (OPA)

(Ovine pulmonary carcinomatosis, Jaagsiekte, Lungenadenomatose des Schafes)

Aetiology and Occurrence
In contrast to the earlier suggestion that *Herpetoviridae* were the causal organisms of OPA, it is now been established that the disease is caused by a retrovirus after typical retrovirus particles have been detected in adenomatous lesions (Perk et al.

1974). Mature virions have an average diameter of 110 nm and, when negatively stained, appear as pleomorphic, enveloped particles with spikes on the surface, the morphology of OPA retrovirus (OPAV) being distinct from that of any other retrovirus. The viral genome consists of single stranded RNA with a sedimentation constant of 60–70 S. All attempts to cultivate the virus *in vitro* have failed so far, severely hampering its characterization and attempts to develop a vaccine (Verwoerd et al. 1994).

OPA was recognized a century ago in South Africa and today is to be considered as occurring worldwide with the exception of Australia. It has considerable economic impact in those countries which have a substantial sheep population, being an especially serious problem in the extensive sheep production systems of the High Andes (Seifert 1992, Verwoerd et al. 1994).

Epizootiology, Pathogenesis, Clinical and Pathological Features
Sheep of all breeds, but especially the highly productive breeds of the traditional tropical sheep breeding regions are susceptible to OPA. Goats are also susceptible, but less so. It is mostly young animals which become infected, the virus probably being introduced into a flock when affected sheep are brought into a clean flock. The virus is excreted by all sheep with lung lesions, however small they may be and inhaled by susceptible animals. In the USA, crowding is considered to be a factor favourable to the disease.

The incubation period is very long; it can last from 4 months up to several years. The inhaled virus penetrates the epithelial cells of the mucous membranes of the respiratory tract where it establishes disseminated foci of infection. The affected cells become transformed by the infection, proliferate and produce neoplastic areas which in the course of months or even years extend and may infiltrate the whole lung. This may also lead to the appearance of metastases in the regional lymph nodes and other internal organs. Because of the tumorous lesions in the lung, the function of the lung is seriously hampered, a development which is of fatal consequences for sheep which are kept on highland pastures. The disease causes serious losses in the sheep populations of the vast sheep ranches of the Cordillera in Central Perú.

The course of the disease is chronic but unstoppable. At first the animals show dyspnoea following physical effort, and finally continuous coughing, progressive emaciation, dyspnoea and nasal secretion appear. In the progressive stage of the disease, the raling respiration can already been heard from a distance. The animals drop their heads, and a mucous, watery secretion drips from the nostrils – a pathognomonic indication. Surprisingly the animals continue to feed and do not show a rise in the body temperature until a bacterial secondary infection leads to a fatal outcome in the final stage.

The pathological lesions are restricted to the thorax. The lung tissue is condensed and about 10 times as heavy as normal. Wherever tumorous lesions are present, the lung adheres to the pleura. The lung tissue is mostly collapsed and filled with liquid; abscesses and pleuritis may be present. The tumours may appear disseminated but also proliferated like nodules which makes the lung appear bacon-like. Histologically, the disease is characterized by a proliferation of the respiratory epithelium in the bronchi and the alveoli. The typical adenomes develop on the wall of the alveoli which becomes thickened and infiltrated with lymphocytes. Through coalescence of singular adenomes, the original structure of the lung is lost (Merck 1991, Rolle and Mayr 1993, Seifert 1992).

Diagnosis
Wherever the disease is enzootic, the typical clinical and pathological features of the disease allow a presumptive diagnosis. With Giemsa or Papanicolaou staining, clusters of macrophages and epithelial cells can be identified in nasal secretion as diagnostic characteristics. The final diagnosis can be established with the histopathological examination of the lung lesions (Seifert 1992). An ELISA test for detecting viral antigens and immune complexes in lung-rinse fluid has been developed, but its use is impracticable because of the problem of sample collection (Verwoerd et al. 1994).

In early literature, the disease was often confused with maedi, which is caused by a lentivirus and characterized by chronic, non-neoplastic inflammatory lesions (Montana progressive pneumonia) (Verwoerd et al. 1994).

Treatment
An efficient treatment of OPA is unknown.

Control
The most important measure of control is to prevent the introduction of animals from infected flocks into clean flocks. Because of the long incubation period quarantining is unfeasible. Culling of evidently sick animals does not lead to the eradication of the disease since infected animals cannot be detected. Iceland has become disease-free by rigourously culling all infected flocks of sheep. Wherever the disease is enzootic, the incidence of acute cases gradually may diminish which is supposed to be the consequence of a gradual build-up of immunity. Trials to prevent the disease through vaccination with vaccines prepared from formolized extractions of organs from infected animals in the USA so far have only brought inconclusive results (Merck 1991).

3.4.11. Maedi-Visna (MV)

(Ovine progressive pneumonia, Montana sheep disease, Zwoegersiekte, Graaff-Reinet disease, La bouhite)

Aetiology and Occurrence
MV is caused by the maedi-visna (MV) virus first isolated in Iceland and is the prototype of a large group of lentiviruses which belong to the family of *Retroviridae*. The morphogenesis and the morphology are distinct from those of OPA. The virions bud with a crescent-shaped core and have no intermediate space

between core and envelope. Mature virions are 80–100 nm in diameter, are membrane bound and have eccentric cores, sometimes surrounded by a clear intermediate layer (Verwoerd et al. 1994).

MV, one of the so-called "slow virus diseases" in sheep, was first described in South Africa and was confused with OPA for a long time. The virus has been isolated in all major sheep-breeding countries and is often closely related to OPA (Verwoerd et al. 1994).

Epizootiology, Pathogenesis, Clinical and Pathological Features

All breeds of sheep are susceptible to the MV virus. Differences in susceptibility between the breeds have been observed. Natural cases of MV in goats have also been reported in various countries. Droplet infection via the respiratory route under close contact is thought to be how the transmission occurs. Transmission is also possible through the colostrum to the lamb presumably by means of infected monocytes. Whether haematophagous insects are involved in the transmission or not is unclear. The virus persists latently, generally in the presence of high antibody titres, and infected animals are therefore carriers for life. Most infected sheep show little or no sign of disease, but remain carriers and can transmit the infection to others. No sex or age predilection has been found in the susceptibility of animals to infection (Verwoerd et al. 1994).

The MV virus replicates in the monocytes where it is contained by the immune response of the host which effectively eliminates the cell-free virus but is unable to destroy infective cells in which the virus genome is latent, the virus however being spread inside the cells to all tissues of the body. It has been shown that similar to the AIDS virus the MV virus can induce mild immunosuppression by affecting macrophage activity thus predisposing infected animals to secondary infections (Verwoerd et al. 1994).

Typical clinical signs of MV, rarely seen in young animals, are dyspnoea, a dry cough and occasionally nasal discharge. In the absence of secondary infections, there is no fever, and the animal eventually dies from weakness and hypoxia 6–12 months after the appearance of clinical signs, pulmonia caused by *Pasteurella* spp. often being the final cause of the fatal outcome. Occasionally, nervous signs like an unsteady gait, stumbling and falling, trembling facial muscles and an inability to extend the fetlock may occur. Gradual progression leads to paresis and eventually total paralysis accompanied by loss of condition. There is usually no loss of appetite. There is no fever and the animal remains alert throughout the course of the disease. The clinical course may last up to a year, but there is no recovery once clinical signs manifest themselves (Verwoerd et al. 1994).

The basic pathological lesions of MV are hyperplasia of lymphoid tissue which may develop into lymphoid nodules, and primary degenerative changes in the brain, joints and walls of blood vessels. At necropsy, in advanced cases of MV, the lungs are found to be enlarged, maybe 2–4 times their normal weight, they have a firm rubbery consistency, lack normal crepitations, and are greyish in colour, the lymph nodes being grossly enlarged. Secondary bacterial pneumonia may be evident. The characteristic histopathological feature in the lungs is a diffuse, often pronounced thickening of the alveolar septa caused by infiltration of mono-nuclear cells and hyperplasia of smooth muscle cells, partly obliterating the alveoli. Cerebral lesions are characterized by chronic demyelinating encepha-lomyelitis. The lesions which are found on the joints are hyperplasia and necrosis of the synovial membrane, necrosis and erosions of articular cartilage, necrosis and fibrosis of subchondral bone, and extensive periarticular fibrosis. The mastitis which some infected sheep develop is characterized by interstitial accumulation of lymphocytes, the formation of periductal lymphoid nodules, and the necrosis at the site of these nodules (Verwoerd 1994).

Diagnosis
A presumptive diagnosis can be based on clinical signs but has to be confirmed by histopathological examination. Virus isolation can be achieved by cultivation of macrophages from peripheral blood or milk, most ovine cell cultures being suit-able for this purpose. The virus can be isolated *post mortem* by explanting affected tissues in which cytopathic effects develop slowly and include the appearance of refractile stellate cells and syncytia. The presence of the virus can be detected by reversed transcriptase assay, electron microscopy or immunolabelling techniques. The demonstration of virus antibodies of the MV virus can be accomplished with virus neutralization, CFT, immunofluorescence, AGID and ELISA (Verwoerd 1994).

Treatment
Since MV invariably has a fatal outcome, affected animals should be slaughtered.

Control
No vaccine for the immunoprophylaxis of MV is available, eradication being the only viable option. This has been achieved on a national basis in Iceland by the implementation of a drastic slaughter-out policy. MV can be eradicated from a flock, or its incidence reduced to an insignificant level by test and slaughter (Ver-woerd 1994).

3.4.12. Vesicular Stomatitis (VS)

(Stomatitis vesicularis, Sore mouth of cattle and horses)

Aetiology and Occurrence
VS is caused by a *Rhabdovirus* which occurs in 4 serotypes (Indiana, Cocal, Alagoas and New Jersey). 3 subtypes are known within the Indiana type. The types and subtypes are serologically distinct but possess common antigenic com-ponents. The virus is bullet-shaped with a length of 178–188 nm and a diameter of 70 nm, it grows in a large variety of tissue cultures and can be multiplied in embryonated hens' eggs. Of the small laboratory animals guinea-pigs, mice, ham-ster, chinchillas and ferrets are susceptible; CNS symptoms appear in these animals following intercerebral infection.

The VS virus is not very resistant to environmental influences and is destroyed by the usual disinfectants even in low concentrations; it is highly sensitive to acids.

VS usually occurs epizootically in temperate regions but only enzootically in warmer climates on the American continent. Under the conditions of intensive cattle production systems (feed-, dairy-lot) it may cause serious economical losses. In South America, VS is of special importance since the lesions which appear in cattle are often taken for FMD.

Epizootiology, Pathogenesis, Clinical and Pathological Features
Equines, cattle and pigs are susceptible under natural conditions. The transmission of VS in South America occurs through direct or indirect contact whereby refuse from kitchen and meat products are supposed to be important vehicles. In the USA, it is supposed that the Indiana serotype is transmitted by phlebotomine sand flies, an observation which is confirmed by the increased incidence of VS during the rainy season. Wild boar, raccoons, deer and antelopes are thought to be reservoirs of infection (Merck 1991, Rolle and Mayr 1993).

VS can spread rapidly, often affecting many animals in one week. A persistent phase of the virus also is likely. The primary route of infection is unknown but may be through the skin or respiratory and alimentary tracts. Generalized lesions occur, but viraemia is unusual. Diseased animals excrete the virus with material from the vesicles and saliva.

The incubation period is 2–8 days, or possibly longer. After as little as 24 hours, fever and inappetence become apparent. Frequently excessive salivation is the first sign, and blanched, raised vesicles in variable size appear, some not larger than a pea, while others may involve the entire surface of the tongue. In horses, the lesions are principally confined to the upper surface of the tongue but may involve the inner surface of the lips, corners of the mouth, and the gums. In cattle, the lesions may also occur on the hard palate, lips and gums and sometimes extend to the muzzle and around the nostrils. Secondary lesions involving the feet of horses and cattle are not exceptional, and cattle in particular in intensive production systems may also show lesions at the teats. In natural infections in pigs, foot lesions are frequent, and lameness is often the first sign observed. Ordinarily there are no complications and the disease is self-limiting, with recovery in about 2 weeks. In dairy herds and feed-lots, loss of production may be serious, and mastitis may be a sequela in dairy cows. Wherever horses are kept together with cattle, horses show symptoms prior to, or at the same time as the cattle. Serum antibodies persist for life, but recrudescence of the disease or reinfection may occur (Merck 1991, Seifert 1992).

Diagnosis
VS, while economically important, is of particular significance because of its similarity to FMD, vesicular exanthema, and swine vesicular disease. VS is easily confused with FMD especially in South America where FMD is enzootic and often shows a benign course. When VS affects horses under natural conditions, there is no diagnostic problem because horses are not susceptible to FMD. This

fact can also be taken advantage of if the disease appears at first only in cattle. It is enough to take a swab of saliva from salivating cattle and infect with it the oral cavity of an equine. Within 24–48 hours, the typical symptoms of salivation and mouth lesions will appear in the equine if VS virus is present. Diagnosis can also be made on the distribution and character of the lesions, and the disease may be differentiated from horse pox by the absence of papules and pustules. In cattle and pigs, diagnosis can be confirmed by CFT and ELISA using a suspension of the epithelial lesions as an antigen. If this is negative, the material is passaged in mice, embryonating eggs, or tissue cultures, and then CFT, ELISA and virus neutralization test are performed. Antibodies in the sera of recovered animals can also be detected by these tests. Electron microscopy may be performed on the original sample or passaged material. In sera of convalescent animals specific antibodies can be detected with ELISA as soon as 4–5 days p.i. (Merck 1991, O.I.E. 1990, Seifert 1992).

Treatment
In intensive production systems, the course of disease can be shortened when mild disinfectants or tetracyclines are added to the drinking water.

Control
Horses and cattle which recover from the disease are immunized at least for one year. In South America, regionally type-specific aluminium-adsorbed vaccines are available which are now produced on tissue cultures.

References:

Adlam, C. and Rutter, J. M. (1989): Pasteurella and Pasteurellosis. Academic Press, London.
Barnard, B. J. H., J. J. van der Lugt and E. Z. Mushi (1994): Malignant catarrhal fever. In: Infectious diseases of livestock (eds: Coetzer, Thomson and Tustin), 2, 946–957, Oxford University Press, Cape Town/R.S.A.
Barnard, B. J. H., E. Munz, K. Dumbell and L. Prozesky (1994): Lumpy skin disease. In: Infectious diseases of livestock (eds: Coetzer, Thomson and Tustin), 1, 605–612, Oxford University Press, Cape Town/R.S.A.
Bastianello, S. S. and J. W. Nesbit (1994): Haemorrhagic septicaemia. In: Infectious diseases of livestock (eds: Coetzer, Thomson and Tustin), 2, 1180–1183, Oxford University Press, Cape Town/R.S.A.
Bergey, D. H. (1984): Manual of Systematic Bacteriology, 1, Williams & Wilkins, Baltimore/London.
Bishop, G. C., P. P. Bosman and S. Herr (1994): Bovine brucellosis. In: Infectious diseases of livestock (eds: Coetzer, Thomson and Tustin), 2, 1053–1066, Oxford University Press, Cape Town/R.S.A.
Bötcher, L. (1988): Zur Pathogenese und Immunprophylaxe von Pasteurella-haemolytica-Infektionen. Berl. Münch. tierärztl. Wschr. **101**, 221–228.
Diallo, A, W. P. Taylor, P. C. Lefèvre and A. Provost (1989): Atténuation d'une souche de virus de la peste des petits ruminants: candidat pour un vaccin homologue vivant. Rev. Élev. Méd. vét. Pays trop. **42**, 311–319.
Domenech, J., Ph. Lucet, B. Vallat, Ch. Stewart, J. B. Bonnet and A. Hentic (1982): La brucellose bovine en Afrique centrale. III. – Résultats statistiques des enquêtes menées au Chad et au Cameroun. Rev. Élev. Méd. vét. Pays trop. **35**, 15–22.

El Bashir, S. (1993): Entwicklung einer Methode zur Herstellung einer hochgereinigten Pasteurella multocida-Vakzine mit Hilfe des Göttinger kontinuierlichen Bioreaktor-Verfahrens. Göttinger Beitr. Land. Forstwirt. Trop. Subtrop. 88.

El Hag Ali, B. and W. P. Taylor (1988): An investigation on rinderpest virus transmission and maintenance by sheep, goats and cattle. Bull. Anim. Hlth. Prod. Afr. **36**, 290–294.

Guillemin, F., M. Mosienyane, T. Richard and M. Mannathoko (1987): Immune response and challenge of cattle vaccinated simultaneously against rinderpest and foot-and-mouth disease. Rev. Élev. Méd. vét. Pays trop. **3**, 225–229.

Häusser, V. (1989): Methodische Untersuchungen zur kontinuierlichen Kultivierung von Mycoplasma mycoides ssp. mycoides im Göttinger Bioreaktor. Göttinger Beitr. Land. Forstwirt. Trop. Subtrop. 45.

Henning, M. W. (1956): Animal Diseases in South Africa. Central News Agency, S.A.

Huchzermeyer, H. F. K. A., G. K. Brückner and S. S. Bastianello (1994): Paratuberculosis. In: Infectious diseases of livestock (eds: Coetzer, Thomson and Tustin), 2, 1445–1457, Oxford University Press, Cape Town/R.S.A.

Huchzermeyer, H. F. K. A., G. K. Brückner, A. Van Heerden, H. H. Kleeberg, I. B. J. van Rensburg, P. Koen and R. K. Loveday (1994): Tuberculosis. In: Infectious diseases of livestock (eds: Coetzer, Thomson and Tustin), 2, 1425–1441, Oxford University Press, Cape Town/R.S.A.

Hunter, P. and S. Herr (1994): Leptospirosis. In: Infectious diseases of livestock (eds: Coetzer, Thomson and Tustin), 2, 998–1008, Oxford University Press, Cape Town/R.S.A.

Kaaden, O.-R., A. Walz, C.-P. Cerni and U. Wernery (1992): Progress in the development of a camelpox vaccine. Proc. 1st int. Camel Conf., Dubai, 47–49, R & W Publ. Newmarket, England.

Kitching, R. P., N. J. Knowles, A. R. Samuel and A. I. Donaldson (1989): Development of foot-and-mouth disease virus strain characterisation – a review. Trop. Anim. Hlth. Prod. **21**, 153–166.

Lefèvre, P. C. (1991): Une maladie en pleine expansion: la peste des petits ruminants. Rev. Mond. Zootec. **66**, 55–58.

MacOwan, K. J. and J. E. Minette (1976): A mycoplasma from acute contagious caprine pleuropneumonia in Kenya. Trop. Anim. Hlth. Prod. **8**, 91–95.

Mahnel, H. (1974): Labordifferenzierung der Orthopockenviren. Zbl. Vet. Med. **21**, 242–258.

Makinde, A. A. and A. O. Ezeh (1981): Primary and secondary humoral immune responses in cattle experimentally infected with Dermatophilus congolensis. Bull. Anim. Hlth. Prod. Afr. **29**, 19–23.

Maré, C. J. (1994): Contagious caprine pleuropneumonia. In: Infectious diseases of livestock (eds: Coetzer, Thomson and Tustin), 2, 1495–1497, Oxford University Press, Cape Town/R.S.A.

Mariner, J. C., J. A. House, A. E. Sollod, C. Stem, M. van den Ende and C. A. Mebus (1990): Comparison of the effect of various chemical stabilizers and lyophilization cycles on the thermostability of a vero cell-adapted rinderpest vaccine. Cited from Mariner et al. 1990: The serological response to a thermostable vero cell-adapted rinderpest vaccine under field conditions in Niger. Vet. Microbiol. **22**, 119–127.

Merck and Co. (1991): The Merck Veterinary Manual, 7th ed., Merck, Rahway, N.J., USA.

Mitscherlich, E. and K. Wagener (1970): Tropische Tierseuchen und ihre Bekämpfung. II. Aufl., Parey, Berlin u. Hamburg.

Munz, E. (1992): Pox and pox-like diseases in camels. Proc. 1st int. Camel Conf., Dubai 43–46, R & W Publ. Newmarket, England.

Munz, E. and K. Dumbell (1994): Orf. In: Infectious diseases of livestock (eds: Coetzer, Thomson and Tustin), 1, 616–620, Oxford University Press, Cape Town/R.S.A.

Munz, E. and K. Dumbell (1994): Pseudocowpox. In: Infectious diseases of livestock (eds: Coetzer, Thomson and Tustin), 1, 625–626, Oxford University Press, Cape Town/R.S.A.

Mutters, R., W. Mannheim and M. Bisgaard (1989): Taxonomy of the Group. In: Pasteurella and Pasteurellosis, 3–34, Acad. Press, London.

Nesbit, J. W. and S. R. van Amstel (1994): Pneumonic pasteurellosis in cattle. In: Infectious diseases of livestock (eds: Coetzer, Thomson and Tustin), 2, 1169–1179, Oxford University Press, Cape Town/R.S.A.

O.I.E. (1990): Recommended diagnostic techniques and requirements for biological products, II, 11, O.I.E. Paris.

Perk, K., R. Michalides, S. Spiegelman and J. Schlom (1974): Biochemical and morphological evidence of the presence of an RND-tumour virus in pulmonary carcinoma of sheep (jaagsiekte). J. Nat. Canc. Inst. **53**, 131–135.

Plowright, W. and R. D. Ferris (1962): Studies with rinderpest virus in tissue culture. Res. vet. Sci. **3**, 172–182.

Plowright, W., G. R. Thomson and J. A. Neser (1994): African swine fever. In: Infectious diseases of livestock (eds: Coetzer, Thomson and Tustin), 1, 568–599, Oxford University Press, Cape Town/R.S.A.

Provost, A. (1981): Rinderpest and C.B.P.P. A new approach to strategic and tactical problems. O.I.E., Paris, Rev. Sc. Tec. **3**, 589–678.

Rolle, M. and A. Mayr (1993): Mikrobiologie, Infektions- und Seuchenlehre. 6. Aufl., Enke, Stuttgart.

Rossiter, P. B. (1994): Rinderpest. In: Infectious diseases of livestock (eds: Coetzer, Thomson and Tustin), 2, 735–757, Oxford University Press, Cape Town/R.S.A.

Rossiter, P. B. and H. M. Wamwayi (1989): Surveillance and monitoring programmes in the control of rinderpest: A review. Trop. Anim. Hlth. Prod. **21**, 89–99.

Rossiter, P. B. and W. P. Taylor (1994): Peste des petits ruminants. In: Infectious diseases of livestock (eds: Coetzer, Thomson and Tustin), 2, 758–765, Oxford University Press, Cape Town/R.S.A.

Rotermund, H. (1980): LTF-130 – eine wirksame Vakzine gegen Rindertrichophytie. Mh. Vet. Med. **35**, 334–335.

Schneider, D. J. (1994): Listeriosis. In: Infectious diseases of livestock (eds: Coetzer, Thomson and Tustin), 2, 1374–1377, Oxford University Press, Cape Town/R.S.A.

Schneider, H. P., J. J. Van der Lugt and O. J. B. Hübschle (1994): Contagious bovine pleuropneumonia. In: Infectious diseases of livestock (eds: Coetzer, Thomson and Tustin), 2, 1485–1494, Oxford University Press, Cape Town/R.S.A.

Schudel, A. A. and A. M. Sadir (1986): Fiebre aftosa. Adel. Microbiol. Enf. Infecc. **5**, 22–36.

Scott, D. B. (1994): Mycoses. In: Infectious diseases of livestock (eds: Coetzer, Thomson and Tustin), 2, 1521–1533, Oxford University Press, Cape Town/R.S.A.

Seifert, H. S. H. (1990): Humanhygienische und wirtschaftliche Bedeutung der Brucellose in der Tierhaltung des Sahel. Göttinger Beitr. Land. Forstwirt. Trop. Subtrop. **51**, 187–192.

Seifert, H. S. H. (1992): Tropentierhygiene. Fischer, Jena.

Seifert, H. S. H. (1995): unpublished empirical findings.

Sharma, B., B. S. Negi, A. B. Pandey, S. K. Bandyopadhyay, H. Shankar and M. P. Yadav (1988): Detection of goat pox antigen and antibody by the counter immunoelectrophoresis test. Trop. Anim. Hlth. Prod. **20**, 109–113.

Spencer, B. T. (1994): Erysipelothrix rhusiopathiae infections. In: Infectious diseases of livestock (eds: Coetzer, Thomson and Tustin), 2, 1378–1383, Oxford University Press, Cape Town/R.S.A.

Taylor, W. P., S. Al Busaidy and T. Barrett (1990): The epidemiology of peste des petits ruminants in the Sultanate of Oman. Vet. Microbiol. **22**, 341–352.

Thomson, G. R. (1994): Foot-and-mouth disease. In: Infectious diseases of livestock (eds: Coetzer, Thomson and Tustin), 2, 825–852, Oxford University Press, Cape Town/R.S.A.

Van Tonder, E. M. and R. F. Horner (1994): Dermatophilosis. In: Infectious diseases of livestock (eds: Coetzer, Thomson and Tustin), 2, 1472–1481, Oxford University Press, Cape Town/R.S.A.

Venter, B. J., J. G. Myburgh and M. L. van der Walt (1994): Bovine salmonellosis. In: Infectious diseases of livestock (eds: Coetzer, Thomson and Tustin), 2, 1100–1112, Oxford University Press, Cape Town/R.S.A.

Verwoerd, D. W., R. C. Tustin and A.-L. Williamson (1994): Maedi-visna. In: Infectious diseases of livestock (eds: Coetzer, Thomson and Tustin), 2, 792–796, Oxford University Press, Cape Town/R.S.A.

Verwoerd, D. W., R. C. Tustin and A.-L. Williamson (1994): Jaagsiekte. In: Infectious diseases of livestock (eds: Coetzer, Thomson and Tustin), 2, 783–792, Oxford University Press, Cape Town/R.S.A.

Wamwayi, H. M., P. B. Rossiter and J. S. Wafula (1991): Confirmation of rinderpest in experimentally and naturally infected cattle using microtitre techniques. Trop. Anim. Hlth. Prod. **23**, 17–21.

Weidmann, H. (1991): Survey of means now available for combatting brucellosis in cattle production in the tropics. Anim. Res. Dev. **33**.

Weiser, A. (1995): Analyse der Tierhygienesituation in mobilen pastoralen Tierhaltungssystemen in der Butana/Ost-Sudan. Göttinger Beitr. Land. Forstwirt. Trop. Subtrop. **101**.

Wernery, U. and O. R. Kaaden (1994): Infectious diseases of camelids, 52–56, Blackwell Wiss. Berlin.

Wilson, O. B. and S. F. Amakiri (1989): Chemotherapy of Dermatophilosis in a herd of German Brown-N'Dama crosses and White Fulani cattle. Bull. Anim. Hlth. Prod. Afr. **37**, 315–317.

Wittmann, G. (1990): Virusträger bei Maul- und Klauenseuche. Berl. Münch. Tierärztl. Wschr. **103**, 145–150.

Yilma, T., D. Hsu, L. Jones, S. Owens, M. Grubman, C. Mebus, M. Yamanaka and B. Dale (1988): Protection of cattle against rinderpest with *Vaccinia* virus recombinants expressing the HA or F gene. Sc. **242**, 1058–1061.

4. Plant Poisoning

4.1 Introduction

Poisonous plants can cause enzootic diseases with serious consequences in cattle and small ruminants in the extensive animal production systems of the tropics. Wherever the intensification, the increase of the stocking rate and the introduction of exotic highly productive breeds have led to management systems which restrict free grazing of the animals or have led to a loss of the innate instinct of the animals to avoid poisonous plants, plant poisoning will occur. At the same time climatic influences, especially a long-lasting dry season may increase the incidence of plant poisoning. Because of their empirical experiences, traditional animal holders have been aware of the connection between losses in grazing animals and the appearance of poisonous plants on the pasture for a long time. During the last 200 years, the incidence of plant poisoning in ruminants on tropical pastures has increased in proportion to the development of the semi-arid pasture ranges by European settlers in the Americas, Australia and South Africa. Last but not least, the uncritical transfer of experience in animal management acquired in Europe to the tropical environment has led to an increase of losses through plant poisoning.

The fact that a potential fodder plant is poisonous does not necessarily imply that it will cause poisoning when ingested. There are ecological, socio-economical and technical circumstances within the production system which eventually force the animal to feed on a poisonous plant during a season of the year when its toxicity is at its peak. Therefore, for the following presentation it is not the systematology of causal plants which has been chosen as a guide. Instead the mechanics which lead to plant poisoning are described, e.g. the toxicological-physiological processes. Based on these, the syndromes are characterized which eventually are produced by particular or groups of poisonous plants. The influence of the production system on the development of plant poisoning also has been taken into account.

True poisonous plants only are those which contain poisonous substance in an intolerable amount under normal conditions. In the USA, about 30000 species of plants are known, but only 700 of those are supposed to be poisonous. It is impossible to systematically classify families of poisonous plants. In contrast, most of the poisonous plants are found between *Leguminosae*, *Solanaceae* and *Gramineae*, to which most of the plants produced for human consumption also belong. It is because of the conditions which induce plants to become poisonous for grazing animals that the real number of poisonous plants is considerably reduced. On the other hand, plants which are primarily unsuspicious may become poisonous under certain meteorological conditions: after excessive fertilization, after the application of herbicides and pesticides, after being affected by pests, or through inappropriate storage and conservation or because of emissions from industrial plants.

Even if the poisonous constituents of the plant are metabolized by the poisoned animal to non-poisonous substances during the process of plant poisoning, the carcass of the slaughtered animal is unfit for human consumption. Because of fast autolytic processes of the tissues putrefaction is imminent and the meat becomes inedible. It has also to be considered that plant poisons and their metabolites are excreted with the milk (aflatoxin).

4.1.1. Definition and Terminology

The definition given by Paracelsus (1493–1541) still holds true: All things are poison and nothing is without poison; it is only the dosage which makes a thing not a poison (*Dosis solum facit venum*). Last but not least it is true for fodder plants that it is not the property of the plant but the amount in which and the way it is ingested by the organism and the condition in which it remains there which cause its toxicity.

During the following presentation by using the precise term **toxin** – produced by bacteria, fungi and parasites – and **poison** – the latter being of plant or animal origin – the origin of the poisonous principle at or in the plant is made clear.

Poisoning is a disease which is caused after the ingestion of an innate substance in a harmful amount. Interrelations between a poison and an organism determine the course of the disease. Resistance and condition of the organism on the one hand as well as dosage and intensity of the poison on the other hand, site and time of affection, conditions of reabsorption and excretion are determining factors of these mechanisms. The intensity of the mechanism of poisoning is identical to the size of the biochemical reaction between poison and organism. Since these mechanisms are interrelated, it is not only the poison which affects the organism but also the organism which acts upon the poison.

4.1.2. The Influence of the Poison

The chemical and physical structure of the poison determine the mechanism of poisoning. The solubility of the poison in water or in lipids is a prerequisite for the enteral, pulmonary or cutaneous reabsorption. A high chemical reactivity and the ability to be adsorbed by other substances are further factors which determine the action of the poison. Further to the dosage, it is the concentration of the poison at its site as well as its time of action which influence the course of the poisoning. To a certain degree (critical dose), the organism is able to metabolize the poison without presenting clinical symptoms. This depends mostly on whether the poison has been ingested once, sporadically or continuously. Poisoning can also appear if the continuously ingested critical dose has accumulated (heavy metals, chlorinated hydrocarbons). Not only may the poison itself accumulate, but it may also accumulate its action in such a way that after repeated short-time contact with the poison, lesions appear which in total lead to a definite adverse affection of the organism. A typical example is fern poisoning. Each time after the plant is ingested, a subclinical inhibition of prothrombin is caused. Only when the supply of prothrombin in the organism is exhausted do clinical symptoms appear. In contrast, poisons which do not accumulate and are eliminated quickly can be ingested over a long time if the ingested amount remains below the critical dose, e.g. subclinical doses of cyanide can be supported by well-fed animals.

With plant poisoning, the relationship between plant and the poison which it contains or which is produced by it also has to be taken into account. Generally, plant poisons have no nutritive value, but they are useful to the plant because they

- protect the plant against herbivores and insects;
- protect plants against other plants through the inhibition of the growth of rival plants;
- are part of its metabolism;
- are residues of its metabolism.

Environmental factors, soil and climate determine the type of vegetation. The local climate depends not only on the precipitation but also the seasonal distribution of rainfall and evaporation. Furthermore, the altitude and topography as well as the type of soil influence the use of the vegetation. In the semi-arid pasture ranges in the tropics, the end of the dry season is the time of the year during which plant poisoning becomes acute. Beside poisonous evergreen bushes and trees, other deep rooted plants also often remain with green leaves for a long time or get new leaves earlier. Wherever the pastures are overgrazed, these plants are eaten by hungry and thirsty animals. Daily climatic differences may also influence the poisonous potential of plants. During hot, humid weather, cyanogenic plants contain more cyanogenic glycosides than when cooler temperatures prevail. On the other hand, cyanogenic glycosides are produced by the plant if it withers abruptly after the sudden onset of a dry period. Nematodes, fungi and bacteria produce toxins in plants predominantly during hot, humid weather.

The type and the quality of the soil are important to the poisonous potential of some fodder plants. Fodder plants growing on soils, rich in copper and selenium, can accumulate these heavy metals. With an increased nitrogen content and after the sudden onset of a dry spell, an increased content of alkaloids may appear in the plants. An imbalance of the soil nutrients phosphorus, sulphur and molybdenum in relation to nitrogen will influence the toxicity of nitrogen-accumulating plants which under those conditions are unable to transform nitrate to ammonia and other organic nitrogen-containing components of the plant.

The topography of the pasture may indirectly become part of the appearance of plant poisoning. On mountainous pastures, cattle may be unable to graze everywhere, thus the animals may eat poisonous plants when crowded together on flat parts which are close to the water place.

4.1.3. Resistance and Susceptibility of the Animal

Under the rather heterogeneous conditions of the extensive animal production systems in the tropics, innate properties of the animals which determine the course of poisoning are especially important. Individual and breed-specific variations of the reaction to poisoning and genetically determined resistance and instinct, as well as the capacity to get accustomed to the poison, along with the condition, health status (latent infection), weight, age, sex and productivity, and perhaps also state of gestation, will determine the course of poisoning. Thus some animals of a population may be more sensitive than others. Abnormal reactions (idiosyncrasy) of some individuals to the poison may occur in one or another way. The individual and genetically determined sensitivity to poisoning probably depends mainly on the variation of the enzyme pattern of the organism.

Generally but not necessarily, the poisoning depends on the relationship dose:body weight of the organism. Under certain conditions, however, well-fed animals may be more sensitive because they reabsorb poisons soluble in lipids more quickly and/or can accumulate these in their body fat. In general, new-born and young animals are the most sensitive to poisoning because the enteral reabsorption of the poison takes place more quickly and more completely while the metabolizing enzymes of the liver (transaminases and hydrolases) are not yet fully functional. Neither do young animals possess the experience to evade poisonous plants, an experience which they will eventually learn from their mother, nor have they got accustomed to the poisonous principle. With progressing age, the resistance increases, which, however, decreases again in older animals which is possibly a consequence of the reduced adaptability of the organism to the mechanism of poisoning.

Numerous poisonous actions are specific to gender and especially persistent and economically important in females. This is true mainly for substances with teratogenic action and those which cause abortion.

Condition and productivity influence the course of poisoning since well-fed animals accumulate the poison in the fatty tissues and are able to eliminate it *ad hoc* this way. In the long term, however, they are exposed to a cumulative poisoning because they ingest more feed and thus more poison. Animals in bad condition as well as those of high productivity are more exposed since they have only a limited detoxifying capacity of the liver and capability of elimination through the kidneys. A slow peristalsis of the intestine enhances the reabsorption of the poison and hyperperistalsis reduces the reabsorption since it increases the passages of the poison through the intestine.

Finally, it can also be the hierarchic position in the herd which determines how much of the particular poisonous plant is ingested by the animal. Animals with a high social position have the first choice, while those on a lower level have to be satisfied with the remains. Thus they may be forced to graze on remaining poisonous plants.

The genetic disposition of a species and breed to plant poisoning depends on the reaction of the ingested poison with the enzyme systems of the organism. Such polymorphisms influence the stability of the cell membrane and its permeability. Naturally, the principle differences of the digestive systems of monogastrics and ruminants are important to their sensitivity for poisoning.

Depending on their resistance, individuals and breeds are able to adapt to the continuous intake of plant poisons. With ruminants, the flora of the rumen is of special importance since it can break down poisonous substances (mimosine from *Leucaena*). Because of the particular importance of this mechanism for the management of ruminants in the tropics, it is referred to in a special chapter (II/4.2.3).

An important species-specific factor which influences the appearance of plant poisoning is the grazing behaviour of the animal. There are principal differences between cattle, sheep and goats. While cattle mainly feed on grasses, sheep and especially goats also graze on leaves from herbs and browse on bushes and trees. Cattle will feed on herbs and browse on ligneous plants only when grass and water become scarce. As a consequence of the selective pressure, last but not least due to the availability of certain fodder plants, the physiology and morphology of cattle, sheep and goats have adapted differently during evolution. The adaptation to a particular feed determines the feeding behaviour of each species. The physiological and morphological differences involved are of rather complicated nature.

It is also specific the way different species compose their diet dependent upon the availability of feed and water. Furthermore, environmental conditions are important in determining the choices of the animal for selection and the attractiveness of the fodder plant. Dependent upon instinct, the animal eventually will ingest replacements uncritically even when essential feed stuffs become scarce. Ruminants graze selectively as long as they are allowed to; they prefer particular stages of growth or parts of a plant species. The selective behaviour is typical for the species, the breed and the individual. Autochthonous breeds in the tropics are much more selective than imported exotic breeds. South American Criollo cattle will refuse poisonous plants even under extreme conditions while imported Hereford cattle and especially Holstein-Friesian on the same pasture eat poisonous plants at the onset of the dry season indiscriminately (Seifert 1995). The pattern of selection is in no way constant, it changes both during the day and from day to day as well as during the year. Older animals are less selective than young animals and hunger reduces the selectivity. Feed selection combines basic nutrients and specific nutrients into an optimal diet for the animal. It is so far still unknown why a particular plant is palatable or attractive to a particular animal. The term "feed wisdom" ("Nahrungsweisheit" in German) describes the ability of many animals to refuse or avoid an unpalatable or poisonous plant. Depending on the breed, cattle are able to test whether a fodder plant is palatable and tolerable or not. The offspring of autochthonous cows sired using A.I. with sperm from highly productive exotic breeds were much more affected by plant poisoning than calves sired by local bulls (Seifert 1995). It is also the adaptation to the locality which is important in this connection. Autochthonous cattle which have a high selective capacity where they are born become sensitive to plant poisoning if they are brought into a region which is ecologically different. In South Africa, many kinds of tulips (*Moraea* and *Homeria* spp.) occur which cause poisoning in cattle, sheep, goats and occasionally in donkeys. The poisonous principle of the tulips is a strong stomach and intestinal irritant causing inflammation of these organs. Animals accustomed to graze on pastures on which tulips grow are seldom poisoned, apparently because they come to know the plant. Animals not accustomed to such pasture, or very hungry animals will eat large quantities and get poisoned. In order to prevent poisoning in newly acquired animals, the farmer assists the "feed wisdom" of imported cattle by orally applying an extract of the bulbs of tulips repeatedly to exposed animals wherever tulip-poisoning is enzootic. The animals will not forget the awful taste of the tulip extract and in the future refuse to feed on the plant (Mönnig and Veldman 1956).

In contrast to the usual pattern of selection, only exceptionally poisonous species are preferred (*Astragalus* spp., *Cestrum* spp.). It may also occur that animals particularly like to eat the plants when they are at a poisonous stage (*Sorghum sudanense*).

Principally the pattern of feed selection depends on the normal hunger-full-cycle whereby, beside an appetite for specific food stuffs, environmental influences and those of management may become relevant. Management of the animal can adversely affect its selective pattern, e.g. with pasture rotation, supplementary feeding and restriction of water provision.

The character of the ecosystem pasture is determined by the interactions between animal and plant. Two of these interactions are relevant: between animal and

– ingested feed, and
– its provision with fodder.

The amount and quality of ingested fodder determine the behaviour of the animal during feed intake. They are decisive factors regarding the further search for fodder but also perhaps for avoiding fodder plants and at the same time determining the degree of satiety. The feed intake as a sequence of continuously and repeatedly ingested food stuffs is repeated daily. The components of the diet produce a feed mosaic which determines the amount of the intestinal content, the activity of the intestine and of those organs which function in connection with it and finally the digestion of the feed, the reabsorption of the digestive products as well as the composition and biochemical capacity of the microorganisms of the intestinal tract. The feed mosaic can cause numerous alimentary diseases. With plant poisoning, the feed mosaic is aetiologically involved. Beside the absolute amount of feed which has been ingested, it also controls the reabsorption rate of the poison. How quickly and in which concentration the poison acts on the susceptible organs of the organism depends on the reabsorption rate of the poison. It makes a big difference whether the poison is ingested continuously and in low concentration or if it has been ingested in a short space of time in a high concentration. The feed mosaic also influences the course of the poisoning through the amount of feed in the intestine and its digestive rate as well as by diluting the poisonous constituents of the plant. Through the intake of water, the excretion of harmful substances via the kidneys is enhanced and at the same time together with the other components of the fodder, the chemical structure of the content of the intestine is regulated. Particular components in the feed mosaic can reduce the poisonous action, e.g. calcium with oxalate or iron with gossypol. The mechanism of detoxication of rumen flora in ruminants is determined through the feed mosaic (II/4.2.3). Hunger, a diet with low content of crude fibre and sudden change of feed may have disastrous consequences for the rumen flora. The microorganisms of the rumen can develop their capacity for detoxication only if they are able to get adapted gradually to the ingested poison; at the same time, their tolerance is enhanced continuously. The rumen flora is unable to adapt to poisons which are quickly metabolized by the organism (cyanide).

The interaction between animal and its provision with feed are part of the overall relationship between animal and environment. In the context of the ecosystem, they have to be understood as part of the cycle soil-plant-animal, taking into account the influence of different pasture managements, as well as the influence of the stocking rate on the growth of the plants and the structure of the soil. The interactions between animal and its feed are of short as well as of long term consequences regarding its productivity and the development of problems. In the short term, the combination of a high stocking rate and lack of fodder leads to a reduction of productivity and eventually an increased intake of poisonous plants. In the long term, the composition of the plant cover will change, and with it, a higher incidence of non-palatable plants and also of those which are poisonous.

4.1.4. The Importance of the Digestive System of the Ruminant to the Course of Plant Poisoning

Ruminants by means of their rumen flora are able to break down ingested poisons but can also synthesize poisonous substances from plant constituents. On the other hand poisonous principles of plants may damage the ruminal flora.

With its rumen flora, the ruminant is able to detoxify various substances, for instance oxalates, pyrrolizidine-alkaloids, etheric oils, cyanides, selenium and mimosine.

Plants which contain **oxalic acid** occur worldwide. They are mostly rather palatable and the ruminants will eat them readily. Those plants which contain over 10% oxalic acid in dry matter are poisonous. This is so for some species of the family *Polygonaceae* (*Rumex* spp.), *Chenopodiaceae* and *Oxalidaceae*. Some grasses also have a high content of oxalic acid, for instance *Setaria sphacelata* (3–8% oxalic acid in dry matter) which is common in Australia. Plants which contain oxalic acid can be divided into two groups, the plant juice-pH of which is about

- 2.0: in these plants the oxalate-ion appears mostly as acid oxalate (mostly potassium oxalate/HC_2O_4), e.g. *Oxalis pes-caprae* and some *Rumex* spp.;
- 6.0: in these plants the oxalic acid is mostly contained as oxalate-ion (C_2O_4), mostly as soluble sodium oxalates and insoluble calcium and magnesium oxalates, as for instance in some *Chenopodiaceae*.

The rumen flora can detoxify oxalate directly or it can be transformed with calcium to the non-reabsorbable calcium oxalate. If not all of the oxalic acid is detoxified by this mechanism oxalate enters the blood stream and binds with the circulating calcium. As a consequence, hypocalcaemia will appear. The insoluble calcium oxalate which has been produced metabolically is deposited in the tissues especially in the kidneys and the ruminal wall. After adaptation, ruminants become able to detoxify large amounts of oxalate ingested with the feed in their rumen and to excrete it with the faeces.

Pyrrolizidine-alkaloids in most animal species cause chronic lesions of the liver but also damage to the lungs, kidneys and other organs and can also cause carcinogenic effects. The species of fodder plants containing pyrrolizidine-alkaloids which are grazed on mainly by cattle and sheep are *Amsinkia*, *Echium*, *Heliotropium*, *Crotalaria* and *Senecio*. The pyrrolizidine-alkaloids heliotrin and lasiocarpin, contained in *Heliotropium europeum*, are reduced in the rumen at first to 1-goreensin and heliotrin acid or angeloxy-methylene-pyrrolizidine as the final metabolite. Goreensin and heliotrin acid are finally also metabolized into this compound. The detoxication is achieved by Gram-negative and Gram-positive cocci (*Peptococcus heliotrinreducans*) which compete in the rumen with methane-producing bacteria for the hydrogen donors H_2 or formiate. If the methane production in the rumen is reduced by applying for instance chloroform or carbon tetrachloride to the animal, the metabolism of heliotrin can be enhanced and thus the time of adaptation of the animal can be shortened.

Etheric oils are also broken down by the microorganisms of the rumen and thus detoxified. Since most plants which contain etheric oils are not very palatable to ruminants this mechanism is only of minor importance.

Cyanides, e.g. prussic acid glycosides, are easily hydrolized in the rumen. The metabolic rate depends on the quantity of ingested poison and the physiological situation of the rumen flora. Well-fed animals with an adequate functioning rumen are in a much better position to metabolize cyanide than meagre ones.

Selenium is transformed by the rumen flora into a less reabsorbable form through bonding to the protein of the bacteria regardless of whether it has been ingested as organic or inorganic matter. This self use by the bacteria leads to detoxication.

The mechanism of **detoxication of mimosine** in the rumen is of particular importance in the semi-arid pasture ranges of the tropics and can contribute to improve the feed supply for ruminants. The bush *Leucaena leucocephala* which during pre-Columbian times was already being used by the Indians as a source of food, has in the meantime been distributed worldwide in the tropics as a fodder plant and has become an important source of nitrogen last but not least for ruminants. If, however, more than 30% of the diet is composed of leaves from *L. leucocephala*, poisoning appears in ruminants and monogastrics which is caused by the free amino acid mimosine. Mimosine is present in all parts of the plant. The ruminant autolytically metabolizes mimosine immediately after ingestion through the alkaline effect of the saliva, but especially by means of the rumen flora to 3.4-dihydroxypyridine (DHP) which has an antithyroid effect of the thiouracil type. The complete metabolism of mimosine takes several days and has a close linear correlation to the amount of ingested mimosine and to the feed mosaic. DHP is mainly excreted with the urine. The detoxication of mimosine in the rumen of sheep is much slower, and therefore they tolerate only lower amounts of *Leucaena* in their diet.

The **production of poisons** from plant constituents by the microorganisms of the rumen appears after certain plant species, for example *Astragalus miser*, *Trifolium* spp., *Brassica* spp., and photosensitizing substances (phylloerythrin) have been ingested.

Astragalus miser (Timber milkvetch) contains miserotoxin, a β-glycoside of 3-nitro-1-propanol. By hydrolizing the glycoside in the rumen, glucose and 3-nitro-1-glucanol result, the latter a poison with a neurotropic effect. *Trifolium pratense* and *T. subterraneum* contain isoflavines, e.g. formononetin, biochanin A and genistein. Formononetin is transformed in the rumen into the oestrogenic acting equol, the primarily oestrogenic compounds biochanin A and genistein being partly inactivated. *Brassica* species (cabbage) contain S-methylcystine-sulfoxide (SMCO). It is transformed by the rumen flora into dimethylsulfide (DMS) which causes the syndrome of so-called cabbage poisoning. During the metabolism of chlorophyll ingested with plant material, the rumen bacteria produce the photodynamic active substance phylloerythrin, which is normally excreted via the gall and the intestine. It may cause a hepatogenous photosensitivity if the liver has already been damaged through other feed poisons or if the excretion of the gall liquid is blocked by mechanical obstacles.

Different types of plant contents can **damage the rumen flora**. This leads through disorders of the rumen function (reduction of cellulose metabolism and production of volatile fatty acids) to a severe diminution of the condition of the animal. The alkaloid perloline contained in *Festuca arundinacea* not only causes so-called rye-grass-staggers poisoning, but also has a serious negative effect on the rumen flora. Etheric oils from plant species like *Artemisia*, *Eucalyptus*, *Pseudotsuga* and *Juniperus* spp. contain monoterpene-alcohols which inhibit the rumen bacteria. Beside their direct poisonous action on the organism, myco-toxins may also lead to pathogenic effects by interfering with the rumen flora (Martinez 1990, Osterhoff 1981).

4.2. Toxicological-Physiological Mechanics of Plant Poisoning

Poisonous plants contain constituents which directly or indirectly cause specific or combined complexes of disease symptoms. It has already been pointed out that it is therefore uncertain if and when and under which conditions a fodder plant or another plant ingested by an animal becomes poisonous. A classification of plant poisoning analogous to the systematology of the aetiologically relevant plant species is impractical. Instead, in the following presentation, the alternatives for diagnosis and treatment of plant poisoning are grouped according to the complexes of symptoms which are caused by plant poisons through their toxicological-physiological mechanics. The division according to medical-clinical criteria takes into account the poisonous action on the organism. A differen-tiation has to be made here between local and general effects. The resorptive effects either affect the target organ where the main action of the poison becomes effective, or the mechanism of the effect itself. Accordingly, a differentiation can be made between blood, nerve, protoplasma and parenchymal (liver, heart, and kidney) poisons. Due to their mechanism of effect mitosis, capillary and haemolytic poisons can be identified.

Apparently there are plant poisons which can inhibit all essential metabolic functions of the organism. There is for almost each organ, each endocrine gland and metabolic function of the organism a correspondent inhibitor perhaps contained in a plant. One may only speculate where the phylogenetic cause for this phenomenon is to be found. In the following a scheme has been developed which may be helpful in finding indications about the poisonous principle of a particular plant (Cheeke and Shull 1985). Plant poisons are listed which affect particular organs and tissues of the organism. The compilation however does not claim to be complete.

Lesions in the **oral cavity** are the consequence of

– proteolytic enzymes: bromelain from pineapples and papain from papaya (silage) may damage the mucosa of the mouth;
– oxalate crystals, e.g. from *Dieffenbachia sequine* which may cause severe inflammation of the mucosae of the mouth and pharynx;
– tannins, which can cause astringent ulcers.

In the **rumen**

– nitrates are reduced to nitrites through the action of the rumen bacteria;
– poison from mesquite and lupins as well as oxalate cause paralysis of the motor nervous system;

- cytoplasmic proteins and saponins cause tympany;
- the mucosa is damaged through the action of oxalates.

In the **intestine**

- saponins, tannins and selenium-containing amino acids cause inflammations and other lesions of the intestinal mucosa;
- enzyme inhibitors block the physiological effect of trypsin and amylase;
- lectins reduce the permeability of the mucosa and subsequently the reabsorption of nutrients;
- nitrates cause diarrhoea;
- pyrrolizidine-alkaloids provoke prolapse of the rectum.

The **liver** is damaged through

- hepatotoxins, as e.g. pyrrolizidine-alkaloids, which cause irreversible centrilobular necrosis, and lupin poisons which cause fatty liver cirrhosis (II/4.2.3);
- pyrrolizidine-alkaloids and lupin poisons lead to an increased concentration of copper and with it cause changes of the intermediate metabolism of zinc and iron.

The **kidney** is damaged by

- pyrrolizidine-alkaloids, oxalates and lactones.

The following can appear in the **circulatory system**

- blockage of the oxygen interchange through inhibition of the cytochrome oxidase after the release of cyanide from cyanogenic glycosides (III/4.2.6);
- haemolysis, caused by the effect of saponins, copper-poisoning and pyrrolizidine-alkaloids;
- disorders of haematopoiesis and anaemia caused by pyrrolizidine-alkaloids and copper poisoning;
- inhibition of blood clotting through dicumarol compounds which are antagonistic to vitamin K (eagle fern, clover) (II/4.2.4);
- vasoconstriction through ergo-alkaloids and other mycotoxins which grow on grasses (fescue foot) (II/4.2.11);
- vasodilatation through *Veratrum*-alkaloids;
- hypercalcaemia through calcinogenic glycosides;
- hyperglycaemia, e.g. through the poisonous amino acid hypoglycine contained in *Blighia sapida*;
- hyperglycaemia through fluoracetate which is contained in numerous poisonous plants in South Africa and Australia;
- hypercholesterolaemia through cyclopropenoid-fatty acid from cottonseed meal.

Damages to the **heart** and of the **heart function** are the consequence of the ingestion of

- eruca acid from rape seed meal and gossypol from cottonseed meal;
- digitonin from monkshood and coniin, a piperidin-alkaloid contained in hemlock (tachycardia);
- *Veratrum*-alkaloids which cause bradycardia;
- fluoracetic acid from *Dichapetalum* spp. which interferes with the metabolism of pyruvate and causes sudden death (II/4.2.4);
- unknown plant poisons in *Cestrum* spp. (alkaloids and glycosides) and *Pachystigma* spp. which lead to severe peracute lesions of the heart muscle and of the coronary vessels (II/4.2.4).

Lesions of the **skeleton** are caused by

- toxins from lupins and *Astragalus* spp. which lead to deformities of the skeleton of the foetus;
- eagle fern poisoning which damages the bone marrow;
- oxalates in tropical grasses which cause hyperparathyroidosis and with it fibrosis of the bones of the head ("big head").

Defects of the **eyes** and interference with the **eye sight** are caused by the ingestion of

- atropine, e.g. from thorn apple;
- *Astragalus* spp. and *Stypandra imbricata* ("blind grass") which cause blindness through lesions to the *Nervus opticus* and eagle fern which leads to disorders of the eye sight through degeneration of the neuroepithelium of the retina (II/4.2.8).

In the **nervous system** the following disorders may appear

- dysfunction of the nerve cells caused by indolizidine-alkaloids which are contained in *Swainsonia* spp. and *Astragalus* spp. They cause accumulation of mannose in the lysosomes of the nerve cells and with it axonal dystrophy (II/4.2.8);
- degeneration of the spinal cord through lathyrism;
- inhibition of the cholinesterase by solanin;
- somnolent effects caused by *Stipa robusta* ("sleepy grass") which can put horses in a somnolent state for days;
- ataxia caused by pyrrolizidine-alkaloids;
- polyneuritis because of lack of thiamine caused by thiaminase contained in eagle fern.

The **skeletal muscles** are damaged through

-   the lack of selenium which appears subsequent to lupin poisoning and causes a myopathia (II/4.2.8).

The **thyroid gland** is damaged through

-   a number of goitrogenic plant contents which inhibit the synthesis of thyroxine. These are for example the glucosinolates in *Brassica* spp., the thiocyanates which are produced subsequent to the detoxication of cyanogenic glycosides and dihydropyridon (DHP) which is produced metabolically in the rumen after the ingestion of *L. leucocephala* (II/4.2.1).

Damages to the **reproductive system** are caused by

-   a number of mycotoxins and phytooestrogens contained in clover and alfalfa which have an oestrogenic effect (II/4.2.10);
-   gossypol from cottonseed meal which leads to infertility in the male animal;
-   teratogenic effects which are produced by a number of poisonous plants such as lupins, hemlock and *Veratrum*;
-   pine needles which contain substances which cause abortion.

The **immune system** can be damaged by

-   lectins, aflatoxin and other plant contents.

Lesions of the skin are caused by

-   photosensitization which is the consequence of the ingestion of constituent elements of plants of the genus *Hypericum* and others as well as clover species. Secondary photosensitization appears subsequent to liver lesions such as those caused by pyrrolizidine-alkaloids. As a consequence, the liver is unable to break down the phylloerythrin contained in fodder plants which subsequently can react photodynamically in the skin (II/4.2.1);
-   mimosine- and selenium poisoning which lead to alopecia;
-   ergot- and other mycotoxins which cause demarcations of the skin through vasoconstriction (II/4.2.11).

Disorders of the **protein metabolism** are the consequence of the ingestion of

-   protease inhibitors e.g. trypsin inhibitors from soy beans;
-   indospecin, an amino acid which is similar to arginine and leads to arginine deficiency when animals graze on a pasture where *Indigo spicata* is endemic. Pyrrolizidine-alkaloids destroy the capacity of the liver to diaminize amino acids and to synthesize urea from ammonium;

- pyrrolizidine-alkaloids which interfere with the DNA- and RNA-metabolism and inhibit the protein synthesis in the cell metabolism.

Disorders of the **metabolism of carbohydrates** are the consequence of

- inhibition of the digestion of starch caused by α-amylase-inhibitors which are contained in several grains and beans;
- mannosidosis which is caused through the accumulation of α-mannose in nerve tissues after the ingestion of *Swainsonia* spp. and *Astragalus* spp. (II/4.2.8).

Disorders of the **fat metabolism** are caused by

- lupinosis in sheep (II/4.2.8);
- aflatoxins which lead to steatorrhoe (II/4.2.11).

Disorders of the **energy metabolism** are caused by

- fluoracetates which are contained in several poisonous plants and inhibit the tricarbon acid cycle through blockage of the aconitase;
- nitropropionic acid, e.g. from crown vetch (*Coronilla varia*) which inhibits the succinate of the dehydrogenase;
- cyanide (HCN) from cyanogenic glycosides which inhibits the cytochrome oxidase (II/4.2.6).

The **cell division** is inhibited by

- pyrrolizidine-alkaloids which block the prophasic stage of mitosis; the metaphase is arrested with lupinosis.

Disorders of the **mineral metabolism** are caused by

- chelators, e.g. phytine acid and oxalates which inhibit the reabsorption of minerals; mimosine binds to aluminium, iron and copper;
- hepatotoxins which inhibit the reabsorption of copper and zinc by the liver. Pyrrolizidine-alkaloids increase the copper content of the liver drastically but reduce the concentration of zinc;
- oxalates which lead to hypocalcaemia and calcinogenic glycosides which cause hypercalcaemia.

Disorders of the **metabolism of vitamins** are caused by

- vitamin antagonists contained in the feed as, e.g., thiaminases, calcinogenic glycosides which inhibit vitamin D, dicumarol which inhibits vitamin K and lipoxidase which inhibits carotene, the pre-vitamin A (II/4.2.4).

As is obvious from the scheme described above, it is difficult to find a correlation between the toxicological-physiological mechanics of the poisonous principle of the plants and the complexes of symptoms and thus to be able to determine particular syndromes as a consequence of the ingestion of a particular poisonous plant. There are nevertheless obvious connections which allow a grouping like it has been attempted in the following. Osterhoff (1981) has compiled known poisonous plants of the tropics and the symptom complexes caused by those in an extensive table which has been part of the first edition of this book but has been omitted in the second edition because of its volume. In order to find practical instructions for the diagnosis and prevention of plant poisoning in the tropics in the following examples of typical plant poisoning for each of the affected organ systems are described as they can appear in extensive animal production systems of the tropics.

4.2.1. Affections of the Skin and the Hair Coat

• **Photosensitization**

Aetiology, Occurrence and Pathogenesis
Photosensitization in animals appears when photodynamic substances hypersensitize the skin to light. The inflammation which mostly appears in the skin is the consequence of the absorption of light by the photosensitive substance which as far as its amount and spectrum are concerned would be harmless to a healthy skin. It is mainly cattle and sheep which are affected in the tropics. Two different principles of photosensitization have to be differentiated:

– **primary** photosensitization and
– hepatogenous or secondary **photosensitization**.

Primary photosensitization: the photodynamic agent is absorbed through the skin, or from the gastrointestinal tract unchanged and reaches the skin in its "native" form. Examples of these are hypericin from *Hypericum* sp. (St.-John's wort) and fagopyrin from *Fagopyrum esculentrum,* (buck-weed). Plants in the families *Umbelliferae* and *Rutaceae* contain photoactive flurocoumarins (psoralens), which cause photosensitization in livestock and poultry. *Ammi majus* (bishop's-weed) and *Chymopterus watsonii* (spring parsely) have produced photosensitization in cattle and sheep respectively. Ingestion of *A. majus* and *A. visnaga* seeds has produced severe photosensitization in poultry. Species of *Trifolium* and *Medicago* (clover and alfalfa), *Erodium* and *Brassica*, have been incriminated, e.g. in "trefoil dermatitis" and "rape scald". Many other plants have been suspected, but the phototoxins have not been identified.

    Secondary (hepatogenous) photosensitivity: This is by far the most frequent type in livestock. The photosensitizing agent, phylloerythrin (a porphyrin), accumulates in the plasma due to impaired hepatobiliary excretion. Phylloerythrin is derived from the anaerobic breakdown of chlorophyll by micro-organisms in the forestomachs of ruminants. Phylloerythrin, but not chlorophyll, is normally absorbed into the circulation and is effectively excreted by the liver into the bile. Failure to excrete phylloerythrin due to hepatic dysfunction or bile duct lesions increases the amount in the circulation. Thus, it reaches the skin where it absorbs and releases light energy, which initiates a phototoxic reaction.

Phylloerythrin has been incriminated as the phototoxic agent in the following conditions: common bile duct occlusion, facial eczema, lupinosis, congenital photosensitivity of Southdown and Corriedale sheep, and poisoning by *Tribulis terrestris* (puncture vine), *Lippia rehmanni*, *Lantana camara*, several *Panicum* spp. (kleingrass, broomcorn millet, witch grass), *Myoporum laetum* (ngaio), and *Narthecium ossifragum* (bog asphodel).

Photosensitization has also been reported with liver damage associated with various poisonings: ragwort and other *Senecio* spp., blue-green algae, *Nolina texana* (bunch grass), *Agave lecheguilla* (lechuguilla), *Holocalyx glaziovii*, *Kochia scoparia*, or *Tetradymia* (horse brush or rabbit brush) and *Brachiaria brizantha*. It is likely that phylloerythrin is the phototoxic agent in many of these (Merck 1991).

Photosensitizing substances which have been deposited in the skin either as primary plant contents or products of disorders of the intermediate metabolism (phylloerythrin) absorb the UV-radiation and lead to sunburn on little pigmented parts of the skin. Black and white cattle are especially susceptible; but also sheep can show symptoms of sunburn in particular in the face and on the ears.

The **clinical picture** of photosensitization is similar regardless of the cause. The first sign is a distinct photophobic reaction of the photosensitized animals; they seek shade and turn their back to the sun; they squirm in apparent discomfort and scratch and rub lightly pigmented, exposed areas of skin, e.g. ears, eyelids and muzzle. The inflammatory lesions which appear are restricted to white or clear pigmented parts of the skin which are exposed to sun-radiation. The course of disease can be divided into three phases:

– during the early stage (1–2 days after the onset of clinical symptoms) an erythema of the skin appears at first which later turns into an oedema; the skin is warm and painful. Due to the oedematous-inflammatory swelling and a seam of bristled hair, the affected parts of the skin are demarcated from the unaltered pigmented regions. If the animals are given shade at this stage, the lesions soon clear up. Otherwise the
– progressive stage (2–3 weeks after exposition) follows; deep necrosis of the skin appears which can be complicated through secondary infections. In black and white cattle, the white parts of the skin can slough off as entire plates and partly dry, partly bleeding or even ulcerative purulent surfaces remain. Even the tongue, the teats, the udder and the scrotum can become affected. In sheep the woolless parts of the face and the ears are most expos-ed. The ears may become enlarged and be covered by crusts in such a way that they hang downwards ("big head"). In the following
– final stage the skin starts to recover from the margins of the lesions, the hair

cover returning gradually but mostly less dense than previously (Merck 1991, Osterhoff 1981, Rosenberger 1970).

The **diagnosis** results from the inflammatory and necrotizing lesions restricted typically to the non-pigmented skin. Primary photosensitivity can be suspected if the disease only appears enzootically on particular pastures and/or after the animals have been feeding on them. Whenever hepatogenous photosensitivity is suspected, the function of the liver has to be controlled. Secondary photosensitivity is usually only sporadic; if it is hepatogenous, icterus may be present. Phylloerythrin can be detected with spectral-photometry in the serum.

The **treatment** has to be concentrated on topical chemotherapeutical and anti-inflammatory measures (ointments) to the skin lesions, perhaps supported by corticosteroids; fly strike has to be prevented. The animals most definitely have to be protected from sun-radiation and perhaps only allowed to graze at night. The prognosis is favourable. With hepatogenous photosensitivity, the cause of the liver damage has to be removed.

• Alopecia as Consequence of Mimosine Poisoning

The prominent feature of mimosine poisoning caused by *Leucaena leucocephala* in ruminants is loss of hair and/or wool. Because of the variable mimosine content of *L. leucocephala*, the amount of the daily intake of mimosine and not of the poisonous plant itself is the decisive criteria of poisoning. Therefore the discussion is about a mimosine and not a *Leucaena* poisoning.

The **pathogenesis** of mimosine poisoning is not fully understood. Because of its structural similarity with tyrosine, mimosine possibly inhibits the tyrosine-decarboxylase and competitively tyrosinease whereby the growth of hairs in the hair follicle is inhibited. It is also supposed that the metabolic pathways of phenylalanine and tyrosine are blocked and the incorporation of these amino acids into the protein molecules is prevented. Furthermore it is presumed that those enzymes which are part of the DOPA metabolism are inhibited. Besides this, mimosine is thought to enter a complex formation with pyridoxalphosphate, the most important coenzyme of the amino acid metabolism, whereby disorders of the transaminization occur. As a strong chelator, mimosine binds with metal ions (mainly iron, aluminium and copper) and competes this way with those enzymes of the organism which require metal ions. Finally mimosine is supposed to block the cell multiplication by inhibiting the incorporation of thymidine into DNA.

The **clinical picture** of mimosine poisoning depends on the amount and the frequency of the ingestion of the poison. The course of disease can be acute as well as chronic. A few days after ingestion of large amounts of mimosine, the following appear together or separately: alopecia, salivation, refusal of feed, loss of weight, increased activity and excitability, paralysis of the hind limbs, swelling of the bucal papillae, ulcera on the dorsal aspect of the tongue and cataracts of the eyes.

After a few days or weeks, death supervenes. If only low doses of mimosine have been ingested after a few months, an accumulative effect becomes manifest which presents similar symptoms. With both alternatives the low thyroxine content of the blood plasma causes goitre.

Pathological manifestations of mimosine poisoning are ulcers and necroses of the oesophagus, the rumen and reticulum, haemorrhages in the membranes and the mucosa of the jejunum, emphysema of the lung, hyperaemia of the liver and hypertrophia of the thyroid gland (Martinez 1990).

The **diagnosis** is unequivocal if a known connection between feeding *Leucaena* and alopecia exists.

The **treatment** of mimosine poisoning is symptomatic. The feeding of *Leucaena* can by no means be continued.

Two alternatives for the **prevention** of mimosine poisoning exist. Either one has to reduce or discontinue the feeding of *Leucaena* completely, or through gradual increase of the proportion of *Leucaena* in the diet, the rumen flora of the animals can be adapted to an increasing amount of mimosine. If animals are available which are already adapted to a higher proportion of mimosine in the diet, ruminal fluid from those can be transferred to animals which have to be adapted. The transferred ruminal fluid will assist in the development of a ruminal flora capable of the physiological metabolism of mimosine (Martinez 1990).

• Chronic Skin Necrosis Caused by *Crotalaria* Poisoning

Aetiology and Occurrence
In production systems of tropical plantations in particular, more than 600 species of *Crotalaria* spp., a nitrogen-fixing leguminosa, are used for mulching and soil improvement. Cattle and horses which are either intentionally or unintentionally grazed on such cultures ingest wild plants or seed in fodder cultures, or with feed-concentrate succumb to a severe poisoning. The typical symptom of *Crotalaria* poisoning is the severe skin necrosis. Though the species *Crotalaria* is usually avoided by animals, the plant may be ingested by some individuals, especially horses, when fodder is scarce. They mainly ingest leaves and stems but seeds and seed stalks only reluctantly. Poultry can be poisoned by *Crotalaria* seeds. Concentrations of >0.05% produce signs of poisoning; 0.3% causes death in 18 days (Merck 1991). All parts of the plant contain the poisonous alkaloid monocrotalin during all stages of vegetation.

The **clinical symptoms** of *Crotalaria* poisoning can appear as either acute or chronic signs. The chronic symptoms often only become manifest months after *Crotalaria* has been grazed on. The coat becomes lustreless, the animals are weak, present slight icterus, ascites, diarrhoea, tenesmus, prolapse of the rectum, lung emphysema, incoordination, perhaps aggressiveness, muscle spasms, they may refuse feed and lose weight. During the course of disease, dry skin necrosis appears in horses in particular (Fig. 67). Without losing the hair, the skin shrinks

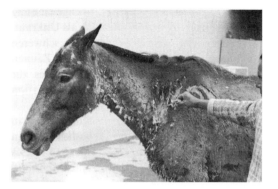

Fig. 67. Skin necrosis of a horse after Crotalaria poisoning

like leather and can be pulled off in hand-sized plaques or it even sloughs by itself. The skin defects reach down as far as the *Stratum germinativum*. In the neck and shoulder region, the animals may lose their skin entirely. In combination with severe general symptoms, these extensive skin lesions lead to death.

The acute course mostly lasts only a week. Blood-tinged tar-like faeces and tenesmus, bloating, blood-tinged nasal secretion, incoordination, colic and icterus, and perhaps also haemoglobinuria appear, and death soon supervenes. The skin lesions rarely appear during the acute course.

Pathological features of *Crotalaria* poisoning further to the skin necrosis lesions are ascites, swelling or cirrhosis of the liver and haemorrhages, parenchymatous hepatitis, proliferation of the periportal connective tissue, megalocytosis, centrilobular liver necrosis, haemorrhages in the spleen, emphysema, oedema, hyperaemia, hepatization of the lung and serous catarrhal-fibrinous pneumonia. Lesions in poultry consist of ascites, swelling or cirrhosis of the liver and haemorrhages.

Wherever there is a connection between the described clinical and pathological features and grazing on *Crotalaria*, a presumptive **diagnosis** can be made. It is difficult to diagnose chronic *Crotalaria* poisoning, the skin necrosis in horses, however, being an important diagnostic feature (Fig. 67). Acute cases can be confused with clostridial enterotoxaemia, last but not least because *Clostridia* can be isolated bacteriologically. The determination of the liver enzymes SGOT (serum-glutamine-oxalic acid-transaminase) and SGPT (serum-glutamine-pyruvate-transaminase) can be helpful in confirming the diagnosis.

For **treatment**, damage to the liver should be taken care of. Chronic poisoning can be treated successfully if it is recognized on time and the cause is eliminated especially by applying methionine in combination with dextrose intravenously. The prognosis however is unfavourable in general.

As a **preventive** measure, grazing on *Crotalaria* spp. should be avoided by all means. Since these plants are regularly cultivated intentionally, this should be possible under all circumstances (Sippel 1964).

4.2.2. Diarrhoea, Gastroenteritis and Constipation

The gastrointestinal tract is the first organ which is affected by numerous plant poisons directly after ingestion of the poisonous plant. Mustard oil glycosides, oxalates, nitrates, etheric oils, lectins and other plant contents can cause inflammation of the gastrointestinal tract with the corresponding disease symptoms.

• **Castor Oil Plant Poisoning in Cattle**

Aetiology, Occurrence and Pathogenesis
Poisoning following the ingestion of leaves, stems or seed of the castor oil plant is a plant poisoning of cattle which is becoming increasingly important in exotic cattle in the tropics, especially in areas of irrigated agriculture. The bush *Ricinus communis L.* is widespread in the arid tropics. The plant is cultivated for oil production but appears often in corn and millet cultures as a weed. Whenever exotic cattle are grazed on corn stubbles, severe outbreaks of poisoning can appear. Even if the animals only take a few leaves from plants which they find on the roadside, along trenches or fences, poisoning will be the consequence. In contrast, South American Criollos would not browse on castor bushes under any circumstances (Seifert 1995).

The poisonous principle of castor beans was already described by Stillmark way back in his PhD-thesis when working with Robert Koch as a postgraduate student. In all parts of the bush, especially in the seeds, the thermolabile toxalbumin is contained in variable concentrations. This lectin is composed of a toxic and a sugar-binding component. Whenever the docking part of the ricin binds to parts of sugars of molecules from membranes of any cell of the organism, the poisonous part is channelled into the interior of the cell where it blocks the protein synthesis.

Prominent features of the **clinical picture** of castor poisoning are severe digestive disorders and colic. Even small amounts of parts of the plant or the seed cause poisoning. After oil extraction, the fat insoluble ricin remains in the seed meal; 200–300 g ricin are deadly to cattle. The symptoms of poisoning usually appear after a latent period of up to 3 days; if large amounts of plant material are ingested, death may appear apoplectically. All animals of a group which have had access to feed containing ricin are usually soon affected with severe symptoms: refusal of feed, salivation, thirst, severe colic with transpiration, roaring, groaning, teeth grinding, signs of a tense and painful abdominal wall, tenesmus, trembling, cramps, choking, and difficulties in swallowing, paralysis of rumination and of the rumen and sometimes tympany; the animals may collapse suddenly. The stage of excitation soon turns into complete apathy. The profuse dark-brown watery and evil-smelling faeces which are mixed with blood, parts of mucous membranes and/or fibrin are all pathognomonic. At first the body temperature becomes raised and later on subfebrile, the heart pounds irregularly. The heartbeat is sometimes tachy-cardiac, sometimes bradycardiac. Dyspnoea, and oedema and emphysema appear in the lung. The field of percussion of the liver is enlarged. With increasing dehydration and recumbency, death supervenes (Seifert 1995).

Necropsy findings are severe abomaso-enteritis with spotty reddenings and thickenings, necroses and ulcers of the mucosa, enlargement of the intestinal lymph nodes, oedema and interstitial emphysema of the lung, splenomegaly, enlarged liver and kidneys. Characteristically dark red, partly tar-like petechiae appear subendocardially and on all serosae of the body cavit.es (Rosenberger 1970, Seifert 1995).

The **diagnosis** of castor oil poisoning is often difficult since small-holders will not believe that their exotic cattle browse on castor oil plants because they know that their native cattle never would touch them. Careful observation is required in order to become aware of the epizootiological connection. The clinical and pathological features are difficult to separate from other toxic causes (clostridioses) of the disease (Seifert 1995).

The symptomatic **treatment** almost without exception has an unfavourable prognosis. Even if the cause of poisoning has been removed immediately after the first symptoms have appeared, all affected animals will usually die. Even days after the poisoning has been diagnosed and the access of the animals to the plant has been restricted, further cases will appear in the herd which again can rarely be treated successfully (Seifert 1995).

As a measure of **prevention**, it is important to guarantee that exotic cattle have no access to castor oil plants or seeds. The stubble can only be grazed on after the removal of all castor oil plants; if straw is harvested, care has to be taken that no parts of castor oil plants are mixed with the straw. The ricin remains active in dry vegetative parts. Cattle are able to become immunized against the toxalbumin ricin by gradual adaptation. Since it is rather easy to prevent cattle from having access to castor oil plants this alternative remains theoretical.

• Oak Poisoning (*Quercus* spp.) in Cattle

Aetiology, Occurrence and Pathogenesis
A number of different species of the genus *Quercus* occurs in the tropics and subtropics. Wherever small-holders take their animals for grazing, particularly into woodlands, poisoning can appear. The animals are often brought into bushy areas when the fields are being worked and also in the subsequent period of vegetation. In north-eastern Mexico, oaks are endemic in the so-called "matorral", a vast tree-savannah which is used as pasture. Apparently all species of oaks are poisonous though there are seasonal differences of toxicity. The green acorns, which the cattle often even crave, are the most dangerous. Oaks contain large amounts of tannic acid (gallotannin 7–9%) both in the leaves and the bark, tannic acid having a severely irritating effect on the mucosa of the abomasum and the intestine after repeated ingestion; it is reabsorbed after hydrolization.

The **clinical symptoms** of oak poisoning appear after grazing the animals for several days under oaks. Almost all animals of the herd become gradually affected presenting severe symptoms especially if exotic cattle are involved. At first, the milk production is reduced or ceases completely; typically, the milk tastes bitter. Further signs are: colic-like unrest, excitation, separation from the herd, seeking shade, wanting to stay at the water place, a stiff gait, progressive apathy, refusal of feed (with exception of acorns and oak leaves), thirst, a dry muzzle which is covered with nasal secretion, paralysis of the rumen and rumination, slight tympany, occasional vomiting, contracted and painful abdominal walls, and as a typical pathognomonic feature, protracted defaecation of black-green and mucous covered balls of faeces which later on turn into a severe diarrhoea with evil smelling mucous- and fibrin-containing faeces. Initial retention of urine is followed by polyuria, the body temperature is subnormal, the cardiac rhythm bradycardiac and irregular, the pulse low and hard, the mucosae are pale, and erythropenia and lymphocytosis appear. The animals are weak, stagger and become recumbent with their legs stretched out, they groan and grind their teeth. Oedemas appear on the neck, the brisket, the chest, the underbelly, the inner aspect of the hind limbs, the perineum and around the anus. With progressing dehydration, death supervenes.

Characteristic **necropsy findings** are the yellowish, gelatineous oedemas of the subcutis, the increased yellowish-red liquid in the body cavities and petechiae in the serosae and also on the mediastinum as well as at the peri-, epi- and endocardium. In the abomasum, lesions of a catarrhal enteritis dotted with haemorrhages are found along with necroses and/or ulcers. The intestinal content varies from dark-brown to blackish and tar-like in colour, the content of the rumen smelling putrefied and perhaps being mixed with parts of oak leaves or the remains of acorns. The kidneys are pale, in acute cases enlarged, and in chronic cases, signs of contraction as consequence of interstitial nephritis are evident; haemorrhages are found present on the capsule and on the cortex. Histologically, the proximal parts of the tubules are enlarged and filled with necrotic masses, hyaline cylinders are present in the collecting tubules (Rosenberger 1970).

The **diagnosis** is obvious if there is a connection between grazing the animals under oaks and the appearance of the severe gastrointestinal symptoms and necropsy findings as described above. The remains of acorns found in the forestomachs and the typical lesions of the kidneys are pathognomonic criteria. Enterotoxaemia caused by *C. perfringens* has to be excluded using differential diagnosis.

Part of the **treatment** has to be an immediate change of pasture. The oral application of linseed slime and sodium hydrogen carbonate or calcium carbonate and the i.v. application of plenty of electrolytes with methionine and dextrose are indicated. The prognosis of the peracute course of oak poisoning is unfavourable. It always has to be taken into account that cases of oak poisoning still can appear even weeks after the animals have been removed from the oak pasture.

To **prevent** oak poisoning, the safest alternative is of course to keep cattle away from pastures where oaks are endemic. It is, however, possible to find out under which conditions, perhaps by supplying additional roughage, the ingestion of limited amounts of oak leaves and acorns is tolerated by the animals. An adapted pasture management is prerequisite for the prevention of the poisoning (Rosenberger 1970).

4.2.3. Disorders of the Liver

Plant poisons which have been reabsorbed enterally are mainly metabolized by the liver. Even after a heavy challenge through poisoning, the liver remains capable of detoxifying surprisingly high amounts of poisonous substances. If, however, excessive doses of poisons are reabsorbed, acute or chronic alterations appear in the liver which mostly become manifest as centrilobular necrosis which after the continuing intake of poison turns into a fatty degeneration. If the poisoning only lasts for a short time, the liver is able to regenerate; if, however, the intake of poisonous substances continues, the fatty parenchymatous degeneration is transformed into cirrhosis of the liver. The lesions of the liver lead to further metabolic disorders which again in a *circulus vitiosus* will have an adverse effect on the function of the organ.

The liver can be damaged by numerous poisonous principles of plants, for example by alkaloids as well as by mycotoxins which are produced on plants. Gossypol which is contained in cottonseed meal is also a severe liver poison.

In the following, a few relevant examples of liver poisoning in tropical animal production systems are described.

• Poisoning by Plants of the Genus *Senecio*

Aetiology and Pathogenesis

Numerous species of the genus *Seneciaceae* which belong to composites are endemic in the tropics. *S. jacobaea* (Groundsel, Ragwort, Stinking Willie), *S. aureus, S. burchelli, S. cunninghami, S. ilifolius, S. isatidus, S. longilobus, S. retrorsis, S. riddellii* are some of the *Senecio* species which appear in the tropics (Australia, South Africa, America) and are recognized as poisonous. The biennial or perennial herbs are poisonous both in a green state as well as when in a dried state. Because they are not very palatable they are only ingested if the alternatives for selection of the animals are restricted, e.g. through intensive pasture rotation. All of the mentioned species contain esters of hydroxypyrrolizidine bases, e.g. jacobine, a toxic alkaloid isolated from *S. jacobaea*, which after either increased short-time or cumulative long-time ingestion causes irreversible lesions of the liver. The poisonous principle is inactivated in silage but not by preparing hay. Cattle seem to be poisoned most frequently, but horses to a lesser extent. Sheep often eat *Senecio* with impunity but can also be poisoned.

Clinical symptoms of acute *Senecio* poisoning appear after animals have ingested the equivalent of 1–5% of their body weight of *Seneciaceae* over a few days. Chronic poisoning occurs if an equivalent of up to 150% of the body weight of the animal as wet weight of the poisonous plant has been ingested for a month. The symptoms of the acute poisoning already appear a few days after the animals have been brought onto the suspect pasture; up to 30% of a herd can be affected. The number of poisoned animals will gradually increase and animals will still show symptoms for weeks after they have been removed from the pasture. First signs are a sudden drop of milk production, unrest and separation from the herd, walking in circles, excitation and aggressiveness with rabies-like signs, later on weakness and staggering, knuckling of the fetlock, eye-sight disorders which can turn into complete blindness. Feed is refused and the animals show signs of increased thirst. After initial constipation, diarrhoea with prolapse of the rectum appears. The muzzle is dry and scaly, the conjunctivae appear dirty and icteric. The bilirubin content of the serum is increased up to 20 mg %, bile pigment is found in the urine and the body temperature remains normal or can be subnormal. During the chronic course of poisoning, oedematous swellings of the skin and perhaps hepatogenous photosensitivity with its typical signs (II/4.2.1) can appear. The milk smells peculiarly sweet. The liver enzymes are notably increased and icterus can appear.

The acute course can lead to death after a few days with typical signs of a highly acute liver inflammation. Chronic poisoning causes infirmity and emaciation which lasts for weeks. Even after the animals have been removed from the suspicious pasture they will not usually recover.

At **necropsy**, the yellowish colour of the serosae and the fat of the carcass is obvious, the liver is enlarged, shows signs of yellowish, fatty degeneration and after the chronic course of poisoning appears tough and cirrhotic. Extensive oedemas of the mediastinum and ascites are apparent. Histologically, the typical picture of the "nutmeg liver", the sign of a centrilobular fatty degeneration, will be found. The small veins are partly obstructed and show perivascular necroses and fibroses (Muenscher 1951).

The **diagnosis** is extremely difficult if there is no information about a connection between symptoms of poisoning and grazing the animals on a pasture on which *Senecio* species which are known to be poisonous are endemic. Since autochthonous cattle under no circumstances will take up *Seneciaceae*, one can only suspect *Senecio* poisoning if exotic cattle have been brought onto a potentially poisonous pasture. Alone the fact that *Seneciaceae* are endemic is not a diagnostic criterion. It has to be made sure that a known poisonous species is present which is ingested under normal management conditions. Feed trials can confirm a presumptive diagnosis (II/4.3.2).

As a **treatment**, the application of methionine and dextrose is effective. The animals have to be put to rest, and good hay as well as enough drinking water have to be provided.

As a **preventive** measure, the mechanical or chemical removal of recognized poisonous *Senecio* species only is feasible where the intensification of pasture management is economically justified. Under ranching conditions, the problem usually can be controlled by reducing the stocking rate to half and changing from rotating to permanent pasture; it may be an advantage to graze the animals on a suspect pasture when the herbs have become overripe.

• Poisoning by Cottonseed Meal

Aetiology, Occurrence and Pathogenesis
Gossypol, the predominant pigment and probably the major toxic ingredient in the cotton plant (*Gossypium* spp.), and other polyphenolic pigments are contained within small discrete structures called pigment glands, found in various parts of the cotton plant. The gossypol content of cottonseeds varies from just a trace to >6%, and is affected by both plant species/variety and environmental factors such as climate, soil type, and fertilization; it is also a natural component of the "glandless" variety of cotton which is all but rarely cultivated. Cottonseed is processed into edible oil, meal, linters (short fibres), and hulls. Depending on the process of extraction, gossypol is contained in cottonseed meal, linters and hulls. Since the pigment has an unfavourable influence on the colour of the oil, the effort is made during the production process to keep as much as possible of the pigment in the seed meal. With heat extraction, gossypol is bound to protein and thus inhibited. Cottonseed meal which has been extracted mechanically (pressed) is poisonous. In the tropics, cottonseed meal is used in dairy production and often mixed with molasses as an energy source and sometimes applied up to 3–5 kg/day in the diet. Seed hulls which contain parts of the untreated seed are used also as a bulk feed for ruminants. Ruminants are able to detoxify large amounts of gossypol by binding it in the rumen to lysine and transforming it into non-poisonous protein. Nevertheless, intensive feeding with the above mentioned amounts will already lead to severe poisoning after 1–2 weeks with symptoms which are similar to those of *Senecio* poisoning, immature ruminants being most frequently affected. Other susceptible species are monogastrics and poultry. As far as differences of cattle breeds regarding their susceptibility are concerned, the Holstein are the most sensitive (Merck 1991, Seifert 1992).

The first **clinical symptom** is a gradual reduction of milk production, lack of appetite and apathy all of which regularly appear mostly with the animals of highest productivity. Subsequently, typical symptoms of icterus become apparent: the conjunctivae are yellowish and dirty, the eyes sunk in, mucosae of the oral cavity, the muzzle, the mucosa of the vagina and the skin around udder and vagina present a yellowish colour. Diarrhoea appears, the urine is dark to red-brown, the rumen motor activity is paralysed with subfebrile temperature, and circulatory decompensation becomes apparent. After dyspnoea and convulsions set in, death often supervenes unexpectedly.

Apart from hepatoxicity, signs may relate to cardiac or reproductive effects. Prolonged exposure can cause acute heart failure. Pulmonary effects and chronic dyspnoea are most likely secondary to cardiotoxicity. Reproductive effects include in males reduced libido with decreased spermatogenesis, which may be reversible, and in females irregular cycling, luteolytic disruption of pregnancy, and direct embryotoxic action. Green discolouration of egg yolks and decreased egg hatchability have been reported in poultry.

Signs of prolonged excess gossypol exposure in many animals are weight loss, weakness, anorexia, and increased susceptibility to stress. Young lambs, goats, and calves may suffer a cardiomyopathy and sudden death, or a more chronic course with depression, anorexia, and pronounced dyspnoea. Adult dairy cattle may also have gastroenteritis, haemoglobinuria, and reproductive problems. Acutely exposed monogastric animals may have sudden circulatory failure, while subacute exposure may result in pulmonary oedema secondary to congestive heart failure; anaemia may be another common sequela. Violent dyspnoea ("thumping") is the outstanding clinical sign in pigs (Merck 1991).

Pathologically, gossypol poisoning is characterized by extensive oedemas, haemorrhages, icterus of the serosae and the fat, nephritis and haemorrhagic abomaso-enteritis. The body cavities including the pericardium are filled with copious amounts of amber-coloured or red tinged fluid with fibrin clumps, the mediastinum in the lung is oedematous, haemorrhages are found on the serosae, and the epi- and endocardium; the heart is enlarged, flabby, pale, streaked and mottled with dilated ventricles, and valvular oedemas may be present. The liver is enlarged, hardened and on outer inspection already shows the typical picture of the "nutmeg liver"; icterus is evident. Histologically, degeneration of the heart muscle and centrilobular liver necroses are typical lesions. The kidney, spleen, and other splanchnic organs may be congested, and may have petechiae. Haemoglobinuria, and oedema and hyperaemia of the visceral mucosa may occur. Further signs are those described with *Senecio* poisoning.

A presumptive **diagnosis** is rather easy because the connection between feeding cottonseed meal and severe disorders of the liver function or even sudden death will be obvious. Determination of the liver enzymes will be helpful. Free gossypol can be detected in the diet of the animals and in the tissues of the carcass. It should not exceed 1000 ppm in the diet and not be above a limit of 10–20 ppm as bound or free gossypol in the tissues of the ruminants (Merck 1991, Seifert 1992).

As with *Senecio* poisoning, the i.v. application of dextrose and methionine and oral application of linseed slime will be helpful as a **treatment**. Removal of the cottonseed meal-containing concentrate and feeding with good hay rich in nutrients as a treatment can be taken for granted.

The **prevention** of cottonseed poisoning can be guaranteed when applying an appropriate diet which is adequate for the productivity of the animal and which does not contain an excessive proportion of cottonseed meal. It is important to find out which technology has been used for processing the cottonseed. A high intake of protein, calcium hydroxide, or iron salts appears to be protective in cattle. Cattle should also be given at least 40% of dry matter intake from a roughage source. Added iron of up to 400 ppm in swine diets and up to 600 ppm in poultry diets has been reported to be effective in preventing signs (Merck 1991, Seifert 1992).

4.2.4. Disorders of the Circulatory System, Including Blood and Haematopoiesis

Poisonous plant contents which are reabsorbed through the portal circulation directly affect the circulatory system and the blood.

• **Gousiekte**
(Quick or sudden disease, Flabby heart)

Aetiology and Occurrence
Gousiekte is an economically important plant poisoning in South Africa, it is characteristic of the mechanism of direct poisoning of the heart after enteral reabsorption of the poisonous plant constituents. Gousiekte is caused by *Pachystigma pygmaeum*, *P. thamnus*, *Pavetta harborii*, *P. schumaniania* and *Fadogia monticola*. The poisonous principle of the plants is unknown; all ruminants can be affected. The first symptoms of poisoning appear 2–6 weeks after grazing animals on a gousiekte pasture. The latent period makes it difficult to diagnose the disease since it is not always known whether or not the animals have been in contact with the plants.

The **clinical symptoms** are determined by the peracute course of the disease which is dominated by symptoms of a severe dyspnoea. The animals cough, show abdominal respiration, bellow anxiously or bleat and become recumbent with head and neck in a characteristically stretched position. Dyspnoea and tachycardia increase and only a few hours after the appearance of the first signs of poisoning the animals die while bellowing or bleating. Symptoms are often not even observed because of the fast course of the disease and the animals are found dead on the pasture.

Characteristic **necropsy findings** are signs of acute heart failure, ascites, hydrothorax, hydropericardium, stasis of the liver and the other abdominal organs. At the apex of the heart, focal myocarditic lesions are present, the heart has lost its characteristic consistency, the heart muscle is soft and tender, and petechiae are found on the endo- and epicardium. It is supposed that the plant poison interferes with the cell metabolism and the capacity of the heart muscle to contract thus causing disorders of the cardiac rhythm and heart dilatation ("flabby heart").

The **diagnosis** is obvious if losses occur on gousiekte pastures, the animals presenting the typical lesions at the heart at necropsy.

An effective **treatment** of gousiekte is unknown.

Gousiekte can be **prevented** by only grazing on gousiekte pastures during winter. The poisonous plants are at their most poisonous after sprouting in spring and are eagerly eaten by the animals in this state. Ploughing of the pastures and cultivation of fodder plants are effective prophylactic measures (Watt and Breyer-Brandwijk 1962).

• **Poisoning by** *Cestrum* **spp.**

Aetiology, Occurrence and Pathogenesis
In the tropical bush savannahs as well as at the edge of agriculture areas on waste land, and on those hills which cannot be used for agriculture but rather only as communal pastures, bushes of the species *Cestrum* spp. which belong to the family of *Solanaceae* are endemic with numerous species. It is known that leaves, stems, flowers and seed of *C. auriculatum*, *C. parqui* and *C. hediondinum* are extremely poisonous for horses, cattle and sheep. *Cestrum* poisoning has been reported from Argentina, Chile, Perú, Cuba and also Italy. The development and the course of *C. auriculatum* poisoning in Perú is characteristic of the *Cestrum* poisoning. *C. auriculatum* is an evergreen bush reaching up to 3 m which mostly grows on moist spots. In Perú, the bush is endemic along the river-beds of the coastal agriculture area and in the surroundings of the connected irrigation systems. But mostly *Cestrum* grows as a member of a typical hill foot vegetation in the surroundings of creeks in the high valleys of the Andes. While Criollo cattle will under no circumstances browse on the bush, exotic cattle ingest all parts of the plant whenever the stocking rate in a ranch has past a critical level. Even when enough drinking water is available during the rainy season, losses will appear under such conditions. Though the leaves of the bush have an evil smell, they are ingested by cattle. The seedlings which sprout on natural pastures are the most dangerous. In a trial carried out in 1960, we found out that when adapted cattle (Criollo-Highland crosses) were inseminated with sperm from Hereford or Sta. Gertrudis bulls, increased losses appeared in the offspring of the A.I. bulls in contrast to the calves sired by local crossbred bulls (Seifert 1960).

The **poisonous principle** of *Cestrum* spp. is not fully understood. An alkaloid (parquina) as well as a glycoside (parquinosidol) have been described for *C. parqui*, and a glycoside (glucerosina) has been described for *C. auriculatum* (Carcamo 1957).

The **clinical symptoms** of *Cestrum* poisoning are characterized by the course of the poisoning which is always peracute. Affected animals are mostly found dead on the extensive pasture. After experimental poisoning, signs of excitation appear at first followed by disorders of coordination and paralysis of the hind limbs, muscle tremor, sometimes diarrhoea and tenesmus, dry muzzle and finally recumbency appear. After only a few hours, the animals enter into a coma still trying to support their heads on any available object. Disorders of the sensory system are obvious and during the following agony, reddish frothy foam appears on the nostrils. Death supervenes with signs of dyspnoea and asphyxia.

Characteristic **pathological features** of *Cestrum* poisoning are dark red almost blackish haemorrhages which are spread over all serosae of the body cavities especially in the intercostal space, the mediastinum and the peri-, epi- and endocardium (Fig. 68 on p. 2 annex). There are signs of a severe heart muscle degeneration, the heart muscle has lost its consistency, is enlarged, tender and flabby, pale with greyish spotty, partly extensive infarcts; the entire heart muscle may be discoloured to greyish yellow; dilated ventricles and valvular oedema are present. The liver is enlarged with the typical signs of a nutmeg liver because of centrilobular liver necroses. In the lung an extensive oedema with large amounts of blood-stained frothy foam in the bronchi and bronchioles is present (Seifert 1960).

Wherever the incidence of *Cestrum* poisoning is known, the **diagnosis** can be made based on the characteristic tar-like haemorrhages on the heart (Fig. 68 on page 2 annex). Poisoning with phosphoric acid esters has to be considered for differential diagnosis.

An effective measure of **prevention** of *Cestrum* poisoning is the drastic reduction of the stocking rate in a ranch. It has been possible to demonstrate that whenever the stocking rate has been reduced to half, even highly problematic pastures can be grazed. Equally, by changing the pasture management from rotating to permanent, the incidence of poisoning will be reduced. Extreme caution is required where exotic highly productive cattle are supplied to small-holders who have no experience in tending those animals. Holstein-Friesian cattle which are tethered to road sides will try to reach the leaves of a *Cestrum* bush nearby and browse on its leaves. It is relatively simple to remove the plant since it is easy to recognize. It is achieved best by mechanical means. *Cestrum* species are extraordinarily resistant to herbicides, even the so-called "brushkillers" (Seifert 1960, see also II/4.4.1).

• *Dichapetalum* Poisoning
(Gifblaar, Poison leaf poisoning)

Aetiology, Occurrence and Pathogenesis
Dichapetalaceae are highly poisonous plants which are endemic to the natural pastures of eastern and southern Africa. Poisoning in cattle, sheep and goats has been reported from Nigeria, Kenya, Tanzania, Mozambique and South Africa. Under experimental conditions, all mammals are susceptible to *Dichapetalum* poisoning. Known poisonous species of *Dichapetalum* are *D. aureonitens*, *D. barteri*, *D. braunii* Engl. & Krause, *D. cymosum* Engl., *D. macrocarpum* Engl. which is also used as an arrow-poison, likewise *D. mossambicense* Engl., *D. ruhlandii* Engl., *D. stuhlmannii*, *D. toxicarum*, *D. venenatum* Engl. & Gilg. The fruits from *D. deflexum* Engl. are considered edible in Tanzania. *Dichapetalaceae* have roots which reach 20 m deep and extensive root-stocks which allow the plant to survive even extremely dry seasons. With the first rain, the lancet-shaped juicy green leaves sprout which have a distinct vascular structure. The bushy plant which is densely covered with leaves does not grow higher than 20 cm, the green leaves having a strong smell. In South Africa, *D. cymosum* is considered

the most poisonous plant of all: with the ingestion of only 3–4, leaves an adult cattle can be killed, and 1–2 leaves are enough to kill a sheep. The toxic principle of *Dichapetalum* is fluoracetic acid (FCH_2COOH); dry leaves contain 15 mg/g. Because of its water solubility, the poison has a fast action, especially if the animals take water after having ingested the plant. Fluoracetic acid interferes as fluoracetate with the pyruvate-metabolism and leads to the accumulation of citric acid, especially in the heart muscle, apparently because of the transformation into fluorcitric acid. The peracute fatal syncope which is the consequence of the poisoning is caused either through ventricular fibrillation or excessive stimulation of the CNS. Both mechanisms can develop differently in the different animal species. The animals may prefer the poisonous plants to good pasture and even hay.

The peracute course of the poisonous process characterizes the **clinical signs**. The animals often die with acute apnoea and staggering a few minutes after drinking water following the ingestion of *Dichapetalum* leaves. During a more protracted course, the animals can be observed staggering; they fall down, try to get up, present signs of acute apnoea and are highly excited. Subsequently the animals appear dazed, and become recumbent until death supervenes. Signs of a protracted course of poisoning which can appear on occasion are an increased but weak pulse, elevated temperature, severe ataxia of the hind limbs and sometimes diarrhoea (Watt and Breyer-Brandwijk 1962).

Characteristic **pathological features** of *Dichapetalum* poisoning are cyanotic mucosae, lung oedema and hypostasis of the liver; the heart is filled with coagulated cyanotic blood.

The **diagnosis** is obvious if signs of the acute fatal syncope can be combined with grazing on *Dichapetalum*-infested pastures.

Mostly there will be no time for applying **treatment**. Cardiac stimulants and laxatives can be helpful. If animals show signs of poisoning they should not be allowed to drink water.

As a **preventive measure**, the poisonous plants can be eliminated selectively from the pasture applying herbicides. Before grazing, suspicious pastures have to be controlled for young sprouts of *Dichapetalaceae*.

• Poisoning by Eagle Fern/Brake Fern/Bracken (*Pteridium aquilinum*)

Aetiology, Occurrence and Pathogenesis
Eagle fern is native on acid soils which are poor in phosphorus with a base of sand or gravel. It is common on upland pastures, abandoned fields and open wood lands. The perennial plant occurs widely in the humid tropics and is especially troublesome where small-holders graze their cattle during the vegetative period of their field crops in brushy areas and forests. Wherever shifting cultivation is carried out, the plant often appears densely on abandoned fields which have become pastures for ruminants following their use for agriculture. Bracken is mostly grazed on at the onset of the rainy season when

the animals search for roughage as compensation to the young grass. Cattle may also consume bracken at the end of the dry season when there is a scarcity of other feed. Hogs dig for the fresh root-stocks especially in autumn and yet they do not suffer from poisoning after having eaten the plants. The leaves and roots of bracken contain a storage resistant thermolabile and alcohol-soluble factor which inhibits the production of megakaryocytes in ruminants after continuous ingestion and along with that, the production of thrombocytes as well as granulocytes. In contrast, the pteridism of horses is caused by thiaminase which is also contained in bracken.

The **clinical signs** of bracken poisoning depend on whether the cause of the disease is acute or chronic.

The symptoms of the **acute bracken poisoning** only appear after the animals have grazed on pastures where bracken is endemic. Cattle have to ingest fern in amounts which is about equivalent to their own body weight. The first symptoms are weakness, reduced feed intake, digestive disorders, emaciation and haemoglubinuria, signs which gradually appear in all animals of the affected herd. As the symptoms progress, the animals start to bleed from the nose and the vagina, petechiae appear on the visible mucosae (eye, nose, mouth, vagina) which at the same time become anaemic; flea bite-like haemorrhages appear on the skin. Together with oedemas along the intermandibular space and of the pharynx, dyspnoea and tachycardia become evident. In young animals, pulmonary symptoms, and in older animals, signs of enteral disorders become manifest. Thrombopenia (< 100000 thrombocytes/mm^3), agranulocytosis and lymphocytosis as well as erythropenia are further clinical signs. The disease symptoms may be complicated through febrile septicaemic secondary infections. Death often supervenes only a few days after signs of diathesis appear. Even weeks after the animals have been taken from the fern-infested pasture, cases of poisoning will appear because of the accumulation of the poisonous principle in the organism of the affected animals. Recovered animals will become emaciated and do not usually regain their previous condition.

Chronic poisoning only appears in adult animals which have grazed for years on pastures with bracken but have only ingested small amounts of the plant (0.5–1.0 kg b.w./day). Signs of chronic fern poisoning are intermittent haemoglobinuria of varying intensity, erythropenia as well as slight thrombo- and leukopenia. The animals try continuously to urinate, the lower corner of the vagina being covered with bloody secretion. The animals become apathetic, refuse feed, their coats become lustreless and staring, and constipation or diarrhoea may appear. With progressing cachexia and oedemas which appear on the neck and brisket, dyspnoea and cardiac decompensation become evident until death supervenes. The prognosis of the disease is extremely unfavourable and cannot be influenced even after changing the pasture.

Pathological characteristics of the acute course of fern poisoning are the spectacular haemorrhages which are found on all serosae of the body cavities and the internal organs as well as on the mucous membranes of the respiratory, excretory and digestive system and in the form of haemorrhagic ulcers in the abomasum and jejunum. The only signs of the chronic poisoning are blackish-red petechiae in the mucosa of the bladder and wine-red stained urine in the bladder (Rosenberger 1970).

The **diagnosis** of acute fern poisoning is easy to establish where a connection between pasturing on bracken-contaminated pastures and signs of fern poisoning is evident. It has to be taken into consideration that the symptoms can still appear after the animals have been removed for weeks from a suspicious pasture. In the tropics, fern poisoning is often confused with babesiosis which however is characterized by high fever and icterus. Chronic fern poisoning is very difficult to identify.

Part of the **treatment** has to be an immediate change of pasture; blood transfusion, application of iron-dextran preparations as well as feeding with good hay can be helpful. Broad spectre antibiotics can prevent secondary infections.

The **prevention** of fern poisoning has to be guaranteed through an adapted pasture management and/or by avoiding pastures where bracken is endemic. Wherever economically feasible, the vegetative conditions for ferns can be removed through cultivation of fodder plants and fertilization of the pasture.

4.2.5. Development of Oedema Caused by Poisonous Plant Principles

Aetiology and Occurrence
Plants of the genus *Setaria* cause oedemas of the skin and subcutis in cattle in southern Africa, Australia and America. *Pimelea* spp. causes this in Australia as well ("St. Georges disease"). The poisoning occurs mostly on ranches where the animals are being kept on improved natural extensive pastures. *Setaria* spp. is cultivated in order to improve natural grass-land in association with gramineas. Pimeleae in principle are not palatable and are avoided by cattle. The plants which have also become dried at the end of the dry season are ingested only by the animals if green feed becomes extremely scarce. *Setaria* becomes poisonous only at the end of its vegetative period. Animals of both sexes and of all ages can become poisoned. Disease symptoms appear when cattle have grazed for 3–10 weeks on pastures where pimeleae are endemic. A daily dose of 60 mg/kg b.w. of dry leaves are enough to cause poisoning.

The **clinical symptoms** of *Setaria* poisoning are extensive oedemas in the intermandibular space, the brisket and belly down as far as the limbs, as well as diarrhoea which in the course of the disease will turn into dysentery. The animals lose weight quickly, show a staring lustreless coat, pale, anaemic and dirty conjunctivae; they are apathetic and become recumbent. In acute cases, death may supervene without the described symptoms becoming evident.

Necropsy findings, apart from this are the oedema of cutis and subcutis hydrothorax and ascites. The liver is enlarged and deep purple-red in colour. In the gastrointestinal tract, lesions of a catarrhal gastroenteritis are evident especially in sheep and horses which occasionally also become poisoned.

The **diagnosis** can be made if a connection exists between the occurrence of pimeleae on the extensive pastures and the appearance of typical symptoms in poisoned cattle. Brisket disease which used to be diagnosed in the Rocky Mountains and on other highland pastures and was supposed to be caused by lack of oxygen is now also thought to be caused by pimeleae poisoning.

An effective **treatment** of the disease is unknown.

As a **preventive measure**, the animals should only be grazed during the rainy season on pastures where pimeleae are endemic (Parmelee et al. 1960).

4.2.6. Disorders of the Respiratory Systems

• Cyanide Poisoning

Aetiology, Occurrence and Pathogenesis
In contrast to common opinion, no plants exist which primarily contain cyanide. There are, however, 250 known genera with more than 1000 species which contain cyanogenic compounds, mostly glycosides of α-hydroxynitriles. These cyanogenes are composed of one or several sugar glycons (mostly containing glucose) and a HCH-containing aglycon. Through enzymatic action, cyanide is released from the cyanogenic glycosides. Cyanogenic glycosides with relevance to the poisoning of animals are

– amygdalin (laetril), which occurs in rosaceae (e.g. cherry laurel, wild cherries) and the stones of almonds, apricots, peaches and apples. Prunasin present in those plants, is of the same structure as amygdalin with the difference that only one instead of two molecules of glucose are bound to the aglycon;
– dhurrin, which occurs both in grain- and fodder sorghum (*S. sudanense, S. vulgare, S. halepense*) and grasses (*Cynodon dactylon*, *Glyceria* spp., *Melica uniflora*) but it can also occur in sugar cane (*Saccharum officinarum*);
– linamarin, which occurs in white clover (*Trifolium repens*), flax, cassava and several species of beans.

Cyanogenic glycosides accumulate in plants whenever their cycle of vegetation is interrupted, for example when they wither after a sudden removal of water, after application of herbicides or under special climatic influences (renewed onset of the dry season after an early beginning of the rainy season). The production of cyanide in the plant is connected to the destruction of the plant tissue whereby the enzymes of the plant itself become effective, or those from other plants which are components of the feed mixture, freeing hydrocyanic acid (HCN) from the cell structure. The cyanogenic glycosides are localized in the vacuoles of the plant tissue and the enzymes in the cytosole or in other plants. The enzymatic split of the cyanogenic glycosides may occur following different forms of cell destruction for example through the effect of freezing, drying, mechanical destruction during harvesting or preparation of the feed, through trampling and finally when chewing during feed intake. HCN is freed from cyanogenic glycoside in the following way:

– cyanogenic glycoside $\xrightarrow{\text{ß-GLYCOSIDASE}}$ sugar + aglycon
– aglycon $\xrightarrow{\text{HYDROXYNITRILE LYASE}}$ HCN + aldehyde or ketone.

The enzymes can also be produced by the microorganisms of the rumen. The optimal pH for the action of the enzymes is around pH 7.0, thus HCN in the rumen of ruminants is more quickly released than in the stomach of monogastrics. The process of further enzymatic toxin release can be as follows:

– at first mandelonitrile + glucose are released from amygdalin and HCN + benzaldehyde are released from mandelonitrile;
– glucose + hydroxymandelonitrile are released from dhurrin and HCN + hydroxybenzaldehyde are released from hydroxymandelonitrile.

The enzymatic split and the reabsorption of HCN occur in the organism of the animal only 5–10 minutes after the cyanogenic plant has been ingested. The poison penetrates the cells of all tissues where it inhibits the cytochromoxidase. The change of the iron contained in the chromoxidase from a valency of 2 to 3, which is essential for all oxidative processes in cells, is inhibited by the bonding of the cyan-radical to the Fe^{+++}. This leads to a blockage of the ATP production and with it to the inhibition of the provision of energy for the cell. The oxygen provided by the blood to the tissues cannot be used resulting in a histotoxic anoxia and internal asphyxiation. The venous blood remains arterial. Through the interruption of the utilization of oxygen, the nerve cells (respiratory centre) especially are damaged and asphyxia results. On the other hand, the liver, kidney and the thyroid gland contain an enzyme, the thiosulphate-sulphur-transferase which converts cyanide to the non-poisonous thiocyanate, the latter being excreted with the urine. Therefore, poisoning only appears if the reabsorbed amount of HCN exceeds this detoxication potential.

The appearance of cyanide poisoning depends, however, very much on the content of the cyanogenic glycoside in the plant. It varies according to the vegetative stage, season and day-time. White clover can be more poisonous in the morning than in the afternoon, and feed sorghum is more poisonous during hot, humid weather than on hot, dry days. It is also important how much of the cyanogenic glycoside is ingested at a time, 2 mg HCN (cyanogenic glycoside)/kg b.w. ingested at one time are potentially lethal for large and small ruminants. If the same dose is distributed over a longer time, the above described mechanism of detoxication can break down the HCN; sheep are able to detoxify up to 2 mg HCN/kg b.w. in the space of 12 hours, and cattle can detoxify 15–50 mg/kg b.w. in the same amount of time. Feeding a diet rich in energy will support the detoxication. Therefore, poisoning only appears if the poisonous limits of HCN are exceeded within a short interval which mostly happens in the case of hungry and greedy animals.

Ruminants are more susceptible than monogastric animals, and cattle more so than sheep.

The **clinical symptoms** of cyanide poisoning are spectacular and the course of disease is peracute almost every time. Signs can occur between within 15–20 minutes and a few hours after animals consume poisonous forage. Excitement can be displayed initially, the affected animals suddenly showing unrest and staggering, and in a few minutes acute severe dyspnoea follows with tachycardia. The respiration becomes dyspnoeic, the cattle stand with their mouths wide open, they salivate, frothy foam appears around the mouth, the animals present nystagmus, dilatation of the pupillae, lacrimation, increased excitability, muscle tremors and finally break down with tonic-clonic convulsions, vomiting at the same time. Remaining recumbent with the fore legs stretched out sideways, the animals gasp and groan, grind their teeth, and present transpiration and tympany. The body temperature drops, unconsciousness, irregular and flat respiration appear, and with typical bellowing, the animals die with their limbs stretched out stiffly. The mortality rate will usually be 100%.

Typical **pathological features** of cyanide poisoning are the bright cherry-red blood, which may only clot slowly or not at all, as well as the typical smell of bitter almonds which escapes when the carcass is opened. Mucous membranes may be pink initially, then become cyanotic after respiration ceases. The muscles of the carcass are of dark colour, the lungs being filled with bloody foam. Typical bright cherry-red petechiae are found on the serosae and also on the epi- and endocardium (Fig. 69 on page 2 annex). The liver, serosae surfaces, tracheal mucosa, and lungs may be congested or haemophagic. The mucosa of the rumen and abomasum is slightly reddened. In the tropics, it is typical that the putrefaction of the carcass only appears slowly (Seifert and Beller 1969).

A presumptive **diagnosis** can be made if a number of animals of an affected herd die with the described spectacular disease symptoms. Often, however, the disease is confused with tympany, last but not least because it mostly appears after the pasture has been changed. The typical smell of bitter almond which escapes from the mouth, nose and perhaps the rumen following piercing with a trochar or at necropsy, as well as the cherry-red watery blood are all important diagnostic criteria. The cherry-red haemorrhages on the heart are rarely to be found in any other disease (Fig. 69). Under field conditions, the diagnosis can be confirmed by the picric acid test: a drop of reagent (5 g hydrogen carbonate, 0.5 g picric acid, 100 mL aq. dest.) and rumen content and/or homogenisate from liver, muscle or heart are mixed on filter paper. If the colour of the reaction changes from yellow to brown or red within 10 minutes, one has to suspect HCN poisoning. The test can also be carried out in a test tube (Seifert and Beller 1969).

Differential diagnoses include poisonings by nitrate/nitrite, urea and pesticides.

The **treatment** of acute cyanide poisoning is more or less hopeless. The repeated infusion of a 10–15 mL 20% solution of sodium nitrite immediately followed by a 40–50 mL 20% solution of sodium thiosulphate to an animal of about 500 kg/b.w. as has been recommended (Merck 1991, Rosenberger 1970) has proven to be of little success. The same is true for the oral application of sodium thiosulphate in order to detoxify the cyanide which still remained in the rumen (Seifert and Beller 1969).

It is extremely difficult to **prevent** the appearance of cyanide poisoning under the conditions of tropical animal production systems. Information about local empirical experiences regarding the seasonal and daily toxicity of fodder plants like sorghum can be very helpful. It may perhaps only depend on the way the sorghum has been cut which determines whether or not feeding the plant leads to poisoning. Normally, cyanide will be released from the glycoside in silage. During the distribution of the silage, the released HCN will escape and the feed will become harmless.

Seifert and Beller (1969) described a typical example of cyanide poisoning and its prevention in Perú. After grazing cattle on wilted sugar cane, cyanide poisoning appeared only if supplement feed with *Prosopis juliflora* (carob) pods was provided. The *Prosopis* pods contained the enzyme required for splitting the cyanogenic glycoside which was contained in large amounts in the sugar cane which had suddenly wilted. As soon as the supplementation with *Prosopis* was stopped, no further cases of poisoning appeared.

4.2.7. Neuromuscular Lesions Caused by Plant Poisoning

Aetiology, Occurrence and Pathogenesis
Poisoning caused by larkspur (*Delphinium* spp.) which especially occurs in the USA is a characteristic representative of this disease complex. In the mountainous areas of the western USA, *Delphinium* poisoning is economically the most important disease of cattle. Sheep in contrast are relatively resistant and only become poisoned after ingesting extremely high doses of the poisonous principle. There are about 80 larkspur species which are commonly divided into the groups of tall and low larkspurs and vary extremely as far as their toxicity is concerned. The most poisonous tall larkspurs are *D. nelsonii* and *D. andersonii*. The poisonous principle are toxic alkaloids, delphinine, delphinoidine, delphisine, and staphisagroine which have been found in certain species of larkspur. Of these, delphinine seems to be the most powerful and most constantly associated with the symptoms produced in cases of poisoning. The alkaloid, deltaline, has been found in *D. occidentale* (Muenscher 1951). The poisonous effect of the alkaloids is complex. Therefore, little is known about which active compounds primarily or exclusively cause poisoning. The DL_{50} for an adult cattle is about 0.5–0.7% of its body weight which as wet weight of the plant has to be ingested in a short time. After sprouting in spring, perennial larkspur contains the highest amount of poison. After flowering, the alkaloid content of the plant decreases.

The **clinical symptoms** of larkspur poisoning are determined by the nervous symptoms which are typical for this type of poisoning. Loss of appetite, general uneasiness, a staggering gait as well as splayed hind legs are the first symptoms. The animals stumble, fall down and lie with their feet extended more or less rigidly. After a varying interval of rest they may perhaps be able to get up again, the signs of weakness however returning quickly in most cases, now accompanied by muscle tremors and convulsions, nausea, bloating and constipation. The animals show signs of pain, look back towards their bellies and groan. Because of the nervous disorders of the pharynx, animals which vomit may die because of pneumonia after having choked on feed, otherwise death supervenes because of respiratory paralysis with signs of convulsions and acute dyspnoea.

At **necropsy**, no characteristic signs are found on the carcass.

No **treatment** is known.

As a **preventive measure**, the recommendation is to graze the more resistant sheep at first on suspicious pastures or to graze cattle only after larkspur has flowered (Rosenberger 1970).

4.2.8. Disorders of the Nervous System Caused by Plant Poisoning

In a large number of plant poisonings the central and/or peripheral nervous system is affected. In the following, however, only those syndromes of poisoning will be discussed which affect mainly the nervous system and which mostly present no lesions at all at necropsy.

• *Astragalus* spp. and *Oxytropis* spp. Poisoning

Aetiology, Occurrence and Pathogenesis

Herbs of the genus *Astragalus* and *Oxytropis* which occur in the mountainous ranges of the USA but also in the highlands of the tropics cause poisoning in horses, cattle and sheep; it is known as "loco poisoning". The small perennial leguminosae which grow upright with spreading stems and with typical white labial flowers occur in humid places in extended meadows, especially on mountainous ranges, for instance in the high Andes. When looked at superficially, the plants which remain a juicy green during the dry season looking like fields of white clover, are especially popular with sheep during this season. It has been observed in Perú that while exotic Corriedale sheep graze eagerly on the herbs, local Criollo sheep will not touch it (Seifert 1995). It has also been found in the USA that whenever exotic animals have become accustomed to ingesting the plant when fodder has become extremely scarce, the animals have become addicted and have continued to feed on *Astragalus* until they have become poisoned. More than 100 species of *Astragalus* occur in the mountainous ranges of the USA and the tropical highlands; in Perú *A. garbancillo* and *A.brackenridgei*, and in the USA *A. bisulcatus, A. campestris, A. diphysis* and others which are locally important poisonous plants. The toxicity of the plant depends on the quality of the soil. Apparently *Astragalus* and *Oxytropis* accumulate selenium, molybdenum, arsenic, tin and other heavy metals. These selenium-bearing loco-weeds are believed to be able to impart their toxicity to other forage plants. The decay of roots and other parts returned annually to the soil becomes a potential source of selenium, which is available to growing crops and forage plants, thus making them somewhat toxic. Selenium replaces sulphur in the essential amino acids which become poisonous this way. On the other hand, glycosides are also thought to be the cause of *Astragalus* poisoning as is miserotoxin for instance. A toxic base, locoine, has been isolated from *Astragalus earlei* grown in Texas. The green as well as the dry plants are poisonous (Muenscher 1951).

The **clinical symptoms** of *Astragalus* poisoning are already described by the name "loco poisoning" (loco span. crazy). In Perú the disease is also called "renguera" (limp). The first symptom in horses, cattle and sheep is dizziness followed by an uncoordinated staggering gait and nervous disorders when feeding. The animals characteristically drag their hind legs when walking and lose control of their motor muscles. As the disease progresses, the sensory system becomes affected, which sensibility becomes reduced and finally the animals become emaciated, recumbent, and do not search for feed right up until the time when death supervenes. During the so-called garbancillo poisoning (*A. garbancillo*) in sheep in the high valleys of the Andes, the sheep walk continuously in circles until the characteristic paresis and finally paralysis of the hind legs appears; the animals typically give way in the joints of their hind legs (Fig. 70 on page 2 annex) and remain for some time in this position (renguera) (Carcamo 1957, Muenscher 1951, Seifert 1992).

Because of the different nervous syndromes which are caused by different *Astragalus* species in the USA, the poisonous plants have been divided into two groups which on the one hand mainly affect the CNS ("loco weed") and on the other hand, the respiratory nervous system ("vetch disease"). The course of the disease can be acute, but mostly it will be protracted for several weeks.

Beside signs of progressive cachexia, no **pathological lesions** will be found during necropsy.

The **diagnosis** results from the connection between grazing the animals on pastures where *Astragalus* spp. are endemic and the appearance of the characteristic nervous symptoms. In sheep, the disease is often taken for coenurosis (oncospheres of *Multiceps multiceps*), since sheep also walk in circles when infested with this parasite.

The i.v. application of 1 g sodium nitrite and 4 g sodium hyposulfite as a 10% solution and of 3 g for sheep and 20 g for cattle respectively is supposed to have a **therapeutic** action.

Preventive measures depend on pasture management and the removal of the poisonous plant. Reduction of stocking rate, the fencing off of highly contaminated pastures where the plants grow in dense meadows and keeping the animals away from suspicious pastures during the dry season will reduce mortality. The pasture has to be changed without fail as soon as the first signs of poisoning appear. It has also been found that by adding 4.2 g sodium arsenite/100 kg salt which has to be made available to the animals permanently, the poisonous effect of the selenium can be compensated. Traditionally in South America, sulphur is mixed with the salt which is provided to the animals. This is thought to prevent the replacement of sulphur by selenium in the essential amino acids. Applying herbicides to the pastures is an efficient and easy measure for removing *Astragalus* from the pastures since the plant is easy to recognize and always grows in large associations (Seifert 1995).

• *Swainsonia* Poisoning

Aetiology, Occurrence and Pathogenesis
A disease similar to *Astragalus* poisoning occurs in Australia caused by the perennial *Swainsonia* spp. which dominate natural pastures especially at the beginning of the rainy season. There are more than 50 poisonous species, the most important being *S. canescens*, *S. galegifolia*, *S. greyana*, *S. luteola* and *S. procumbens*. The toxicity of the plant depends on their stage of growth; the poisonous principle is thought to be indolizidine-alkaloids. Young exotic cattle are especially vulnerable while sheep are relatively resistant.

The **clinical symptoms** are more or less similar to those of *Astragalus* poisoning and are identical in cattle and sheep. Initially the animals lose weight until gradually nervous symptoms appear; disorders of the sensory system, a staring look, trembling, shaking of the head which is held in a hanging position, a difficult gait with a stiff, high tread, uncoordinated movements and hypersensitivity are typical symptoms. During hot weather or after a physical challenge to the animals through rounding up for instance, the symptoms increase until death supervenes with panting, salivation and collapse. Pregnant cows abort between the 5th and 7th months of pregnancy. Poisoned animals are especially susceptible to secondary infections, like pneumonia, foot rot and pink eye which is why it is thought that the poisoning also affects the immune system.

At **necropsy**, no other signs are found apart from progressive emaciation. Histologically, cytoplasmic vacuoles are present in the tissues of the organs and nerves; they are thought to be enlarged lysosomes. The lysosomes contain large amounts of mannose which is why it is supposed that the indolizine-alkaloid swainsonin contained in *Swainsonia* inhibits the α-mannosidase which leads to an accumulation of mannose in the lysosomes. If the supply of poison is interrupted, the lesions may regenerate completely provided no permanent degeneration of the nerves has taken place yet.

No effective **treatment** is known.

The **preventive** measures are similar to those described for *Astragalus* poisoning.

• **Lupinosis**

Aetiology, Occurrence and Pathogenesis
The genus *Lupinus* includes annual as well as perennial herbs and a few shrubs.
It is well-known that several of the European lupins contain alkaloids which are
poisonous. In the case of European lupin poisoning, a hepatoxin has also been
made responsible which is thought to be developed by microorganisms which
grow on lupins. This poisonous principle, however, seems of no relevance to the
problem of lupin poisoning in the tropics. The so-called American lupin
poisoning is caused by most of wild *Lupinus* spp. which occur widely in the
tropics. They cause poisoning through one or several of the alkaloids lupinine,
lupinidine, lupanine, hydroxyllupanine, angustifuline, spathulatine and sparteine.
The alkaloids are contained in all parts of the plants especially in the pods and
seed. The bitter wild *Lupinus* spp. are almost never ingested by cattle and only by
sheep when fodder has become scarce or when their pastures are overstocked;
poisoning can also appear if the animals are brought onto lupin cultures which
are grown for fertilizing.

The **clinical symptoms** are highly characteristic within the complex of
syndromes described in this chapter. Muscle tremors, excitation, aimless
wandering around, running against obstacles, pushing against each other and
staggering, are typical symptoms in sheep until the animals become recumbent
and soon die in a comatose state with convulsions. Cattle separate themselves
from the herd, seek shade, refuse feed, pearls of secretion appear on the muzzle,
and they salivate and lacrimate; paralysis of the rumen, constipation, severe
icterus and reduction of milk production become evident; they become
somnolent, weak and emaciated; diarrhoea, oedema of the brisket, and
sometimes unrest and rabies-like symptoms can also be observed. Mostly the
disease lasts no longer than a week. Cattle which have ingested lupins during the
2nd to 3rd month of pregnancy give birth to calves with bent limbs and a bent
spine ("crooked calf syndrome") (Rosenberger 1970).

During **necropsy**, no specific signs of disease will be found.

An effective **treatment** is unknown.

As a **preventive measure** it is of course paramount to keep the animals away
from natural pastures infested with lupins or from fields with lupin crops. Again
there are the exotic breeds which are especially exposed. Lupin shrubs which are
widespread on the highlands of the tropics can be removed mechanically.
Attention has to be paid to the new growths from seeds which easily can be
overlooked on a pasture.

4.2.9. Disorders of the Urinary System Caused by Plant Poisoning

Perfectly working kidneys are the prerequisite for maintaining the electrolyte
balance and for the excretion of harmful body wastes. During the process of
excretion of poisonous substances, higher concentrations of poison may appear

in the kidneys as in the rest of the organism. Especially during secretion or partial reabsorption in the tubulus cells, the poison which is to be excreted reacts with the tissues of the urinary system. Depending on the intensity and the time of the poisonous effect, increased excretion of urine because of the inhibition of reabsorption due to damages to the epithelium of the tubules will appear at first. With progressing poisoning, however, it is followed by retention of urine (oliguria, anuria), with severe lesions to the glomeruli being evident. Albuminuria is the consequence. Mechanical obstruction of the renal tubules or crystallizing substances (oxalic acid) which primarily have been dissolved in blood may also cause oliguria and/or anuria. Because of the retention of substances which have to be excreted from the organism, uraemia with ulcers on the cutaneous and serous mucosae results which can be recognized because of a foetid smell from the mouth.

As it has been pointed out already with the description of the different syndromes of poisoning, several poisonous principles of plants, beside their specific organotropic action, eventually also lead to lesions of the urinary tract such as *Quercus* spp. (oak), *Allium* spp. (onion), *Pteridium aquilinum* (bracken) and others. Plants which contain oxalates are particularly damaging to the kidneys because they cause a haemorrhagic to serofibrinous glomerulonephritis with precipitation of calcium salts and triplephosphate in the tubules and herewith albuminuria.

4.2.10. Abortions and/or Other Effects on the Reproductive System Caused by Plant Poisoning

Aetiology, Occurrence and Pathogenesis
The content of phytooestrogens in numerous forage plants in particular may cause considerable economic loss to animal production in the tropics because of reduction of fertility. In Australia, the losses due to missing lambings and temporary infertility of female sheep are calculated with approx. 1 million missing lambings per year. Small amounts of phytooestrogen however may have a favourable effect on the fertility of the animal since they will stimulate oestrus and milk production, e.g. at the onset of the rainy season because of the higher oestrogen content of the young grass. Increased ingestion, however, of oestrogenic principles of plants as they have been identified only partly, e.g. like isoflavines (genistein, biochanin A, daidzein) and cumoestrol, cause hyperoestrogenic infertility. The phytooestrogens are prooestrogens which develop their oestrogenic effect only after transformation inside the organism. They occur especially in clovers (*Trifolium* spp.) and alfalfa (*Medicago sativa*) but also in grasses, including cereals, their content depending much on the vegetative stage, soil quality, fertilization and the climate. Clover has the highest oestrogen content shortly before and when blooming, and in grasses it is during the stage of fast growth at the onset of the rainy season. In hay and silage, the phytooestrogen content of the fodder is only slightly reduced. Hyperoestrogenic

symptoms in cattle and sheep appear only if considerable amounts of roughage with an increased content of phytooestrogens have been ingested as it is often unavoidable on a tropical pasture when the rainy season starts after a long dry season.

The **clinical symptoms** of the hyperoestrogenic syndrome are most evident in young females. 3 weeks after the onset of the rainy season the animals will show oestric behaviour, a lustrous hyperaemic enlarged vulva, perhaps prolapse of the vagina, and a bull-like development of the head and withers in cattle. The juvenile udder of heifers becomes enlarged, the milk production of lactating cows is reduced, irregular and repeated oestrus appears, and pregnant animals become oestric during the first third of pregnancy and abort. The fertility rate may drop down below 25%.

Characteristic **necropsy** findings are a small cystic degeneration of the ovaries, hydrosalpings, thickening of the uterus with cystic dilatation of the glands and inflammation of the endometrium.

The **diagnosis** is obvious when irregular spontaneous oestrus appears after the onset of the rainy season especially in young animals presenting the above described symptoms.

As a **preventive** measure, an adapted pasture management is important, e.g. by reserving preserved pastures of ripe fodder for the time when the rains start. In production systems with fodder cropping such species of clover should be chosen which are known for their low content of phytooestrogens.

4.2.11. Mycotoxicoses

Intoxications caused by mycotoxins produced on forage plants do not fit all that easily into the scheme of organo-specific symptoms as it has been chosen for the previous presentation. Therefore, the mycotoxicoses of economical importance to tropical animal production are described together in a separate chapter.

Mycotoxicoses are acute or chronic intoxications due to the exposure to feed or bedding contaminated with toxins which may be produced during growth of various saprophytic or phytopathogenic fungi or moulds on cereals, straw, pasture or any other fodder. Mycotoxicoses are characterized by the following principles:

-   the cause might not be immediately identifiable;
-   they cannot be transmitted from one animal to another;
-   treatment with drugs or antibiotics has little effect on the course of the disease;
-   outbreaks are usually seasonal because particular climatic sequences may be favourable to fungal growth and toxin production;
-   the intoxication is mostly associated with the particular feed;
-   the presence of fungi in feed does not necessarily indicate that toxin production has occurred. Because several mycotoxins may often be associated

in the cause of an intoxication, it can be difficult to establish clear clinical pictures. Several mycotoxins are immunosuppressive allowing other pathogens to create secondary infections (Merck 1991).

• Aflatoxicosis

Aetiology, Occurrence and Pathogenesis
Aflatoxins are produced by toxigenic strains of *Aspergillus* spp. (*A. flavus, A. clavatus, A. fumigatus, A. chevalieri, A. niger, A. parasiticus*) which develop luxuriant especially in the humid tropical climate on feed stuffs. Peanut straw and oil cake, soybeans, soybeans straw and oil cake in particular are often contaminated along with cottonseed meal, either in the field or during storage when the moisture content and temperatures are sufficiently high for mould growth. The fungi may also multiply on inappropriately prepared silage made from maize, sorghum and sugar cane. *A. flavus* produces aflatoxin on a substrate which is appropriate for its development which even in a dose of 0.2–15.0 ppm is toxic for calves after feeding for 6–12 weeks. In the USA, the disease caused by feeding mouldy corn is known as "mouldy corn poisoning". Aflatoxicosis occurs in many parts of the world and affects young poultry, especially ducklings and turkey poults as well as young pigs, pregnant sows, calves and even dogs. Adult cattle, sheep, and goats are relatively resistant to the acute form of the disease, but show susceptibility to toxic diets fed over long periods. Metabolites of aflatoxin (aflatoxin M_1 and M_2) are excreted in milk.

Aflatoxin poisoning which occurs in all animal production systems of the tropics, but especially in the case of small-holders, is mostly caused by feed contaminated with *A. flavus*. In particular peanut- and soybean straw which has been harvested and stored inappropriately can cause severe poisoning in ruminants, monogastric animals and poultry. Peanut, soybean and cottonseed oil cake might also lead to poisoning.

Aflatoxins bind to macromolecules, especially nucleic acids and nucleoproteins. Their toxic effects include mutagenesis due to alcylation of nuclear DNA, carcinogenesis, teratogenesis and immunosuppression. The liver is the principle organ affected. High doses of aflatoxins produce severe hepatocellular necrosis; prolonged low dosage leads to a reduced growth rate and liver enlargement.

The **clinical symptoms** of the acute mouldy corn poisoning caused by corn contaminated with *A. flavus, A. fumigatus, A. chevalieri* and *Penicillium rubrum* are characterized by multiple haemorrhages on the visible mucosae and perhaps also on the skin as well as in the anterior eye chamber. Nosebleeds, haemoglobinuria and anaemia are also characteristic. The animals are depressed, lacrimate, refuse feed and have diarrhoea. They become emaciated and die with signs of dehydration. Death may also supervene suddenly following shock with acute apnoea and lung oedema. Characteristics of the chronic course are lacrimation, hyperkeratosis of the skin on the head, neck and trunk as well as emaciation and icterus. There is frequently a high incidence of concurrent infectious diseases, often respiratory ones, which respond poorly to chemotherapy.

The clinical symptoms of aflatoxin poisoning caused by *A. flavus* alone in tropical animal production systems can be similar to *Senecio* poisoning (II/4.2.3). Ataxia, twitching with the ears, grinding of the teeth, apathy and breaking down followed by diarrhoea with prolapse of the rectum, and subsequent recumbency until death supervenes are typical signs of the course of disease in calves. The milk yield will be reduced in adult cattle, the toxin being excreted with the milk. Loss of weight will also become evident.

Pathological features of acute aflatoxicosis are widespread petechiae on the serosae, degeneration and necroses of liver and kidneys and distinct icterus of all serosae. Microscopically, the liver shows marked fatty accumulations and massive centrilobular necrosis and haemorrhage. In subacute cases, the hepatic changes are not so pronounced, but there is some liver enlargement and an increased firmness. There can be oedema of the gall bladder. Microscopically the liver shows proliferation and fibrosis of the bile ductules and an increase in the size of hepatocytes and their nuclei (megalocytosis). The gastrointestinal mucosa may show glandular atrophy and associated inflammation. There may be tubular degeneration and regeneration in the kidneys. Prolonged feeding of low concentrations of aflatoxins may produce diffuse liver fibrosis (cirrhosis) and cholangio- or hepatocellular carcinoma (Merck 1991).

Special pathological features of the *A. flavus* poisoning are severe oedemas in the intestinal system as well as exudates in the body cavities as well as pale colour, enlargement and increased firmness of the liver. Histologically, proliferation and fibrosis of the bile ductules, obliterating endophlebitis of the centrilobular liver veins, diffuse fibrosis and polymorphy of the liver cells will be found (Rosenberger 1970).

The **diagnosis** of the aflatoxicoses is extremely difficult. Poisoning with organic phosphoric acid esters and with *Seneciaceae* will cause similar clinical and pathological signs. The presence and levels of aflatoxin in the feed should be determined. Under field conditions, extracts of suspicious feed can be given to ducklings which die within 7 days with characteristic signs. Aflatoxin M_1 can be detected in the urine of affected animals if intakes of the toxin are high.

The **treatment** of aflatoxicosis is almost hopeless. One can try applying protective treatment to the liver with methionine and dextrose.

As a **preventive measure**, avoidance of contaminated feeds by monitoring batches for aflatoxin content is suggested. Young, newly weaned and pregnant and lactating animals require special protection from suspected toxic feeds. Dilution with non-contaminated feed stuff is one possibility for coping with the problem. Ammoniation of grain reduces contamination but is not currently approved in the USA for use in food for animals. Hydrated sodium calcium aluminosilicates (HSCAS) have recently shown promise in reducing effects of aflatoxin when fed to pigs or poultry; at 5 kg/ton they provide substantial protection against dietary aflatoxin. HSCAS reduces, but does not eliminate, residues of aflatoxin M_1 in milk from dairy cows fed aflatoxin (Merck 1991).

In small-holder animal husbandry, it is important to prevent the use of contaminated peanut- and soybean oil cake and straw as feed and to advise the farmers on appropriate harvesting and storage of straw.

• Fescue Poisoning

Aetiology, Occurrence and Pathogenesis

Tall fescue (*Festuca arundinaceae* Schreb.) is used increasingly to improve tropical pastures in association with leguminosae. In Latin America in particular, the tough grass which typically grows in tussocks and is especially resistant to grazing and trampling has become common in the lowlands. Dense populations of *Festuca* as well as other grasses are invaded during hot, humid weather by *Acremonium coenophialum* which endophytically parasitizes the invaded grasses without presenting any signs. The hyphae are found in the stalk as well as in the sheath and in the seed. *A. coenophialum* produces a complex of ergot alkaloids, mainly loline and perloline. Perloline *in vitro* inhibits the cellolytic activity of the rumen flora. Together, the toxins produced by *A. coenophialum* cause growth of the intima and media of the peripheral arterioles which leads to the symptoms described below. Two other fungi from toxic pastures have been implicated in causing fescue foot. *Fusarium sporotrichioides* produces a butenolide *in vitro*, which is capable of inducing some of the lesions of the disease. In addition, *Balancia epichloe*, like *A. coenophialum*, a clavicipitaceous endophyte, can synthesize ergot alkaloids in culture. Unequivocal evidence that any of these fungi can cause fescue foot, however, is lacking. The aetiology of fescue foot remains unresolved (Merck 1991).

Some reports indicate an increased incidence of fescue lameness as the age of the plants increases, and also as a consequence of severe droughts. Strains of tall fescue vary in their toxicity due to variation in infection level with the fungus, and high variability within a strain. High nitrogen applications appear to enhance the toxicity. The susceptibility of cattle is subject to individual variation. Low environmental temperatures are thought to exacerbate the lesions; however, high temperatures increase the severity of other toxic problems (Merck 1991).

The **clinical symptoms** of fescue poisoning become manifest in different clinical pictures:

– fescue foot (fescue lameness, pie de festuca, Schwingelgraslahmheit);
– summer fescue toxicosis (summer syndrome, epidemic hyperthermia, dysthermic syndrome);
– a developmental disorder and/or cachexia with necrosis of the fatty tissues.

Fescue foot, a dry gangrene of the extremities has been reported from many countries in the tropics as well as in temperate climates. The first signs develop within 10–14 days of grazing on tall fescue grass. There is local heat, swelling, severe pain and lameness of one or more feet. A hind foot is usually affected first. With continued feeding on fescue, an indented line appears at some point, usually between the hock and the claws. Dry gangrene affects the distal part, which eventually may slough. Low environmental temperature is thought to contribute to the severity of the lesions. Affected animals usually have a fever, seek shade in warm weather, and stand in water if it is available (Merck 1991). In sheep, the lesions may also appear on the distal parts of tail and ear.

Summer fescue toxicosis is characterized by a reduction in feed intake, weight gain or milk production in cattle, sheep and horses during the rainy season when grazed on tall fescue. In addition to reduced performance, other signs may appear within 1–2 weeks after fescue feeding has started, and include fever, tachypnoea, a rough coat, lower serum prolactin levels, and excessive salivation; the animals seek wet spots or shade. Lower reproductive performance also has been reported. Agalactia can appear in horses and cattle. Thickened placentae and the birth of weak foals have been observed in horses. The severity increases when environmental temperatures are above 24–27 °C and if high nitrogen fertilizer has been applied to the grass. The breeds which are affected the most are those with a dark coat (Aberdeen Angus); Cebus and their crosses are relatively resistant (Merck 1991, Renner 1987).

Developmental disorders and peritoneal fat necrosis occur in cattle which are about 2 years old after prolonged grazing of tall fescue infected with *A. coenophalium*. The condition is seen in the tropics where tall fescue is used as a primary pasture. Hard masses of necrotic fat form in the omentum, mesentery, and perirenal fat and may cause clinical disease when they compress the intestine, obstruct the birth canal, or ureters (Merck 1991). The condition may also lead to cachexia, and developmental disorders may appear in the foetuses of affected pregnant cows.

At **necropsy** there are few, if any, lesions other than those associated with the swelling or dry gangrene in fescue foot. The pathological lesions of summer fescue toxicosis and peritoneal fat necrosis correspond to those described with the clinical signs.

The **diagnosis** of fescue foot results from the characteristic demarcation of distal parts of the limbs in connection with grazing on festuca pastures. In contrast, the summer fescue toxicosis and peritoneal fat necrosis are much more difficult to diagnose. The connection with grazing on infested festuca pastures always has to be looked for.

An effective **treatment** for festucoses is unknown.

The **prevention** of festucoses can be achieved through fodder cropping and pasture management. The association of festuca with leguminosae may prevent the outbreak of festucoses. Since maturing festuca grasses become infested with *A. coenophalium*, the pastures should be grazed while they are still new, and old and overripe fodder should either be cut or burned. Fescue pastures are especially dangerous if they are reserved for the dry season being populated with overripe and highly contaminated grass under these conditions.

- **Ergotism (Ergotismus)**

Aetiology, Occurrence and Pathogenesis
During the rainy season in humid tropical environmental conditions, on pasture grasses, ryegrass (*Lolium rigidum*) and bluegrass (*Festuca pratensis*) along with other forage plants may develop in particular the sclerotia of *Claviceps purpurea*. The hard, black, elongated sclerotia can contain varying quantities of ergot alkaloids of which the levorotatory alkaloids, ergotamine and ergonovine (ergometrine) are pharmacologically most important. The pharmacologic effect of these alkaloids causes vasoconstriction by directly affecting the muscularis of the arterioles and injure to the vascular endothelium following continued intake. These actions initially result in reduced blood flow and eventually complete stasis with terminal necrosis of the extremities due to thrombosis. A cold environment predisposes the extremities to gangrene. In addition, ergot has a potent oxytoxic action and also causes stimulation of the CNS, followed by depression. Cattle, sheep, pigs and even poultry may be affected by ergotism (Merck 1991).

About 2–6 weeks after cattle start grazing on infested pastures the first **clinical signs** appear, depending on the concentration of alkaloids in the ergot and the quantity of ergot in the feed. The first sign is lameness, the hind legs being affected before the fore legs, but the extent of involvement of a limb and the number of limbs affected depends on the daily intake of ergot. Body temperature and pulse and respiration are increased, the animals refuse feed, and diarrhoea, trembling and muscle convulsions appear. A stiff, stilted atactic gait, sudden break down, and occasional jumping as well as aggressiveness towards fellow animals are further symptoms. Swelling and tenderness of the fetlock joint and pastern are associated with the lameness. Within about a week, sensation is lost in the affected part, an indented line appears at the edge of normal tissue, and dry gangrene affects the distal part. Eventually, one or both claws or any part of the limbs up to the hock or knee may slough off. In a similar way, the tip of the tail or ears may become necrotic and slough off. Exposed skin areas, such as teats and udder, appear unusually pale or anaemic (Merck 1991, Rosenberger 1970).

In sheep, signs of ergotism are similar to those of cattle. Additionally, the mouth may be ulcerated. A convulsive syndrome has been associated with ergotism in sheep.

In pigs, ingestion of ergot-infested grains may result in reduced feed intakes and reduced weight gains. If fed to pregnant sows, ergotized grains result in lack of udder development with agalactia at parturition, and the birth of small litters of weak, undersized piglets of which few survive (Merck 1991).

At **necropsy**, the only constant lesions in cattle are in the skin and subcutis of the affected extremities. The skin is normal down as far as the indented line, but beyond that it is cyanotic and hardened in advanced cases. Subcutaneous haemorrhages and some oedema occur in close proximity to the necrotic area. In sheep, a marked intestinal inflammation may be seen.

A presumptive **diagnosis** can be made if the characteristic signs can be associated with grazing on ergot-infected pastures. When fescue foot occurs in cattle, identical signs and lesions of lameness, and sloughing of the hooves and tips of ears and tail are seen. Extraction and detection of ergot alkaloids may be done in the case of suspect ground grain meals.

No efficient **treatment** is available. If the disease is recognized in time, the affected animals can be brought to a non-contaminated pasture where the lesions may heal.

As a **prevention**, pastures prone to ergot infestation should frequently be grazed and flower head production should be prevented. Association of pasture grasses with leguminosae can reduce the incidence of ergotism.

• Paspalum Staggers

Aetiology, Occurrence and Pathogenesis
The fungus *Claviceps paspali* develops on the ears of dallis- or bermudagrass (*Paspalum dilatum* and *P. notatum*) especially in a hot, humid tropical environment. The yellow-grey sclerotia, which mature in the seedheads after the rainy season, are round and 2–4 mm in diameter; they contain lysergin acid α-oxyethylamide. The life-cycle of this fungus is similar to that of *C. purpurea* (see ergotism). Ingestion of the sclerotia causes nervous signs in the affected animals, the toxicity not being ascribed to ergot alkaloids; paspalinine and paspalitrem A and B, tremorgenic compounds from the sclerotia, are thought to be the toxic principles. Horses, cattle and small ruminants are susceptible to the intoxication. Symptoms of intoxication already appear after the ingestion of 100 g sclerotia, a single dose causing signs that persist for several days. The mature ergots are also toxic, they are most dangerous just when they are maturing into the hard, black sclerotic stage (Merck 1991).

The **clinical signs** start with the animals showing excitation and continuous trembling of the large muscle groups. If they attempt to move, their movements are jerky, and limb movements are uncoordinated. If they attempt to run, they fall over in awkward positions. Condition is lost after prolonged exposure, and complete paralysis can occur. Milk production is reduced and diarrhoea may be present. After continued grazing on contaminated grass, the animals become recumbent until death supervenes (Merck 1991).

At **necropsy**, no specific signs are found.

No causal **treatment** is known, recovery follows removal of the animals to feed not contaminated with sclerotia of *C. paspali*.

As a **preventive measure**, suspicious pastures should be grazed before they come into bloom, or the pasture should be topped to remove affected seed heads (Merck 1991).

• Ryegrass Toxicity

Perennial Ryegrass Staggers

Aetiology, Occurrence and Pathogenesis
In parts of the USA, Europe, Australia and New Zealand, a neurotoxic condition appears in livestock of all ages grazing on pastures where perennial ryegrass (*Lolium perenne*) or hybrid ryegrasses are the major components. Horses, cattle, sheep and farmed deer of both sexes are susceptible, the individual susceptibility, varying greatly. This trait can be inherited.

Tremorgenic neurotoxins, lolitrems, mainly lolitrem B make up the toxic principle. These indole toxins are produced in perennial and hybrid ryegrasses infected with the endophytic fungus *Acremonium loliae*. The amounts of fungal hyphae and lolitrem B in infected plants increase to toxic levels during the rainy season and decrease again to safe levels in the cooler season. Mycelia of the fungus are present in all the above-ground parts of infected plants but are especially concentrated in the sheath of the leaf, flower stalks, and seed. Infected plants exhibit no signs, and the fungus is only spread through infected seed. Viability of the endophytes gradually declines when infected seed is stored at ambient temperatures and moderate humidity so that few seeds contain viable endophytes after 2 years. Neurotoxic tremorgens are believed to produce incoordination of movement by interference with neuronal transmission in the cerebral cortex through the production of a reversible biochemical lesion (Merck 1991).

Clinical signs develop gradually over a few days. Fine tremors of the head and nodding movements are the first signs noted in animals if they are approached quietly and watched carefully. Noise, sudden exercise, or fright elicits more severe signs of head nodding with jerky movements and incoordination if they are forced to move. Running produces stiff, bouncing movements with marked incoordination, and often results in collapse and lateral recumbency with opisthotonus, nystagmus, and flailing of stiffly extended limbs. The attack soon subsides, and within minutes the animal regains its feet and rejoins the group. If forced to run again, the episode will be repeated. The signs are most severe when animals are heat stressed. In outbreaks, morbidity may reach 80–90%, but mortality is low (0–5%). Deaths are usually accidental or due to the inability to forage for food and water (Merck 1991).

At **necropsy** no specific signs are found.

A presumptive **diagnosis** can be made if the seasonal occurrence of characteristic tremors, incoordination, and collapse in several or many animals can be associated with grazing on predominantly perennial ryegrass pastures (Merck 1991).

No specific **treatment** is known, since movement and handling of animals exacerbates clinical signs, individual treatment is generally impractical. Recovery is spontaneous in 1–2 weeks if animals are moved to non-toxic pastures or crops (Merck 1991).

Because the endophytes and lolitrems are not uniformly distributed within the ryegrass plants, control by grazing management can help **prevent** the disease. Well controlled leafy pastures, which are neither allowed to bolt to seed, nor are overgrazed down to the sheath of the leaf, are likely to provide safe grazing during the dangerous season, even when a high proportion of ryegrass plants are infected with endophytes. Encouragement of the growth of other grass species and clover in established swards also reduces the intake of toxic ryegrass. Safe new pastures can be established using ryegrass seed with little or no endophyte infection (Merck 1991).

• Annual Ryegrass Staggers

Aetiology, Occurrence and Pathogenesis
In Australia and South Africa, an often fatal neurotoxic disease occurs in livestock of any age and sex which graze on pastures with annual ryegrass (*Lolium rigidum*) which is in the seedhead stage of growth. The responsible corynetoxins are produced in seedhead galls induced by the nematode *Anguina funesta* and colonized by *Clavibacter* sp. These bacteria-infested galls are present in infected annual ryegrass pastures from the beginning of the rainy season onward but animals show no signs of intoxication until the peak of the vegetative period. The spread of bacteria-infested nematodes to adjacent healthy annual ryegrass pastures is slow. The corynetoxins are highly toxic glycolipids that inhibit specific glycolysation enzymes and therefore deplete or reduce activity of essential glycoproteins. Experimentally, the corynetoxins deplete fibronectins and cause failure of the hepatic reticuloendothelial system.

Outbreaks occur 2–6 days after animals graze on a pasture which contains annual ryegrass infected at a toxic level; deaths occur within hours, or up to a week after the onset of signs.

Clinical signs are those of neurological disorders similar to those of perennial ryegrass staggers (see above). Tremors, incoordination, rigidity, and collapse when stressed, with animals often becoming apparently normal again when left undisturbed. When animals are severely affected nervous spasms supervene, and convulsions in recumbency are soon followed by death. Mortality from annual ryegrass toxicity is commonly 40–50%, occasionally greater (Merck 1991).

Pathological lesions include congestion, oedema and haemorrhage of the brain and lungs, and degeneration of the liver and kidneys.

The **diagnosis** is based on the characteristic neurological signs as described above. Close regard to the history and contents of the pastures grazed will assist in differentiating between staggers caused by perennial ryegrass, phalaris, and the ergots of paspalum and other grasses; polyencephalomalacia and enterotoxaemia are other differential diagnoses (Merck 1991).

No specific **treatment** is practical.

The intoxication can be **prevented** or at least minimized by early recognition of signs and removal to safe grazing areas or by reducing the stocking rate. Grazing on hay remains should be avoided. Burning toxic annual ryegrass paddocks in the dry season destroys most of the galls colonized by bacteria and minimizes the risk of toxicity in the following season (Merck 1991).

4.3. Diagnosis of Plant Poisoning and Detection of Poisonous Plants

4.3.1. Diagnosis of Signs of Plant Poisoning in the Diseased or Dead Animal

The way particular plants cause characteristic syndromes of poisoning has been discussed in chapter II/4.2. Though the syndromes cannot be grouped precisely to a determined plant species, they give presumptive diagnostic hints. Osterhoff (1981) has compiled a correlation of syndromes to presumptive causal poisonous plants. TOXLINE is a data base which is available in the USA and describes the poisonous plants of North America and their poisonous principles. By means of appropriate key words related to the clinical and pathological features found with the poisoned animals in the field, information about the presumptive causal poisonous plant can be obtained from the data bank (Wagstoff et al. 1989). Plant poisoning should be suspected whenever animals fall ill sporadically or enzootically after a change of pasture, or after they have grazed the pasture down to the stalks of the fodder plants, presenting symptoms of low or subfebrile temperature, diarrhoea, and being dazed. They can also show icterus and perhaps petechiae on the visible mucosae as well as haemoglobinuria. It must, however, always be taken into account that autochthonous animals will not usually show symptoms whereas exotic animals become poisoned and die presenting severe symptoms.

4.3.2. Detection of Poisonous Plants or of the Poisonous Principle

For the diagnosis of plant poisoning, it is necessary to correlate the symptoms of poisoning in the animal with an aetiologically relevant poisonous plant. Only then can preventive effective measures be applied. Parmelee et al. (1960) developed a key of symptoms of plant poisoning which has been structured like a botanic key for the identification of unknown plants. Based on the symptoms of poisoning step-by-step, the responsible poisonous plant is determined. In order to confirm a presumptive diagnosis and to identify the poisonous plant(s) the following diagnostic tools will be helpful:

- observation on the pasture (grazing behaviour of the animals): which plants are grazed or can be reached across the fence of the pasture;
- feeding trials on the pasture or in a pen under normal grazing conditions or as forced feeding;
- trials to detect the poison inside the gastrointestinal tract of the animal (cyanide) or gross and/or histological identification of rests of plants in the rumen.

• Observations on the Pasture

An important prerequisite for the recognition of plant poisoning is a diary of pasture management. After the appearance of presumptive plant poisoning, it has to be reconstructed when the animals were moved to another pasture and how long the animals have grazed on a particular pasture. It is furthermore important to know which breed/cross, category of animals and age group has been affected significantly by the poisoning. Poisoned animals in the extensive tropical production system (ranch) are often only found when they are dead. Therefore, anamnestic observations are especially important. Experienced animal owners and herdsmen pay good attention to their animals, so they will notice changes in the behaviour of the animals and symptoms of disease early on. They are also mostly aware of potentially poisonous plants which occur in the region. The situation is completely different where exotic highly productive breeds have been distributed to traditional animal holders. The Indians in the Andes know by experience that their Criollo cattle will on no account browse on *Cestrum* spp. Therefore, they do not suppose that exotic cattle which have been given into their care will willingly if not greedily ingest leaves of these bushes. Similar circumstances have been observed in India where small-holders were supplied with German Holstein-Friesian cattle. When the animals were tethered along the roadside, the owners did not mind observing the animals browsing on *Cestrum* or castor oil bushes (Seifert 1995).

It will be rather difficult to identify the relevant poisonous plant if the investigation is started from the botanical view point of the problem. Known poisonous plants which grow on the pasture by no means have to be the cause of poisoning. In order to screen a pasture carefully for its poisonous potential, it is advisable to walk along the paths the animals have made on the pasture, and to inspect the botanical composition at irregular intervals of some square meters on both sides of the path and in the surrounding of the water places and salt licks. One should also observe whether or not potential poisonous plants have been grazed on heavily or only a little.

• Feeding Trials

If it is impossible to identify a poisonous plant on the pasture, it has proven valuable to divide the suspicious pasture into allotments, with electric fences for example, each representing a characteristic plant community, such as a wooded pasture, a bushy pasture, a meadow, a natural pasture, or a seeded pasture. 20–30 cattle or sheep in the most affected category put on areas of about 100 × 100 m will often present symptoms of poisoning either soon or a few days after being exposed to the poisonous plant in the allotment where it grows. The trial can be varied by supplying the animals either with sufficient or limited drinking water. If in one of the allotments symptoms of poisoning or even deaths appear systematically, the suspicious plant can be looked for.

If a suspicious plant has been identified it should be fed to target animals preferably mixed with good, palatable feed. Calves are most suitable since they are usually those most susceptible and represent a relatively low economic value. Even when a poisonous plant has been mixed with good palatable feed, it will be refused if it is not ingested under natural conditions on the pasture; it will probably be refused under all circumstances by autochthonous animals.

- **Detection of Poisonous Principles by Chemical Analysis**

Under the conditions of a tropical animal production system, only the picric acid test for the detection of cyanide can be carried out in the field. Apart from the fact that the poisonous principle of many poisonous plants still is unknown, the analysis of poisonous compounds which are mostly reabsorbed from the gastrointestinal tract soon after ingestion requires a lot of effort which is rarely technically possible.

4.4. Prevention of Plant Poisoning

4.4.1. Removal of Poisonous Plants

As already mentioned several times, the incidence of plant poisoning correlates directly with the intensity of grazing and the management of the pasture. It is extremely difficult to control poisonous plants on overgrazed pastures. With progressive degeneration of the pasture, poisonous species will become dominant. Even after a resting phase, these pastures will not recover and turn back into good grassland. But even with appropriate pasture management, poisonous plants gradually may also dominate the plant community. Such changes in the plant community will appear especially in areas with a long, dry season. It is only possible on occasion to remove poisonous plants with mechanical or chemical means from the pasture, such an approach only being feasible in an intensive production system applying considerable financial resources. Normally one should try to manage the vegetation and the pasture in such a way that the animals do not ingest critical amounts of poisonous plant material.

In order to stabilize a favourable botanic composition of the pasture as far as the provision of an optimal diet for the animal is concerned, unfavourable influences of plant selection by the grazing animal have to be minimized. Therefore, it has to be decided from case to case whether a permanent or rotating pasture is the most appropriate for the particular production system and the particular ecological region. Stocking rate and perhaps fertilization are factors which have to be evaluated economically; the same is true for fodder cropping and the implementation of seeded pastures. Because of long time experience, it is not the fodder potential of a favourable year but of an unfavourable year which is the criterion applied to the stocking rate of a ranch in Texas and New Mexico. This prevents a varying degree of productivity of the pasture from leading to a degradation of the pasture which is favourable to the development of poisonous plants.

Measures to **remove poisonous plants** are required where it is impossible to keep the vegetation of the pasture in balance and where the economic means are available to tolerate a higher stocking rate. Poisonous plants can be removed from the pasture through mechanical, chemical or biological means and can eventually be eliminated definitely. The pasture can also be burned. Burning the pasture is only efficient if the optimal time for the destruction of the poisonous plant is chosen. Most poisonous plants are bushes or herbs. Therefore they may be eradicated selectively with herbicides. Chemical compounds for the removal of poisonous plants are relatively easy to handle and can be applied selectively. For the treatment of singular plants as well as for areal application, herbs can be destroyed with 2,4-D-compounds, wood species with 2(2,4,5-trichlorphenoxy)-propionic acid and 2,4,5-T-compounds. 3,6-dichlor-2-methoxybenzoeacid is applied to remove herbs and bushes. Herbicides have to be applied during the vegetative phase in order to be efficient.

Bushes can be removed mechanically, and if this is not possible, the root-stock should be dug out, and the surface of cut branches or stems can be treated with herbicides. In most cases, one single treatment will be enough.

St. John's wort (*Hypericum perforatum*) can be destroyed with the beetle *Chrysoline gemellata* and *C. hyperici*; ragwort (*Senecio jacobaea*) is destroyed on the west coast of North America by applying the larvae of the cinnabar moth *Tyria jacobaea*. Sheep, which are relatively resistant to *Seneciaceae*, can be used to reduce the population of ragwort before the contaminated pasture is grazed by cattle.

The direct eradication of poisonous plants will be economically feasible where the poisonous plants grow in clusters because of the particular soil quality, for instance in depressions (*Astragalus* spp.). Under those circumstances they may even be fenced out and the contaminated area can be reforested. Poisonous plants are often found at the edge of the pasture, close to the water place, in the surroundings of the pen and along the paths of the animals where the soil has become especially compacted. Careful observation and removal of potential poisonous plants from those areas are prerequisites for each further measure for the intensification of pasture management.

Only rarely will one single measure alone be sufficient for removing a poisonous plant from a pasture; all measures for the removal or eradication of poisonous plants have to be accompanied by those for improving the pasture and the management of the pasture. After treatment of a pasture, the area has to be left to rest at least for 90 days in order to allow the fodder plants to regenerate and, if possible, to dominate the plant community. One has to make sure that non-perennial plants get a chance to multiply and that the ecological balance on the pasture has recuperated.

It is most difficult to monitor the effect of the measures of removal of poisonous plants from the pasture. Often the plants sprout from remaining roots or birds bring back the seeds of poisonous bushes (*Cestrum* spp.). Shoots which often go unnoticed in the pasture grass are usually especially poisonous (*Cestrum* spp.).

4.4.2. Management Rules for the Prevention of Plant Poisoning

One of the most important preconditions for the prevention of plant poisoning is the provision of enough feed and water to the animal because

- some poisonous plants only develop on an overgrazed pasture; these are especially the so-called invading plants (*Senecio* spp. and *Geigeria* spp.). Gradually they will replace the original plants almost completely;
- most poisonous principles depend on the ingested dose as far as their toxicity is concerned. If enough feed is provided, it will dilute the poisonous effect. Natural pastures of the tropics have a large diversity of species and are therefore good for animals with a good ability for selection;
- animals in bad condition are more exposed; their host defence mechanism is reduced and with it their ability for detoxication.

Knowledge of and attention to the following management rules may protect almost all animal production systems of the tropics against economical losses caused by poisonous plants:

- Recognize locally occurring poisonous plants and their poisonous principle.
- Adapt the stocking rate to the conditions of the pasture.
- Avoid areas with a high incidence of poisonous plants by either fencing them out or keeping the animals away from suspicious spots.
- Never hurry the animals when they are driven through an area which is highly contaminated with poisonous plants; the animals only select their feed when left alone.
- Change the pasture under pasture rotation before it is completely grazed down. Animals should not be used to remove weeds.
- When the animals are brought to the pasture after the rainy season, one has to make sure that enough fodder is available (stocking rate); poisonous herbs and bushes are especially attractive during this time of the year.
- The animals should be moved from dry pasture with a high percentage of crude fibre to newly grown grass only gradually, and perhaps supplementary crude fibre (hay, straw) should be provided. Animals newly brought to the pasture should be adapted to it gradually.
- Provide enough water.
- Provide enough mineral supplements.
- Programmes for the up-grading of local breeds towards highly productive breeds should be implemented only as far as pastures free of poisonous plants are available.

- Destroy poisonous plants if they become dominant on the pasture (Osterhoff 1981).

Specific management rules for the prevention of plant poisoning in the different animal production systems of the tropics are presented in chapter III.

References[1]

Carcamo Vega, J. (1957): Estudio Botánico de las Principales Plantas Tóxicas y Daninas para el Ganado en el Perú. Escuela Nacional de Agricultura, Lima/Perú.

Cheeke, P. R. and L. R. Shull (1985): Natural Toxicants in Feed and Poisonous Plants. AVI Publishing, Westport, Connecticut.

Martinez Muñoz, A. (1990): Untersuchungen zu Möglichkeiten und Grenzen des Einsatzes von *Leucaena leucocephala* als Ergänzungsfutter für Ziegen im Nord-Osten Mexikos. Göttinger Beitr. Land. Forstwirt. Trop. Subtr. 55.

Merck Veterinary Manual (1991): A Handbook of Diagnosis and Therapy for the Veterinarian. Merck, Rahway, N.J., USA.

Mönnig, H. O. and F. J. Veldman (1956): Handbook on Stock Diseases. Nasionale Bookhandel, Johannesburg, S.A.

Muenscher, W. C. (1951): Poisonous Plants of the United States. The Macmillan Company, New York.

Osterhoff, F. (1981): Das Problem der Pflanzenvergiftungen in der Tierhaltung der Tropen und Subtropen unter besonderer Berücksichtigung der Rinder- und Schafhaltung in semiariden Gebieten. Diss. FB Agrarwiss., Göttingen.

Parmelee, G. W., C. L. Gilly and W. T. Gillis (1960): Key to Symptoms of Plant Poisoning. Michigan State Univ. Veterinarian **20**, 121–127.

Renner, J. E. (1987): Die Festukose (Schwingelgrasvergiftung) – eine wichtige Intoxikation bei Rindern in Südamerika. Dtsch. tierärztl. Wschr. **94**, 281–282.

Rosenberger, G. (1970): Krankheiten des Rindes. Parey, Berlin und Hamburg.

Seifert, H. S. H. (1960): Die Bekämpfung eines "Rindersterbens" in der Cordillere Nord-Perus. Zbl. Vet. Med. **10**, 929–1020.

Seifert, H. S. H. (1992): Tropentierhygiene. Fischer, Jena.

Seifert, H. S. H. (1995): Unpublished empirical findings.

Seifert, H. S. H. und K. A. Beller (1969): Blausäurevergiftung beim Rind, verursacht durch Abweiden von Zuckerrohr (Saccharum officinarum) und Zufütterung von Früchten des Algarrobobaumes (Prosopis juliflora). Berl. Münch. Tierärztl. Wschr. **5**, 88–91.

Sippel, W. L. (1964): Crotalaria Poisoning in Livestock and Poultry. Annals New York Academy of Sc. **111**, 2.

Watt, J. M. and M. G. Breyer-Brandwijk (1962): The Medicinal and Poisonous Plants of Southern and Eastern Africa. Livingstone, Edinburgh and London.

---

[1] Where not expressly mentioned, the information is mainly from Osterhoff (1981) and Seifert (1992). Osterhoff also cites further references.

Part III – Animal Health Management Adapted to the Tropical Animal Production System

1. Introduction

The scientific basis for understanding the epizootiology and prophylaxis of tropical diseases has been explained in part I; tropical diseases, divided into epizootiological complexes, have been described in part II. In part III, taking for granted that the first two parts have been read, a definition of the respective production systems will be presented along with the relevant relative advantages and disadvantages of the prevention of animal diseases within those structures. After this, based on these framework conditions, suggestions will be made for developing schemes for the prevention of animal diseases which take into consideration both, the socio-economic situation of the animal holder, and the ecological and political environment of the production system. In so doing, an effort will be made to reach a cross-margin for the business results of the respective animal holder through measures of animal health. The particular resistance of autochthonous breeds in the tropics as described in part I will be taken into account, and it is considered very important not to disturb the balance which may exist in autochthonous animal populations between pathogen and host with measures directed against the pathogen which are only effective short-term (chemoprophylaxis). The connection between the development of the immune response of the organism following natural infection or vaccination, and the influence of the environment explained in part I will also be given consideration. Without wanting to discredit the governmental veterinary infrastructures of tropical countries, it will be made clear how, without government help and in view of economic criteria, efficient disease control is feasible in each production system. In so doing, suggestions will be made which might contradict official regulations which are currently valid in some countries. The suggestions, which are based on distinct scientific knowledge, ought be taken note of by the responsible government bodies. The reason for such obvious discrepancies is usually animal disease legislation which accords with the relevant legislation of the highly developed industrial countries and ignores the socio-economic as well as, above all, ecological environmental conditions of animal production in the tropics.

2. Nomadism

2.1. Definition of the Production System

Nomadism is a production system which is only to be found rarely in its original form nowadays. The word nomadism comes from the Greek and means "looking for pasture". In order to be able to describe the animal health problems of pastoral animal husbandry, one ought none the less go back to its traditional framework. In order to roughly define the various nomadic groups, the UN Economic and Social Council decided the following:

- **nomadism** is the migration of the entire group with animals, whereby "pure nomads" concentrate on pastoral animal husbandry and are not involved in farming;
- **semi-nomadism** is likewise predominantly oriented towards pastoral animal husbandry; however farming may be done by a part of the group;
- **transhumance** is migration with a herd with the necessary herdsmen and parts of their families, while the rest of the group remains in more or less settled camps or villages.

Pastoral animal husbandry is as ever, relatively speaking, the best form of human utilization of the Sahel. It is an optimal active symbiosis between man and the ecology, and the only alternative for obtaining economic benefits without big investments. Expensive projects for altering this production system have failed without exception. Therefore it is necessary to keep nomadism as mobile and varied as possible in order to be able to make human utilization of the Sahel possible in the future (Janzen 1991). Adapted animal health measures must make a particular contribution to this.

In the dry regions of Northwest India and Pakistan, in the Near East, in North Africa and the Sahel zone south of the Sahara as well as in parts of Madagascar, nomadism is, in one or other of the forms described here, presumably the best system of utilization, relatively speaking, for dry areas with a yearly precipitation level of 100 up to a maximum of 700 mm. Traditional migration serves the transition between the seasonally determined availability of feed and water, although at the same time, seasonal high points of diseases and/or their vectors (tsetse flies) are avoided where possible.

Pure nomadism occurs predominantly on the steppes and savannahs of East Africa (Maasai, Samburu, Turkana, Galla, Pokot, Karamojong, Bahima, Watussi, Baggara etc.) and in the Sahel of West Africa (Fulani, Shuwa, etc.). The percentage of nomadic animal holders in the entire population is reducing everywhere. It fluctuates between 70–75% in Somalia, 60% in Mauritania, 20–30% in the Sudan, 20% in Niger and is only marginal in Tanzania for example (Berlal et al. 1981).

The nomads choose their main herd animals according to the various ecological conditions of the region they are living in. As as rule, it is cattle and camels added to which come sheep and goats. It can happen that small ruminants are the only remaining animals after a dry period. The size of the herd, the extent and the direction of the migration as well as the connection with farming are extensively dependent upon the availability of water for people and animals as well as the type and amount of pasture. Apart from this, the incidence of disease and spread of the disease vectors (tsetse flies) as well as the appearance of endoparasites, and finally access to salt, either as a natural occurrence or as trading article, determine the path of migration. The migration route, the main herd animals, the structure of the herd as well as the choice of pure or semi-nomadism are not static, but rather are subject to a dynamic adaptation process of interdependent influencing factors. This leads to yearly seasonal as well as long-term (dry season) fluctuations in herd stocks, and the main herd may have to be changed if ecological influences demand it (Baggara in the Sudan from cattle to camel husbandry). One of the results of a comprehensive study our group has carried out in eastern Sudan (Butana) was that the nomads will change the quality of their stock according to ecological conditions. Whenever the rain fail and the pasture becomes extremely scarce the first animals sold are cattle, while small ruminants are kept as long as possible. As soon as grazing conditions improve, efforts are made to upgrade the stock with cattle again (v. Schutzbar 1994, v. Schutzbar et al. 1994). The diversification of the herds and the choice of changing main herd animals are survival strategies and at the same time ensure adaptation to environmental conditions which maintains the balance between animal production and ecology. The diversification of herd stocks allows efficient utilization of pasture land and the reliable provision of nourishment. Mixed stocks do not compete for fodder because of their different grazing behaviour. Besides, camels and goats are more resistant to drought and can still produce milk when cattle are no longer able to do so. On the other hand, the latter produce more milk than goats and fetch a higher sales price than camels; beside this, they are more reproductive than camels (Reckers 1991, v. Schutzbar 1994).

The priority goal of nomadic and semi-nomadic animal husbandry is being able to feed large families. The less crop produce which is available, the stronger the dependence upon the produce of the animals. The well being of the animals is thus the middle point of an economic lifestyle, dependent upon nature and endangered in its existence, which primarily serves the needs of subsistence. The animals deliver milk for daily nourishment (for children about 80% of the food, and adults about 60%) meat, skins and leather for household goods and tents (Berlal et al. 1981, Holter 1994, Holter and Kirk 1994). Beef is only eaten on special occasions, unless the animal has died and has to be utilized immediately. In some nomadic tribes (Maasai, Samburu, Turkana), the blood from the cattle, taken from the jugular vein of the living animal, is a staple part of their diet. The health of the animals with regard to the spread of zoonoses (brucellosis, anthrax, tuberculosis, Q-fever) is of extraordinary importance above all where milk and blood are consumed raw (Seifert 1988).

Dry periods must be bridged with animal stocks, and a relatively high number of female animals (50–60%) is necessary to compensate for reduced fertility and to secure the ability of the herd to regenerate itself after losses incurred either through these dry periods or through disease (rinderpest). Next to this, the possession of a large number of animals determines the social regard of a family within a tribal community and builds a "social network". Animals serve as nuptial gifts and are slaughtered for cultural festivities. The question is, however, whether or not the position of animal ownership amongst the nomads outways the necessity for having a large number of mother animals for providing the required food basis for the group and as regeneration reserves of the stocks. It is also known of the East Pokot in Kenya that they sell their steers at the right time, and only retain cows as long as they are fertile (Reckers 1991, v. Schutzbar 1994, v. Schutzbar et al. 1994).

Depending on the position of the market and the mentality of the nomads, animal products are exchanged for crops products and above all for salt, all of which can be used to provide for the family and the animals. While the Turkanas in Northwest Kenya are almost self-sufficient, the Maasai practise intensive exchange with the farmers and get bananas and other agricultural crops in exchange for sour milk and meat, for example. The Fulani in West Africa virtually have a contract with the farmers in Senegal to be able to graze their animals on the ground-nut fields (ground-nut straw with high protein content) after the harvest which means that the fields are fertilized and provided with valuable organic material at the same time. Generally, only enough animals are sold to make it possible to satisfy those needs which are not satisfied by nomadic production itself (e.g. tea, sugar and household goods). The economic goal of the nomads is not as a rule the best marketing possible, but rather the security of subsistence. Thus the hypothesis exists that higher prices for nomads only mean that they have to sell less animals in order to be able to cover their production needs (Berlal et al. 1981). So, it is contradictory that the nomads are developing a clearly economic interest in selling animals ready for slaughter in countries with favourable export markets for cattle and sheep in the neighbouring Gulf States (Somalia, Sudan) (Janzen 1991, Kirk and Mai 1994).

In contrast to private ownership of animals, pasture land and watering points are used communally and cannot be bought and sold. All tribe and clan members have an equal right to use them. Thus the group with the largest number of animals has the greatest advantage. Migration is also connected to social divisions: in some tribal areas (e.g. Turkana), the situation is completely open, but in other clan or family areas (e.g. the Maasai), the process is traditionally determined. Some communities indirectly settle pasture rights through private ownership of watering points (Kirk 1994, Lewis 1961). The pure nomads change their location when pasture land has been exhausted and when they have information about better grazing areas, or reason to believe that there are better grazing grounds for the animals elsewhere. The less favourable the climatic conditions are, the more often migration occurs. However, migration is mostly undertaken between traditional pastures in dry and wet periods, as with the Fulani in West Africa. This can mean that perhaps only the younger members of the groups will look for pastoral areas which are farther removed (semi-nomads). Large herds are divided up and are grazed by different group members in order to avoid passing on diseases (dispersion). In the process, herds are also formed into animal categories (mother cows, steers for slaughter). Water conditions also influence migration. The Cebus in the Sahel need water at least every 3 days and so cannot be removed any further away than 30–40 km from a watering point. The animals are usually penned at night in the camp to protect them from theft and preying animals. Providing for and holding calves is problematic for most nomads. For one thing, calves only have a low status – whereby heifers as future mother cows are better treated – and for another, the group is dependent upon the milk of the mother cows. As a rule, the calves are only allowed what remains of the milk in the morning and the evening after the milking has been done, and during the day are locked up in a pen which may be muddy and filthy. It is only when they can get by on raw feed and water that they are grazed separately from the cows. Rates of loss of 20–25% for heifers and up to 50% for calves have been observed (Wilson 1963). The fertility of the mother cows is extremely low because of shortages of fodder, minerals and disease. It can be as low as 40% (Oyenuga 1966). Brucellosis plays a big part in causing infertility; applying the Coombs test we found an infection rate in cattle of 36.5% in Senegal and 22% in the Sudan, in sheep of 13.6% in Sudan and 32.8 or 27% in camels in the Sudan (Seifert 1990, Weiser 1995).

The communal rights of utilization adequately comply with nomadic production under the conditions of an extensive economy. Where access to wells is controlled and their condition maintained by tribes and clans through traditional structures, there is a certain guarantee that the production system stays in balance, and the destruction of the environment through overstocking is prevented. This is of particular importance where new water points are created by well or reservoir construction. The East Pokot make use of such foreign-built installations as these according to traditional rules: the water is reserved for humans and young animals in dry periods. Thus the area surrounding the watering point does not get overgrazed; the dam is closed to all for some time during periods rich in rain if there is sufficient water available. A regimented system of dam management was introduced which regulates water usage in a sensible way (Reckers 1991). The fluctuating productivity of the pasture, bush fires etc., require alternating pasture possibilities. On the other hand, unlimited exploitation for all – first come, first served – leads, under increasing external and internal pressure, to a quickly growing deterioration of ecological conditions. Stock density usually exceeds the pasture potential. In times of drought, it is not rare that a part of the herd starves to death which is why families with large cattle herds do not get into difficulties so easily – this is also a reason for maximizing animal stocks. In this context, the documented empirical experience which comes from observations in West Africa is important: soil-borne diseases, primarily botulism, only appear when pasture is degraded and minerals get used up through the creation of additional wells and the attempts to make nomads sedentary ("maladie de forage"/"well-disease").

The yearly extraction rate in nomadic herds is relatively slight under normal conditions if no sales have been forced upon them by drought. For pure nomads it fluctuates around 9% and can rise to 14% for semi-nomads. The quality of the meat of the cattle which has been sold is poor, the animals are very old (6–9 years), and the profits from slaughter are low (45–50%) (Meyn and Münsterer 1971).

The relative advantages of the nomadic production system are diverse from an animal health point of view. The exploitation of autochthonous breeds adapted to a location, principally the Sanga cebus in Africa, guarantees a high-degree relative resistance to vector-borne diseases. As a rule, the calves become gradually premunized under the protection of the maternal colostrum through natural infection. Although babesiosis and anaplasmosis certainly belong to those infections which are enzootic to nomadic production systems, clinical cases are usually only observed when the animals are already suffering from extreme feedshortages. Trypanosomosis is, on the other hand, a problem where tsetse regions cannot be left quickly enough, or political influences and/or measures force the nomads to remain in tsetse regions when this region increases in size in the rainy season. The Sanga cebus have no notable relative resistance to trypanosomosis (1/2.2.3.3). Nomadic animals have apparently not been selected by the prevalence of trypanosomiasis for resistance, last but not least because nomads have known for centuries to avoid tsetse regions. East Coast Fever is also a disease which creates losses especially amongst calves if nomadic migration takes the herds into areas in which carrier ticks are endemic and the disease is enzootic. In the case of ECF, it is the pathogenetic property of the pathogen which also overcomes the distinctive relative unspecific resistance of the autochthonous breeds. Those herds which only enter ECF regions temporarily are particularly endangered (II/1.2.2.2).

The advantages of nomadic production systems which are determined by the exploitation of resistant breeds are complemented by the traditional knowledge and experience which the nomads have about the seasonal occurrence of disease vectors and pathogens including those of parasites. Part of this knowledge and experience is traditional protection and defence measures like, for example, grazing cattle at midday and late in the evening during the inactive time of the tsetse flies, as well as keeping the cattle around the smoke of a grass fire in the evenings to protect the animals from haematophagous flies and mosquitoes (Batutsi in Burundi), along with looking for the presence of natural minerals, and the relative optimization of rations using extensive pasture rotation (Seifert 1988). In this respect, the knowledge of the nomads about edible and poisonous plants is remarkable. The people as well are able to make do on fruits and leaves both in hard times and during migration.

The relatively late maturity of the autochthonous animals used by the nomads and their low production level are further factors which support the hardiness of the animals. Sticking to the ecosystem of the adapted breeds (perhaps camels) makes sure that the animals in the area chosen by the group find sufficient feed and water.

The relative disadvantages of the nomadic production system from an animal health and technical point of view concerns the apparent and at the same time lack of knowledge or lack of interest about the connection between the keeping and feeding of calves as well as the infectious causes of infertility. The poor feeding and keeping of calves as well, for example, as a lack of interest in regularly removing the ticks from the calves may well only lead to an insufficient premunity from the protection from the maternal colostrum. Just as difficult to understand in the case of some tribes is that mother cows which have not calved for years still remain in the herd. Experienced traditional herdsmen elsewhere know that a cow which has not been pregnant for two seasons is unlikely to become pregnant again. Selection targeted towards productivity is also hindered by this.

From a technical point of view, it is disadvantageous that the nomads employ principally aesthetic and phenotypical characteristics in the selection of breeding animals. The selection of longer horns (Watussi cattle of the Batutsi in Burundi) or the choice of the best tempered and smallest bulls (Antandroy in Madagascar) are not criteria which serve to increase productivity.

2.2. Disease Prevention

Attempts for systematic vector control or chemoprophylaxis of vector-borne diseases are both untenable and contraindicated with the nomadic production system. No governmental infrastructure in either Africa or Asia would be in the position to organize or finance such a programme. This is also true for the application of pyrethroids, even if they are dermally applied. If such measures are only sporadically carried out, the balance between vector and host is disturbed. It is possible that the calves do not become premunized at the right time and can be lethally infected during migration in a heavily infested vector area. Wherever ECF is enzootic, one should attempt to optimize the keeping and feeding of the calves through extension. In this context, the application of LA-tetracycline may be helpful for supporting the organism of the young animal in building up its premunity (II/1.2.2.2). Apart from that, it makes sense to teach the children, for example, to regularly collect ticks, above all from the young animals, on the one hand to prevent a heavy tick burden, and on the other to lessen natural infection so that the calf is in the position of being able to develop a premunity.

Protection against soil-borne diseases should first of all be achieved by maintaining and supporting a grazing rotation which is as extensive as possible. Principally, migration away from extremely dry areas during the dry season will contribute to relieving the incidence of disease. It is also perhaps helpful to provide the animals with minerals (phosphorus) at watering points. This means, however, both a long-lasting interference with the ecosystem as well as the creation of uncontrolled wells. The application of vaccines for protection against botulism and gas gangrene, as is already being carried out as a governmental measure in West Africa, would have the same consequence as abandoning of pastoralism in favour of exploitation of the savannah in a ranching system. It also creates the problem of overstocking and thus the destruction of the ecosystem. In this regard, the occurrence of soil-borne diseases, in particular botulism, is an indicator of the beginning of the collapse of the balance between the exploitation and the maintenance of the ecosystem. Where the immunoprophylaxis of soil-borne diseases, in particular anthrax and gas gangrene, is unavoidable because of a high incidence of the diseases, an effort should be made to achieve long-lasting protection with booster vaccinations for calves which gradually turns into a lifelong immunity boosted by natural infection (II/2.2.1. and II/2.3.1).

Amongst the contact diseases, the big epizootics rinderpest and CBPP are of long-lasting and economical significance for pastoral production systems; protection against brucellosis is also necessary from an economic and human health point of view. Vaccination against FMD is not of much use at the present time. When the disease appears, it has a benign course as a rule. The short-term vaccination interval necessary for protection is economically untenable for the nomadic production system.

When making efforts to stabilize nomadism, one should not try to start with the feeding and maintenance of the young animals, but rather with the immunoprophylaxis. The thing to do would be somehow to encourage private initiatives amongst the nomads so that they get their calves vaccinated a number of times against rinderpest and brucellosis between the ages of 6 and 12 months. At the same time, the adult animals could be vaccinated against CBPP. The application of several vaccines at the same time is contraindicated. The appropriate advice is that vaccinations are carried out in sufficient intervals which then excludes centrally planned blanket vaccination. Such an adapted immunoprophylaxis has the advantage of being more effective and can be carried out at lower costs.

Endoparasites are without question the cause of considerable losses amongst calves in a nomadic production system. On the other hand, the animals are in the position to premunize themselves after an infection, much as they are in the case of vector-borne diseases, if the condition of the animals allows it. All measures for improving the keeping and feeding of calves therefore support the mechanism of natural premunization of young animals against endoparasites. Sporadic chemotherapeutic or chemoprophylactic procedures would thus only get in the way. And apart from that, they are scarcely tenable in a pastoral production system. There would only be a point to anthelmintic treatment of calves if the animal holders were in the position to acquire and understand the product themselves and thus be able selectively to treat young animals which are particularly infected by parasites. In contrast to this, it must be taken into consideration whether or not overstocking is reduced through the natural selection of weaker young animals.

Under normal conditions, plant poisoning does not play a part in nomadic production systems.

Mineral deficiency is a considerable problem in pastoral animal husbandry if the traditional migration routes cannot be kept to, and/or the nomads do not have the products which they can exchange for salt. The degradation of the pasture leads to extreme conditions of lack of fodder, the consequence being botulism. The preparation of mineral licks at the watering points is an indispensable aid for maintaining the health and productivity of the herds.

3. Small-holders

3.1. Definition of the System

Small-holder farming is to be found in the tropics and subtropics, principally in the humid savannah which borders onto the savannah, or in humid tropical areas. As as rule, such areas are more heavily settled and the farms are extremely small. Their main income is derived from crops which support subsistence and help to create cash-crops to varying degrees. In competition with farming for the little land available, animals can only continue to exist through high output. Another point of view may however be relevant: the desire for prestige which dominates the economic way of thinking. The economic functions of animal production are

– the self-supply of milk and meat;
– draught when preparing fields;
– perhaps milk production and sale of dairy products as cash crops;
– producing dung for the improvement of soil fertility;
– the production of dung as fuel if necessary;
– increasing the farm income to a modest degree.

The size and structure of the farms can vary from purely subsistence farms, as for example amongst the Indians in the High Andes, all the way up to integrated crop-animal production farms.

The relative advantages of animal production integrated with farming are the availability of by-products or harvest remains from the crops, although a proper harvest and appropriate exploitation of such feed resources are not yet very developed. Functioning infrastructures for carrying out measures for animal health are often more or less available from organizations which are responsible for the collection of milk or for artificial insemination. Daily supervision of the individual animals is relatively easy to carry out, the size of the herds is limited, and the animals are easy to reach without having to be restrained (chute).

The relative disadvantages of the production system by far outweigh the advantages from an animal health point of view. The motivation for animal production is minor, the main purpose being targeted at crops for which the financial resources are made available virtually without exception. A certain economic contribution towards animal production can only be expected where milk is produced and sold. If, however, as in India for example, the production of milk is in the hands of the women, who are also in charge of all income earned from dairy products, considerable limitations are made upon this contribution. Next to the only slight motivation towards animal production within the small-holder production system, there is an extremely low level of knowledge about keeping of animals and how to control animal diseases. This can even extend to the farmers' not knowing when a cow is in season. Where feeding, keeping and health for these breeds are concerned, these small-holders are absolutely helpless especially when high productivity animals are distributed to them through credit or governmental measures. A considerable relative animal health disadvantage of the small-holder production system has developed worldwide through the exploitation of unadapted breeds. Above all, the indiscriminate use of artificial insemination has contributed to the suppression of the resistance potential of autochthonous breeds without a noticeable increase in production having been able to be reached with genetic upgrading. On the contrary, the expected economic advantages are ruined as a rule through higher morbidity and mortality rates.

The integration of animal production into crop production in the small-holder production system usually works to the disadvantage of the animal production and does not exploit the resources of farming for animal production optimally. After the harvest, the animals are herded onto the stubble in order to graze on it and to fertilize it at the same time. It would be better to harvest the straw with the crops before the nutrients get returned to the roots when the plants die. During the period of vegetation, the animals may have to work without sufficient feed supply. They are dependent on roughage which grows on the edge of the fields. During the vegetation period, the animals are frequently kept on a communal pasture which is overstocked and degraded. The results of such practices are mutual infection with contact diseases and the outbreak of soil-borne diseases.

The relative disadvantages of small-holder animal production systems integrated with farming can be summarized as follows. They are

- little traditional knowledge and little interest in animal husbandry since the animals are not the main business interest;
- the replacement of robust autochthonous breeds with exotic ones thanks to unplanned and indiscriminate use of artificial insemination;
- poorly targeted utilization of available resources;
- no knowledge of how to produce balanced feed rations;
- community keeping of animals during the vegetation period on degraded communal pastures;
- inadequate care of the young animals.

Disease complexes which inevitably arise out of the structure and practices of small-holder animal production systems are

- vector-borne diseases which have a high incidence where exotic breeds have replaced autochthonous ones;
- soil-borne diseases which become enzootic when animals are kept on overstocked communal pastures and are given insufficient minerals, or when enterotoxaemia is provoked by a sudden change in feed;
- contact diseases, above all FMD, brucellosis and tuberculosis as well as CBPP in West Africa, the spread of which is aided by the structure of the production system;
- endoparasites in young animals;
- plant poisoning if exotic breeds are grazed on stubble where, for example, *Ricinus* is growing, or if these cattle are tethered on roadsides which are covered with bushes (*Cestrum* spp.);
- symptoms of mineral deficiencies as a result of unbalanced feed and/or communal pasture through which soil-borne diseases are in turn provoked.

3.2. Prevention of Diseases

Under consideration of the relative advantages of the small-holder production system described here, it would be possible to set up an animal health and extension service which could take advantage of the existing infrastructure for milk collection or artificial insemination. Such a system has been organized by the National Dairy Development Board in India in an exemplary way. This means that there is the possibility of constructing the measures required for preventing disease which are farm-specific and are adapted to the natural cycle of the farm which in turn is determined by season and vegetation period. Here the emphasis should be put on advice.

For protection against vector-borne diseases, a breeding programme must be chosen where possible which maintains the autochthonous resistance potential and/or to restore it using artificial insemination, by crossing with Cebu dairy breeds for example (Sahiwal, Gyr). With the introduction of acaricides and pesticides for controlling vector-borne diseases the questionability of such methods for human health must be taken into account. The use especially of pour-on/spot-on techniques may only take place if compounds of a non-systemic nature are used (pyrethroids I/4.3.1.4). First and foremost one should attempt to build up a life-long unsterile immunity through the natural premunization of properly maintained and nourished young animals when the calf is still protected by the maternal colostrum. Where the utilization of high productivity breeds can no longer be avoided, the animals should be stabled and fed on crop remains or feed taken from the edges of fields or forests. Examples of simple and appropriate cowsheds, adapted to ecological conditions and which are affordable through private means can be found above all in East Asia. To prevent the penetration of ticks into the sheds, there is the possibility of siloing the raw feed which is possible under small-holder conditions and on a small scale. After only 10 days in the silo, the feed is fermented in the tropics; one can be sure that the tick larvae have been killed off. With the emerging resistance of the ticks, as well as to pyrethroids, the following are essential for dairy production in the mid-term:

- the keeping of adapted breeds which can tolerate exposure to the permanent challenge of infection after their premunization as young animals, or
- stabling (zero-grazing) and feeding of high productivity breeds in such a way that they remain tick-free, and other biting arthropods are prevented from developing through appropriate cleanliness and hygiene in and around the cowshed.

Protection against soil-borne diseases is possible through improvement in the keeping and feeding of young animals as well as through farm-specific booster vaccinations for the calves. A concept which is described as follows for ranch systems has also proven its worth on small-holder farms. It should always be designed and carried out specifically for each farm (III/4.2). Above all, the harvesting of fodder, storage, and the production of balanced feed rations with the required percentages of minerals and trace elements can contribute considerably to prevent the appearance of soil-borne diseases.

For legal reasons, a collaboration with the governmental infrastructure must be made when implementing schemes to prevent contact diseases. It is, however, advisable, to coordinate vaccination programmes with the governmental veterinary service, so that they take into consideration the best time of year for creating maximum immunity in the mother animal. Apart from this, as on ranches, the double immunization of the young animals between the ages of 6 and 12 months makes a large contribution towards disease prevention. The basic concept of ranches can be taken over by small-holder production systems (III/4.2)

Endoparasites must be prevented with general measures for keeping, feeding and if necessary, through chemotherapeutic and chemoprophylactic treatment of the young animals. Medication should only be applied where it is understood by the animal holder and can be maintained in combination with improved keeping in private initiatives.

It has to be guaranteed when applying appropriate extension that exotic breeds are kept away from poisonous plants and are supplied with sufficient minerals.

In order to develop the personal interest of the farmers with health interventions and to guarantee the continued existence of preventive measures in the long term, the idea would be to divide the costs and tasks involved between the governmental veterinary service and the private initiatives. In that case, the government veterinary service would be limited to

- the coordination of vector control, and if necessary to create facilities for the control of the dip liquid; where required and indicated, it should also organize tsetse control in the environs of the animals;
- financing and organizing vaccination for protection against contact diseases – FMD, rinderpest and CBPP;
- implementing diagnostic services for disease incidence.

The private initiatives, whether by the farmers themselves or through cooperatives or small-holder organizations, would be left with the following:

- planning farm management and suitable breeding programmes which would minimize the incidence of vector-borne diseases and to integrate natural and artificial methods of premunization into them (III/4.2);
- to prevent local contact and soil-borne diseases through decentralized vaccinations which especially cater to villages or farms, similar to the ranch system (III/4.2);
- reducing calf losses caused by endoparasites through the targeted use of anthelmintics;
- preventing diseases caused by deficiency and thus at the same time soil-borne diseases using supplementary feed and minerals (Seifert 1989).

4. Ranching

4.1. Definition of the Production System

Ranches are extensive farms with cattle and/or sheep which are designed to produce meat, milk, wool or skins. In Northeast Mexico, there is a special form of production which is goat and lamb meat. Ranches are to be found predominantly in the semi-arid pasture belt of the southern hemisphere as well as in the USA, Central and South America, southern and southeastern Africa, and in Australia where either the amount of precipitation is not adequate for crops, or the lack of convenient transport does not allow their marketing. An exception to this are the farms which have been established in clearings in the humid tropics. It is the low fertility of the soil here which gives rise to animal production instead of crop farming.

Characteristic for ranches is grazing carried out on the land available on the ranch without any additional purchase of feed, or the laying on of feed reserves using hay and silage. Feed reserves for the dry season are created by sparing pastures or by irrigating natural or seeded grassland. The costs of intensive feed production and silage, for example would put the economic results of the ranch into question. The stocking rate is so determined that the amount of feed grown in a bad year is taken as the standard. In contrast, it is customary in Northern Mexico to buy more animals when a lot of fodder is available after a good period of rain, and to sell them when feed is low after an extended dry period. Such diverse management methods have considerable animal health consequences.

The intensity of exploitation of a ranch is dependent upon environmental prerequisites, its capital base and the production target of the farm. The choice of breed, from autochthonous to adapted crossbreds or high productivity breeds follows accordingly. In Table 41, the stages of farm intensity are presented with the individual technical factors and the hygienically relevant consequences which arise from them (Seifert 1978). From an overview of the table, it becomes clear that with the introduction of rotating pasture, fencing, increase of the stocking rate, purchase of breeding animals and the use of taurine breeds, the loss of calves as well as the incidence of disease in adult animals increases rapidly, and demand a considerable expenditure for their control. With intensification,

Table 41. Management characteristics of ranch structures according to criterias relevant to animal health

Intensity level	Farm organization measures in				
	Breeding	Rearing	Feeding	Stock	Use
1	No breed selection.	Animals of all categories and age groups are kept together. Only special measure: branding. No super-vision of the animals.	Permanent grazing the whole year round; usually very low stocking rate (S. America: 1 animal/ 10–20 ha, possibly 50–100 ha); natural grassland, frequently in large part wooded grassland.	Autochthonous breeds or breeds which through natural selection have become highly disease-resistant and less productive. Criollos, Cebus.	Meat
2	Selection of bulls according to pheno-type; year-round calving season; castration of males or separation of bulls selected for future slaughtering.	Separation into herds of suckler cows, young stock and commercial animals. No super-vision of the animals.	As in 1.	As in 1.	Meat
3	Buying in of bulls from a breed with a higher performance potential, frequently products of a cross between the native breed and Cebus, or pure Cebus; still no use of taurine high-performance breeds; castration; year-round calving season.	Separation into categories; weaning of calves; supervision of the individual herds by herdsmen; morbidity control; sporadic therapeutic measures for individual and general animal health.	Beginning of grazing rotation; no grassland improvement; no stockpiling of fodder, sporadic supplying of minerals; higher stocking rate than in 1 and 2.	Native breed and/or Cebu crossbreds. No taurine high-performance breeds.	Meat meat a milk
4	Use of bulls from taurine or Cebu high-performance breeds; cross-breeding pro-grammes; buying in of female breeding stock.	As in 3. In addition at least applying herd brand or that of the degree of hybrid-ization; setting up of fences and of facil-ities for marking and vaccination; organized animal health scheme.	Grazing rotation; provision of watering places; regular or continuous supply of minerals; establishment of fodder reserves by protecting fodder grown during the rainy season; grass-land improvement; sowing; irrigation of some parts during the dry season.	Crosses between native breeds and taurine or Cebu high-per-formance breeds; sometimes development of regionally stable combination crossbreds.	Meat meat a milk

Vector borne diseases	Incidence of disease				
	Soil-borne diseases	Contact diseases	Parasitoses	Deficiency diseases	Intoxications, in particular from plants
	–	Low incidence and only when new diseases have been introduced into the neighbourhood.	–	–	Possibly snake bites
	–	As in 1	–	–	As in 1
	Losses in weaned young stock	As in 1 and occasionally sudden introduction of disease with high losses after bulls have been bought.	Sporadic in calves.	Dependent on stocking rate.	As in 1
High incidence	Losses in young stock and cows during the dry season, esp. with high stocking rate and intensive grazing rotation; anthrax and entero-toxaemia on irrigated grasslands and in some circumstances where steers are fattened on pasture	High losses in the event of disease outbreak; part-icular risk to bulls, suckler cows and sucking calves in the rainy season; disease outbreaks in weaned animals.	Calf losses through liver flukes in fattening steers and young cows	Dependent on management and economic state of the farm	In imported breeds and with problems after introduction of new breeds through artificial insemination

Table 41. Management characteristics of ranch structures according to criterias relevant to animal health (continued)

Inten-sity level	Farm organization measures in				
	Breeding	Rearing	Feeding	Stock	Use
5	Pure bred (pedigree) and commerical herds (crossbreds); cross-breeding programmes; limitedcalving season (3–4 months); purchase of bulls only to introduce new blood or to set new breeding goals.	As in 3 and 4. In addition, at least in purebred herds, individual marking (branding of bulls, marking of suckler cows); in some cases artificial insemination; pregnancy test.	As above. Sometimes still more cultivated grassland for suckler cows with calves; supplementary feeding, particularly in the dry season.	Taurine and *Bos t. indicus* high-performance breeds; combi-nation cross-breds; native breeds only in commercial herds.	Meat

conditions arise which are favourable to the singular relevant disease complexes in specific ways:

- The incidence of vector-borne diseases increases in the same way that the resistance of the autochthonous animals becomes lost through the use of high performance animals, and demands vector control accordingly.
- Soil-borne diseases increasingly cause losses amongst young animals as soon as the challenge to the animals increases during the dry season through intensification of the grazing rotation and the increase of stocking rate; when the fodder plants sprout at the beginning of the rainy season, the perfect conditions for enterotoxaemia are created.
- Contact diseases result from the purchase of young animals, stress caused by increased productivity, an increase in the stocking rate, and failures of health management which are either put up with or unconsciously made.
- Plant poisoning increases in the same degree that grazing rotation is intensified and exotic breeds are introduced.
- Parasites as well as deficiency diseases result from inadequate farm management.

The step from stage 2 to stage 3 (Table 41) is thus only economically tenable when the capital base of the farm is big enough and the price for products covers costs. This may not be the case for many highly developed ranches, and the mounting losses are then written off on tax by the owners against profits made from other activities. These connections are also important for planning an adapted animal health concept for such an operation. The advantage of large and financially well equipped farms is that they can put a farm-specific animal health scheme with a private veterinary service into action independent of the governmental infrastructure. Often problems in collaboration between measures

| ctor borne eases | Incidence of disease | | | | |
	Soil-borne diseases	Contact diseases	Parasitoses	Deficiency diseases	Intoxications, in particular from plants
gh risk, o in *Bos t. icus* high-formance eds	As in 4	As in 4	As in 4 highly dependent on management and ecological conditions	Insignificant where there is good management	As in 4

of screening for disease prevalence and prevention of diseases carried out by the government and those organized by the ranches themselves arise because of the inability of such governmental interventions to be able to adapt to the individual ecological, technical and economical features of the individual farms (Seifert 1969).

4.2. Animal Health Management

The prerequisite for an effective disease prophylaxis on a ranch is the integration of the management of animal health into the normal running of the ranch. If in contrast to this the immunoprophylaxis is carried out as an individual or perhaps imposed action, the costs will be disproportionately high, lead to negative effects and not achieve the protection which is expected. As a rule, a ranch is a large self-contained farm, often with borders which are natural or geographically determined. In this respect, such a farm is in the position to be able to control animal transportation and to be able to put an independent animal health concept into action on its own. Thus there is little point where animal production is dominated by ranches to attempt to combat FMD, for example with blanket vaccinations in the form of "campaigns". If such campaigns are enforced when the animals are under particular stress at the peak of the rainy season, at a time when rounding up creates an unreasonable burden on the cattle, or when the calving season has set in and losses amongst calves during mass vaccination would be unavoidable, or if fresh-milking cows would lose a considerable amount in production on a dairy-ranch, the farm manager would be more or less forced to avoid carrying out such a procedure in the usual style of the given country. On the other hand, there is the expectation that vaccines protect the animals from the threat of disease at its highest yearly peak. That can only work when the animals are vaccinated at a point in time which guarantees the development of the best possible immunity.

Below is an example of how an animal health concept for a given location can be integrated into the running of a ranch under prescribed ecological conditions (mountain valleys in the Andes in northern Perú). The animal health scheme is consciously conceptualized for a beef-ranch. Another concept for milk-producing farms will be suggested in III/5.2. Figure 71 shows how the best time for calving takes place between from the middle of February until the middle of May if the rainy season is from the middle of October until the middle of April. This means that the calves will be weaned at the end of the dry season from the middle of October until the middle of November and on into the beginning of the rainy season. The animal health scheme of a ranch which operates under such ecological and economical conditions can be organized as follows:

- measures for vector control depending on the intensity of the production system and the genetically determined resistance of the animals can be carried out in the rainy season with diping, spraying, or with dermal application in the dry season (pour-on/spot-on);
- he calves can be premunized naturally through a light tick invasion in the dry season if the genetic quality of the animals permits it. This is possible without any complication under the protection of maternal antibodies, and thanks to the good condition of the young animals resulting from the good feed conditions at the onset of the dry season. They are then optimally protected against the consequence of increasing vector activity in the rainy season. Measures for vector control ought to be carried out only sporadically on such farms in the rainy season and perhaps not at all in the dry period;
- the calves can be protected against soil-borne and contact diseases which are enzootic in the area with a booster vaccination during and after weaning. But all the necessary vaccinations must not be combined with one another. The vaccine for anthrax should never be combined with the one for FMD, for example. By applying highly concentrated antigens, if possible intracutane-ously, it is possible to protect the calves for a life time with a booster vac-cination against anthrax and gas gangrene (II/2). Subsequently occurring field infections maintain the immunoprotection through a natural booster effect;
- mother cows, replacements and bulls can likewise be protected against contact diseases which are locally enzootic with a booster vaccination before the rainy season at the time of weaning. Protection against FMD can be more certainly guaranteed this way than when the vaccination is carried out on the mother cows which have just calved and these have a low immuno-competence, for example. Once the calves are weaned and the cow has ceased to lactate the organism is able to place its entire resources at the disposal of effective immunoprotection. This way, it can be guaranteed that the calves receive a maximum in maternal passive antibodies which protect against infections which occur at a given location. Cows calving for the first time are given a booster vaccination against rabies in vampire areas;

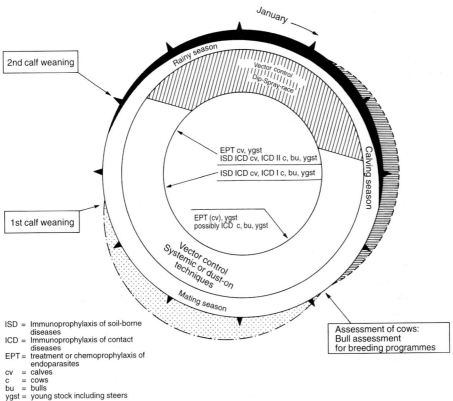

Figure 71. Diagram showing the integration of breeding and management measures with the animal health scheme taking into account local climatic conditions.

— where required, the adult animals and replacement can be vaccinated once against soil-borne diseases at the beginning of the dry season at the time when the cows are classified for the approaching mating season;

— if it is necessary and economically tenable, the calves can be chemo-prophylactically treated against endoparasites at the time when the mother animals are being classified, as well as when they are being weaned.

The concept described integrates the ecological and operational cycles in such a way as to be able to utilize the point in time when the animals are best disposed towards an immune response, and to protect them optimally during the periods when the challenge of infection is at its highest. This example can be adapted to the ecological conditions of any given location. The opportunity also presents itself for transferring this strategy to the small-holder production systems of any geographic region. Prerequisite for carrying this out, however, is that a private or para-governmental veterinary service is available on the ranch or at village level which can plan, coordinate and carry out the necessary measures. In so doing, care must be taken that, at the outbreak of disease, a precise diagnosis can be made, if necessary in collaboration with a governmental diagnostic laboratory. Apart from this, the necessary diagnostic examinations must be made by the veterinary service, and the owners must be assisted when purchasing stock so as to guarantee that latently sick animals are not introduced onto the farm; it might be worth recommending that a quarantine should be set up at the farm level in an isolated region where breeding animals can be carefully examined and observed before they are introduced into the herds.

Apart from controlling vector-borne, soil-borne and contact diseases with chemo- and immunoprophylaxis, the available resources of the ranch must be utilized for removing the breeding and hiding places of the vectors and endoparasites. This includes

- draining humid areas and building suitable drinking troughs so that survival is difficult for the said parasites during the dry season;
- the introduction of a pasture management which takes into account the most unfavourable habitat conditions of the parasite on the pasture. Wet pastures must be grazed in the dry season, and dry ones in the rainy season;
- taking the weaned animals to dry, high-lying pastures which are low in parasites (Seifert 1969, 1978).

The necessary principal operational measures for protection against plant poisoning have been described in II/4.5.2.

5. Dairy Farming

5.1. Definition of the Production System

Keeping cattle for dairy production is an integrated part of farms of all sizes in areas with farming, irrigation and permanent crops. Milk can also be produced alongside meat on dairy-ranches. Highly specialized and intensive operations also produce milk in the tropics along with feed or even feed which is for external use (dairy-lot). Independent of their structure, most farms are connected to a milk-collecting organization and/or processing plant or an organization for artificial insemination.

For the extensive and as well intensive dairy farm, it is characteristic in both tropical and temperate regions that

- the animals which graze relatively near to the milking area or on the dairy-lot are given feed purchased from the surrounding area;
- only those female calves which are needed as replacements are raised with milk-exchangers;
- the lactating cows are subject to individual control during daily milking;
- it is difficult to isolate the animals from contact diseases because of daily milk collection, and the purchase of breeding materials and feed;
- milk production must be guaranteed the whole year round, independent of the seasonally conditioned availability of fodder thanks to feed reserves (silage, feed crops and the purchase of feed).

Only the calves on dairy-ranches and perhaps small-holder operations can be allowed to feed on what remains after milking, a practice which prevents mastitides in the long run. Amongst small-holders who keep autochthonous animals, the calves are usually tethered up next to the cow during milking in order to stimulate milk secretion.

On a dairy-ranch, cows are given feed concentrate according to the amount of milk they produce.

The genetic quality of the animals corresponds to the level of intensity of the farm. Autochthonous animals are seldom still used for milk production, even by small-holders. High-productivity breeds are usually crossed with the autoch-thonous breeds, and the tendency of absorption crossing over to the exotic breeds, Holstein-Friesian as a rule, is barely stoppable worldwide. Better conditions exist from an animal health point of view where Cebu combination crossbreds specialized in milk production (Sahiwal, Gyr) are kept.

The incidence of disease on dairy farms is dependent on

- the structure, the socio-economic conditions of the owner/manager of the farm, the geographical location, as well as political and economic conditions;
- the genetic quality of the animals;
- the environmental conditions.

The incidence of vector-borne diseases is first and foremost dependent on the resistance of the breed of cattle being used in this production system. In parts of Africa where East Coast Fever is enzootic and exotic breeds are used for milk production, heavy losses cannot be avoided without the weekly application of acaricides. The same is the case for anaplasmosis in South America. The hygienic consequences of such interventions for humans have been described in I/4.3. On the other hand, farmers in South America, for example in the humid tropical regions as in the environs of Lake Maracaibo in Venezuela, are able to produce milk and meat without problems and virtually without the use of acaricides using a triple cross of Criollo x Cebu x Brown Swiss.

With sensible management, soil-borne diseases are seldom a problem on dairy farms. Of course, in relation to this disease complex, the same goes for dairy ranches as for beef ranches. Enterotoxaemia must be reckoned with where there are feed crops and pasture rotation is applied.

Contact diseases are a significant health problem for dairy operations. The combination of challenge to the organism during the peak production period after calving and climatic stress, in particular during the hot, humid season, are factors which lead to the outbreak of contact diseases, principally FMD. Besides that, there is a high risk of introducing the infection through animal transportation and other unavoidable outside contact. Apart from FMD, it is brucellosis and IBR/IBV above all which can be introduced into the stocks on the pastures.

Poisoning from pasture plants only plays a role on dairy ranches. In intensive dairy operations, feed hazards created by protein concentrates must be reckoned with (II/4.2.3).

Parasites do not usually appear on dairy farms. They are principally relevant for small-holder production systems.

Mineral deficiencies are due to serious errors of management in the dairy production sector.

5.2. Concept of Animal Health for the Dairy Production Sector

Independent of its structure, the socio-economic conditions of the owner, the geographic and climatic environment of the farm, and the type of animals used, a dairy farm must produce milk all year long. A seasonal calving time is thus out of the question, and coordination of animal health measures, as suggested for a beef-ranch, is impossible. On the other hand, it is this very vaccination, uncoordinated in its form, which leads to heavy production losses as well as perhaps, to doubtful immunoprotection. In order to be able to stimulate the cows to an immune response at the time when their immunocompetence is at its best, a suggestion would be to undertake the coordination of a health programme in dependence of the productivity curve of the individual animals contrary as it is the case on ranches. To simplify the organization of such a concept on a farm, groups of cows at a similar stage of gestation can be formed. The same thing can happen at the village level with small-holder operations. Since the animals are usually individually registered with the organization which is in charge of artificial insemination, there are no problems in this regard.

As Fig. 72 shows,

- in the 5th/6th month of pregnancy, the required vaccination can be carried out;
- in the 6th/7th month, the booster vaccinations can be administered when drying off the cows.

The prerequisite for this is that all necessary vaccines are applied at once, and only vaccines which are harmless can be used. On the other hand, the animals are far less stressed with individual or group treatment, and abortions caused by mistreatment during blanket vaccination are prevented.

If the mother cows are vaccinated at the end of the lactation period, not only is maximum protection for the organism of the cow achieved, but also for the calf because of the optimizing of the antibody level in the colostrum. Infectious diseases must always be controlled in the case of young calves through the immunization of the mother animal. The calves may only be actively immunized when the passive immunization from the maternal colostrum has subsided. This does not take place as a rule until the calves are 6 months of age (Fig. 72). The animal health measures which accompany the weaning of the calves must first and foremost be aimed at optimizing the conditions for keeping, if possible by raising the calves in the open in the tropics. Whether this means single stalls, an open cowshed, or only under a roof which keeps off sun and rain, depends on the structure of the farm operation. In Argentina, the calves are tethered and are protected with a transportable roof which at the same time has a set-up for the drinking trough attached to the base of the structure. Just keeping the calves appropriately, ought to ensure that endoparasitoses are prevented. Where this is not the case, chemotherapeutic and/or other prophylactic measures can be undertaken during weaning and again with the herds of young animals (Fig. 72;

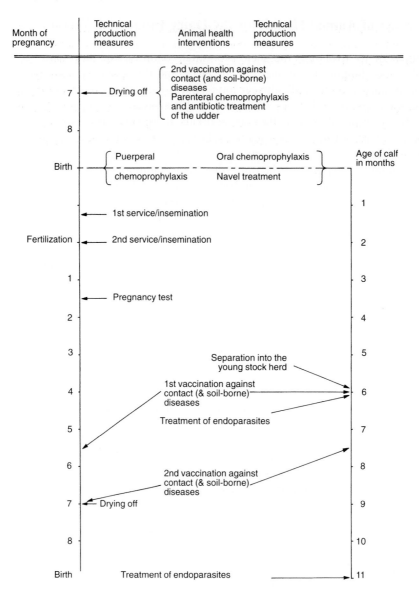

Figure 72. Animal health scheme of a dairy unit in the tropics.

Seifert 1969, 1978).

The owner of the operation will usually agree to such a concept without reservation since any sort of loss of production can be avoided, and hardly any additional work is necessary. In putting this concept into action, a functioning private, or perhaps cooperative veterinary service is required. The concept consciously contradicts that of disease control using blanket vaccination – a strategy which is impossible to realise completely in the tropics and which also has the described production losses as a result. If it is possible with the operation-specific animal health concept described here to motivate more farm operations to cooperate, it will be possible in the long term to control FMD, in Argentina for example. Thanks to the current concepts which are centralistic and work consciously against the interests of the farms, half a century has passed since the end of World War II without success, and little success is likely in the future foreseeable.

6. Feed-lot

6.1. Definition of the Production System

A feed-lot is an intensive fattening operation based on the North American model where cattle or sheep are penned in groups of varying sizes for fattening. The animals which are kept in the restricted space of a pen go through a fattening period of up to a year in which they are fattened according to market demand. When applying such intensive fattening, young animals can be removed from the grazing areas, perhaps as soon as they have been weaned, and can also be economically fattened using agro-industrial by-products. Fattened animals bring higher profits and can open up a particular market for animal production in the tropics. By reducing the numbers of young animals on a ranch, the number of mother cows can be increased and thus productivity as well.

In order to have easy access to feed sources, a feed-lot is usually operated in a farming area and may also be relatively far removed from the grazing lands which mostly are in another geographical and ecological region. It is an exception to the rule that feed-lots have their own feeder-calf production. The cooperative owners of a feed-lot may also be the proprietors of ranches in the dry savannah where they raise young animals (Venezuela). As an intensive fattening operation, a feed-lot can be an independent production unit under communal, private or semi-governmental management on the one hand. On the other hand, it can also be a branch of a larger farming operation, or be integrated into a production cooperative (Seifert 1979).

In the semi-arid grazing areas of the world, large numbers of cattle are grazed on extensive pastures, which only permits the production of mediocre animals to be slaughtered unfattened. When fattening with agro-industrial by-products, as happens in feed-lots, the beef becomes qualitatively and quantitatively improved. This means not only that the potential for feeder-calves/animals can be exploited, but also that additional feed resources of farming areas are opened up. Aside from that, areas for cattle fattening can be made accessible in which there is a high incidence of vector-borne diseases. The intensive production techniques of a feed-lot allows for the technical and economic application of disease prevention schemes which are not feasible on extensive farms.

Since fodder cropping for cattle fattening is only economically tenable in the tropics in exceptional case because of yields which are usually low, by-products from the agro-industry present themselves in the main as a food source for feed-lots, and these are:
as roughage

- cane tops;
- by-products from pineapples, bananas etc.;
- straw from ground nuts, rice, barley and wheat and from other crops;
  as concentrates
- molasses,
- secondary products from mills (rice bran and others);
- oil cakes (cottonseed meal, soya cakes and so on).

The quality of the feed is determined by the energy content: the more intensive the fattening operation, and the more productive the animals, the higher the energy content must be. The use of agro-industrial by-products also leads to disease problems.

6.2. Enzootic Diseases of Feed-lot Stock

Feeder-calves which are brought from differing environmental regions to a feed-lot in a farming region are subject to many challenges. These are:

- stress during transportation with possible subsequent infection;
- a change in climate;
- confrontation with an environment which is unfamiliar, and against which they have neither passive maternal protection, nor have they developed their own active immunity;
- intensive contact in the feed-lot with animals from foreign stocks, which come from environments with other infections and are possibly disease carriers ("crowding");
- intensive exposure to infection on the feed-lot through the feed, the drinking water, the floor of the pen, and intensive contact with fellow animals;
- challenge to the organism through an abrupt change in feed, and being fed with high-energy rations;
- management measures of the intensive production system, such as vaccination, marking, weighing, separation, etc.

The potential spectrum of disease in a feed-lot takes in the site-specific epizootiological situation corresponding to the complexes of

- vector-borne diseases: trypanosomosis, babesiosis, theileriosis, anaplamosis, heartwater, rabies in vampire regions;

- soil-borne diseases: enterotoxaemia, anthrax, and botulism. Gas gangrene and tetanus only occur occasionally;
- contact diseases: CBPP, FMD, pasteurellosis, rinderpest, stomatitis vesicularis, IBR/IBV;
- diseases caused by spoiled or poisoned feed: mycotoxicosis and poisoning caused by poisonous principles from oil cakes.

Endoparasitoses and deficiency diseases can be prevented with appropriate management in feed-lots.

The causes for the occurrence of these individual disease complexes in tropical feed-lots are particularly varied and include

- with vector-borne diseases
  • the introduction of premune, latently infected animals;
  • insufficient vector control in the feed-lot;
- with soil-borne diseases
  • an abrupt change in feed to an high-energy ration (molasses) with a high percentage of dry matter;
  • feeding with contaminated bone-, carcass-, fish- and blood meal (anthrax and botulism);
  • contamination of the soil of the pen with *B. anthracis* spores;
- with contact diseases
  • transport and adaptation stress ("crowding") (pasteurellosis, FMD);
  • inadequate selection and diagnostics (tuberculosis, CBPP, paratuberculosis);
  • inadequate immuno- and chemoprophylaxis;
  • use of contaminated feed (FMD, brucellosis);
  • hazards through contaminated or toxic feed;
  • using oil cakes containing poisonous plant principles (cottonseed meal, soya cake) as feed;
  • mouldy concentrates, straw and silage with, for example, high aflatoxin content. Sugar cane tops can also be contaminated with mycotoxins.

No concessions with regard to measures of animal health can be made for autochthonous animals in feed-lots such as letting them remain in a state of premunity because of their higher resistance level. Since they are kept under conditions unfamiliar to them in the feed-lot, their premune balance may collapse; and apart from that, they could be a source of potential infection for susceptible highly productive animals.

The low relative resistance of taurine highly productive beef breeds means that rather than showing symptoms of typical feed-lot diseases in the tropics (CBPP, anthrax, enterotoxaemia, anaplasmosis) within the intensive production system of the feed-lot, they will usually die apoplectically often shortly after feeding. Disease outbreaks of this kind can mean the financial ruin of a farming operation. If beef is to be exported to industrialized countries, additional effort is needed to keep the farm free of notifiable diseases (rinderpest, CBPP, FMD). In this context, the introduction of bovine tuberculosis must be prevented which is clinically inapparent in numerous tropical regions were the feeder-cattle are produced extensively on the savannah (II/3.3.2.1).

6.3. Animal Health Management of the Feed-lot

The economic success of a feed-lot is dependent upon whether or not the daily gain of the animals cover their maintenance costs per day, and above and beyond that, whether or not a profit can be returned. A feed-lot is economically an extremely sensitive production system which runs up considerable costs for feed, interest which has to be paid on animal capital, amortization of the equipment, running costs for machinery etc., as well as personnel costs, even when using agro-industrial by-products as a feed resource. Therefore the main prerequisite for the successful running of such an operation is that the animals gain weight from the day they arrive, and that there is neither a stagnation in nor any loss of weight because of disease or management procedures. Apart from this, the value of individual animals is so high that total losses must be excluded. For this reason, the animal health scheme is responsible for not allowing diseases to occur (keratoconjunctivitis, stomatitis vesicularis) which only reduce daily weight gains, even if they are only the result of the animals refusing feed, something which could be tolerated in an extensive operation. Outbreaks of disease with high morbidity and a high mortality rate must not occur under any circumstances.

The production principle of an intensive fattening operation is the optimizing of production factors and the minimizing of costs, last but not least through organizational measures. The animal health concept must likewise fit into this principle. The economic viability of the feed-lot is connected with the minimizing of production-reducing factors which are determined by influences of animal health. The operational concept of the feed-lot must therefore include an animal health scheme which rigorously ensures water-tight and absolute disease control, even under the ecological and socio-economic conditions of the tropical environment. The targets of the animal health concept must be:

– cleaning up and isolating the site of the feed-lot from vectors and other disease reservoirs;
– prevention of the introduction of diseases;
– control of vector-borne and infectious diseases during the fattening cycles;
– prevention of diseases caused by management and feeding errors and of factorial diseases.

The choice of the site for the feed-lot must be made so that the operation is infrastructurally open, however, not in the immediate vicinity of centres of civilization, so that air pollution can be avoided and straying pets can be kept out. In Northeast Mexico, feed-lots are set up in valleys of the Sierra Madre which are uninhabited and are of no use for farming.

The positioning of the feed-lot should be on a hill if possible (incline 3–10%) so that there is natural drainage and so that the animals are cooled by the breezes which occur as a rule in the afternoon in the tropics. A proven set-up for pens is one which is square with sides 30 m long, and with room for 100–150 animals. Two sides of the pens border onto neighbouring pens, and the other two are equipped with troughs for feeding (fence-feeding). In view of the intensity of the operations, feeding takes place mechanically. A floor composed of gravel and rocks guarantees that mud cannot develop which means avoiding hoof damage in the rainy season. Mud on a feed-lot can strongly reduce productivity and creates favourable conditions for vector reproduction. The excrement must be drained off in such a way that no muddy spots remain. With the recommended animal density, the ground is trodden under foot so much that fly larvae have no chance of developing. Such a system is also no habitat for ticks. Flies can only breed along the troughs and fences where there is loose feed or dung lying around (Seifert 1971). A feed-lot can be kept practically vector-free with regular cleaning and removal of dung. If there should be a build-up of mud in the rainy season in spite of this, the dung can be piled up in the middle of the pen. The flies' eggs are then destroyed in the dung heap by the self-heating of the dung.

Where multi-host ticks exist, a double fence should surround the area and the lower parts of the fence should be impenetrable to rodents. Thorough inquiries or perhaps investigations must establish whether or not the location is contaminated with spores of pathogens of soil-borne diseases which may be the case on waste-lands which used to be used as waste disposal sites and where dead animals have perhaps been disposed of. The nesting and resting places of carrion birds (vultures) can be highly contaminated with spores of pathogens of soil-borne diseases.

Where there is a favourable habitat for disease vectors, the vegetation must be cleared in an appropriate radius and it might also be necessary to treat tsetse fly resting places with a selective application of insecticides (I/4.2.1.1).

Apart from hygienic and technical factors, it is the quality of the animals which determines the success of the feed-lot. The animals must be dehorned and castrated. The selection of the animals therefore requires particular care with regard to quality. Since this means paying more as a rule, animal health criteria (state of health) can be used when selecting the animals, and animal health interventions can also be applied in the process of selection.

By introducing preventive animal health measures into the process of selecting the animals, a three-tiered concept of animal health can be realized which starts at the location of the origin of the animals (Fig. 73).

Since the animals find themselves in a familiar environment on the pasture of origin, and are not yet challenged by external influence, it might be a good idea to leave them at first on the farm of origin after selection, or to leave them on an extensive pasture (holding ground) in the same region, and to carry out the first

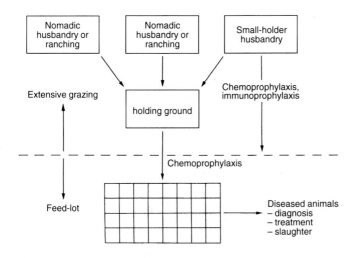

Figure 73. Animal health scheme for a feed-lot.

stage of the necessary animal health interventions. These are

- careful selection following external inspection;
- diagnosis in the field of latent infections with the assistance of feasible serological and allergic reactions;
- chemoprophylactic treatment against endoparasites, ticks and haematozoarias;
- immunoprophylaxis against soil-borne and contact diseases which are relevant at the site of the feed-lot;
- chemoprophylaxis before transportation.

This means that vectors and carriers of infection are prevented from being introduced to the site of the feed-lot. Animals latently infected with CBPP (II/3.3.1.1) and tuberculosis (II/3.3.2.1) can be recognized and eliminated. By applying oxytetracycline, or Imidocarb and Berenil (10 mg/kg b.w. respectively), a *sterilisatio magna* can be achieved as far as trypanosomes, babesiae, theileriae, anaplasma and rickettsia are involved, and the animals can be introduced to the lot uncontaminated. If long-acting tetracycline is used, the animals can also be protected at the same time from pasteurellosis during transportation (shipping fever). Vaccination against soil-borne diseases (anthrax, enterotoxaemia) and contact diseases (CBPP, FMD, rinderpest) on the location of origin of the animals, done best as a booster vaccination, ought to guarantee a massive immunity which lasts for the whole of the fattening period and requires no repetition during the fattening period. The animals finally become releaved of the burden of endoparasites through a targeted application of anthelmintics.

Once the first stage of the animal health scheme has been realized, quarantining is superfluous. If after being brought to the site of the feed-lot, the animals were to be put into quarantine where the described chemo- and immunoprophylactic would be carried out, they would not be able to immunize themselves the same way in the ecologically different surroundings and under the challenge of transportation and adaptation. Apart from that, daily costs for the fattening operation per animal would arise which could not be covered by the corresponding daily gains.

The second stage of the animal health scheme is carried out at the moment of the arrival of the animals on the feed-lot. When the animals are weighed and marked on arrival, an additional treatment against vectors using spray or pour-on takes places; a second application of oxytetracycline/Berenil to ensure for *sterilisatio magna* is also possible. Tetracycline LA should be applied in any case in order to avoid the consequences of "crowding" (pasteurellosis). The application of paraimmunity inducers is not economically tenable in the tropics especially since the climatic problems of a closed shed which are favourable to IBR are not relevant with an open pen in the tropics (Hofmann and Schulz 1982).

The third stage of the animal health scheme accompanies the fattening period and consists simply of careful vector control through the regular removal and destruction of the breeding places of the vectors. Keeping the feed troughs and their surroundings clean is the most important rule. It will only be necessary in exceptional cases to spray insecticides in the area of the fattening pens. Chemo- or immunoprophylactic measures for the animals themselves should no longer be required. If infections should occur in spite of this, oxytetracycline (2 mg/kg b.w./animal) can be mixed into the feed concentrate. By doing this, those mistakes in feed which provoke enterotoxaemia can be made up for. Where the soil of the pen is heavily contaminated with *B. anthracis* spores, such chemoprophylaxis will back-up the immunoprophylaxis applied previously at the same time. The bacterial infections which favour IBR are also controlled in this way, and the occurrence of IBR is prevented wherever it is enzootic. The experience of intensive animal production systems in Europe can be transferred to this situation (Hofmann and Schulz 1982). Chemoprophylaxis against IBR is particularly indicated wherever type- and site-specific vaccines are not available. If the medication is withdrawn 4 weeks before the end of the fattening, there will be no concerns in the question of human health.

A lot of attention must be paid to morbidity control. The animals must be observed daily while feeding, and cattle which do not immediately come to the feed troughs when the feed is distributed must be separated. They must be taken to the sick bay, carefully examined and treated if necessary. After being restored to health, such animals as these must be slaughtered, and the carcasses must be closely inspected. This prevents diseases which are difficult to recognize from spreading within the herd (CBPP, tuberculosis, liver flukes). The treated animals may under no circumstances be reintroduced into the fattening paddock. They would not be accepted by the other animals which would cause unrest, and this in turn would disturb the daily weight gain of the animals.

During the entire fattening period, the animals ought not be brought all at once into the chute either for weighing or for treatment. However, if this is unavoidable they should be treated with an acaricide, because if the animals all have the same smell, unrest and new hierarchical clashes are avoided.

The animal health scheme described is fully integrated into the management course of the purchase, delivery and fattening of the animals and requires a minimum of personnel and complicated technical procedures. The success of this strategy is based on

- the complete and utter elimination of carriers and vectors of pathogens;
- an effective immunity against the infectious diseases which occur in a given location;
- additional chemoprophylactic control of potential disease carriers.

In so doing, as experience shows, losses of less than 0.1% can be reached, even in tropical locations with animals of 5000 or more in number (Seifert 1969, 1979).

References

Berlal, A.-R., B. Bös and H. Mayer (1981): Nomadische Viehhaltung in Afrika. Entw. u. Ländl. Raum **3**, 7–9.

Hofmann, W. and W. Schulz (1982): Verhütung und Bekämpfung infektiöser Bestandserkrankungen in der Bullenmast. Prakt. Tierarzt **8**, 698–704.

Holter, U. (1994): Domestic Herd Economy. Anim. Res. Develop. **39**, 68–81.

Holter, U. and M. Kirk (1994): Access to Resources and Social Networks. Anim. Res. Develop. **39**, 177–187.

Janzen, J. (1991): Mobile livestock keeping – A survival strategy for the countries of the Sahel? – The case of Somalia. Appl. Geogr. Dev. **37**, 7–20.

Kirk, M. (1994): Changes in Land Tenure in the Butana. Anim. Res. Develop. **39**, 82–95.

Kirk, M. and D. Mai (1994): Socio-Economic Differentiation in Animal Keeping Societies. Anim. Res. Develop. **39**, 188–202.

Lewis, J. M. (1961): A pastoral democracy. A study of pastoralism and politics among the northern Somali of the Horn of Africa. Oxfort Univ. Press, London.

Meyn, K. and F. Münsterer (1971): B Rinder i. Handbuch d. Landw. u. Ernähr. Entwicklungsländern 2, Pflanzl. u. tier. Prod. Trop. Subtrop. 729–775, Ulmer, Stuttgart.

Oyenuga, V. A. (1966): The level of Nigerian livestock industry. World Rev. Anim. Prod. **1**, 91–104.

Reckers, U. (1991): Verbesserung der Lebensbedingungen nomadischer Viehhalter – Ein Projekt der Deutschen Welthungerhilfe. Entw. u. ländl. Raum, **2**, 11–14.

v. Schutzbar, R. (1994): Effect of Declining Fodder Resources on Herd Management Strategies. Anim. Res. Develop. **39**, 123–134.

v. Schutzbar, R., H. S. H. Seifert and A. Weiser (1994): The Tale of the "Cattle Complex", or Human Reactions to the Risk of Losing Animals through Diseases and Environmental Influences and contrary to the Improvement of Animal Health. Anim. Res. Develop. **39**, 173–176.

Seifert, H. S. H. (1969): Unpublished findings as chief veterinarian and director of the Div. Ganadera of the Empresa Agrícola Chicama Ltda., Hda. Casa Grande, Trujillo/Perú.

Seifert, H. S. H. (1971): Die Anaplasmose. Schaper, Hannover.

Seifert, H. S. H. (1978): Hygiene management on extensive and intensive cattle units in the Tropics and Subtropics. Anim. Res. Develop. **7**, 49–102.

Seifert, H. S. H. (1979): Hygiene und Gesundheitsprobleme, in Expertengespräch: Einrichtung und

Betrieb von Feed-lot am tropischen Standort. DSE-Bericht, 31–36.

Seifert, H. S. H. (1988): Ranching – Eine Alternative zum Nomadismus Die Erde **119**, 269–274.

Seifert, H. S. H. (1989): An integrated preventive animal health scheme for dairy production in the Tropics. 4th. Intern. DLG Symp. on "Modern Cattle Production", Giessen.

Seifert, H. S. H. (1990): Tierhaltung im Sahel. Göttinger Beitr. Land. Forstwirt. Trop. Subtrop. **51**, 187–192.

Weiser, A. (1995): Analyse der Tierhygiene-Situation in mobilen pastoralen Tierhaltungssystemen in der Butana/nordost-Sudan. Göttinger Beitr. Land-Forstwirt. 101.

Wilson, F. (1963): Calvings and calf mortality in Ankole long-horned cattle. E. Afri. agric. For. J. **29**, 178–181.

Index

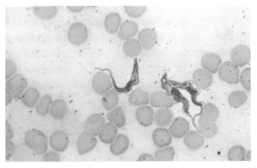

Figure 42. *Trypanosoma brucei.* Giemsa-stain

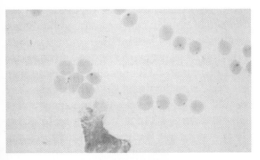

Figure 48. *Anaplasma marginale.* Giemsa stain.

Figure 46. *Theileria parva. Merozoites inside lympho-cytes of spleen.* Giemsa stain.

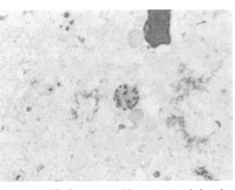

Figure 50.
*Bluetongue. Stripey
degeneration of
heart muscle.*
(photo: Weiss,
Onderstepoort)

Figure 51.
*Characteristic
paralysis of hind
quarters with
rabies transmitted
by vampires.*
(photo: Käthler)

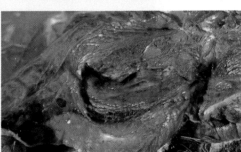

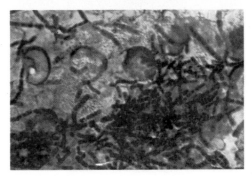

Figure 52. *B.anthracis bacilli in the blood of a bovine.*

Figure 53. *Gas gangrene of
the musculature of a sheep*
(strain 335 Madagascar)

Figure 54. *Carcass of a heifer, dead as a result of
blackleg*

Figure 55. *Tonic-clonic convulsions in a foal with
tetanus.*

Figure 56. Epizootic Lymphangitis in a cow.

Figure 58. Groupous pneumonia with CBPP.

Figure 59. Dermatophilosis in Fulani-Cebus.

Figure 60. Brucellosis. Characteristic hygromas in a N'Dama cow.

Figure 61. Orf: Balanoposthitis.

Figure 68. Characteristic haemorrhages on the heart and degeneration of the heart muscle after poisoning with cestrum auriculatum.

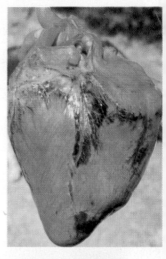

Figure 70. Paresis after Astragalus poisoning

Figure 69. Characteristic haemorrhages on the heart after cyanide poisoning